The Value of Outcrop Studies in Reducing Subsurface Uncertainty and Risk in Hydrocarbon Exploration and Production

Geological Society books refereeing procedures

The Society makes every effort to ensure that the scientific and production quality of its books matches that of its journals. Since 1997, all book proposals have been refereed by specialist reviewers as well as by the Society's Books Editorial Committee. If the referees identify weaknesses in the proposal, these must be addressed before the proposal is accepted.

Once the book is accepted, the Society Book Editors ensure that the volume editors follow strict guidelines on refereeing and quality control. We insist that individual papers can only be accepted after satisfactory review by two independent referees. The questions on the review forms are similar to those for *Journal of the Geological Society*. The referees' forms and comments must be available to the Society's Book Editors on request.

Although many of the books result from meetings, the editors are expected to commission papers that were not presented at the meeting to ensure that the book provides a balanced coverage of the subject. Being accepted for presentation at the meeting does not guarantee inclusion in the book.

More information about submitting a proposal and producing a book for the Society can be found on its website: www.geolsoc.org.uk.

It is recommended that reference to all or part of this book should be made in one of the following ways:

BOWMAN, M., SMYTH, H. R., GOOD, T. R., PASSEY, S. R., HIRST, J. P. P. & JORDAN, C. J. (eds) 2016. *The Value of Outcrop Studies in Reducing Subsurface Uncertainty and Risk in Hydrocarbon Exploration and Production*. Geological Society, London, Special Publications, **436**.

RACEY, A., FISHER, J., BAILEY, H. & ROY, S. K. 2016. The value of fieldwork in making connections between onshore outcrops and offshore models: an example from India. *In:* BOWMAN, M., SMYTH, H. R., GOOD, T. R., PASSEY, S. R., HIRST, J. P. P. & JORDAN, C. J. (eds) 2016. *The Value of Outcrop Studies in Reducing Subsurface Uncertainty and Risk in Hydrocarbon Exploration and Production*. Geological Society, London, Special Publications, **436**, 21–53. First published online May 23, 2016, http://doi.org/10.1144/SP436.9

GEOLOGICAL SOCIETY SPECIAL PUBLICATION NO. 436

The Value of Outcrop Studies in Reducing Subsurface Uncertainty and Risk in Hydrocarbon Exploration and Production

EDITED BY

M. BOWMAN
Texas A&M University, Qatar

H. R. SMYTH
Neftex Exploration Insights, UK

T. R. GOOD
ExxonMobil, UK

S. R. PASSEY
CASP, UK

J. P. P. HIRST
Sedimentology Consultant, UK and

C. J. JORDAN
British Geological Survey, UK

2016
Published by
The Geological Society
London

THE GEOLOGICAL SOCIETY

The Geological Society of London (GSL) was founded in 1807. It is the oldest national geological society in the world and the largest in Europe. It was incorporated under Royal Charter in 1825 and is Registered Charity 210161.

The Society is the UK national learned and professional society for geology with a worldwide Fellowship (FGS) of over 10 000. The Society has the power to confer Chartered status on suitably qualified Fellows, and about 2000 of the Fellowship carry the title (CGeol). Chartered Geologists may also obtain the equivalent European title, European Geologist (EurGeol). One fifth of the Society's fellowship resides outside the UK. To find out more about the Society, log on to www.geolsoc.org.uk.

The Geological Society Publishing House (Bath, UK) produces the Society's international journals and books, and acts as European distributor for selected publications of the American Association of Petroleum Geologists (AAPG), the Indonesian Petroleum Association (IPA), the Geological Society of America (GSA), the Society for Sedimentary Geology (SEPM) and the Geologists' Association (GA). Joint marketing agreements ensure that GSL Fellows may purchase these societies' publications at a discount. The Society's online bookshop (accessible from www.geolsoc.org.uk) offers secure book purchasing with your credit or debit card.

To find out about joining the Society and benefiting from substantial discounts on publications of GSL and other societies worldwide, consult www.geolsoc.org.uk, or contact the Fellowship Department at: The Geological Society, Burlington House, Piccadilly, London W1J 0BG: Tel. +44 (0)20 7434 9944; Fax +44 (0)20 7439 8975; E-mail: enquiries@geolsoc.org.uk.

For information about the Society's meetings, consult *Events* on www.geolsoc.org.uk. To find out more about the Society's Corporate Affiliates Scheme, write to enquiries@geolsoc.org.uk.

Published by The Geological Society from:
The Geological Society Publishing House, Unit 7, Brassmill Enterprise Centre, Brassmill Lane, Bath BA1 3JN, UK

The Lyell Collection: www.lyellcollection.org
Online bookshop: www.geolsoc.org.uk/bookshop
Orders: Tel. +44 (0)1225 445046, Fax +44 (0)1225 442836

British Library Cataloguing in Publication Data

A catalogue record for this book is available from the British Library.
ISBN 978-1-78620-140-9
ISSN 0305-8719

Distributors
For details of international agents and distributors see:
www.geolsoc.org.uk/agentsdistributors

Typeset by Nova Techset Private Limited, Bengaluru & Chennai, India
Printed and bound by CPI Group (UK) Ltd, Croydon CR0 4YY

Contents

Acknowledgements

The Geological Society thanks the following companies for their generous sponsorship of the Petroleum Group and the conference that led to this volume. The Petroleum Group is thanked for its contribution towards the colour printing in this volume.

Petroleum Group corporate sponsors at the time of the conference:

Sponsors of the 'Reducing Subsurface Uncertainty and Risk Through Field-Based Studies' conference:

Reducing uncertainty and risk through field-based studies

M. B. J. BOWMAN[1]* & HELEN R. SMYTH[2]

[1]*Texas A&M University, Education City, Al Rayyan, Qatar*

[2]*Neftex Exploration Insights, Halliburton, Milton Park, Abingdon OX14 4RW, UK*

**Corresponding author (e-mail: mbjbowman@live.co.uk)*

Abstract: Field-based studies, be they regional or basin-scale fieldwork, or the examination of outcrop analogues, have been an important component of petroleum system studies in the upstream oil and gas industry for many years. Recent technological advances, founded on traditional field methodologies, whether they are in digital outcrop characterization or developments in processing and computational capabilities, have fuelled a resurgence in such fieldwork. These studies enable the building of realistic models that allow the exploration or production geoscientist to navigate effectively from plate-to-play-to-pore. The interrogation of field data and resulting models provide a more informed means of benchmarking the subsurface, as well as a more realistic view of subsurface uncertainty and the management of associated risk in subsurface description and prediction.

This Special Publication records 12 papers given at a conference titled 'Reducing Subsurface Uncertainty and Risk Through Field-Based Studies' organized by the Petroleum Group of the Geological Society of London, and held in London from 4 to 6 March 2014. The observations drawn in this introductory paper reflect the authors' experiences, presentations at the conference and manuscripts within this volume.

Reducing uncertainty and risk through field-based studies

This volume showcases the value of field-based studies in reducing uncertainty and risk in hydrocarbon exploration and production, and, as such, it is a companion to the volume *Sediment Body Geometry and Heterogeneity*, edited by Martinius *et al.* (2014). The papers include examples of stratigraphic and structural studies, outcrop analogues, field data developing play analogues, and the application of modern techniques such as LiDAR (Light Detection And Ranging) and 3D photogrammetric analysis. The studies were undertaken in geographically diverse locations such as the Faroe Islands, Italy, Algeria, India, the USA and Trinidad (Fig. 1), representing a range of tectonic settings and a wide geological time frame.

The integration of field-based studies into the exploration or production workflow is not a new concept; the conference that preceded this volume 'Reducing Subsurface Uncertainty and Risk Through Field-Based Studies' was convened in response to the resurgence in these studies and the adoption of new technologies. The conference was held over 3 days at the Geological Society in Burlington House, when 40 papers were presented (Bowman *et al.* 2014). The 12 papers that comprise this volume offer an excellent reflection of the range in scope and topics offered at the event. The meeting reflects the upturn in interest in field-based studies over recent years, and how these are being used to better reflect subsurface uncertainty and to manage risk for subsurface description and prediction in the upstream oil and gas industry. This is, in part, a reflection of the major advances in capturing and visualizing outcrop data through the advance of techniques such as LiDAR technology combined with a more 'user-friendly' ability to create meaningful numerical and digital characterizations of these data which are of more direct value to subsurface practitioners, specifically in the upstream oil and gas industry.

The papers in this volume have an applied focus for the upstream oil and gas industry. They mainly relate to applications in the exploration, appraisal and development phases of the business. In addition, there are a limited number of technique-based papers that reflect some of the advantages available to geoscientists today in terms of digital technology, data manipulation and visualization. Every field-based study carries the expectation of safety management, an area addressed by Gill & Lynn (2015); it is a critical topic that has appropriately received greater attention in recent years, reflecting the oil and gas industry's increased focus on this aspect of their risk management.

From: Bowman, M., Smyth, H. R., Good, T. R., Passey, S. R., Hirst, J. P. P. & Jordan, C. J. (eds) 2016. *The Value of Outcrop Studies in Reducing Subsurface Uncertainty and Risk in Hydrocarbon Exploration and Production*. Geological Society, London, Special Publications, **436**, 1–8.
First published online August 11, 2016, http://doi.org/10.1144/SP436.13

Fig. 1. World map illustrating the range of papers offered in this volume.

Context

Outcrop analogues and field studies have been an important part of petroleum-system-related studies in the upstream oil and gas industry for many years. This need to understand the entire system harks back to the very earliest days of our industry in such places as East Texas, the Middle East and the South Caspian, informing the decisions on where to drill exploration wildcat wells in frontier areas that led to the birth and growth of the upstream industry we know today.

In these early days, the use of field data was largely based on qualitative observation, maps and diagrams that offered general insights into the petroleum system of an area, particularly focusing on improved exploration for working source rocks and potential reservoir systems. It was the advances in computational technology in the late 1980s that enabled a step change in how and why geoscientists collected and used the data. This change resulted in more informed and, consequently, better business decisions, and an enhanced subsurface understanding. The development of numerical modelling and characterization techniques enabled geologists and other subsurface practitioners to transfer their previously qualitative field observations and descriptions into numerical data. Stochastic modelling, the application of statistical and more deterministic techniques enabled insights into the size, shapes and architecture of different depositional bodies. This is reflected in some of the seminal and classical works at the time, such as Weber & van Guens (1990), Begg & Williams (1991), Haldersen & Damsleth (1993) and Hirst *et al.* (1993). This created a new branch of geoscience and saw the first recognized applied computational geologists, able to transfer the traditional qualitative observations into digital data. These new numerical static descriptions could be upscaled and transferred into simulations and predictions of reservoir performance by reservoir and petroleum engineers including Haldersen *et al.* (1987), Dubrule (1989) and Geehan & Pearce (1994). The data, associated models and interpretations have enabled a deeper understanding of the nature and range of subsurface uncertainties, and how these impacted risks to delivering promised resource assessments, reservoir production profiles, production forecasts and well rates (Fig. 2) (Fox & Bowman 2010).

Since about 2000, developments in computational capability, data capture, visualization and uncertainty assessment have lead to major changes in the way geologists worked, placing greater demands on numerical and computational skills. These advances have also lead to a massive explosion in fieldwork as academia and industry revisited classic outcrops and sought out new field areas to deliver detailed descriptions and maps that could be readily quantified in digital format. The classic works in areas such as the Book Cliffs (Hampson *et al.* 1999; Hodgetts & Howell 2000), Karoo deepwater and slope systems (Flint *et al.* 2011), carbonates (Ruppel 2012), and elsewhere (e.g. Reynolds *et al.* 1998; Hampson *et al.* 2008) reflect this new dawn in applied data collection. It was largely focused on reservoir systems, where the digital static description of the outcrops became the initial

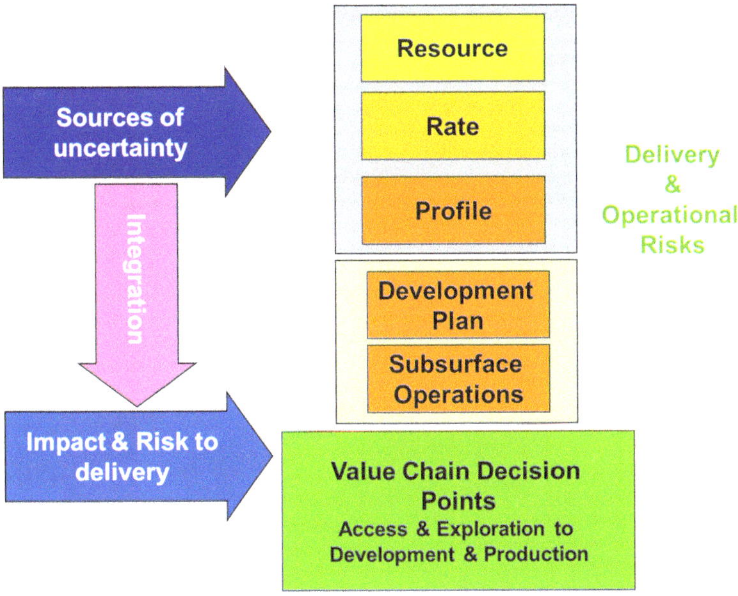

Fig. 2. Subsurface uncertainty and risk-management high-level workflow illustrating the impact of fundamental uncertainty sources (resource size, well rate and production profile) on key E&P value chain decision points.

and core part of a subsurface workflow focused on permeability architecture description, modelling and prediction (Fig. 3).

The industry driver behind this new growth in subsurface description and understanding was the move to new, more challenging areas for exploration and production. Offshore developments in more challenging regions, such as the North Sea, shifted focus to drilling fewer, often high production-rate wells from typically space-constrained offshore production and drilling facilities in increasingly deeper water, with complicated geology and challenging metocean conditions. All of this put pressure on the subsurface team to focus more on optimally placing the wells drilled, deeper consideration of recovery mechanisms and their impact on reservoir drainage, sweep efficiency, and recovery factors (Fig. 4). The arrival of numerical characterization enabled the geologist to transfer their historically qualitative descriptions into more useful numerical models that also facilitated this more rigorous approach to understanding subsurface uncertainty and risk management. The early work was largely focused on the exploration, appraisal and development phases of the business where major strategic and commercial decisions are made, driven largely by the subsurface description, particularly of the reservoir system and its impact on permeability architecture (Weber & van Guens 1990; Gluyas & Swarbrick 2004; Jahn *et al.* 2008).

Following these advances and changing business demands, a whole range of field-based research efforts began to describe more rigorously the outcrops around the globe as analogues for subsurface prospects, discoveries and fields. This boom also lead to a number of significant advances in our understanding of reservoir depositional systems, focusing more on siliciclastics initially (Reynolds *et al.* 1998; Richards *et al.* 1998; Richards & Bowman 1998) but also including some seminal works on carbonates (Lucia 2007; Ruppel 2012). These exercises by research groups, such as the North Africa Research Group (NARG) at the universities of Manchester, Heriot-Watt and Delft, and the University of Aberdeen and many others besides, were almost exclusively funded by industry.

Today, the hydrocarbon industry finds itself once more at a tipping point, with returns on investment under pressure: explorers are required to resolve ever-more challenging reservoir issues with greater precision, while the business demands for lower costs has led to increased focus on an understanding of the subsurface, which, in turn, results in better informed business decisions. The resolution, speed of accessibility of techniques such as LiDAR (Hodgetts 2013), photogrammetry and GIS (geographical information systems), along with supporting software, have evolved significantly in recent years; combined with major advances in computational capability and high-performance computing

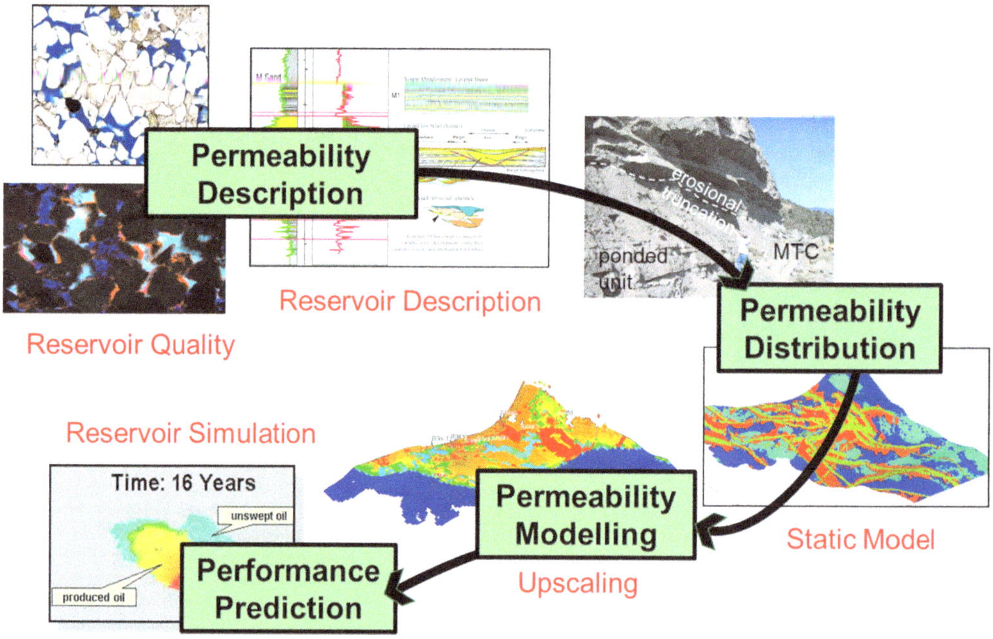

Fig. 3. Fundamental subsurface workflow for a reservoir description that illustrates the foundation that outcrop studies have as an input to describing, distributing and modelling permeability (adapted from a figure provided by A.J. Pulham).

centres becoming the norm in the industry, subsurface visualization and digital description has now become routine (Seers & Hodgetts 2014; Pyrcz & Deutsch 2014). With a requirement for sound input data, these advances have led to a resurgence in outcrop description and characterization. At the same time, the demands of industry for better numerical descriptions, characterizations and integration of the subsurface were being driven by stepouts into deeper water, more complex geology, deep high-pressure and high-temperature environments, unconventional reservoirs such as shale gas and

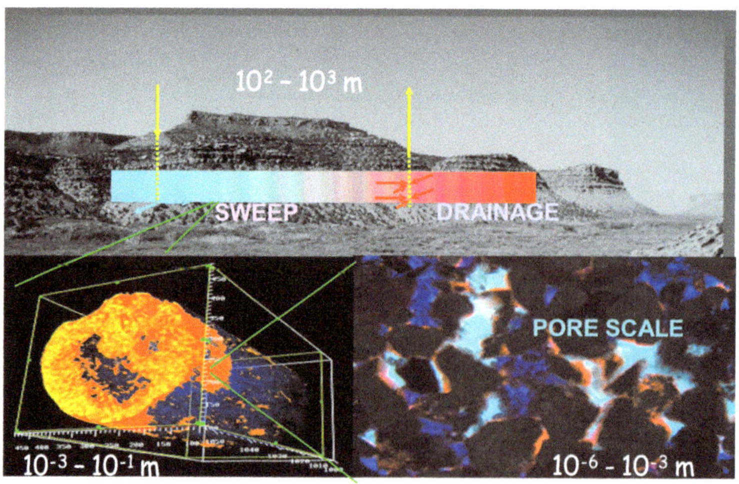

Fig. 4. Scales of dynamic focus using outcrop studies and their relevance to subsurface performance: outcrop scale is the focus of this Special Publication.

oil, and a high cost base. Together, these have meant that the time was right to take another step forward in reducing subsurface uncertainty and risk through field-based studies as a basis for better and more informed business decision making.

This volume, and the conference that led to its compilation, was instigated in response to the latest advances in capability and demand for high-quantity field-based studies. It covers most, if not all, of the latest techniques and how they are applied. The papers in this volume cover the full range, from frontier basin petroleum system assessment (Racey *et al.* 2016) to field appraisal and development (Bowman *et al.* 2014; Hirst 2016), covering structure, stratigraphy and also unconventional resources.

This volume is structured to present the papers in four sets around the following themes:

- health and safety in the field;
- fieldwork in frontier and mature basins;
- fieldwork techniques, current application and new developments;
- uncertainty characterization.

Health and safety in the field

The opening paper of this Special Publication is a timely and thought-provoking paper on health and safety in the field (**Gill & Lynn 2016**). This paper considers fieldwork safety practices and their management from a Shell perspective. Field safety rightly occupies a more prominent place in our thinking for fieldwork today (e.g. Oliveri & Bohacs 2005). The best practices, often common-sense approaches, are not always adhered to; this paper presents an opportunity to reflect and learn skills that are relevant to those in our industry who are responsible for the safety of participants, be they students or colleagues in the field.

Fieldwork in frontier and mature basins

The paper by **Racey *et al.* (2016)** offers an example of fieldwork applied to frontier exploration informing the understanding of an integrated petroleum system. This shows how field data are used to help aid basin entry and the early stages of exploration access decisions.

Vosgerau *et al.* (2015) present results from fieldwork on the Faeroe Islands Tertiary volcanic system. This paper explains the application of photogrammetry and how it is being used in a complex, intra-volcanic exploration target. The work serves as an excellent analogue for the area exploration focus of West of Shetlands in the UK Continental Shelf (UKCS) and the Norwegian Offshore Continental Shelf (NOCS), which has resulted in the discovery of the Rosebank Field that is currently under development by Chevron (Duncan 2015).

The papers by **Hurst *et al.* (2015)** and **Rossi & Craig (2016)** offer examples of how field-based studies are being used in more mature exploration modes. Hurst *et al.* (2015) summarize and update consortia work on injectites as possible reservoirs, applying them to the appraisal of the Volund Field in the NOCS. Rossi & Craig (2016) show how outcrop data can be successfully integrated with subsurface data to enable new insights into hydrocarbon plays and reservoir types through the rigorous application of sequence stratigraphic techniques. They look at the impact of relative sea-level changes on reservoir prediction in the NW Italian Alps.

Reservoir and field development

Moving along the exploration and production (E&P) value chain into reservoir appraisal and field development, **Hirst (2016)** and **Bowman (2016)** offer two very detailed analyses of shallow-marine and wave-dominated shoreline systems in Algeria and Trinidad, respectively. Both are focused on analogues for gas field reservoirs in their areas. Hirst (2016) presents the results of fieldwork in shallow-marine sandstone reservoirs analogues: these sands exhibit complex, non-intuitive architecture that is subseismic and difficult to predict. The paper shows how field studies have helped in understanding the facies, developing a depositional model for such reservoirs and explaining the varied outcomes in the subsurface. Bowman (2016) offers a synthesis of a major effort to describe outcrop analogues of the subsurface reservoir in the Columbus Basin, offshore Trinidad. The paper integrates recent publications on specific aspects of these outcrops (Bowman & Johnson 2015) to offer novel insight into the controls, architecture and predictability of these reservoirs and other analogous high-sedimentation and high-subsidence rate aggradationally dominated shoreline reservoir systems.

Donovan (this volume, in press) offers an insight into how fieldwork and reservoir analogues have materially impacted the understanding and prediction of unconventional source-rock-based reservoir systems (shale oil and shale gas). Using a reservoir-scale outcrop of the Eagleford Group in West Texas, USA, the paper illustrates how detailed outcrop observations and field descriptions can help in predicting sweet spots in subsurface shale reservoir systems. Applying traditional stratigraphic description and prediction techniques to these systems has advanced the understanding of how to drill and develop these unconventional reservoirs more effectively.

Fieldwork techniques current application and new developments

During the conference, several presentations focused on new developments or adaptations of existing techniques such as LiDAR and photogrammetry (see Vosgerau *et al.* 2015). The other papers in the volume are more diverse, ranging from specific analytical techniques to the value and application of digital mapping to non-oil and gas projects which could be readily transferred to subsurface prediction and description.

The keynote paper by **Sanderson (2015)** offers an overview of how field-based structural analogues have helped in better understanding the geometry and predictability of post-depositional structural deformation. These techniques have helped constrain the understanding of structural compartmentalization and trap definition.

Papers by **Jordan & Napier (2015)** and **Newell & Shariatipour (2016)** illustrate the step changes in digital outcrop mapping that are transforming the mapping work carried out by the British Geological Survey (BGS) in the UK. These papers provide excellent illustrations of the major advances created by using digital techniques, both software and hardware, in traditional mapping. The BGS is at the cutting edge for applying these approaches and actively transferring this capability to other government-funded geological surveys around the world. These papers provide details on the challenges of data management and validation caused by the advance in techniques such as GIS systems, together with numerous other new and diverse data sources.

Uncertainty characterization

The final manuscript by Peel & White (2015) focuses on a key theme of the conference – uncertainty characterization. This paper offers a challenging and thought-provoking discussion into the use of, and abuse of, uncertainty descriptions in the subsurface.

Conclusion

This volume demonstrates the value of field-based studies in reducing uncertainty in hydrocarbon exploration and production. The variety and range of papers is testament to the value and integral position that fieldwork occupies within the hydrocarbon industry. They mark a major milestone in the advances made in using outcrop data to better constrain subsurface understanding and uncertainty management. Future advances are likely to involve computational/processing capabilities, the reduction in cost of technology such as LiDAR, and moves into remote and predictive mapping, as well as photogrammetry from unmanned aerial vehicles (UAVs).

The authors wish to thank reviewers Dave Hodgson and Phil Hirst who improved the content of this review. Thanks are also extended to the remainder of the editoral team: Tim Good, Simon Passey, Phil Hirst and Colm Jordan, who have helped to shape the whole volume and provide context for the introductory paper.

References

Begg, S.H. & Williams, J.K. 1991. Algorithms for generating and modeling sand-body distributions. *In*: Lake, L.W., Carrol, H.B. & Wesson, T.C. (eds) *Reservoir Characterization II*. Academic Press, San Diego, CA, 613–639.

Bowman, A. 2016. Outcrop analogues for hydrocarbon reservoirs in the Columbus Basin, offshore east Trinidad. *In*: Bowman, M., Smyth, H.R., Good, T.R., Passey, S.R., Hirst, J.P.P. & Jordan, C.J. (eds) *The Value of Outcrop Studies in Reducing Subsurface Uncertainty and Risk in Hydrocarbon Exploration and Production*. Geological Society, London, Special Publications, **436**. First published online February 24, 2016, http://doi.org/10.1144/SP436.8

Bowman, A.P. & Johnson, H.D. 2015. Storm-dominated shelf-edge delta successions in a high accommodation setting: the palaeo-Orinoco Delta (Mayaro Formation), Columbus Basin, South East Trinidad. *Sedimentology*, **61**, 792–835.

Bowman, M.B.J., Smyth, H.R., Good, T., Hirst, P., Passey, S. & Jordan, C. 2014. *Reducing Subsurface Uncertainty and Risk through Field-based Studies*. Geological Society, London, Abstracts Books.

Donovan, A.D. In press. Making outcrops relevant to the subsurface: an example from the Eagle Ford Group of West Texas. *In*: Bowman, M., Smyth, H.R., Good, T.R., Passey, S.R., Hirst, J.P.P. & Jordan, C.J. (eds) *The Value of Outcrop Studies in Reducing Subsurface Uncertainty and Risk in Hydrocarbon Exploration and Production*. Geological Society, London, Special Publications, **436**. First published online August 11, 2016, http://doi.org/10.1144/SP436.13

Dubrule, O. 1989. A review of stochastic models for petroleum reservoirs. *In*: Armstrong, M. (ed.) *Geostatistics, Volume 2*. Kluwer Academic, Dordrecht, 493–506.

Duncan, L. 2015. The influence of volcanic rocks on the characterisation of Rosebank Field – new insights from ocean-bottom seismic data and geological analogues integrated through interpretation and modelling. *In*: *Abstracts of the 8th Petroleum Geology Conference of NW Europe 2015*. Geological Society, London.

Flint, S.S., Hodgson, D.M. *et al.* 2011. Depositional architecture and sequence stratigraphy of the Karoo basin floor to shelf edge succession, Laingsburg depocentre, South Africa. *Marine and Petroleum Geology*, **28**, 658–674.

Fox, R.J. & Bowman, M.B.J. 2010. The challenges and impact of compartmentalization in Reservoir

Appraisal and Development. *In*: JOLLEY, S.J., FISHER, Q.J., AINSWORTH, R.B., VROLIJK, P. & DELISLE, S.D. (eds) *Reservoir Compartmentalization*. Geological Society, London, Special Publications, **347**, 9–23, http://doi.org/10.1144/SP347.2

GEEHAN, G.W. & PEARCE, A.J. 1994. Geological reservoir heterogeneity data bases and their application to integrated reservoir description. *In*: AASON, J. (ed.) *North Sea Oil and Gas Reservoirs III*. Norwegian Petroleum Institute, Trondheim, 131–140.

GILL, C.E. & LYNN, G.J. 2015. Keeping safe in the field: what, how and why? *In*: BOWMAN, M., SMYTH, H.R., GOOD, T.R., PASSEY, S.R., HIRST, J.P.P. & JORDAN, C.J. (eds) *The Value of Outcrop Studies in Reducing Subsurface Uncertainty and Risk in Hydrocarbon Exploration and Production*. Geological Society, London, Special Publications, **436**. First published online December 18, 2015, http://doi.org/10.1144/SP436.1

GLUYAS, J. & SWARBRICK, R. 2004. *Petroleum Geoscience*. 3rd edn. Blackwell, Oxford.

HALDERSEN, H.H. & DAMSLETH, E. 1993. Challenges in reservoir characterization. *American Association of Petroleum Geologists Bulletin*, **77**, 541–551.

HALDERSEN, H.H., CHANG, D.M. & BEGG, S.H. 1987. Discontinuous vertical permeability barriers: a challenge to engineers and geologists. *In*: KLEPPE, J. (ed.) *North Sea Oil and Gas Reservoirs*. Graham & Trotman, London, 127–151.

HAMPSON, G.J., DAVIES, S.J., ELLIOTT, T., FLINT, S.S. & STOLLHOFEN, H. 1999. Incised valley fill sandstone bodies of Southern North Sea analogues. *In*: FLEET, A.J. & BOLDY, S.A.R. (eds) *Petroleum Geology of Northwest Europe: Proceedings of the 5th Conference*, 771–788.

HAMPSON, G.J., RODRIGUEZ, A.B., STORMS, J.E.A., JOHNSON, H.D. & MEYER, C.T. 2008. Geomorphology and high-resolution stratigraphy of progradational wave-dominated shoreline deposits: impact on reservoir-scale facies architecture. SEPM (Society for Sedimentary Geology), Special Publications, **90**, 117–142.

HIRST, J.P.P. 2016. Ordovician shallow-marine tidal sandwaves in Algeria – the application of coeval outcrops to constrain the geometry and facies of a discontinuous, high-quality gas reservoir. *In*: BOWMAN, M., SMYTH, H.R., GOOD, T.R., PASSEY, S.R., HIRST, J.P.P. & JORDAN, C.J. (eds) *The Value of Outcrop Studies in Reducing Subsurface Uncertainty and Risk in Hydrocarbon Exploration and Production*. Geological Society, London, Special Publications, **436**. First published online May 26, 2016, http://doi.org/10.1144/SP436.11

HIRST, J.P.P., BLACKSTOCK, C.R. & TYSON, S. 1993. Stochastic modelling of fluvial sandstone bodies. *In*: FLINT, S.S. & BRYANT, I.D. (eds) *The Geological Modelling of Hydrocarbon Reservoirs and Outcrop Analogues*. International Association of Sedimentologists, Special Publications, **15**, 237–251.

HODGETTS, D. 2013. Laser scanning and digital outcrop geology in the petroleum industry: a review. *Marine and Petroleum Geology*, **46**, 335–354.

HODGETTS, D. & HOWELL, J.A. 2000. Synthetic seismic modelling of a large-scale geological cross-section from the Book Cliffs, Utah. *Petroleum Geoscience*, **6**, 221–229, http://doi.org/10.1144/petgeo.6.3.221

HURST, A., HUUSE, M., DURANTI, D., VIGORITO, M., JAMESON, E. & SCHWAB, A. 2015. Application of outcrop analogues in successful exploration of a sand injection complex, Volund Field, Norwegian North Sea. *In*: BOWMAN, M., SMYTH, H.R., GOOD, T.R., PASSEY, S.R., HIRST, J.P.P. & JORDAN, C.J. (eds) *The Value of Outcrop Studies in Reducing Subsurface Uncertainty and Risk in Hydrocarbon Exploration and Production*. Geological Society, London, Special Publications, **436**. First published online December 16, 2015, http://doi.org/10.1144/SP436.3

JAHN, F., COOK, M. & GRAHAM, M. 2008. *Hydrocarbon Exploration & Production*. 2nd edn. Elsevier, Amsterdam.

JORDAN, C.J. & NAPIER, B. 2015. Developing digital fieldwork technologies at the British Geological Survey. *In*: BOWMAN, M., SMYTH, H.R., GOOD, T.R., PASSEY, S.R., HIRST, J.P.P. & JORDAN, C.J. (eds) *The Value of Outcrop Studies in Reducing Subsurface Uncertainty and Risk in Hydrocarbon Exploration and Production*. Geological Society, London, Special Publications, **436**. First published online December 16, 2015, http://doi.org/10.1144/SP436.6

LUCIA, F.J. 2007. *Carbonate Reservoir Characterization*. 2nd edn. Springer, Berlin.

MARTINIUS, A.W., HOWELL, J.A. & GOOD, T.R. (eds). 2014. *Sediment-Body Geometry and Heterogeneity: Analogue Studies for Modelling the Subsurface*. Geological Society, London, Special Publications, **387**, http://sp.lyellcollection.org/content/387/1

NEWELL, A.J. & SHARIATIPOUR, S.M. 2016. Linking outcrop analogue with flow simulation to reduce uncertainty in sub-surface carbon capture and storage: an example from the Sherwood Sandstone Group of the Wessex Basin, UK. *In*: BOWMAN, M., SMYTH, H.R., GOOD, T.R., PASSEY, S.R., HIRST, J.P.P. & JORDAN, C.J. (eds) *The Value of Outcrop Studies in Reducing Subsurface Uncertainty and Risk in Hydrocarbon Exploration and Production*. Geological Society, London, Special Publications, **436**. First published online January 13, 2016, http://doi.org/10.1144/SP436.2

OLIVERI, S.R. & BOHACS, K. 2005. *Field Safety in Uncontrolled Environments: A Process Based Guidebook*. AAPG, DEG and ExxonMobil Upstream Geoscience.

PEEL, F.J. & WHITE, J. 2015. Do technical studies reduce subsurface risk in hydrocarbon exploration: and if not, how do they add value? *In*: BOWMAN, M., SMYTH, H.R., GOOD, T.R., PASSEY, S.R., HIRST, J.P.P. & JORDAN, C.J. (eds) *The Value of Outcrop Studies in Reducing Subsurface Uncertainty and Risk in Hydrocarbon Exploration and Production*. Geological Society, London, Special Publications, **436**. First published online December 16, 2015, http://doi.org/10.1144/SP436.4

PYRCZ, M.J. & DEUTSCH, C.V. 2014. *Geostatistical Reservoir Modeling*. Oxford University Press, Oxford.

RACEY, A., FISHER, J., BAILEY, H. & ROY, S.K. 2016. The value of fieldwork in making connections between onshore outcrops and offshore models: an example from India. *In*: BOWMAN, M., SMYTH, H.R., GOOD, T.R., PASSEY, S.R., HIRST, J.P.P. & JORDAN, C.J. (eds) *The Value of Outcrop Studies in Reducing*

Subsurface Uncertainty and Risk in Hydrocarbon Exploration and Production. Geological Society, London, Special Publications, **436**. First published online May 23, 2016, http://doi.org/10.1144/SP436.9

Reynolds, A.D., Simmons, M.D. *et al.* 1998. Implications of outcrop geology for reservoirs in the Neogene productive series; Apsheron Peninsula, Azerbaijan. *American Association of Petroleum Geologists Bulletin*, **82**, 25–49.

Richards, M.T. & Bowman, M.B.J. 1998. Submarine fans and related depositional systems ii: variability in reservoir architecture and wireline log character. *Marine and Petroleum Geology*, **15**, 821–839.

Richards, M.T., Bowman, M.B.J. & Reading, H.G. 1998. Submarine Fan Systems I: characterisation and stratigraphic prediction. *Marine and Petroleum Geology*, **15**, 689–717.

Rossi, M. & Craig, J. 2016. A new perspective on sequence stratigraphy of syn-orogenic basins: insights from the Tertiary Piedmont Basin (Italy) and implications for play concepts and reservoir heterogeneity. *In*: Bowman, M., Smyth, H.R., Good, T.R., Passey, S.R., Hirst, J.P.P. & Jordan, C.J. (eds) *The Value of Outcrop Studies in Reducing Subsurface Uncertainty and Risk in Hydrocarbon Exploration and Production*. Geological Society, London, Special Publications, **436**. First published online June 29, 2016, http://doi.org/10.1144/SP436.10

Ruppel, S.C. (ed.). 2012. *Anatomy of a Giant Carbonate Reservoir: Fullerton Clear Fork (Lower Permian) Field, Permian Basin, Texas*. American Association of Petroleum Geologists, Studies in Geology, **63**.

Sanderson, D.J. 2015. Field-based structural studies as analogues to sub-surface reservoirs. *In*: Bowman, M., Smyth, H.R., Good, T.R., Passey, S.R., Hirst, J.P.P. & Jordan, C.J. (eds) *The Value of Outcrop Studies in Reducing Subsurface Uncertainty and Risk in Hydrocarbon Exploration and Production*. Geological Society, London, Special Publications, **436**. First published online December 17, 2015, http://doi.org/10.1144/SP436.5

Seers, T.D. & Hodgetts, D. 2014. Comparison of digital outcrop and conventional data collection approaches for the characterization of naturally fractured reservoir analogues. *In*: Spence, G.H., Redfern, J., Aguilera, R., Bevan, T.G., Cosgrove, J.W., Couples, G.D. & Daniel, J.-M. (eds) *Advances in the Study of Fractured Reservoirs*. Geological Society, London, Special Publications, **374**, 51–77, http://doi.org/10.1144/SP374.13

Vosgerau, H., Passey, S.R., Svennevig, K., Strunck, M.N. & Jolley, D.W. 2015. Reservoir architectures of interlava systems: a 3D photogrammetrical study of Eocene cliff sections, Faroe Islands. *In*: Bowman, M., Smyth, H.R., Good, T.R., Passey, S.R., Hirst, J.P.P. & Jordan, C.J. (eds) *The Value of Outcrop Studies in Reducing Subsurface Uncertainty and Risk in Hydrocarbon Exploration and Production*. Geological Society, London, Special Publications, **436**. First published online December 17, 2015, http://doi.org/10.1144/SP436.7

Weber, K.J. & van Guens, L.C. 1990. Framework for constructing clastic reservoir simulation models. *Journal of Petroleum Technology*, **42**, 1248–1297.

Keeping safe in the field: what, how and why?

CAROLINE E. GILL* & GWILYM J. LYNN

Shell UK Limited, 1 Altens Farm Road, Nigg, Aberdeen, AB12 3FY, UK

**Corresponding author (e-mail: caroline.gill@shell.com)*

Abstract: Geological field trips are a fundamental part of the development of geoscientists both in academia and in industry. The principles learned in the field are often directly translated into every-day work and therefore maintenance of field time is critical. Across the world, the focus on health, safety and the environment for geological field trips is increasing and it is no longer acceptable to pile onto a bus and go out for an afternoon with no prior planning. This paper will address why field trip safety is of increasing importance, considerations that should be taken in order to keep geoscientists safe in the field and practical guidance on how to enable planning for safe field trips in the future.

Rationale

Geological field trips play a critical role in the education of all geoscientists, from initial forays at school and university through to independent research and continued professional development in academia and industry. They provide a way to physically explore the results of ancient processes and understand the scale and interactions of concepts, and allow a picture of the geological evolution of a region to be determined. Fieldwork gives the geoscientist evidence that can then be used to understand an area or process or as an analogue.

Today, when the majority of us conduct fieldwork, we do so as either students affiliated to an academic institution or as professional geologists, and are therefore bound by the appropriate health and safety at work legislation for the country in which we are operating. A safely executed field trip is therefore not simply a nice-to-have but a legal requirement. It is the responsibility of the field trip organizer and participants to ensure that the provision of health, safety and environment (HSE) factors on a given trip meet the appropriate legislation. In the UK, companies are governed by regulations from the Health and Safety Executive on the basis of the Health and Safety at Work etc Act 1974 (Health & Safety Executive 2003). The act is governed by the principle of 'so far as is reasonably practicable' (Health & Safety Executive 2003), requiring companies to look at the risks and take sensible measures to mitigate them where it is sensible to do so. For further details on the legal requirements in the UK at least, the reader is pointed to the Health and Safety Executive website for up-to-date information on the regulations (www.hse.gov.uk).

Field trips are often considered as a good opportunity to get away from the restrictions of the office environment and there can be a tendency to act as if you were at a location in your own time, with your own perceptions of risk and safety. Field trips in a professional context are not in this category and activities or perceptions of risk that you may do in your own time do not fall within the remit of acceptability for a field trip. A significant amount of safety on field trips is basically about common sense; however, with a bit of additional planning and preparation, most hazards can be avoided and incidents prevented (Oliveri & Bohacs 2005).

Planning and preparation

HSE plan

A key requirement for any field trip is the development of an HSE plan. This provides the basis for all aspects of how HSE will be managed throughout the trip, providing an invaluable resource for leaders and participants alike. Common elements of an HSE plan should include logistics, participant details, risk assessment, risk mitigation measures including safety equipment and planning in the event of an incident. These are each discussed below.

Logistics

An understanding of the logistics of the field trip contributes significantly to the safe execution of a trip. It ensures that all participants are in the right place at the right time and sets expectations on the length of field days. A clear plan for how trips hold together, particularly multi-day trips, with associated travel, enables efficiency as well as safety. It

From: Bowman, M., Smyth, H. R., Good, T. R., Passey, S. R., Hirst, J. P. P. & Jordan, C. J. (eds) 2016. *The Value of Outcrop Studies in Reducing Subsurface Uncertainty and Risk in Hydrocarbon Exploration and Production*. Geological Society, London, Special Publications, **436**, 9–20.
First published online December 18, 2015, http://doi.org/10.1144/SP436.1

also provides information on where the group is at any one time in the event that they need to be contacted in an emergency.

Participant details, roles and responsibilities

Knowledge of participants' personal details together with their roles and responsibilities is a key component to a safe field trip. Participants should provide personal data, in particular any medical conditions and medicines associated with the condition. They should also provide an emergency contact name, the relationship of the emergency contact to the participant and a range of telephone numbers. This named individual needs to be willing to take decisions on behalf of the participant in the event of an incident that incapacitates the participant.

Information on participant health and pre-existing medical conditions is very important, particularly for long trips where days need to be structured to provide an optimum learning experience and where region-specific health risks need to be factored in. Careful planning and pre-briefing of participants on likely physical demands is necessary to avoid incidents on long walks in arid or mountainous areas, where temperature, humidity and altitude can all increase the level of risk. For certain, more remote excursions, a brief summary of participants' previous field experience can also prove useful.

Medical and emergency information should be made available and carried by designated first aiders on the trip so that in the event of any incident this critical information can be passed on to the healthcare professionals. If preferred by participants, it may be carried in a sealed envelope that is opened only in the event of an incident.

Risk assessments

All HSE plans should include a risk assessment as required by law. The purpose of a risk assessment is to identify all possible hazards that may be encountered on the field trip. Hazards in this context can be defined as 'the potential to cause harm, ill health or injury, or damage to the environment'. A generic risk assessment is not pertinent for fieldwork. Only a trip-specific risk assessment is appropriate. For example, a location-specific risk assessment will need to be updated with the physical abilities of participants, the expected weather given the timing of the trip and any other relevant information such as tide times, military firing range times or travel schedule. When undertaking a risk assessment for a field trip the value of local knowledge cannot be underestimated. Exactly where and when can the tide cut the group off? Where is it safe to enter and exit with a coach? What are the indoor options in the event of inclement weather? How long does it take to get to the nearest hospital in the event of an incident?

Figure 1 shows an example of part of a risk assessment highlighting a few hazards common to most field trips. Once risks and their associated causes have been identified, measures can be put in place to reduce the severity of such threats. Measures are designed to reduce them, either through avoidance, for example, missing out a particular outcrop or walking a longer way around to reach it, or through additional mitigation, for example, wearing helmets under a particularly loose outcrop and restricting the time spent underneath such cliffs. By breaking the day down and considering all possible outcomes in advance, the chances of a safer trip are greatly improved.

Reconnaissance trips

If new or unfamiliar field areas are being visited, then a reconnaissance trip prior to the main trip has significant value. It gives an opportunity to visit the area and assess possible hazards without the exposure of a large field party. If coastal outcrops are to be visited, then reconnaissance trips should aim to visit at similar tidal states as for the planned excursion. Reconnaissance trips provide a valuable opportunity to check logistical issues such as approach times to outcrops, mobile phone reception at outcrops in case of an incident, emergency parking provision and local hospital locations, for example. During these trips, it is often good to consider 'What if?' scenarios, for example, if a participant slips here, what is the consequence; how would you handle the incident in the location; how long would it take for help to arrive; and where are the evacuation routes from each locality?

HSE leads

A dedicated person to deal with HSE issues and be responsible for group safety is considered a sensible precaution. The HSE Lead has the authority to evacuate the group where necessary, for example owing to a rising tide or incoming electrical storm. They should be capable of exercising dynamic risk assessment, continually assessing the group's situation and taking decisions as appropriate. The HSE Lead should have a primary focus on safety, rather than on the geology, and for this reason in more hostile environments the HSE Lead may need to be professionally qualified in outdoor leadership or wilderness medicine rather than a geoscientist participating in the technical aspects of the trip. A shared technical and HSE Lead is a significant safety exposure as experience shows that as geoscientists we have a tendency to get carried away whilst

THREAT	CAUSE	CONSEQUENCE	PROBABILITY	EFFECT	RISK	CONTROL MEASURES	ACTION	RESIDUAL RISK
Road Accidents	Trave to and from field destination	1) Minor Injuries	L	M	L	Transport by Coach, Ensure Seat belts used. Ensure loading and offloading localities are in safe environs.	HSE Lead/ Whole Party	L
		2) Severe Injuries/Fatalities	L	H	M			
Slips, Trips etc.	Wet rocks, poor underfoot conditions, vegetation/seaweed/algae.	1) Cuts and Bruises	M	L	L	Ensure Suitable Footwear is used. Take time and care when walking on uneven surfaces/steep ground. Prevent rushing/assist groups over tricky steps.	Whole Party	M
		2) Twisted Ankles	M	M	M			
		3) Broken Bones	M	M	M			
		4) Fatalities	L	H	M			
Traffic	Injury through cross ng roads/railways or by walking alongside roads/railways	Road/Railway Accident	M	H	H	Special Attention required when crossing or working alongside travel infrastructure. Use traffic wardens with high-visibility vests to assist and ensure visibility.	HSE Lead/ Whole Party	L
Exposure to Adverse Weather	Changing Weather, Unsuitable Clothing, Time Overruns	1) Cold	M	L	L	Ensure Suitable Clothing Carried. Ensure Familiarity with Location, duration of approaches. Ensure Weather Forecast checked before departure. Be prepared to adapt programme in light of changing weather. Carry Spare equipment including Group Shelters. Consider Backup localities.	HSE Lead/ Whole Party	L
		2) Hypothermia	L	H	M			
		3) Lightning Burns	L	H	M			

Fig. 1. Example of part of a Risk Register for a field excursion.

interpreting rocks in the field and a technical lead, waxing lyrical on the detail of a coastal section, may not notice the approaching tide.

Permits and access

Certain field areas require either working permits or lie on private land, where permission is required prior to access. Identification of these in the planning stage will avoid disappointment on the day if access is refused. Other areas have very strict access regulations where access is only permitted at certain times of year, for example military firing ranges and areas where protected birds nest. Some field areas around the globe may also become subject to travel warnings or advice, from, for the example the UK Foreign and Commonwealth Office, which are important to heed prior to visiting a region.

Contingency planning

In the event of inclement weather or an unplanned incident when it is not safe to access the planned outcrops, it is sensible to have a backup plan. This may consist of alternative outcrops, such as when coastal sections are impassable owing to sea state, or may be an indoor option for lectures or a group exercise.

Common field trip hazards

Hazards associated with fieldwork can generally split into four categories:

- travel;
- fieldwork;
- hotel residence;
- personal health.

Hazards related to travel

The most dangerous part of any field trip is usually travelling to the destination and then to and from the field localities. As a consequence, understanding the key hazards associated with travel is important when putting together a mitigation plan. Road conditions, weather, local driving behaviours, the condition of vehicles, the use of intoxicating substances, driver fatigue, route conditions/choice, pedestrians and animals all act as hazards that need to be considered. Mitigations include ensuring drivers are adequately trained for the specific environment and vehicle, allowing adequate time for journeys and using journey management plans.

Hazards related to fieldwork

Geoscientists are exposed to a number of specific hazards when out in the field environment. Every locality has a different subset of hazards, depending on the landscape, time of year, climate and other factors. The discussion below considers some of the more common field hazards but is not exhaustive.

In the outdoor environment the most common injuries are related to slips, trips and falls. These may be triggered by external factors such as fallen rocks or uneven ground and may be exacerbated by sharp objects such as rocks, coral or vegetation. Temperature extremes whilst working in the field constitute a hazard to participants and can cause heat stroke, dehydration or sunburn, or hypothermia in the opposite case. Extreme weather including wind, snow, fog and rain may present physical hazards in the form of flash floods, freak waves, avalanches or bush fires, or personal hazards such as exhaustion, which may then trigger illness or cause an accident through tiredness. Any factor, such as tiredness, which elongates the length of the field day may result in the additional hazard of being benighted or working in low light.

The fauna and flora of the field area may also present hazards to the field scientist. Snake bites are usually serious. Insects may bite or sting, which is a particular problem to allergic individuals, and in certain locations may transmit diseases such as malaria or Lyme disease. In addition vegetation may produce toxic sap (e.g. giant hogweed, poison ivy or oleander) or pollen, to which some members of a trip may be allergic. Contact with wild or domestic animals is also a hazard in the field environment, particularly where animals may carry diseases such as rabies.

Finally, the geology itself may present very specific hazards such as active volcanoes or seismically active areas at risk from earthquakes. In these circumstances regular updates to the evacuation plan need to reflect the current risk profile at the time of the field trip.

Hazards related to overnight accommodation

Residential field trips carry an additional set of risks when compared with day trips. Overnight residence in an unfamiliar environment introduces hazards such as fire, lack of knowledge of evacuation routes and emergency procedures, blocked fire exits and theft of personal or company property. Mitigations can include prior vetting of accommodation and vigilance by participants on fire escapes and personal belongings.

Hazards related to personal health

The field environment can vary from being very benign to having significant risks to personal health depending on location and time of year and the composition of the group that is visiting the area.

The primary risks relating to personal health are rooted in participants being unfamiliar with the field environment. For example, lack of fieldwork experience, poor physical condition, improper or unfamiliar fieldwork gear or equipment failure all are risks associated with the occasional field geologist. In addition, a lack of awareness of the environment by some members places the remaining field trip participants at risk.

Poor leadership and organization can result in participants being exposed to risks including a lack of guidance in the field, improper documentation or missing vital information. These may be exacerbated by careless behaviour, pre-existing medical conditions, language or cultural differences and unfamiliarity, separation of individuals from a group and simple human error.

Finally participant non-adherence or lack of procedures and guidelines results in hazards. For example consumption of alcohol or drugs, including medicines, in the field, or dropped objects such as rucksacks from a coach luggage rack.

Risk mitigation

Risks, once identified, can be mitigated using a wide range of strategies including the most basic of simply modifying the itinerary to not visit a specific locality. For any trip, with a specified itinerary, there will always be a degree of residual risk, some of which can be mitigated through thorough HSE training for staff, others through use of correct equipment and some through simple procedural measures. These aspects of risk mitigation are discussed below.

HSE training

Travel. Driving is the single most dangerous activity undertaken on field courses. The risk can be minimized with the employment of professional drivers and by using a coach or minibus with a local driver for group transport. Local drivers are always preferred. Their knowledge of geography, traffic conditions, traffic laws and language are by far the best way to prevent accidents, provided the local driver is competent. Where local drivers are not available, participating drivers should familiarize themselves with the driving regulations and local conditions, including speed limits, of the country in which they are driving. All drivers should have a valid driving licence for the type of the vehicle being driven. Drivers should take time to familiarize themselves with the controls of the vehicles and remember that, if they are driving a four-wheel-drive or minivan, they respond differently to a car. Appropriate driver training can also reduce the risk of a driving-related incident. For all trips, defensive driver training is appropriate. On specific trips, off-road and skid-pan training may be beneficial. Strict enforcement of basic rules such as a zero tolerance policy on drink driving, a commitment to the use of seatbelts and the restriction of the use of mobile telephones and hand-held radios by the driver also add significantly to a safer trip. When driving in convoy an individual in each vehicle must know the proposed route. A safe driving policy can be built into the trip HSE plan.

The provision of appropriate vehicles is a key risk mitigation for all field trips. In an urban setting, coach travel with a professional driver is often the safest approach. In a remote setting, where travel is on dirt roads, a four-wheel-drive vehicle with extra ground clearance may be more appropriate. In particularly remote areas with dirt roads, risk mitigation may include carrying additional spare wheels and fuel and having extra passenger space to evacuate the group in the event of a breakdown.

First aid. All field trips should have first aid-trained personnel to deal with incidents, with the minimum level of training dictated by the relevant regulatory body (e.g. the Health and Safety Executive for the UK). The field environment can be a relatively hostile environment and very different to that simulated in basic first aid training, which often assumes an indoor environment and that professional medical help will arrive in a few minutes. First aid providers for the outdoor industry should always be considered for field trip first aid training. These organizations deliver training that is specific to giving first aid in more remote environments, focussing on skills that better prepare staff for dealing with medical emergencies. It should be stressed, however, that these are still first aid scenarios, where the core aims of preserving life, preventing worsening and promoting recovery are still adhered to, with a focus on immediate and temporary care, with the acceptance that medical assistance will be available in the near term, rather than requiring secondary medical care.

Scenario-based training, with regular refreshers, has been shown to have led to an improvement in staff competency in dealing with incidents. It has been shown to reduce the likelihood of panic, and most adequately prepares leaders and first aiders for execution of plans in the event of an emergency. For example, first aiders who are practically trained using scenarios are much less likely to panic during an incident and are more likely to remember the correct sequence of actions and their training on the relevant treatment plans given the situation they find themselves in. Figure 2 shows some example of outdoor-specific first aid training with the inclusion of scenarios.

Fig. 2. Examples of outdoor specific first aid training with inclusion of scenario-based training.

A ratio of one first aider to 10 participants is considered a reasonable ratio, with a minimum of two first aiders per group. If more remote field locations are considered, far from any possible assistance, then a more advanced medical provision and associated training should be considered, guided by the principle that any evacuation will be challenging and not immediate.

Each individual first aider should carry a first aid kit appropriate to the location being visited and be familiar with the contents. In addition, every vehicle used for a trip should have a first aid kit. In most countries first aid kits may not contain medicines and with specific exceptions first aiders are not trained to administer any medicines.

Water margin safety. Geology is often viewed most optimally on coastlines or in river beds, particularly in the UK and Ireland. The introduction of water, particularly a fast-moving river or a stormy sea, as a hazard into a field environment is a very serious risk and the best mitigation of this is avoidance. However, water margins still present a risk even in more benign conditions and basic training in the use of throw-lines is an option to better mitigate this, with a focus very much on prevention, rather than cure.

Staff competencies. Fieldwork can often take staff away from their comfort zone into a very different environment to their normal day jobs. Participants look to field trip leaders for guidance in both a conscious and unconscious manner. Leadership behaviours and competencies are therefore a critical and often overlooked part of field trip planning and preparation, which is often focussed around the technical geological aspects rather than HSE and outdoor leadership roles.

How leaders behave in the field environment will influence how participants act and contribute to a safer overall trip. People are strongly influenced by basic heuristics (Tversky & Kahneman 1974), so a person designated or appointed as leader has a large degree of influence over a group of individuals. A leader's behaviour begins with leading by example, from wearing the correct clothes and equipment to demonstrating good field etiquette. When provided with clear descriptions of reasons and rationale for why things are being done, participants are more likely to oblige in the majority of cases, for example, 'We are using helmets here as the cliff is loose and don't go under that overhanging section of cliff as you can see the evidence of recent rock falls', rather than 'Wear your helmets at all times and don't go over there'. A leader's behaviour can make the difference between a successful trip, where HSE is effectively managed but the participants do not feel bogged down by the HSE aspects and are focussed on the geology, to one of ineffective HSE management where the HSE aspects are thrust upon the participants at all times. In this latter case, guidance from the HSE Lead is more likely to be ignored and the technical aspects of the trip become blurred with factors such as keeping helmets on and being told where and where not to step.

Leadership competencies are crucial: research has demonstrated that people are willing to follow leaders who they rightly or wrongly consider experts and switch off their own decision-making abilities, often leading to catastrophic consequences (e.g. McCammon 2004). These competencies in leadership in the field environment can only be gained by experience, but can be supplemented and improved through training, for example, that provided by Mountain Training in the UK. Staff involved in HSE-critical roles should feel comfortable and confident in the outdoor environment in order to make effective decisions. Scenario-based training, such as simulated emergency response events, can provide those participants in HSE critical roles with competency and confidence in the event of an incident.

There is a wealth of literature on this subject in the outdoor press, much of which is applicable to the field environment, such as publications such as *Mountaincraft and Leadership* (Langmuir 2013), *Hillwalking* (Long 2014) and *International Mountain Trekking* (Plas y Brenin Instructional Team 2013).

Safety equipment

Personal protective equipment. In the offshore environment or in any industrial facility Personal Protective Equipment or PPE is part of life and an expected requirement; the same is also true in a field environment as part of an organized field excursion. PPE requirements will vary depending upon the region where the trip is conducted and the season in which the locality is visited. Here a few common themes are explored.

One of the simplest pieces of PPE that should be mandatory for all participants on all excursions is a robust pair of boots, with good tread and adequate ankle support. One of the biggest causes of injury in the outdoor environment in the UK is injuries from simple slips, trips and falls (Sharp 2007). Any risk assessment of a field locality will always carry this as one of the risks and often it remains as one of the highest after mitigations have been put in place as it is difficult to control. One of the mitigations for this injury mechanism is ensuring the participants have suitable footwear, reducing the likelihood of a slip and possibly decreasing its severity.

In many field localities, for example in the UK and Ireland, waterproof and warm clothing is essential to ensure that groups can cope with all weather conditions. Long days away from transport can mean significant periods potentially exposed to the elements with the effect of loss of morale/interest in the learning component of the trip and hypothermia in more extreme cases. Keeping participants warm and dry enables them to fully take advantage of the trip and to get the most educationally from the experience.

On trips where safety requires the group or an individual to be seen easily, high-visibility jackets should be worn. Examples of this include: when working on road or railway sections; when crossing a golf course; or when working in quarries or other industrial sites. Different or distinctive high-visibility jackets or vests are also a useful tool for identifying key HSE staff, particularly in a group when members do not know one another.

Other than these basic pieces of PPE, other items are very much location specific. These are likely to be driven by location, climate, remoteness and length of field day. For example, operating in desert and semi-arid regions items such as sun cream, hat and sunglasses may be considered essential items.

Good exposures of rocks are often found on cliffs and steep slopes. This often introduces a quandary into risk management and mitigation. In a simplistic risk assessment, all such outcrops would be avoided and studied only from a distance. In reality, it is important for participants to be able to investigate some of these outcrops more closely in a safe fashion. Risk assessment should be used to identify those cliffs or sections of cliffs for which there is no safe route of approach. Safe approach is time limited and dynamic risk assessment should be applied where a cliff is made more unstable by recent wet weather or storm activity. Where approach is possible, time at the base of the cliff or steep slope should be minimized and restricted to direct observation. Discussion and recording of observations should be conducted away from the fall zone. For working in these environments, the use of a helmet is advised as a method to partially mitigate the risk. It will not remove all risk of injury and careful group management at the outcrop is still necessary. Climbing helmets have been seen to be most effective for use in the field, with a good firm fit, staying on when moving around and robust enough to be used in the field environment. Helmets will rarely be required for all outcrops on a given trip, with prior and dynamic risk assessments aiding in their pragmatic use.

Safety cards are a good way for participants to have easy access to key information on the safety aspects of a trip, contained on a pocket-sized laminated piece of card. Contact phone numbers for the trip leader, HSE lead and first aiders and contact hotel information are useful both for day trips where a participant may become separated from a group and on longer trips where out-of-hours contact may be required.

Non-compliance in the use of PPE in the field is unfortunately a common problem on field trips, often not helped by overzealous HSE leaders enforcing procedures at all times, for example, helmet use at all times, even at the smallest of outcrops. A clear distinction needs to be made, however, between a field trip as part of employment and what one would do in one's own time. Employers are bound in the UK by the Health and Safety at Work etc. Act 1974 (Health & Safety Executive 2003), and as such need to take precautions guided by their risk assessments as to what PPE is required for the localities in question. Participants need to follow this guidance in order to safely allow participation in the activity as determined by the employer's risk assessment. HSE leads have a large role to play here, helping to keep the participants safe and following the risk assessment pragmatically through informed communication and group management.

Figure 3 shows a number of examples of Personal and Group Personal Protective Equipment in use in the field.

Group equipment. Some additional equipment should also be carried by the group to cater for possible incidents. First aid kits are a pre-requisite for all trips in conjunction with trained first aiders. Kit contents will be dictated by the legal framework of the country being operated in and it should be borne in mind that items legal in one country may not be in another. Group shelters, essentially large windproof bags that up to 12 people can get into in an emergency, are useful tools for incidents in inclement weather or even for a warmer lunch stop at an outcrop! Group shelters can also be used as a stretcher in the event that a field party needs to evacuate an ill or injured participant. In coastal or water margin environments a throw-bag can be carried if suitably trained staff or participants are available.

Communication devices are an important tool in the case of any emergency. All field parties should have a clear plan of what external help is available and how it should be contacted in the event of an incident. Mobile phone coverage is gradually reaching most parts of the world; however, there are some regions where a satellite phone may be more appropriate. Choice and selection of a communication medium will certainly vary based on world location and the remoteness and seriousness of the environment being worked in. It will vary from a mobile or satellite phones to VHF radios or Emergency Personal Locator Beacons.

Fig. 3. Examples of fieldwork Personal Protective Equipment: (**a**) boots with good soles and ankle support; (**b**) waterproofs – jacket and trousers providing protection against elements; (**c**) use of throwline at coastal outcrops where a slip could result in a participant in the water; (**d**) use of reflective vests to identify key HSE personnel – useful in large groups, where participants may not know one another; (**e**) use of hard hats when working under potentially unstable cliff sections; (**f**) group shelter for use in emergencies for isolating casualty and group from elements.

Regional differences may also influence the choice of equipment, for example, trips to study sedimentology in Svalbard may require equipment to deal with polar bears or trips to study structures in the Canadian Rockies may need to carry bear spray. It is important to identify the appropriate risks and translate the risks into mitigation using these items in the planning phases of a trip in order to be adequately prepared.

Group items such as spare warm or waterproof clothing, sun cream, spare hats and additional food and fluids should be carried at the discretion of the HSE Lead as dictated by group make-up and environment.

Procedures

Some risks associated with fieldwork can be mitigated through having procedures in place to reduce the likelihood of an event. One of the biggest risks in all fieldwork is driving to the outcrop or field area. As previously discussed, driver training is part of the mitigation, but keeping convoys of vehicles together is another good mechanism. In areas where a coach is not appropriate for transportation and multiple vehicles need to be used, keeping the group together helps to prevent individual vehicles becoming separated and keeps help close by in the event of any incident.

Outcrop management is important, from keeping participants away from dangerous areas and ensuring correct PPE is used (e.g. helmets) to maintaining vigilance for other previously unidentified risks.

On the field trip

HSE briefing

Once on the field trip, the focus for the participants should be on learning (and enjoyment!). A successful trip will be one where the safety aspects are effectively managed, but not in such a way as the participants feel they have been thrust upon them. In order to keep a focus on safety, an HSE briefing at the start of the trip, complemented by additional daily or outcrop specific briefings if required, can help to reinforce important messages. The objective of these briefings is to highlight the key risks, identify the mitigation measures, for example wearing helmets under certain outcrops, and to reinforce any company-specific or location-specific rules or guidelines. The outcome of the briefing should be a common and shared perception of risk for the whole group. Points should not be laboured, but participants should have a clear understanding of what is expected of them to safely execute the trip.

Dynamic risk assessments

Once on the trip, both the HSE Lead and participants should be vigilant of the hazards in the field area and should use careful judgement throughout the trip. An element of dynamic risk assessment should be used if new hazards have appeared since previous trips or reconnaissance visits or if the nature of a given hazard changes during the visit. For example, a bridge may have been washed away since the last visit or there may have been recent landslips or rock falls. Changing weather conditions such as the increasing threat of electrical storms will also require dynamic risk assessment. Modification of plans during the trip should be considered if it is deemed that conditions have significantly changed. Figure 4 shows some examples of where a dynamic risk assessment may need to be applied.

Incidents

Clear planning in the event of any incident is essential to enable a rapid and positive conclusion. In the event of an incident all participants need to know their role and how to execute emergency procedures safely and quickly. Clear procedures to follow, including emergency contact numbers, knowledge of local hospitals and emergency service provision should all be embedded in the safety plan and familiar to the HSE staff. A clear line of command starting from the HSE Lead is required in the event of any incident. With any incident there is always a focus on the casualty, but it is important that the rest of the group is also well managed and reassured to prevent a bad situation from getting worse and additional casualties occurring.

Health risks

Participation in field excursions in various parts of the world exposes geoscientists to a range of health risks that may not be present in the places where they live, for example, malaria, Lyme disease and dengue fever. The distribution of these health risks is constantly changing and as a result up-to-date information is always needed before leaving for a field trip. Detailed information can be found from the Foreign and Commonwealth Office in the UK and from dedicated travel clinics who can also provide any necessary vaccinations and medication needed prior to the trip.

The environment

Localities with good geological outcrops suitable for teaching are found in some of the world's most beautiful and scenic areas. Geological groups

Fig. 4. Examples of where dynamic risk assessment may need to be applied: **(a)** path collapse since previous visits; **(b)** late season snow prevents access to outcrops; **(c)** recent rockfalls may make access to outcrops difficult; **(d)** road washouts makes access difficult; **(e)** avalanche debris/landslides blocking access routes; **(f)** weather including sea state can make access to outcrops inadvisable/impossible

often will have to share the outcrops with other tourists and outdoor enthusiasts and it is important to balance the needs and expectations of different users. The old adage of 'take only photographs and leave only footprints' is important to follow. As such, a group's impact on the environment should be minimized, taking consideration of obvious mitigations such as not leaving rubbish, as well as considering group size, which at some localities should be limited in order to minimize environmental impact. Hammering should not be permitted, as aside from the safety implications, there are recent cases in UK where indiscriminate hammering at a Site of Special Scientific Interest has led geology groups into conflict with other outdoor users over outcrop damage. Each field group has a responsibility to preserve access for future trips and generations and not jeopardize such activities through abuse of the privilege of being in these areas.

Conclusions

This paper describes why staying safe in the field is required and offers guidelines on how this can be best achieved. Much of the discussion has focussed on common sense, but the importance of good planning and preparation is highlighted as a measure to reduce the likelihood of incidents. The development of an HSE plan is the bare minimum for any field trip, including a risk assessment and incident plan. Input into this plan, focussing on proper planning and reconnaissance, can significantly reduce risks and enables mitigation or avoidance plans to be put in place. The use of HSE training for staff and the use of a dedicated HSE Lead can all contribute to a safer overall trip. Finally, the other letters in HSE should not be forgotten – participant health and environmental impact can be equally important to the safety aspects of any trip.

In conclusion, the safe execution of field trips in the future will enable us as a geoscience population to continue working in the field and developing our knowledge of the subject through detailed analysis of field analogues.

The authors would like to acknowledge the discussions held with numerous colleagues involved in field trip safety over the years and for Shell UK Limited for permission to publish this paper. We would also like to acknowledge constructive and helpful advice from two anonymous reviewers.

References

Health and Safety Executive 2003. *Health and Safety Regulation – a short guide*. HSC13(rev1). http://www.hse.gov.uk/pubns/hsc13.htm

Langmuir, E. 2013. *Mountaincraft and Leadership*, 4th edn. Mountain Training Boards of England and Scotland, UK.

Long, S. 2014. *Hillwalking: The Official Handbook of the Mountain Training Walking Schemes*, 3rd edn. Mountain Training, UK.

McCammon, I. 2004. Heuristic traps in recreational avalanche accidents: evidence and implications. *Avalanche News*, **68**, 42–50.

Plas y Brenin Instructional Team 2013. International mountain trekking: A practical manual for trekkers & leaders. *In*: *The Official Handbook of the International Mountain Leader Scheme*. Mountain Training, UK.

Oliveri, S. R. & Bohacs, K. 2005. Field Safety in uncontrolled environments: a process-based guidebook. American Association of Petroleum Geologist. Division of Environmental Geosciences The American Association of Petroleum Geologists, Division of Environmental Geosciences, ExxonMobil Upstream Geoscience, Houston, TX.

Sharp, R. H. 2007. *Scottish Mountaineering Incidents (1996–2005)*, Research Report no. **109**. Sport Scotland.

Tversky, A. & Kahneman, D. 1974. Judgement under uncertainty: heuristics and biases. *Science*, **185**, 1124–1131.

The value of fieldwork in making connections between onshore outcrops and offshore models: an example from India

ANDY RACEY[1]*, JOHN FISHER[2], HAYDON BAILEY[3] & SANDIP KUMAR ROY[4]

[1]*Andy Racey Geoscience, 5 Norris Close, Bordon, Hampshire GU35 9EG, UK*

[2]*Genel Energy, 1 Grafton Street, London W1S 4FE, UK*

[3]*Network Stratigraphic Consulting, Harvest House, Cranborne Industrial Estate, Cranborne Road, Potters Bar, England EN6 3JF, UK*

[4]*A-503, Radhe Tower, Pashabhai Park, Baroda-39007, India*

**Corresponding author (e-mail: andy@andyracey.com)*

Abstract: Fieldwork provides an efficient and economical way of obtaining genuine geological data that can be used to constrain, test and develop exploration models in frontier areas. This study of the Cretaceous outcrops of NW India has been used to characterize models for petroleum systems postulated to have developed beneath the Deccan volcanic sequence, offshore western India, where well data are not available. To date, exploration of 'sub-Deccan plays' has mainly been limited to the NW, where the presence of a working Cretaceous petroleum system has been proven in the offshore Kutch Basin.

A revised chronostratigraphic framework has been defined for the Cretaceous section as a result of new biostratigraphic data collected during this study and a thorough revision of existing palaeontological data. The revised framework, together with lithofacies observations from the field, has provided a basis for regional sequence stratigraphic correlation, which has been used to link data from onshore outcrops to models of offshore petroleum systems.

A Lower Cretaceous clastic package that, in part, corresponds to the Lower Goru and Bhuj Formation reservoirs of the Indus Basin of Pakistan and the Kutch Basin in India has been correlated across the region. Towards the top of this package, a transition from fluvial to shallow-marine sedimentation marks the onset of a marine transgression that reaches its peak in the Turonian. Marly limestones and claystones associated with this regionally mapped event are a potential topseal to the Lower Cretaceous clastic package. Geochemical evaluation of several outcrop locations has identified potential source-prone facies that, if present downdip in the offshore, may provide a mechanism for charging a pre-Deccan play.

The data for this study were collected over three field seasons between 2010 and 2011, during which some 137 localities were studied and a total of 586 samples were collected for follow-up laboratory analyses. This work was undertaken in support of BG Groups' offshore exploration effort in western India, with the main intention of using onshore outcrops to identify and help de-risk potential offshore plays. The focus of the fieldwork was on the Cretaceous sections exposed in Narmada, Saurastra, Jaisalmer and Kutch, and, to a lesser extent, the Jurassic in the Jaisalmer and Kutch basins (Fig. 1).

The ages and environments of outcrop sections were analysed, and this information was extrapolated downdip into the offshore within a new sequence stratigraphic framework. This permitted potential reservoir and seal units as far west as Pakistan to be correlated with the outcrops, and also permitted the identification of new potential source intervals in the Saurastra and Cambay areas. The workflow was simple and included an initial literature review to identify areas/sections of potential interest, followed by fieldwork and selected laboratory analyses during which the following were noted:

- The published lithostratigraphic schemes contain many inconsistencies in terms of age, environment and terminology. Moreover, there is additional confusion owing to the inconsistent use of basic lithostratigraphic terms such as Group, Formation, Member and Beds. BG therefore constructed an in-house lithostratigraphy, unifying the plethora of published formation and member names into one simple scheme that broadly fitted field observations.
- A standardized chronostratigraphy was established using various macrofossils (echinoids,

From: Bowman, M., Smyth, H. R., Good, T. R., Passey, S. R., Hirst, J. P. P. & Jordan, C. J. (eds) 2016. *The Value of Outcrop Studies in Reducing Subsurface Uncertainty and Risk in Hydrocarbon Exploration and Production.* Geological Society, London, Special Publications, **436**, 21–53.
First published online May 23, 2016, http://doi.org/10.1144/SP436.9

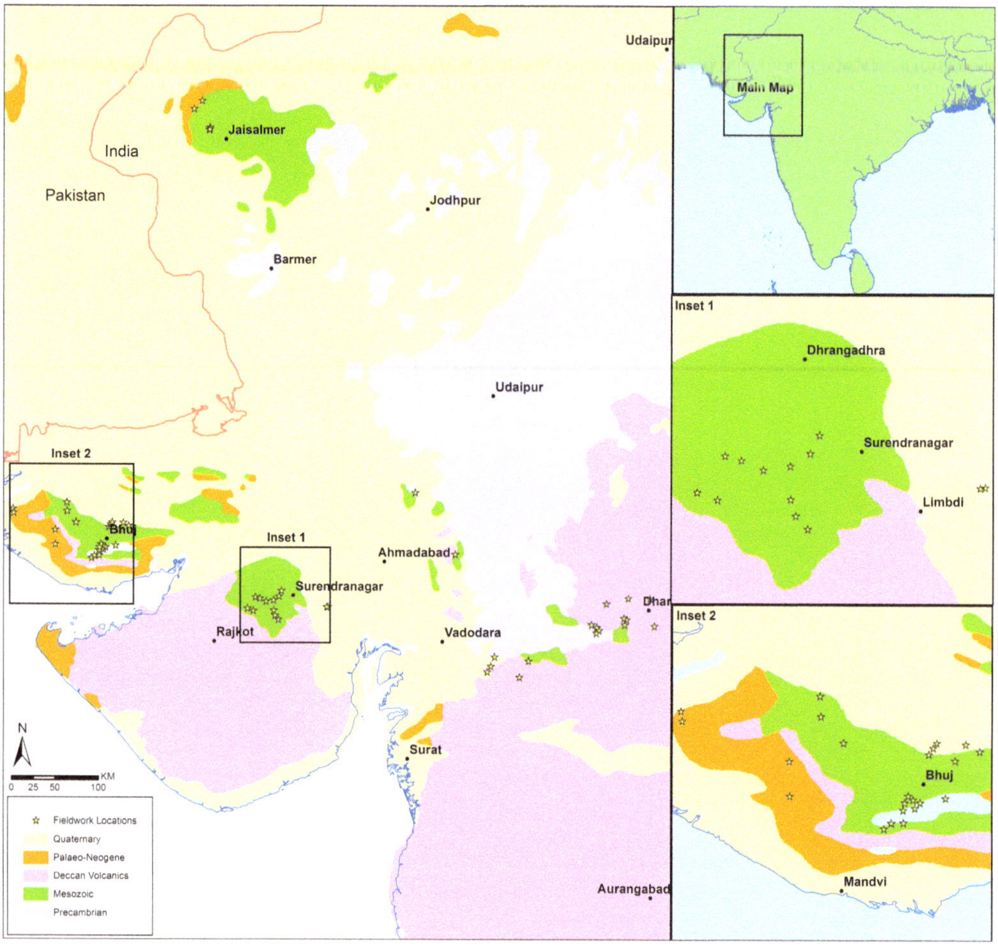

Fig. 1. Geological map highlighting the main studied areas of Mesozoic outcrop in green. The yellow stars all denote studied localities within the Cretaceous (some of which are mapped as Deccan Traps by the Indian Geological Survey). The studied localities marked in the Kutch Basin (inset 2) also include some in the Jurassic and Tertiary, which are not discussed in this paper. Samples localities in the Saurastra (inset 1) and Kutch and Jaisalmer basins are shown on subsequent figures. The blue basement is wholly of Precambrian age and mainly comprises quartz-rich metasediments.

bryozoans and ammonites) and microfossils (foraminifera, nannofossils and palynomorphs).

- The environments of deposition were established through studying the faunas and sedimentary structures in the field, coupled with microfacies analyses of thin sections. Petrographical modal analysis of sandstone samples was used to identify the likely provenance source areas using published geological maps and outcrop data.
- Potential petroleum source intervals were screened for total organic carbon (TOC), and samples containing >1% TOC were analysed further using Rockeval. Potential reservoir intervals were assessed using thin sections and selected plug porosity–permeability measurements.

The overall objectives of this study were to improve the understanding of the pre-Deccan petroleum potential in the offshore, focusing on the Cretaceous by studying the ‘updip’ onshore basins that may have fed sediment into offshore western India. From a company perspective, BG Group – who funded this fieldwork – also hoped to gain an improved understanding of their recently acquired offshore acreage and the pre-Tertiary potential beneath existing fields, such as Panna, Mukta and Tapti, which produce hydrocarbons mainly from the Eocene to the Miocene. Exploration activity in the Kutch offshore area had established the presence of a working petroleum system (source and reservoir) in the pre-Deccan in three wells in the

Jurassic, but none of these were deemed commercially viable.

Mesozoic basins are prolific hydrocarbon producers on a global scale: yet, in India, where they occupy around 400 000 km^2, they remain markedly underexplored. This is especially the case where such basins are overlain by the Deccan Traps, which cover an area of 273 000 km^2 and where proprietary seismic and gravity data studies often indicate an underlying Mesozoic sediment thickness of up to 2.5 km.

The western margin of India has evolved in three main phases: Early Jurassic rifting of India from Africa (Besse & Courtillot 1988), which influenced the tectonic evolution of NW India; mid Cretaceous separation of India from Madagascar (Morgan 1981); and Late Cretaceous separation of India from the Seychelles – the latter evidenced by the Reunion Hotspot. Biswas (1982) suggested that rift basin development in western India commenced in the Early Jurassic in the Kutch and Jaisalmer basins, in the Early Cretaceous in the Cambay Basin and mainly in the Late Cretaceous in the Narmada Basin. The location of these rift basins is strongly controlled by Precambrian basement tectonic trends: the Aravalli Trend (Kutch Basin), the Dharwar Trend (Cambay Basin) and the Satpura Trend (Narmada Basin). The main rift systems of western India are shown in highly simplified form in Figure 2, which shows the Cambay Rift

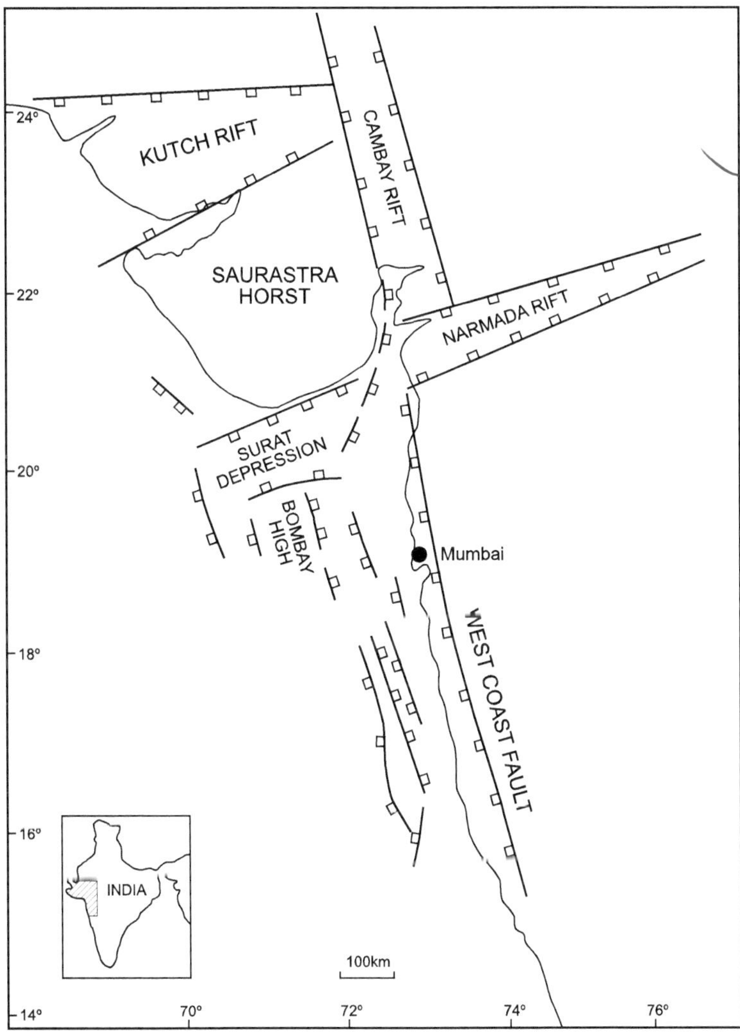

Fig. 2. Simplified map showing the main rift basins in the study area. Note the cross-cutting relationships of the different rifts, the timing of which is discussed in the main text.

truncating the older Kutch Rift and itself being later truncated by the younger Narmada Rift. The role of these Precambrian belts in controlling subsequent basin development is discussed in Raval (2003) and Raval & Veeraswamy (2003).

Database

A total of 475 thin sections were made and analysed to assess reservoir quality, diagenesis and sand provenance. The carbonate thin sections were logged for fossil content (macro- and microfossils) in order to interpret the environment of deposition and age. Altogether, 179 biostratigraphic analyses (100 palynological, 38 micropalaeontological and 41 nannofossil) were undertaken, along with the identification and review of selected macrofossil material collected from the outcrops. Sixty-six mudstone–shale samples were analysed to assess their hydrocarbon source potential using TOC and Rock-eval analysis, whilst 66 sandstone samples were plugged and analysed for porosity and permeability.

The Cambay Basin

The Cambay Basin forms a north–south-trending depression located between the Deccan Trap outcrops of the northern peninsula of India and Saurastra (Kathiawar), and includes the offshore Gulf of Cambay area. It is approximately 400 km long and 40 km wide, and may have originally been more extensive, possibly connecting with the Kutch and western Rajasthan (including Jaisalmer) basins to the NW. The basin is subdivided into several discrete tectonic blocks by broadly east–west-trending faults. The northern part of the basin contains a thin Cretaceous sedimentary sequence resting unconformably on Precambrian basement, and thus affords limited Cretaceous prospectivity. In the southern part of the basin, this Cretaceous sequence is over 1000 m thick (Mukherjee 1983).

Himatnagar Sandstone Formation

Lithofacies. The Lower Cretaceous Himatnagar Sandstone Formation is exposed on the NE edge of the Cambay Basin (Fig. 1), and consists of a sequence of grey to red sandstones, siltstones and claystones. The sandstones generally form fining-upwards sequences, and are predominantly medium grained and moderately well sorted. Sedimentary structures are dominated by large-scale planar and cross-stratification, whilst ripples (symmetrical and asymmetrical) and convolute lamination are also noted.

Sandstone samples of the Himatnagar Sandstone Formation were collected from quarries around Himatnagar and in road cuttings in the easternmost part of the basin where 30–60 m of the formation is exposed. To the east of the town of Himatnagar, the Himatnagar Sandstone Formation rests unconformably on the Erinpura Granite. In thin section, the sands are mainly moderately sorted quartz arenites with average porosities of 14%. Exploration wells in the Cambay area have encountered significant thicknesses of Himatnagar Formation sandstones: for example, >229 m was encountered in the Viramgam-1 well, located some 45 km east of Dhrangadhra on the western side of the basin, and 350 m resting on granite basement in the Serau East-1 well to the immediate north of the Cambay Basin.

Depositional environment and age. Fining-upwards cycles and associated sedimentary structures are interpreted to represent meandering channels deposited in an overall fluvial system. Channel lags, point bars and floodplain facies are also recognizable within this mainly fluvial sequence. However, a marine or estuarine influence has also been noted (see Akhtar & Aquil 1984). Plant fossils (ferns) from this formation are reported to have a Neocomian age (Sahni 1936).

Lameta Formation

Upper Cretaceous fluvial sandstones and calcretes of the Lameta Formation also outcrop in the area, and these have been dated as Maastrichtian, based on their dinosaur and palynofloral content (Tandon *et al.* 1995; Joshi & Ganapathi 2000).

The Narmada Basin

The Narmada Basin is confined to a narrow, westwards-opening, ENE–WSW-trending graben. The basin is bounded by a series of sub-parallel strike-slip faults that form the Narmada–Son Lineament. This lineament represents a major feature that extends into the Moho to a depth of around 40 km (Kaila *et al.* 1981, 1985). The Narmada Basin is around 1200 km long in a broadly west–east direction and extends as far east as Jabalpur. Precambrian basement is exposed along its northern margin. The Narmada–Son Lineament forms the boundary between Precambrian and Palaeozoic rocks to the north and Gondwana deposits to the south. During the Early Cretaceous, fault reactivation resulted in rifting along the Narmada–Son Lineament, with regional uplift to the north and south forming an intervening graben that, in Late Cretaceous times, was temporarily open to the sea at its western end (Biswas 1987). The Narmada Basin did not extend into the Bombay offshore region where Cenozoic sediments and Paleocene Deccan volcanics are seen to rest unconformably on Precambrian basement, but did extend into the Surat

Depression, offshore Saurastra (Babu 1984). Uplift of the Saurastra Arch (Horst) to the NW may have caused the cessation of Cretaceous sedimentation in the Narmada Basin (Biswas 1987). Early Tertiary Deccan Trap basalts (around 900 m thick) overlie the Cretaceous sediments in the Narmada Basin. The Narmada Rift intersects the NNW–SSE-trending Cambay Rift at the western end of the Narmada Basin to create a Cenozoic depocentre.

Cretaceous subsidence within the Narmada Basin was mainly restricted to the western end, with the depocentre both widening and deepening towards the west. The basin fill rests unconformably on Precambrian basement. Over 1800 m of fluvial deltaic Lower Cretaceous Nimar Group sandstones occur in the western part of the basin (based on seismic and borehole data): at outcrop, however, the group thickens from 15–30 m in the east to >150 m in the west (Biswas & Deshpande 1983), although the base and/or top of the group are rarely exposed. In the west, the mainly fluvial portion of this interval is also referred to as the Himatnagar Sandstone Formation or Songir Group, whilst the upper marine portion is commonly assigned to the Uchad (Umrali) Flagstone (Murty *et al.* 1963).

In this study, the Upper Cretaceous sequence is assigned to the Bagh Group and comprises 'transgressive' argillaceous nodular limestones, which become less nodular and more shaley towards the west. These nodular limestones are overlain by 'regressive' shallow-marine bryozoan–echinoderm–molluscan limestones, which are unconformably overlain by calcrete-dominated non-marine deposits of the Maastrichtian Lameta Formation.

The fieldwork in Narmada focused on the Zeerabad, Bagh and Dhar areas in central Narmada, on Jabalpur in eastern Narmada, and around Rajpipla, Kawant and Naswadi in western Narmada (eastern Cambay). A geological map showing the main localities mentioned in the text is provided in Figure 3.

Stratigraphic discussion

The Bagh and Nimar groups outcrop in a series of isolated outcrops along various rivers in the Narmada Valley and are intermittently exposed from Barwaha in the east to Naswadi in the west, a distance of some 275 km. The most complete and most fossiliferous outcrops occur in the vicinity of Zeerabad and Bagh (Fig. 3). The southern limit of the Nimar and Bagh groups is defined by the Narmada Fault and the Satpura Horst, with the latter overlain by the Deccan Traps.

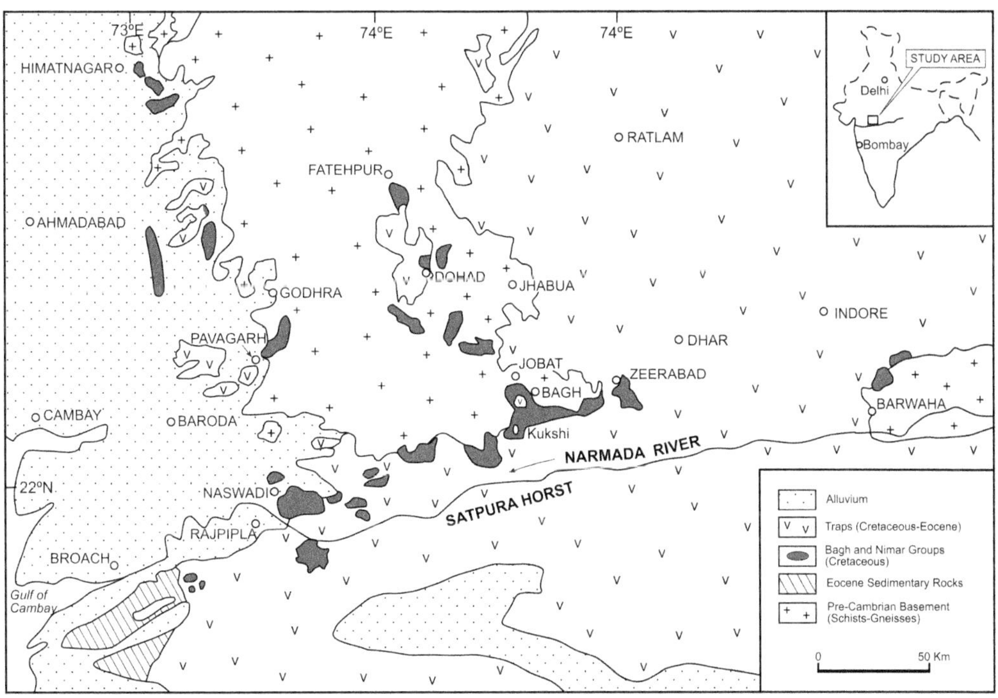

Fig. 3. Location map for Narmada and eastern Cambay showing the main outcrop areas of the Bagh and Nimar Group sediments in grey.

The pre-Deccan sequence in the Narmada Basin comprises three main units: the Lower Cretaceous Nimar Group, the Upper Cretaceous Bagh Group and the Lameta Formation (Fig. 4). The Bagh Group, as applied herein, comprises the Nodular Limestone (including the Deola Marl of some workers) and Coralline Limestone formations. The Bagh Group is unconformably overlain by the Maastrichtian Lameta Formation. The shallow-marine calcareous sandstones, referred to as the 'Upper Nimars', have been included as the basal unit of the Bagh Group by some authors (Dassarma & Sinha 1975). These sandstones show increasing marine influence towards the west and our field observations indicate that they form the uppermost part of the fluvial-dominated Nimar Sandstone of the Nimar Group.

The marine fossiliferous part of the Nimar Formation has been assigned to the 'Upper Nimars', the Amplipura Oyster Bed (or, just simply, the Oyster Bed) and may also be equivalent to the Umrali Flagstone of other authors (e.g. in Murty *et al.* 1963).

The published lithostratigraphic terminology applied to the Cretaceous in the Narmada Basin is complex and varies markedly depending on which publication is applied. For a summary and comparison of the various early schemes of Blandford (1869), Bose (1884), Rode & Chiplonkar (1935), Roy Chowdhary & Sastri (1962), Murty *et al.* (1963) and Poddar (1964), see Dassarma & Sinha (1975). Subsequent additional modifications and additions are noted in Sahni & Jain (1966), Pal (1971), Ramasamy & Madhavaraju (1993), Taylor & Badve (1995), Tripathi (1995, 2000) Kennedy *et al.* (2003) and Smith (2010).

Although the literature is much confused as to which 'units' comprise the Bagh and Nimar groups, the localities we studied comprised a basal sandstone-dominated fluvial sequence that becomes marine influenced at its top (sometimes with associated oysters), which we assign to the Nimar Group. This is overlain by a sequence comprising nodular limestones that are, in turn, capped by bryozoan ('coralline' of older studies) limestones, both of which we assign to the Bagh Group (Fig. 1).

Narmada		
Turonian	Bagh Group	**Coralline Limestone Formation** (Temperate shallow marine)
		Nodular Limestone Formation (Deep to shallow shelf)
Albian-Cenomanian	Nimar Group	**'Upper Nimars'** (Fluvial to shallow marine)
Hauterivian-Albian		**Nimar Sandstone** (Fluvial)

Fig. 4. Narmada basin stratigraphy.

Our field observations indicate that the Bagh Group shows marked lateral variation in facies and includes multiple hardgrounds and hiatuses, with the Nimar Sandstone and Nodular Limestone Formation separated by a significant flooding surface and possible stratigraphic break.

Nimar Sandstone Group

The following account of the Nimar Sandstone Group includes the 'Upper Nimars' of earlier workers. The sandstones, which comprise the bulk of the Nimar Sandstone Group, are a locally important aquifer and, consequently, have reasonable porosity and permeability in the 'near' subsurface.

Lithology. The group is well exposed in the Bagh area around the Bagh Caves, where it reaches thicknesses of 30–40 m (Fig. 5). The unit expands towards the west, reaching a thickness of at least 100 m. The upper and lower contacts of the group were not observed in any single continuous section, so the true thickness could be considerably more. It is unconformably overlain by the Nodular Limestone Formation and rests unconformably on Precambrian acidic basement. The Nimar Sandstone Group (particularly in the eastern part of the basin) is dominated by thickly bedded, red, ferruginous sandstones, with occasional interbedded conglomerates, common siltstones and palaeosols (Fig. 5). Cross-bedding and ripples are common, with occasional carbonaceous clays, plant fragments and roots in interbedded mudstones and siltstones (Fig. 5). The sandstones in thin section often show poor sorting and are texturally immature (Fig. 6f). The sands are often coarse grained and quartz dominated, and are likely to have been derived from local acidic metamorphic and, to a lesser extent, granitic basement, which crop out across the study area.

The uppermost part of the Nimar Group consists of thinly bedded, fine-grained, calcareous sandstones and interbedded siltstones, with common lenticular bedding and ripples. This upper part is bioturbated, and contains macrofossil debris including echinoderms, oysters and bryozoans (Fig. 6e). In thin section, the detrital clastic component is similar to that noted in the underlying fluvial portion indicating a similar sediment provenance.

Depositional environment. The lower part of the Nimar Sandstone Group is interpreted as having been deposited in a fluvial setting. The sandstones represent discrete sandy periods with a lateral switching of metastable, mainly meandering channels, which suggests low accommodation space. The

Fig. 5. (**a**) Outcrop of Nimar Group sandstones in the Man River area near Bagh Caves. Note the person in the bottom right for scale. The lower part comprises coarse to pebbly cross-bedded sheet and channelized sandstones. The upper part of the section is shallow marine and contains bryozoan and echinoderm debris, and is locally bioturbated. (**b**) Pebbly and conglomeratic lags within fluvial channels; (**c**) ripples; (**d**) calcified tree roots; (**e**) fluvial sheet sandstones; and (**f**) fluvial channel incising sheet sandstones.

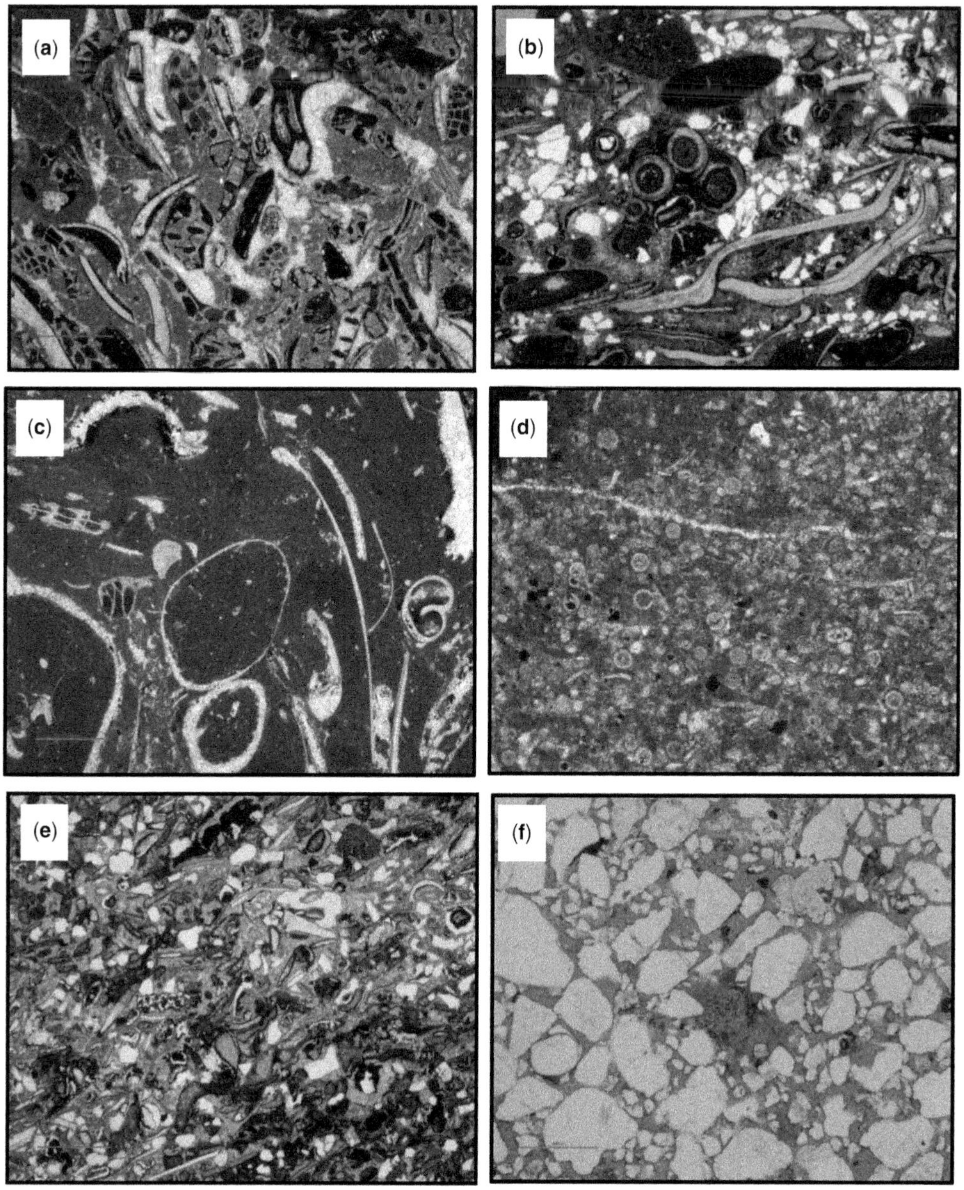

Fig. 6. Photomicrographs. (**a**) & (**b**) Coralline Limestone Formation showing the dominance of bryozoan (a), molluscan and echinoderm debris as packstones and grainstones with common reworked limestone clasts. Quartz grains and metamorphic quartzite fragments are locally common. Note rare serpulid worms (b, just left of the centre). The general absence of micrite envelopes on many of the grains, absence of coral, larger foraminifera and calcareous algae, and the dominance of bryozoans, suggests deposition in a very-shallow-marine temperate setting. (**c**) & (**d**) Nodular Limestone Formation: (c) upper Nodular Limestone showing common bivalve, gastropod and bryozoan debris in a micritic groundmass representing shallow-marine inner-shelf (?lagoonal) deposits; (d) lower Nodular Limestone showing common calcispheres and rare smaller foraminiferal fragments in a partially recrystallized micritic groundmass representing deep-marine, low-energy outer-shelf deposits. (**e**) Upper Nimar Sandstone containing common shallow-marine fragmentary fauna including molluscan, bryozoan and echinoderm debris. (**f**) Poorly sorted, texturally immature fluvial sandstone from the lower part of Nimar Sandstone with a porosity of 16% (blue stained in online version of this figure). Minor kaolinite (pale green in the online version of this figure) (?after feldspar) occurs near centre of image. Porosity is stained blue in the online version of this figure. The red scale bar is 1 mm in (a)–(b), 0.5 mm in (e) and (f) and the field of view in (d) is 1 mm.

channels show little evidence of incision and are separated by palaeosols. Palaeocurrents are mainly directed to the SW and west.

The Nimar Sandstone Group has been interpreted as representing a variety of shallow-marine settings from shallow sublittoral (Chiplonkar *et al.* 1977) to intertidal/subtidal (Singh & Srivastava 1981) to a transgressive storm and fairweather wave-dominated shelf (Bose & Das 1986). Ahmad & Akhtar (1990), based on studies in the western part of the basin, interpreted it as representing an overall macrotidal estuarine complex consisting of shallow meandering channels and mud flats. Overall, the degree of marine influence is observed to increase from east to west along the basin. The sandstones are often calcareous, locally show bioturbation and contain macrofossil debris (echinoid, bryozoan and molluscan), indicating deposition in a shallow-water, open-marine, normal salinity setting.

Age. Jafar (1982) dated the uppermost marine part of the Nimar Group as Turonian, based on nannofossils, and assigned it to the *Eiffellithus eximius* Zone (i.e. late Turonian–early Coniacian), which, based on the older age of the overlying Nodular Limestone (discussed below), must be considered questionable. The presence of the bryozoan *Chiplonkarina dimorphopora* as fragments within its uppermost part suggests an age no older than upper Cenomanian, based on its known stratigraphic distribution (Taylor & Badve 1995). Kundal & Sanganwar (1998) collected a diverse calcareous algae assemblage from the uppermost part of the Nimar Sandstone near Pipaldehla, which included several species with a composite age range of Cenomanian–Turonian.

Plant fossils from the lower fluvial portion of the group suggest a Lower Cretaceous age within the range Hauterivian–Albian and are broadly similar to those of the Wealden (Murty *et al.* 1963). Kumar (1994) recorded a palynomorph assemblage in carbonaceous clays from the lower part of the fluvial portion of the Nimar Sandstone at Umrali village, Jhabua District. The palynomorphs listed (but not illustrated) comprised *Cyathidites australis*, *C. minor*, *Haradisporites mineri*, *Stereisporites* sp., *Cingutriletes* sp., *Callialisporites dampieri*, *C. trilobatus*, *Klausipollenites* sp., *Alisporites ovalis*, *Abiespollenites triangularis*, *Podocarpidites ellipticus*, *P. multesimus*, *Araucariacites australis*, *Cycadopites gracilis*, *C. sakrigaliensis* and *Striatites* sp. This palynomorph assemblage would indicate a Late Jurassic–Early Cretaceous age. Similar palynofloras have been recorded from the Jabalpur Group of eastern Narmada (Kumar 1973, 1992; Kumar & Kulshreshtha 1979), the Katrol/Jhuran and Umia/Bhuj formations of Kutch, and the Dhrangadrha Group of Saurastra.

Petrographical analysis. Both the fluvial and shallow-marine sandstones are dominated by quartz arenites, with detrital compositions dominated by mono- and polycrystalline quartz, metamorphic rock fragments and rare alkali feldspar. Quartz varieties comprise mainly metamorphic sutured and strained grains, and rare composite plutonic quartz grains. The marked dominance of quartz is partly due to feldspar dissolution during diagenesis (Fig. 6f). In the upper part of the formation, calcite cement and fragments of oysters, bryozoans (*Chiplonkarina*) and echinoderms are noted (Fig. 6e).

Akhtar & Ahmad (1991), in a study of 176 sandstone thin sections from the western part of the Narmada area (from Himatnagar in the north to Naswadi in the south, and including Pavagarh and Fatehpur in-between), noted that the textural maturity varied markedly from immature to supermature, but was dominated by immature (33%) and submature (56%) sandstones, with a clay matrix in the immature sandstones varying from 6 to 24% (see Fig. 3 for the locations). Monocrystalline quartz was the main detrital component, averaging 79% around Himatnagar, 59% around Fatehpur, 82% around Pavagarh and 82% around Naswadi (see Fig. 3 for locations).

The Fatehpur samples had a higher feldspar (average 7%) and higher rock fragment (average 26%) content than the other localities, and this may be because they were located closer to the Precambrian acidic basement source area. Rock fragments include metamorphic (phyllites, quartzites and schists) and sedimentary clasts (including shale and siltstone). The feldspars comprise orthoclase, plagioclase and microcline. Akhtar & Ahmad (1991) recorded a heavy mineral assemblage of rounded tourmaline, staurolite, garnet, rutile and euhedral zircon, which together suggest derivation from a metamorphic basement terrain with nearby granite intrusions to account for the euhedral zircon.

Akhtar *et al.* (1992), based on modal analyses of 286 samples, noted porosity ranges of 0–37% (mean 6%). In this study, 33 samples were analysed to give a range of 0.5–20%, with a mean of 7.4% porosity. The detrital grains in Akhtar *et al.* (1992) comprised 97% quartz, 2% feldspar and 1% rock fragments compared to our analysis, which show an average detrital composition of 90% quartz, 4.5% feldspar and 5.5% lithic fragments. The dominance of point and long grain contacts in the thin sections suggests limited pressure solution/burial, with an average pre-cementation porosity of 25% indicating a burial depth of 1000–3500 m according to Akhtar *et al.* (1992). Since the preserved overlying Deccan sequence is generally only around 250 m thick in the area, there must have been a large amount of uplift and erosion prior to the eruption of the Deccan Traps (possibly related to the uplift of the western Ghats). A substantial uplift pre-Deccan most probably resulted in the introduction of meteoric waters, causing the dissolution of the original feldspars, the formation of iron-rich clays and the precipitation of blocky calcite cement.

Sediment provenance. Precambrian rocks of the Aravalli and Dehli supergroups, comprising mainly quartzites and phyllites, occur to the immediate north and NE of the study area and are intruded by granites (leading to the formation of garnet and staurolite). These provide the most likely sediment source area for the Nimar sandstones. In the Pavagarh area, meta-sediments of the Precambrian Aravalli Supergroup, with associated phyllites, schists and quartzite, crop out over a large area and extend southwards to Naswadi and eastwards to Fatehpur, with a large area of gneiss exposed near Naswadi. An abundant source of quartz is therefore available from Precambrian basement across the area. The low textural maturity of many of the sandstones tends to suggest local derivation, with limited long-distance transport and reworking, whilst the general paucity of feldspar in the sandstones (usually $<1\%$), despite its fair to common occurrence in the Precambrian sediment source area, is possibly due to prolonged weathering in a humid tropical climate, as suggested by Akhtar & Ahmad (1991), coupled with its dissolution during burial.

Bagh Group

A typical outcrop section of the Bagh Group and an accompanying sedimentary log are provided in Figures 7 and 8.

Nodular Limestone Formation

Lithology. The formation is best exposed in the Zeerabad, Bagh and Man River Valley areas, where it attains a thickness of around 15 m (Fig. 8). It is conformably overlain by the Coralline Limestone Formation and rests unconformably on shallow-marine sandstones of the Nimar Group (Fig. 7). The lower part of the formation comprises thinly bedded nodular marly limestones, containing fine molluscan debris, smaller benthic, and rare planktonic foraminifera and calcispheres (Fig. 6d). This unit is separated by a thin iron-stained marl band, around 20 cm in thickness, which contains rare ammonites

Fig. 7. Bagh Group outcrop at the cement quarry in Zeerabad. The base of the Nodular Limestone represents a major Turonian transgression, whilst a thin muddy, ammonite-bearing band separating the Lower and Upper Nodular Limestone probably represents a maximum flooding surface separating a deep-marine outer shelf below from shallow open-marine shelf above. The boundary between the Coralline Limestone and the Nodular Limestone represents a sequence boundary above which there is marked emergence and reworking.

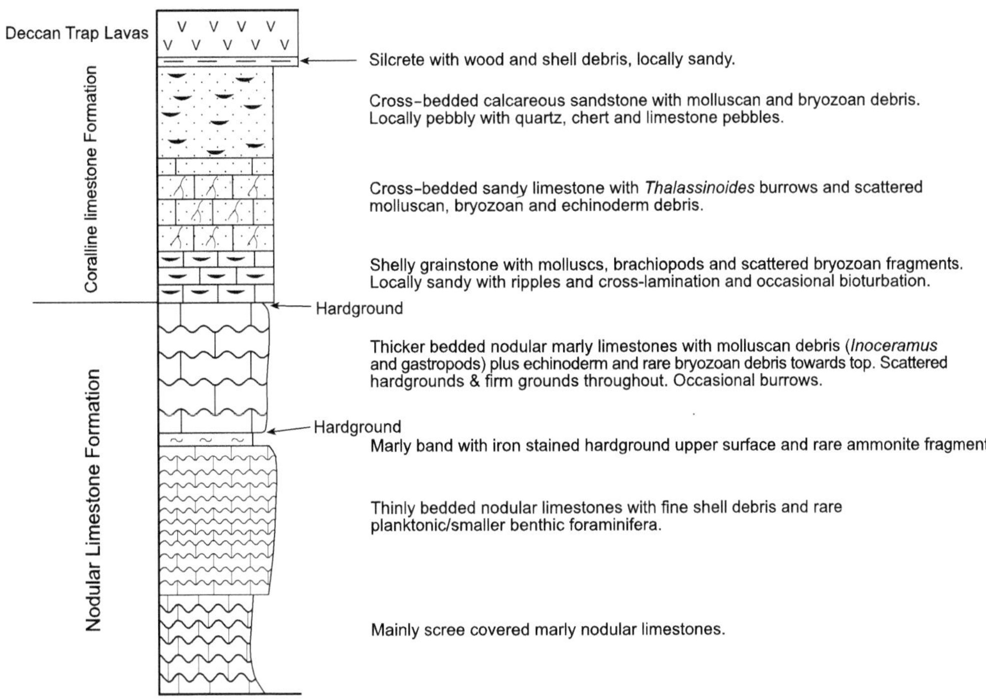

Fig. 8. Simplified summary log of the Bagh Group sequence exposed in quarries around Zeerabad. The sequence is about 12 m thick at this locality. Patches of Nimar Group sandstone outcrop less than 1 m below the base of this section, but the actual contact is not exposed.

and is capped by a hardground before passing into a thicker bedded nodular limestone interval containing more obvious and coarser macrofossil debris, including bivalves, gastropods, bryozoan and echinoderm material (Fig. 6c). A hardground surface with associated *Thalassinoides* burrows occurs at the top of the formation, whilst less obvious hardgrounds/firmgrounds occur in the upper part of the formation.

Around Zeerabad (east), the formation is more bedded than nodular, and contains marine fossils and trace fossils (*Thalassinoides*): further west, in the Bagh area, however, it appears more nodular and lacks obvious fossils, which could be either because it is deeper water and/or it is more distal.

A distinctive thin (20 cm) recessively weathering unit subdivides the Nodular Limestone around Zeerabad. This may approximate with an ammonite flood of '*Placenticeras*' noted in the literature and probably represents a maximum flooding surface (MFS) separating the older calcisphere-dominated interval from the younger and environmentally shallower marine macrofossil-dominated interval in the Nodular Limestone.

Depositional environment. The formation comprises a deeper-water shelf lower portion dominated by micritic nodular limestones containing locally common to abundant calcispheres, rare planktonic foraminifera and rare macrofossil debris (*Inoceramus*, echinoderm and ostracods). The upper part consists of thicker-bedded nodular limestones with hardgrounds, and contains a shallow-water, open-marine fauna of echinoids, bryozoa, gastropods and bivalves, and locally common *Thalassinoides* burrows. The absence of typical warm water taxa such as 'larger' foraminifera, calcareous algae and corals tends to suggest deposition in temperate waters.

Age. There are several records of ammonites and inoceramid bivalves from the formation. Chiplonkar & Ghare (1976) and Chiplonkar *et al.* (1977), on the basis of the presence of various Placenteritid ammonoids, assigned a Coniacian–Turonian age to the Nodular Limestone–Coralline Limestone interval. Bardhan *et al.* (2002) recorded *Placenticeras kaffrarium* from the middle part of the nodular limestone in the Man River area, and noted its occurrence in the Turonian–Coniacian of Madagascar and the Middle Coniacian of Zululand. Bardhan *et al.* (2002) also recorded the inoceramid *Volicceramus involutus* from the uppermost part of the Nodular Limestone below a hardground separating it from the overlying Coralline Limestone, and

Inoceramus cf. *rotundatus* in the lower part of the Nodular Limestone below another hardground. They interpreted both as indicating a Coniacian age and concluded that the overlying Coralline Limestone was Middle Coniacian and the Nodular Limestone was Early–Middle Coniacian in age.

Subsequently, Kennedy *et al.* (2003) found the Late Turonian marker ammonite *Prionocyclus germari* in the upper Nodular Limestone in the Hatani River near Kukshi, and concluded that previous ammonite and inoceramid records were compatible with a Late Turonian age for the upper Nodular Limestone and a Late Turonian–Coniacian age for the overlying Coralline Limestone. However, what they refer to as the upper Nodular Limestone they show as a 1 m-thick unit both over- and underlain by a 0.5 m-thick Coralline Limestone with a 3 m-thick lower Nodular Limestone below resting on the Nimar Group sandstones. It is therefore possible that their upper Nodular Limestone may either be a marly unit within the Coralline Limestone or, more probably, part of the Nodular Limestone, which is known to have bryozoans associated with hardgrounds in its uppermost part. Kennedy *et al.* (2003) also noted that the *Placenticeras kaffrarium*, recorded from the upper part of the Nodular Limestone elsewhere in the area, should be placed in synonomy with *Placenticeras mintoi* and includes a distinctive morphotype, *Parastantonoceras*, which is known from the Turonian of Madagascar. Additional support for a Turonian ammonite age comes from the presence of *Namadoceras*, a probable synonym of *Coilopeceras*, which was recorded from an oyster bed near the base of the Nodular Limestone and was assigned a Middle–Late Turonian age (Kennedy *et al.* 2003).

Kennedy *et al.* (2003) observed that over 30 species of inoceramid bivalves were recorded from the Nodular Limestone Formation by Chiplonkar & Badve (1976*a*, *b*), but that these have been excessively split taxonomically. A fauna from near the top of the formation, reinterpreted by Kennedy *et al.* (2003), comprises *Inoceramus pedalionoides* (equivalent to *Inoceramus heinzei* and *I. patootensis* of Chiplonkar & Badve 1976*a*, *b*), *Mytiloides incertus* (equivalent to *Inoceramus crippsi* and *Inoceramus subtrigonoides* of Chiplonkar & Badve 1976*a*, *b*) and *Mytiloides* ex gr. *striatoconcentricus* (equivalent to *Inoceramus subsulcatus* and *Inoceramus lamarcki var. indicus* of Chiplonkar & Badve 1976*a*, *b*), all of which would indicate a Late Turonian age. Kennedy *et al.* (2003) also mentioned a late Middle–early Late Turonian age for an inoceramid fauna from the lower part of the Nodular Limestone, but did not list the fauna.

Jaitly & Ajane (2013) concluded that the Nodular Limestone is Turonian, based on the occurrence of the morphologically highly variable ammonite *Placenticeras mintoi*. They suggested that the two associated taxa (*P. germari* and *P. kaffrarium*) might represent ecophenotypic variants of *P. mintoi*, indicating a very-low-diversity ammonite fauna.

Smith (2010), in a study of the echinoderm faunas, identified *Euporophyma lefebvrei* (a taxon restricted to the Turonian in Egypt) from the uppermost part of the Nodular Limestone and abundant *Mecaster mutabilis* (previously known from the Upper Turonian of Madagascar). He also noted occurrences of *Leptosalenia keatingei* (a form known from the Upper Turonian of Madagascar as *S. trigonopyga*) and *Hemiaster integer*, also known from the Upper Turonian of Madagascar.

Our studies of the micropalaeontology have yielded a rare and impoverished planktonic foraminiferal fauna from the lower half of the Nodular Limestone Formation. Specimens assigned to *Praeglobotruncana stephani* and tentatively identified specimens of *Helvetoglobotruncana helvetica* in the lower part of the section in the Zeerabad quarries indicate an age no younger than early Middle Turonian (the *Helvetica* Zone of Ogg *et al.* 2008). The presence of *Gavelinella tourainensis* in the immediately overlying samples also supports a Turonian age, although the taxon can range upwards into the Early Coniacian. The abundance of calcispheres, especially in the lower two-thirds of the formation, is similar to blooms noted elsewhere close to the Cenomanian–Turonian boundary in general and more specifically within the *H. helvetica* Zone, as noted here. This flood occurrence of calcispheres may be a useful stratigraphic marker for correlation across the region. The very high calcisphere content is considered typical of upwelling along basin/rift margins and is recognized from latest Cenomanian–Early Turonian sections across northern Europe, North Africa and into Iran (Adams *et al.* 1967; Wilkinson 2011; Omaña *et al.* 2014), all of which lie on a palaeo-latitudinal belt between 20° and 40° N during this period. The NW shelf area of the Indian subcontinent, where the present samples originate, was situated at 30° S at this time, suggesting that there was a parallel palaeo-latitudinal belt of calcisphere abundance in the southern hemisphere. Some authors (Bramwell *et al.* 1999) regard levels of high calcisphere abundance as indicating maximum flooding events. The high abundance levels in our samples may equate with the Early Turonian (93 Ma) MFS. Nannofossil recovery is very poor to absent and is dominated mainly by long-ranging forms. However, the presence of an assemblage comprising *Quadrum eneabrachium*, *Prediscosphaera* sp., *Watznaueria barnesae* and *Q. gartneri* indicates an age within the range Santonian–Turonian (Burnett *et al.* 1998).

Based on the above evidence, we assign a Turonian age to this formation.

Coralline Limestone Formation

Although this formation name is confusing on the basis that it contains very few corals and only rare calcareous algae, it is retained herein on the basis that it is well ingrained in the literature and is easy to identify in the field. Various authors have attempted to replace the name with the term Bryozoan Limestone Formation (on the basis that it is rich in the bryozoan *Chiplokarina* (*Ceriopora* of earlier (i.e. pre-1995) workers), which in the early days of field mapping was probably 'mistaken' as a coral), whilst others have suggested the new name Barwah Formation, which is even more confusing as the town of Barwah is located over 50 km away from the outcrops of this unit. Rode & Chiplonkar (1935) noted the presence of two Coralline Limestone horizons bracketing what was then referred to as the Deola-Chirakhan Marl, and defined these as the Lower and Upper Coralline Limestones. However, it is difficult to separate these owing to the soft nature of the marl and its frequent absence due to either erosion or non-deposition. It is clear that the unit referred to as the Lower Coralline Limestone by these authors is, in fact, due to the presence of bryozoans in the upper part of the underlying Nodular Limestone Formation associated with hardgrounds: in the Coralline Limestone Formation *sensu stricto*, however, the bryozoans are abundant and are scattered throughout the formation.

Lithology. Thickness is highly variable, from a few metres to around 10 m. The formation rests disconformably on the Nodular Limestone Formation and is unconformably overlain by either Maastrichtian Lameta Formation calcretes or Paleocene Deccan volcanics. Where the Lameta Formation is absent, the Deccan Traps are separated from the Coralline Limestone Formation by a thin (< 1 m) silcrete that often contains large wood fragments. The formation shows marked lateral variability in terms of fauna and facies. It comprises a sequence of mainly detrital carbonates, with a broad range of textures from wackestone to grainstone, with occasional rudstones and framestones. The macrofauna is diverse, and includes echinoids, gastropods, oysters, bivalves, brachiopods, serpulid worms and shark teeth, but is overwhelmingly dominated by the bryozoan *Chiplonkarina*. The macrofossils are often bored and sometimes encrusted. Planar and trough cross-bedding are common, as are burrowed and cemented hardground horizons, which exhibit only minor amounts of erosion. Bryozoan fragments and bryozoan limestone intraclasts, together with rounded detrital quartz grains and metamorphic rock fragments (mainly quartzite), are common. Although this formation is generally carbonate dominated, detrital clastic material is often locally common and mainly comprises metamorphic (quartzite and schist) and plutonic quartz grains (Fig. 6b). Locally, around Zeerabad, the formation contains cross-bedded, calcareous sandstones containing common molluscan (mainly oyster) and bryozoan debris, together with pebbles of quartz, chert and limestone.

Depositional environment. The presence of hardgrounds indicates periods of limited deposition, which are often associated with *in situ* bryozoan colonies. The abundance of bryozoan fragments and limestone intraclasts is indicative of multiple phases of erosion and redeposition typical of such very-shallow-marine settings situated above storm wave base (Fig. 6a). However, erosion associated with the hardgrounds is limited, indicating that they are likely to represent short-lived events. Post-Palaeozoic bryofaunas with significant quantities of erect species (such as *Chiplonkarina dimorphopora*) are almost entirely extra-tropical in occurrence and therefore tend to characterize cool-water deposits (Taylor pers. comm. 2014). A cool-water interpretation is supported by the paucity of warm-water taxa such as corals, larger foraminifera and calcareous algae in this formation. The composition of the detrital clastic material observed locally within this formation suggests that it was most probably derived from reworking of the underlying Nimar Sandstone and/or derivation from the nearby Precambrian basement hinterland. The Coralline Limestone Formation is interpreted to have been deposited in a shallow-marine, moderate- to high-energy, temperate setting most probably in a tidal complex.

Age. The presence of the benthic foraminifera *Rotalia* cf. *algeriana* in the lowermost part of the formation in the Zeerabad area may be of importance as this species is common throughout much of the southern Tethys in shallow-water Turonian carbonates.

Bardhan *et al.* (2002) noted the occurrence of the ammonite *Platyceramus mantelli* in the basal part of the formation. They observed that, in Europe, this taxon indicates a Middle Coniacian age. In the upper part of the formation, they recorded the ammonite *Barroisiceras onilahyense*, which they also consider to be indicative of a Middle Coniacian age. However, Jaitly & Ajane (2013) pointed out that this taxon shows wide intraspecific variability and overlaps with the late Turonian species *B. haberfellneri*. Gangopadhyay & Bardhan (2000) also recorded the ammonite *Prionocyclus germari* from the basal part of the Coralline Limestone, which they noted occurs in the Coniacian in Madagascar. However, this contradicts the Kennedy *et al.* (2003) assignment of this taxon to the Late Turonian and may support the view that the reported occurrence referred to by Kennedy *et al.* (2003) as upper Nodular Limestone may, in fact, have been a marly facies variation within the Coralline Limestone. Smith (2010) recorded an echinoid fauna similar to that recorded from the Upper Turonian of Madagascar, of which the important age diagnostic species were *Leptoslania keatingei*, *Tetragramma besairei*, *Hemiaster subsimilis*, *Hemiaster integer* and rare *Mecaster mutabilis*. Taylor & Badve (1995) described common to abundant occurrences of the bryozoan *Chiplonkarina dimorphophora*, which they noted had also been recorded from the Turonian and ?Coniacian of Turkmenistan, Uzbekistan and Tadzhikistan, being especially abundant in the Upper Turonian of Turkmenistan. The palaeontological data therefore tend to suggest an Upper Turonian age for this formation.

The Saurastra Basin

The Saurastra Basin is located in central western India to the west of the Narmada–Cambay basins and consists of a Mesozoic rift sequence overlain by Early Cenozoic Deccan volcanics (Fig. 9). It covers an area of around 52 000 km^2, with an additional offshore area of 20 000 km^2. The basin is bordered to the north by the Kutch Basin, to the east by the Cambay Basin and to the south by the Surat Depression (part of the Bombay Shelf), whilst to the west it opens into the Arabian Sea. The area comprises a large horst block of Lower Cretaceous clastic sediments overlain by Deccan Trap basalts and a thin veneer of Neogene–Quaternary sediments. The area remained a high separating the Kutch and offshore Narmada basins until the Early Cretaceous, when it became a depocentre. It was subsequently subjected to extensive regional uplift and associated Deccan volcanism in the Late Cretaceous–Early Cenozoic (Biswas 1987). The basin structure is dominated by an underlying NNW–SSE-trending Precambrian arch (Sahay 1984), which on the present day coast is expected to be at around 3000 m depth (Biswas & Deshpande 1983). In Saurastra, the main focus areas for the fieldwork were around Wankaner, Chotila, Surajdeval, Than and Ranipat in the north of the basin. A simplified geology and locality map is provided in Figure 10, together with a summary of the onshore exploration wells drilled in the area to date.

Saurastra		
Turonian	Wadhwan Formation	**Bhaduka Limestone Member** (Temperate shallow marine)
Turonian	Wadhwan Formation	**Navania Limestone Member** (Deep shelf)
Albian-Cenomanian	Wadhwan Formation	**Suredranagar Sandstone Member** (Shallow marine)
Unfossiliferous	Dhrangadhra Group	**Ranipat Formation** (Coastal beach to subtidal)
Unfossiliferous	Dhrangadhra Group	**Surajdeval Formation** (Distal delta plain/coastal)
Barremian-Aptian	Dhrangadhra Group	**Than Formation** (Delta Plain)

Fig. 9. Pre-Deccan stratigraphy of the Saurastra Basin.

Dhrangadhra Group

The Dhrangadhra Group rests unconformably on Precambrian basement and is unconformably overlain by Deccan volcanics. It was deposited in an Early Cretaceous failed rift that trended from NE to SW, and consists of Lower Cretaceous (Hauterivian–Cenomanian) fluvial to shallow-marine transgressive sandstones, siltstones and coals (from bottom to the top: the Than, Surajdeval and Ranipat formations). These are unconformably overlain by shallow-marine clastics of the Upper Cretaceous Wadhwan Formation in the north of Saurastra. In the south of Saurastra, the Wadhwan Formation is subdivided into three members: shallow-marine sandstones (Surendranagar Sandstone) overlain by shelfal, partly restricted nodular limestones (Navania Limestone) and capped by Late Turonian shallow-marine bryozoan-rich carbonate grainstones (Bhaduka Limestone), as shown in Figure 9.

The clastic part of the Dhrangadhra Group outcrops in NE Saurastra (NE of Rajkot), where it is up to 500 m thick at outcrop (Chiplonkar & Borkar 1971) and forms part of a large domal feature with gentle bedding dips that show a radiating pattern (Biswas & Deshpande 1983). The overlying Wadhwan Formation is restricted to the structural low along the southern outcrop limits of the Dhrangadrha Group.

The Lodikha-1 borehole located to the SW of Rajkot (Fig. 10) penetrated 1280 m of Deccan volcanics and 2150 m of pre-Deccan. The pre-Deccan sequence comprised 240 m of Upper Cretaceous Wadhwan Formation sandstones, 1100 m of Lower Cretaceous–probable Upper Jurassic Dhrangadhra Group sandstones, claystones and conglomerates, and a 600 m-thick sequence of Upper Jurassic basalts and tuffs representing the Lodhika Formation (Singh *et al.* 1997). Singh *et al.* (1997) reported the following palynoflora from the upper part of the Dhrangadra Group in the Lodihika-1 well: *Microchryidites antarcticus*, *Spheripollenites subgranulatus*, *Spheripollenites psilatus* and *Cornabispora valdensis*, to which they assign a Lower Cretaceous age. They also record *Gliscopollis* sp. and *Exesipollenites tumulus* from the older part of the 'Upper' Dhrangadrha, to which they assign a Jurassic age. Singh & Venkatachala (1988) recorded the palynomorphs *Appendicisporites* and *Imparde-cispora* from outcrops of the Dhrangadrha Group (formation not known), to which they assigned an Aptian (i.e. Lower Cretaceous) age.

Further to the west near the Cambay River, the Dhanduka borehole (Fig. 10) penetrated a 1200 m-thick Deccan sequence followed by 100 m of undated sandstones resting on granite basement. Biswas & Deshpande (1983) estimated that the Cretaceous sediments are around 2000 m thick on the flanks of the Saurastra Arch and around 1000 m thick over the crest. A borehole drilled offshore in the Raju Depression to a total depth (TD) of 3500 m encountered coarse alluvial clastics in the lower Dhrangadhra, and fluvial sandstones and siltstones in the upper Dhrangadhra (Singh *et al.* 1997). They noted poor to fair source rock potential (0.2–1.48% TOC, dominated by gas-prone kerogens) in the Dhrangadhra Group, whilst the overlying delta-plain deposits of the Wadhwan Formation had low TOC values (average 0.6%). A borehole drilled onshore in 1915 is reported to have encountered some gas in Cretaceous sandstones (Johri & Kandpal 1983). The Dhrangadhra Group thins to the NE towards Ahmadabad and the Precambrian Aravalli Craton. The few wells drilled in Saurastra (Viramgaom-1, Dhandhuka-1 and Lodhika-1) had Deccan Trap thicknesses of 175–1300 m and Mesozoic sediment thicknesses from 175 to >2100 m (Fig. 10).

The sandstones of the Dhrangadrha Group are dominated by detrital quartz, with rare feldspar and locally common metamorphic rock fragments.

Than Formation

Lithology. The base of the formation is not exposed and the main exposures occur around the village of Than, where it has been mined by locals for its coal in a number of open pits (Fig. 11). The maximum thickness is around 120 m and the formation broadly coarsens upwards. Three distinct lithofacies are observed:

- Medium- to coarse-grained grey sandstones exhibiting planar and trough cross-bedding form bedsets with thicknesses varying from 1.5 to 3 m. Lateral accretion surfaces are noted in some of the thicker sandstone beds. Palaeocurrent taken from cross-beds suggest a transport direction towards the west and SW. The sandstones

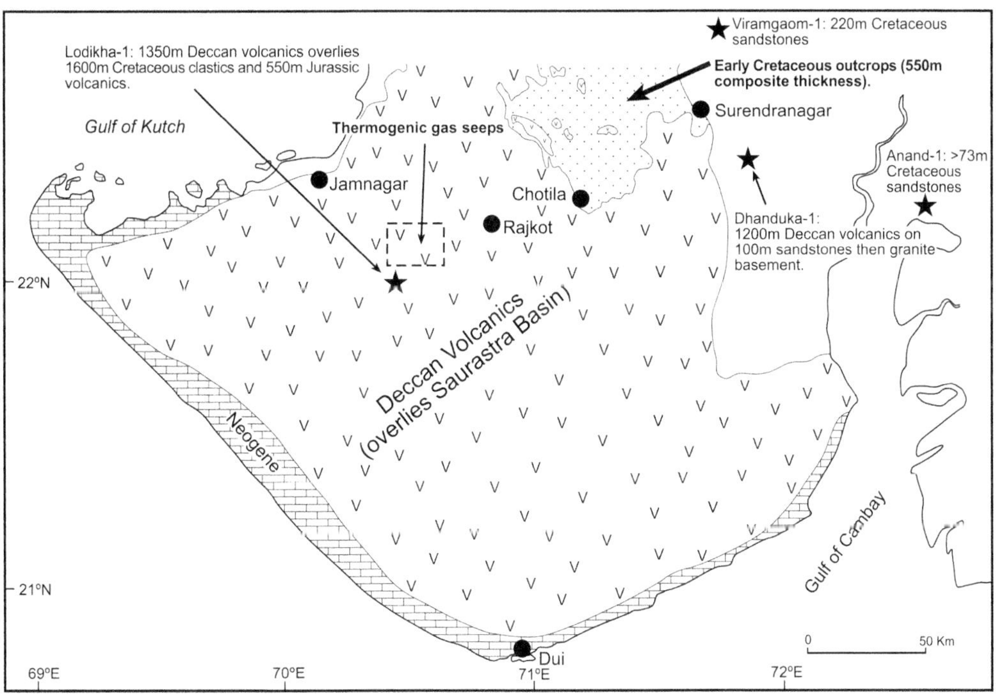

Fig. 10. Locality map of Saurastra showing outcrops of studied Cretaceous sections (stippled area) and also wells that have penetrated the pre-Deccan sediments. The limestone symbol around the coast represents Miocene–Pliocene shallow-marine to aeolian carbonates.

Fig. 11. Dhrangadhra Group, Saurastra. (**a**) Than Formation outcrop showing marked bioturbation. (**b**) A Than Formation sandstone photomicrograph dominated by moderately to strongly compacted quartz grains (with some quartz overgrowths) has a porosity of 16.5% (stained blue). (**c**) Surajdeval Formation part ferruginous trough cross-bedded, shallow-marine sheet sandstones. (**d**) Surajdeval Formation thin-section image showing moderate sorting and compaction, with a porosity of 10% (blue stained). (**e**) Ranipat Formation outcrop showing massive, clean cross-bedded channel sandstones. (**f**) A Ranipat Formation photomicrograph with 14% porosity and which is compositionally similar to the Than Formation: both being quartz arenites, with this sample showing minor kaolinite (top left) and scattered recent dark coloured haematite/limonite cement.

are predominantly quartz arenites and show moderate to strong compaction, yet retain porosities in the range 5–20% (Fig. 11b, d, f).

- Horizontally laminated and ripple-laminated sandstones and siltstones with occasional cross-bedding and interbedded claystones containing

thin layers of plant debris and burrows (Fig. 11a).
- Carbonaceous siltstones and claystones from 0.1 to 18 m thick with thin (10 cm) coal seams often occur in the lower part of the formation and are associated with plant fragments and siderite nodules. Individual coal seams, 2–4 m in thickness, also occur but are laterally discontinuous, extending over a few tens of metres. Interbedded, fine-grained sandstones commonly exhibit root traces. This lithofacies is more common in the lower part of the formation.

Depositional environment. The planar and trough cross-bedded sandstones with lateral accretion surfaces are interpreted as representing mainly meandering channels, whilst the more laminated siltier and claystone rich lithofacies are interpreted as being deposited in a low-energy, delta-plain environment. The more carbonaceous lithofacies in the lower part of the formation is interpreted as being deposited in a back-swamp or interdistributary bay setting.

Overall, the Than Formation is interpreted as comprising a prograding delta distributary complex with back-swamp deposits in its lower part (lower delta plain), and lower to upper delta-plain sediments in its upper part (Aslam 1992; Casshyap & Aslam 1992).

Age. The formation has yielded plant floras that are considered to be similar to those of the Umias Member of the Kutch Basin. The macroflora is said to have a 'Wealden' affinity, and includes *Matonidium geopperti*, *Weichselia reticulata* and *Onpchiopsis psilotoides* (Verma & Rawat 1964; Chiplonkar & Borkar 1971). Verma & Rawat (1964) also recorded the following palynomorph assemblage: *Cyathadites* spp., *Osmundacidites wellmanii*, *Polypodiaceoisporites speciosus*, *Gleichenites* sp., *Trilobatus apiverrucatus*, *Lygodiumsporites* sp., *Acanthotriletes* sp., *Lycopodiumsporites* sp., *Callialasporites trilobatus*, *C. dampieri*, *Callialasporites* sp., *Cicatricosisporites cooksoni*, *Callispora* sp., *Ginkgocycadophytus deterius* var. *majus*, *Pityosporites* spp., *Podocarpidites* sp., *Classopollis pflugi*, *Classopollis* spp., *Monosulcites* sp., *Araucariacites* spp., *Granatricolpites* sp., *Granatriporites* sp. and *Punctatriporites* sp. The presence of angiosperm taxa *Granatricolpites* sp., *Punctatriporites* sp. and *Granatriporites* sp. suggests a Cretaceous age, whilst *Classopollis pflugi* is known from the Early Cretaceous and *Trilobatus apiverrucatus* is common in the Wealden, supporting an overall Early Cretaceous (possibly Hauterivian) age according to Verma & Rawat (1964).

Our interpretation of this palynoflora is that it is older than Turonian, with the presence of angiosperm pollen indicating an age no older than latest Barremian and the presence of *Trilobosporites* tending to support a Barremian–Aptian age. A Late Barremian–Aptian age appears to be most likely for this assemblage.

Surajdeval Formation

Lithofacies. The formation mainly consists of white to pale buff-coloured, lenticular, cross-bedded and trough cross-bedded channel sandstones with local lenses of conglomerate and interbedded red brown laminated and ripple-laminated siltstones stacked in a series of fining-upwards cycles and overall is around 170 m thick (Fig. 11c). The siltstones are often ripple laminated, whilst the sands occasionally show dewatering structures. Individual channel thicknesses vary from 1 to 3 m, and commonly coarsen upwards with thicker channels and larger-scale cross-bedding more common in the upper part of the formation. The sandstones are medium to coarse grained, and are dominated by almost pure quartz arenites (Fig. 11d).

Depositional environment. The formation is interpreted as having been deposited in an overall distal delta-plain and coastal nearshore tidal setting (Casshyap & Aslam 1992), and mainly comprises delta distributary channels that show west- and SW-directed palaeocurrents and are interbedded with thin-bedded coastal-plain deposits.

Ranipat Formation

Lithofacies. The Ranipat Formation mainly comprises a 200 m-thick sequence of light-coloured (mainly white to tan) fine- to coarse-grained, cross-bedded and trough cross-bedded sandstones (Fig. 11e) with bi-directional ripples, herringbone cross-bedding and locally common bioturbation. These are interbedded with thin shales and stacked into fining-upwards cycles of up to 10 m in thickness. Aslam (1991) recorded a range of trace fossils comprising *Ophiomorpha*, *Planolites*, *Skolithos* and *Thalassinoides* from this unit. The sandstones are dominated by almost pure quartz arenites (Fig. 11f).

Depositional environment. The overall depositional environment is nearshore coastal, varying from beach through tidal/intertidal to subtidal. Palaeocurrents are dominantly directed towards the west and SW.

Wadwhan Formation

The Wadhwan Formation comprises three members: from the bottom to the top, these are: Suredranagar Sandstone, Navania Limestone and Bhaduka

Limestone. Based on the age of the underlying Than Formation (Barremian–Aptian), the Wadwhan Formation must be younger.

Suredranagar Sandstone Member

Lithofacies. This member comprises tan-coloured massive, pebbly to fine-grained sandstones with intercalated marls and has a total thickness of 5–20 m. It contains common, often fragmentary, molluscs, echinoids and bryozoans, as well as some calcareous algae. Ammonites are recorded (Chiplonkar & Borkar 1971) from this member, but none have been named or illustrated. The basal part of the formation is pebbly and rests with angular unconformity on the underlying Dhrangadhra Group near the village of Kukda (Chiplonkar & Borkar 1971).

Depositional environment. The Suredranagar Sandstone Member was deposited in a shallow-water, open-marine, normal salinity environment as indicated by its macrofauna.

Age. Samples from the formation from depths of between 1490 and 1535 m in the Lodhika-1 well to the west of the studied outcrops yielded a palynomorph flora including *Classopollis classoides* and *Perinipollenites elatoides*, which is similar to that noted in the Parh and Upper Goru formations in Rajasthan, and which would tend to indicate a broad Late Jurassic–Neocomian age. The macrofauna shows overall similarities with the upper part of the Nimar Sandstone and Bagh Group, whilst the Aptian–Barremian age for the underlying Than Formation supports a post-Aptian age.

Navania Limestone Member

Lithofacies. This argillaceous limestone has yielded echinoids (including *Hemiaster*) and oysters (*Phygraea*).

Depositional environment. The depositional environment is probably a fairly deep shelf or lagoon. The limited outcrops of this member around the village of Navania could not be located and it is now believed that they have been removed or built on.

Age. The presence of *Hemiaster* in beds, which is also recorded in the Nodular Limestone and Coralline Limestone formations of the Bagh Group of Narmada, date this as Turonian.

Bhaduka Limestone Member

Lithofacies. The Bhaduka Limestone varies significantly in thickness from 5 to 20 m and has a macrofauna similar to that of the Coralline Limestone Formation of Narmada, including bryozoa (dominated by *Chiplonkarina*), echinoderms (six species common to both Narmada and Saurastra) and a similar molluscan fauna, although the fauna is often silicified. Amongst the echinoids, *Hemiaster subsimilis* is important for age dating as it was recorded in the Coralline Limestone Formation of Narmada by Smith (2010) and is known from the Late Turonian of Madagascar. Ammonites have also been recorded and include *Placenticeras*, which also occurs in the Nodular Limestone of Narmada (Bardhan *et al.* 2002). Oysters are common towards the top of the member.

Depositional environment. The depositional environment is interpreted as a shallow-marine shelf, based on its contained macrofauna.

Age. The macrofauna is similar to that of the Coralline Limestone Formation from Narmada, which is dated as Upper Turonian.

The Kutch Basin

This Mesozoic rift basin is located on the western margin of the Indian Craton and covers an area of around 71 000 km^2, of which around 60% is located offshore. The basin comprises a series of fault blocks and half-graben aligned along underlying Precambrian structures, and contains around 3000 m of Mesozoic and 900 m of Cenozoic sediments. The Mesozoic sequence was deposited in two megacycles: a transgressive Middle Jurassic cycle and a regressive Late Jurassic–Early Cretaceous cycle (Biswas 1982). The basin is bounded by the Radhanpur Arch to the NE, the Nagar–Parkar Ridge to the north and by the Kathiawar Horst–Saurashtra Arch to the SE. The development of the basin was strongly influenced by the NE–SW-trending Delhi–Aravalli Precambrian faults, which were passive during the Mesozoic and were later reactivated during Cenozoic uplift (Biswas 1982).

The oldest marine sediments in the basin are Middle Jurassic and were deposited in the north of the basin as part of a transgressive megacycle dominated by carbonates and shales. Tectonism resumed in the Late Jurassic at the onset of the deposition of the regressive megacycle, during which the western part of the basin subsided more rapidly than the eastern part and so created the Median High (Biswas 1982). West of the Median High, the sequence is both of deeper water origin and thicker: to the east, however, a thinner sequence of shallow-marine sediments was deposited.

During the Early Cretaceous, the basin filled up as the sea receded, during which time the SE end of the Precambrian Aravalli Range to the east of the Kutch Basin was peneplained (Biswas 1987). A major fault to the east of the Cambay Basin became active at this time and formed a hinge zone, such that the area to the west of this zone, including the Kutch Basin, subsided as part of an extensive platform, with large volumes of sediment input from the erosion of the Precambrian Aravallis located to the east. The Late Jurassic–Early Cretaceous regressive cycle is dominated by deltaic clastics and is up to 1800 m thick. In the Late Cretaceous, differential uplift along five main faults in the basin resulted in the development of a series of east–west basins separated by ridges, whilst other periods of uplift associated with Late Cretaceous plutonic intrusions are also noted.

The field localities are shown in Figure 1, with the main Cretaceous sections studied located in and around the city of Bhuj.

Stratigraphy

The Umia Formation comprises four members that are, from bottom to top, the Umia (35 m), the Ghuneri (400 m), the Ukra (35 m) and the Bhuj (300 m) members, and these rest on the Jhuran Formation (Fig. 12). The Bhuj Member was the main interval studied and sampled in this study, with published information mainly being used to define the age.

Bhuj Member (aka Bhuj Formation)

Lithofacies. The Bhuj Member comprises around 300 m of mainly medium- to coarse-grained sandstones with large-scale cross-bedding and densely bioturbated intervals (Fig. 13a). A diverse trace fossil fauna has been identified (Mude *et al.* 2012), including *Ophiomorpha*, *Thalassinoides*, *Skolithos*, *Paleophycus*, *Planolites*, *Chondrites*, *Rhizocorallium*, *Aulichnites* and *Cylindricus*. Soft-sediment deformation and ripple lamination are locally common (Fig. 13b, c). A short-lived marine incursion with the trigonid bivalve *Pisotrigonia ventricosa* occurs at the top of the underlying Ukra Member.

Kutch		
Middle - Late Albian	Umia Formation	Bhuj Member (Shallow marine-Fluvial)
Late Aptian - Early Albian		Ukra Member (Open marine shelf)
Hauterivian-Barremian		Ghuneri Member (Open marine shallow shelf)
Tithonian		Umia Member (Open marine shelf)

Fig. 12. Stratigraphy of the Cretaceous in the Kutch Basin.

Depositional environment. The Bhuj Member contains very few body fossils, especially in its upper part, and was often considered by earlier authors to be mainly non-marine. However, the diverse trace fossil assemblage indicates deposition in a shallow-marine environment. Overall, it appears to be a regressive package, with deposition above wave base, with some channel sands and local estuarine/lagoonal influence.

Age. Tithonian ammonites including *Virgatosphinctes* (mostly *V. densiplicatus*) and *Micracanthoceras* have been recovered from the Umia Member (Krishna *et al.* 1983). Late Aptian ammonites (*Autraliceras* and *Tropaeum* aff. *australis*) have been recorded from the lower part of the Ukra Member and Early–early Middle Albian ammonites (*Lemuroceras* and *Cleoniceras*) have been recovered from the upper part of the Ukra Member (Krishna *et al.* 1983). Rathore *et al.* (1999) reported radiometric (K–Ar) ages of glauconites from the Ukra Member of between 103 and 107 Ma (i.e. Albian). The Ghuneri Member has been dated, on the basis of the presence of the ammonite *Argentriceras*, as Hauterivian–Barremian.

Rai (2006) reported the presence of nannofossils of Middle Albian age (the CC8 *Prediscosphaera columnata* Zone of Perch-Nielsen 1985 or the NC8 *Tranolithus orionantus* Zone of Bown *et al.* 1998) from grey shales within the Bhuj Member. Singh & Venkatachala (1988) recorded the palynomorphs *Appendicisporites* and *Ephedrites* from the Bhuj Formation, to which they assign an Aptian age.

The most likely age of the Bhuj Member is Albian.

Petrographical analysis. The Bhuj sandstones form an important regional aquifer, and thus have good porosities and permeabilities in the near subsurface. Representative photomicrographs of these sandstones are provided in Figure 13d, e. The sandstones are dominated by detrital quartz and feldspar, with rare metamorphic rock fragments. They contain noticeably more feldspar and fewer metamorphic rock fragments than the sandstones from Saurastra and Narmada.

The Jaisalmer Basin

The Mesozoic outcrop area is shown in Figure 1. The main studied sections were located to the immediate south and west of Jaisalmer, although the Cretaceous outcrops were often of poor quality and limited vertical extent. A table summarizing the stratigraphy of the Cretaceous interval is provided in Figure 14.

Fig. 13. Bhuj Sandstone photographs, Kutch Basin. The Bhuj sandstones contain sedimentary features indicative of both fluvial and shallow-marine settings. Bioturbation is locally common and moderately diverse, but marine body fossils are rare and include the bivalve *Trigonia*. Wood and plant fragments are locally present. (**a**) Thick-bedded amalgamated sandstones overlain by thinner-bedded sandstones separated by siltstones, bioturbated in its upper part as shown in the photograph insert top left (note the person for scale in the bottom left of the photograph); (**b**) dewatering structures; (**c**) ripple lamination; and (**d**) & (**e**) photomicrographs of moderately to sorted quartz arenites with varying amounts of alkali feldspar and minor muscovite. Kaolinite is locally common, whilst recent brown haematite cements are also observed. Porosities (stained blue) are often high (27% in d and 22% in e) and the unit forms an important regional aquifer.

Pariwar Formation

Lithofacies. Also referred to in older literature as the Parihar Formation, the Pariwar comprises a 350 m-thick sequence of sandstones, siltstones and claystones (based on unpublished subsurface data), but is generally poorly exposed. Minor marine incursions are marked by bioturbated fine sandstone

Jaisalmer	
Turonian-Coniacian	Parh Formation (Open marine mid to outer shelf)
Albian-Cenomanian	Goru Formation (Shallow marine)
Aptian	Habur Formation (Shallow marine)
Neocomian	Pariwar Formation (Fluvial with some marine influence)

Fig. 14. Stratigraphy of the Cretaceous in the Jaisalmer Basin, based on outcrops and subsurface data.

and siltstone horizons (Fig. 15a). The Pariwar Formation is broadly subdivided into a lower sandy–pebbly unit *sensu* Roy & Jakhar (2002) (Fig. 15b, c) and an upper argillaceous unit, and contains fossil wood (*Mesemborioxylon* sp.) and plants (including *Ptilophyllum acutifolium*, *Otozamites imbricatus*, *Elatocladus conferta*, *Phlebopteris* sp., *Pagiophyllum* sp. and *Gleichenites* sp.: Das Gupta (1975) reported the additional taxa *Anomozamites* sp., *Actinosporites* sp. and *Cladophlebes* sp.).

Depositional environment. The environment of deposition is overall regressive and fluvial, with intermittent marine influences.

Age. The basal sandy–pebbly part of the sequence, as described by Roy & Jakhar (2002), rests on the Bhadasar Formation. The latter is dated as Tithonian based on the presence of an ammonite fauna comprising *Virgatosphinctes densiplicatus*, *V. communis*, *Aulacosphinctites occultefurcatus*, *Hibolites* sp. and *Perisphinctes* sp. (Das Gupta 1975; Singh 1996). The plant taxa identified within the upper argillaceous unit are ascribed a broad Late Jurassic–Early Cretaceous age. The most likely age for the Pariwar Formation is therefore Early Cretaceous.

Fig. 15. Pariwar Formation, Jaisalmer. (**a**) Coarse-grained, low-angle cross-bedded sandstones overlying bioturbated siltstones (marine incursion); (**b**) mudstone intraclasts within ferruginous, fluvial sandstone; (**c**) coarse quartz pebbly gravels of mainly metamorphic quartz: note the bimodal grain-size distribution; (**d**) photomicrograph of quartz arenite comprising moderately sorted subrounded to subangular quartz grains with minor kaolinite (green). Porosity stained in blue is 22%.

Habur Formation

Lithofacies. This formation, which is around 100 m thick and overlies the Pariwar Formation, was not sampled during the fieldwork. It comprises a basal sandy limestone with common molluscan material overlain by locally ferruginous, mainly calcareous grits and sandstones overlain by a thick fossiliferous limestone containing ammonites, bivalves and belemnites, which are capped by ferruginous grits and sandstones.

Depositional environment. The depositional environment was shallow marine, with the ammonite material introduced during storm surges.

Age. The formation is assigned a Lower Aptian age, based on the occurrence of the ammonite *Deshayesites bodei* from the lower part of the formation, whilst an older Valanginian age was inferred by Roy & Jakhar (2002) based on the presence of the ammonites *Sarasinella subspinosa* and *S.* aff. *trezanesis*.

Goru Formation

This is only identified in the subsurface in the Jaisalmer Basin, and is subdivided into a Lower and an Upper Goru Member: it is 565 m thick in the Kharatar-B well. The Lower Member comprises shales with intercalations of glauconitic sandstone and siltstones, which have yielded a microfauna of *Favussella washitensis*, *Biticinella breggiensis* and *Planomalina buxtorfi*, indicative of a Late Albian age. The Upper Member is dominantly argillaceous, and consists of shales, siltstones and marls, which have yielded a microfauna that includes *Rotalipora cushmani* and *Praeglobotruncana stephani*, indicative of a Cenomanian age (Bhandari 1999). The formation was deposited in an overall transgressive shallow-marine environment with occasional short-lived regressions and, based on its age, its lower member is probably age equivalent to the Bhuj Member on the Kutch Basin. Further to the west, in the Lower Indus Basin of Pakistan, the Lower Goru Member is extended up to the Late Cenomanian, and the Upper Goru has a Turonian–Santonian age range proven by a well-developed planktonic foraminiferal fauna.

Parh Formation

This has only been encountered in the subsurface in the Jaisalmer Basin in the Kharatar-B well, where it was around 350 m thick and had a gradational contact with the underlying Goru Formation. The Parh Formation comprises a sequence of thick argillaceous limestones with interbeds of calcareous clays, marls and siltstones, and was deposited in an open-marine middle- to outer-shelf setting. A diverse planktonic foraminiferal microfauna including *Helvetoglobotruncana helvetica*, *Marginotruncana coronata* and *M. sigali* indicates a Turonian–Coniacian age (Bhandari 1999): that is, broadly time equivalent to the Nodular and Coralline Limestone formations of the Bagh Group of the Narmada area. The age of the base of the Parh Limestone is regionally diachronous, as this formation is dated as Campanian–Early Maastrichtian, based on the planktonic foraminiferal content of wells drilled in the Lower Indus Basin in Pakistan.

Petroleum systems

Reservoir environment and provenance

During the Early Cretaceous, clastic deposition dominated NW India, with fluvial systems eroding large parts of the craton and carrying sediment out to delta systems forming on the 'developing' western coast. In the onshore Indus Basin of Pakistan and the shallow waters of the offshore Kutch Basin in India, the Lower Cretaceous Lower Goru and Bhuj formations are proven reservoirs for hydrocarbons. Regional correlation of these reservoirs with field localities from Saurastra, Cambay and Narmada suggest that similar reservoir potential may exist for the offshore portions to the south of Saurastra, in and around the present-day Mumbai Basin.

Petrographical analysis of sandstones from the Cretaceous of the studied basins shows little variation in overall detrital mineralogy (Fig. 16; Table 1). Most of the sandstone samples studied are classified as arenites. Samples from the Dhrangadhra Group of Saurastra are wholly quartz arenites, those from the Nimar Sandstone Group of Narmada are a mixture of quartz arenites and sublithic arenites, whilst the Bhuj Member sandstones of Kutch and the Pariwar Formation sandstones of Jaisalmer are a mixture of quartz arenites and subfeldspathic arenites.

The sandstones from all the studied areas are dominated by monocrystalline quartz grains that are generally weakly to strongly strained with plutonic and vein quartz varieties noticeably more common in the Nimar sandstones of the Narmada Basin. Polycrystalline quartz grains are mainly of metasedimentary (quartzite) origin. Feldspars vary from minor to abundant, are most common in the Bhuj sandstones of Kutch and are dominated by alkali feldspar. Microcline also occurs rarely in the Nimar and Bhuj sandstones, and may indicate a partially granitic source area for these sandstones. Metamorphic rock fragments are minor to moderately common (locally abundant) in most of the sandstones from all basins, and are normally quartz rich (mainly quartzites) and occasionally schistose. A summary

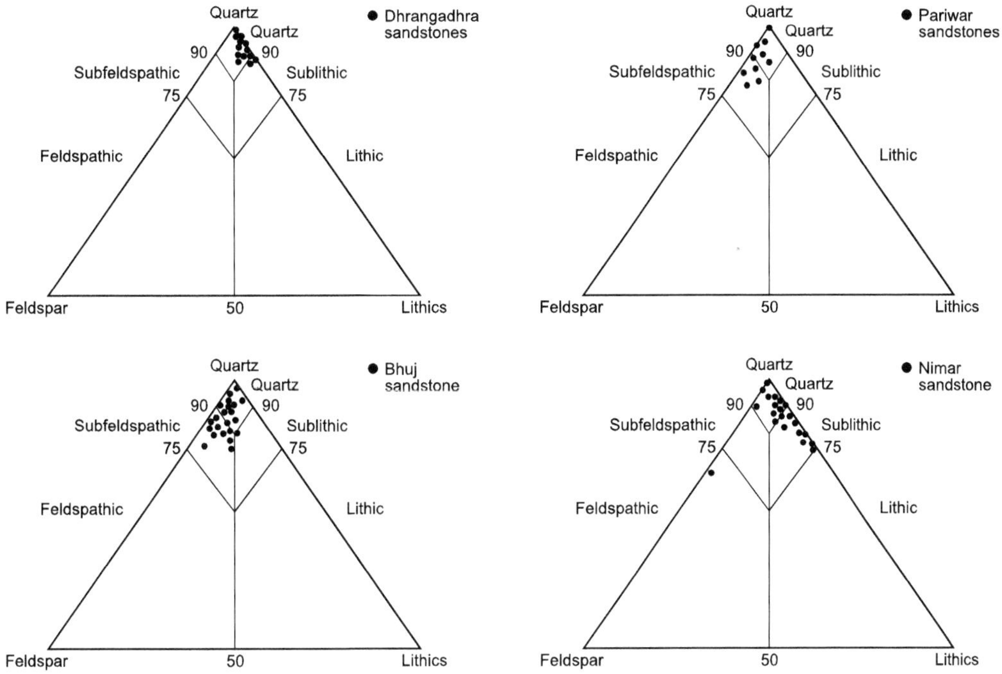

Fig. 16. Sandstone classification and detrital composition for each of the studied basins. The Saurastra area Dhrangadhra sandstones show the greatest degree of mineralogical maturity being quartz dominated and show the closest compositional affinity with the Nimar sandstones of Narmada, although the latter contains more lithic fragments (possibly because it was located closer to the sediment source area). The Jaisalmer (Pariwar sandstones) and Kutch (Bhuj sandstones) basins are compositionally similar and have a higher feldspar content.

of the average detrital compositions is provided in Figure 16 and Table 1, showing the overall compositional similarities between the Drahngadhra (Saurastra Basin), Pariwar (Jaisalmer Basin), Bhuj Sandstone (Kutch Basin) and Nimar Sandstone (Narmada Basin). The Saurastra area shows the greatest degree of mineralogical maturity being quartz dominated and shows closest compositional affinity with Narmada, although the latter contains more lithic fragments (possibly because it was located closer to the sediment source area). The Jaisalmer and Kutch basins are compositionally similar and have a higher feldspar content, including both alkali and plagioclase feldspar (the latter being rarer in Narmada and Saurastra). The Jaisalmer sandstones have a markedly higher alkali feldspar content than the other studied basins, whilst the Kutch and Jaisalmer basin samples show the highest plagioclase content (Table 1). A summary of the heavy minerals recorded in thin sections for these Cretaceous sandstones is provided in Table 2, and show that zircon, tourmaline and, to a lesser extent, garnet are especially common in the Saurastra, Narmada and Kutch basins, and rarer (with garnet

Table 1. *Detrital grain percentage averages for Early Cretaceous sandstones from the studied basins (see the text for the discussion)*

Basin	Monocrystalline quartz	Polycrystalline quartz	Alkali feldspar	Plagioclase feldspar	Microcline	Muscovite mica rock fragments	Plutonic rock fragments	Muscovite mica
Saurastra	82.9	4.4	0.1	1.5	0	3.7	tr	0.19
Narmada	72.3	8.5	2.9	1.3	tr	5.0	0.1	0.10
Kutch	67.2	3.0	2.0	5.9	tr	1.3	0.4	1.0
Jaisalmer	65.3	2.2	11.3	4.3	tr	0	0.6	tr

tr, trace.

Table 2. *Heavy mineral presence/absence in Early Cretaceous sandstones from the studied basins. Precambrian sandstones from the Aravalli Group are included for comparison (numbers refer to the number of samples in which the heavy minerals were observed)*

Basin	Tourmaline	Rutile	Zircon	Apatite	Garnet	Monzanite	Staurolite
Saurastra	12	5	18	1	2	1	2
Narmada	10	3	13	3	3	3	1
Kutch	14	46	23	0	7	0	0
Jailsamer	6	4	6	0	0	0	0
Precambrian	3	2	2	0	0	0	0

absent) in the Jaisalmer Basin, whilst rutile is especially abundant in the Kutch Basin. Potential sediment source areas for the Cretaceous sandstones are shown in Figure 17, and are dominated by Precambrian metasediments and rarer granites.

Reservoir quality

Compactional porosity loss varies and is moderate to strong in the Nimar sandstones of Cambay and Narmada, weak to moderate in the Bhuj sandstones of Kutch, and strong in the Dhrangadrha sandstones of Saurastra. Compaction Index values (the fractional ratio between compaction porosity loss and the sum of the compactional porosity loss plus the cementation porosity loss) were calculated using the method of Lundegard (1992), using a starting porosity of 45%: whereby, values of 1 equate to porosity loss solely by compaction and values of 0 equate to porosity loss by cementation. The Nimar Sandstone values ranged between 0.5 and 1.0, with rare values of 0.0 and 0.1 where carbonate cements are abundant. The Bhuj sandstones had a total range of −0.6 to 0.9 (the negative values being due to an inability to discriminate whether abundant haematite cements were replacive or whether they were primary or secondary pore-filling cements). The Dhrangadrha sandstones had values from 0.6 to 1.0, indicating a dominant porosity loss through compaction. Precambrian sandstone samples from the Aravalli Group near Jaisalmer had values of 0.6–0.9 and it was noted that Precambrian sandstones in the area formed important regional aquifers. The generally poorer reservoir quality of many of the Nimar sandstone samples is, in part, facies related: that is, it is more fluvial dominated and also shows greater porosity loss due to recent diagenesis, especially calcite precipitation. The tighter sandstones in the Bhuj and Pariwar sandstones are often associated with common to abundant recent haematitic and limonitic cements.

Modal porosity estimates from thin-section point-counted data are provided in Figure 18, showing that approximately 50% of the samples have porosities >10%, whilst several (mainly from the Bhuj Sandstone) have values >20%. Comparing the different studied basins, some 27% of the Nimar, 80% of the Dhrangadhra and 50% of the Pariwar sandstones have modal point-counted porosities >10%, whilst 42% of the Bhuj, 2% of the Dhrangadhra and 28% of the Pariwar sandstones have porosities >20%. The average porosities and permeabilities from plug measurements are 16.1% and 271 mD ($n = 17$) for the Nimar, 24.1% and 666 mD for the Bhuj ($n = 22$), 15.95 and 406 mD for the Dhrangadrha ($n = 18$), and 19.1% and 406 mD for the Pariwar sandstones ($n = 8$). Porosity and permeability values from plug samples are provided in Figure 19, and show, as expected based on the modal analyses, that the Bhuj sandstone samples generally show the highest permeabilities, although high permeabilities are also noted in the other three basins. Over half the samples analysed show permeabilities >200 mD, whilst nine had permeabilities >800 mD.

Source rock potential

The evidence for a pre-Deccan petroleum source rock across the study area is summarized in Figure 20 and is discussed in more detail below.

Upper Jurassic

The main known petroleum source rock downdip in the Indus and Kutch areas is the Jurassic, which, although fairly areally restricted in the onshore, could be extensively developed in the offshore. The Lower Cretaceous clastic reservoir of the Lower Goru and Bhuj formations overlie proven Jurassic–Early Cretaceous 'prodeltaic' source rocks of the Sembar and Juhran formations, respectively.

Lower Cretaceous

In the Narmada Basin, the Nimar Sandstone Group is expected to show increasing marine influence downdip and could potentially interfinger distally with source-prone sediment. In addition, the amount of accommodation space is also likely to increase

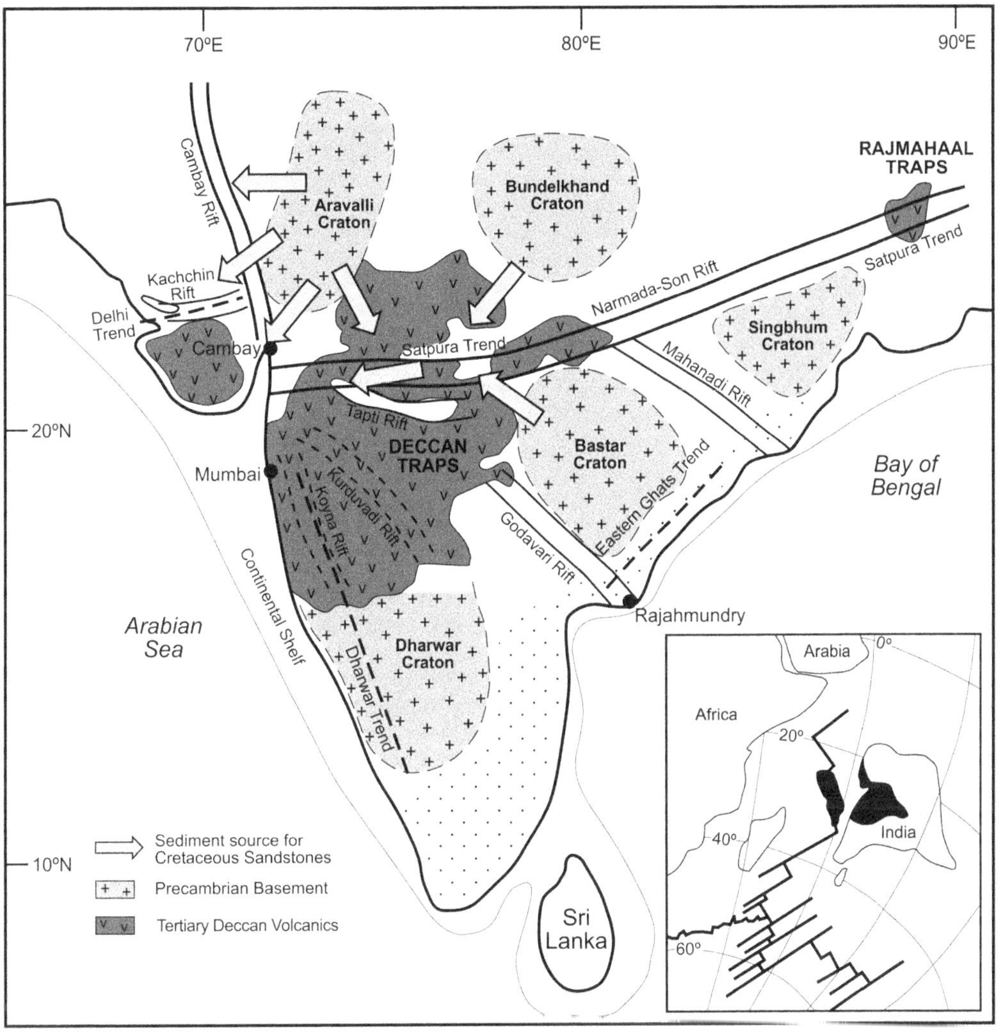

Fig. 17. Map showing the locations of the main rifts, Precambrian basement, Deccan volcanics and the main sediment source areas for the studied Cretaceous sandstones. The grey arrows indicate the inferred Cretaceous sediment-transport directions. Note that the Deccan outcrop area conceals significant areas of Cretaceous sediments and Precambrian basement.

markedly downdip, allowing the preservation of a thicker marine sandstone reservoir sequence.

The source potential of the Than Formation (Saurastra) was evaluated using a portable source rock analyser and also later in the laboratory using Rockeval. Samples were collected from the immediate outskirts of the town of Than. The source rock analyser produced hydrocarbon yields of 1.6–6.7 kg t^{-1} (an average of 3 kg t^{-1} based on analysis of 10 samples), indicating fair to moderate source rock potential. The TOC values varied from 1 to 4.9% (an average of 3% based on 10 samples), but the Rockeval indicated a high maximum temperature (T_{max}) (550–609°C), a low hydrogen index (HI) (9–28) and low $S_1 + S_2$ yields of 0.2–1.35, suggesting the samples were over-mature and that any remaining hydrocarbon potential had been exhausted at the study locations near the village of Than. Owing to the coaly nature of the Than Formation, gas-prone Type III kerogens predominate: however, downdip (offshore), the interval may pass into a more marine oil-prone sequence that may have less overburden, thus creating a possible oil mature source rock. Mani *et al.* (2012) noted

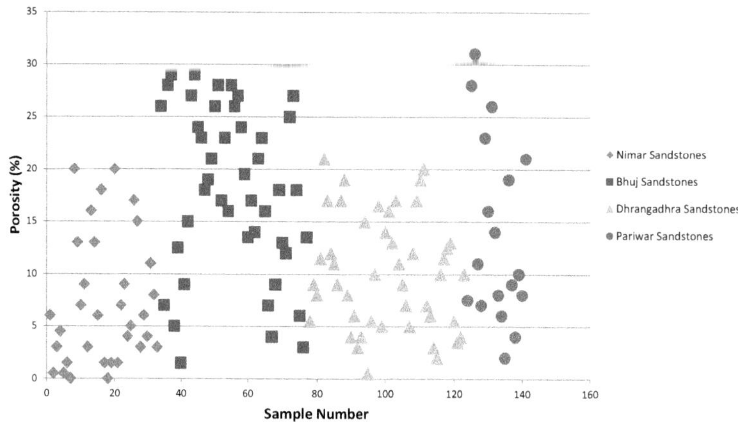

Fig. 18. Porosity estimates (y-axis) from point-counted thin-section data. Notice that the poorer reservoir quality in Nimar Sandstones is, in part, facies related (i.e. it is more fluvial and shows greater porosity loss due to recent diagenetic processes, especially calcite precipitation). The tighter sandstones in the Bhuj and Pariwar sandstones are often associated with recent haematitic and limonitic cements.

gas anomalies in surface soil samples in the area between the Lodikha well location and Rajkot, which most probably came from a pre-Deccan source, possibly the Than Formation.

Potential source rock samples were also collected from a faulted and folded sequence exposed in a road cutting in easternmost Cambay, which comprised shales and interbedded sandstones unconformably overlain by Deccan lavas. A very poorly preserved palynoflora, comprising *Classopollis* spp., *Cicatricosisporites* cf. *australis*, *Cicatricosisporites* sp., non-striate bisaccate pollen and several indeterminate miospores, was recovered from these shales, which together would indicate a Late Jurassic–Early Cretaceous age. The absence of angiosperm pollen may indicate an age older than Barremian. The shale sequence is around 5 m thick and has an average TOC of 2% (max 4%). The high T_{max} values from the Rockeval (>500°C) indicate that these samples have no remaining hydrocarbon-generating potential at our studied location. However, these shales may possess some hydrocarbon-generative capacity elsewhere if not too deeply buried.

Upper Cretaceous

The boundary between the Nimar Sandstone Group and the Nodular Limestone Formation straddles the Cenomanian–Turonian boundary. Source rocks were well developed globally during this timeframe due to oceanic anoxicity: however, this time interval

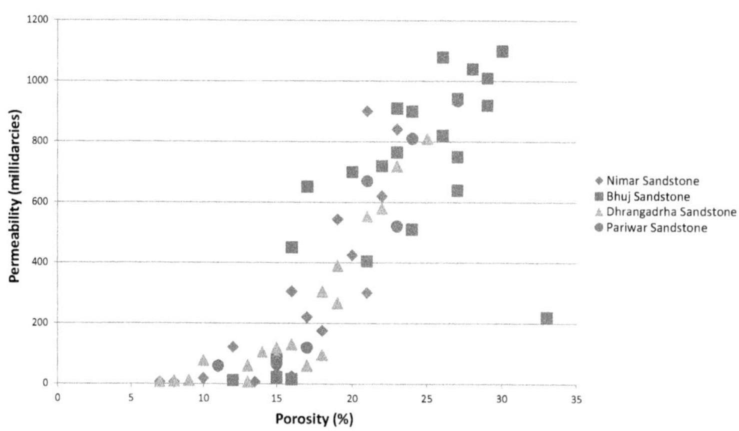

Fig. 19. Plug porosities (x-axis) and permeabilities (y-axis) from selected outcrop samples (see the text for the discussion).

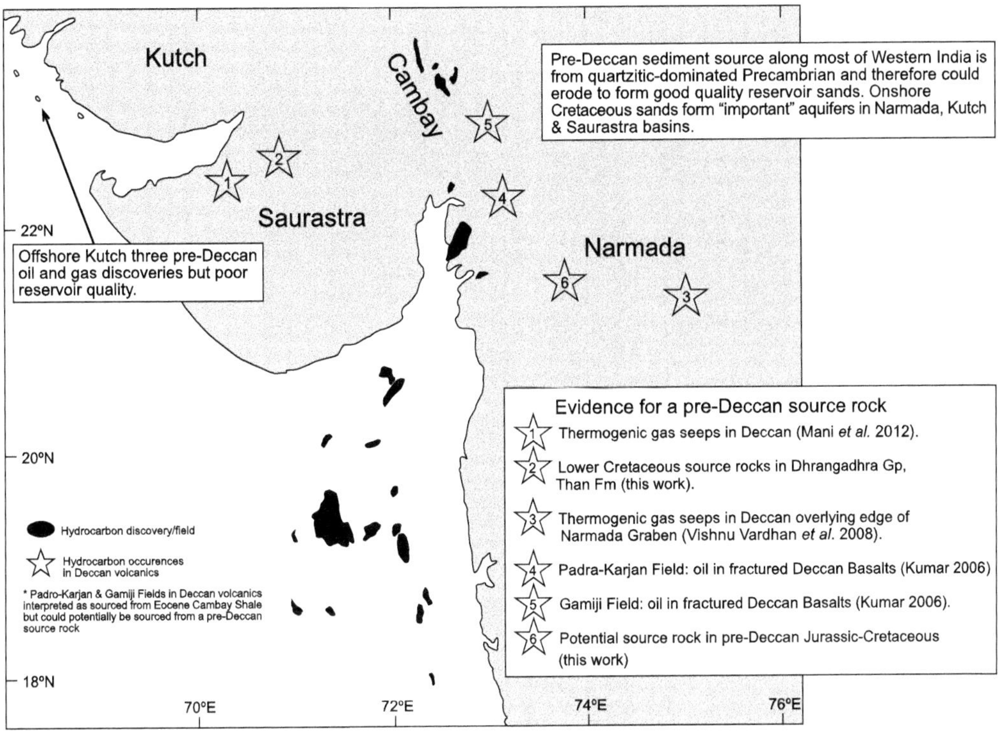

Fig. 20. Regional evidence for a working pre-Deccan source rock in the study area.

downdip is unlikely to have had major source potential along the west coast of India as the western Indian Ocean opened in the Jurassic and by Cenomanian–Turonian time was wide (Scotese 2011) and, therefore, probably oxic.

Hydrocarbon shows

Surface geochemical surveys (Abdul Azeez *et al.* 2011) have been undertaken in the western part of the Narmada area along a north–south transect from just south of Indore to just south of the Tapti River: that is, in an area where the Cretaceous does not crop out and is buried beneath Deccan basalts. This survey has identified microseepages of thermogenic hydrocarbons in areas where the Mesozoic sediment thickness is interpreted to be at least 2 km thick, based on geophysical data. Similar studies to the south of the Narmada River in the Nandurbar region by Vishnu Vardhan *et al.* (2008) showed up to 2.5 km of Mesozoic sediments beneath thick Deccan volcanics where light gaseous hydrocarbons with a thermogenic isotopic signature were observed in soil samples. These geochemical results suggest that a thermogenic petroleum play might exist beneath the Deccan volcanics in this area.

Surface geochemical studies in the onshore western part of the Saurastra Basin indicate the presence of a thermogenic source rock (Fig. 10). Kumar *et al.* (2009), in a study of adsorbed soil gas in the Jamnagar sub-basin, NW Saurastra, noted thermogenically derived gases and suggested the presence of 1000–2000 m of Mesozoic sediments below the Deccan Traps in this area that may have provided sufficient burial (especially when the additional overlying Deccan sequence is added) to generate hydrocarbons. Mani *et al.* (2012) undertook a surface geochemical study east of the town of Rajkot near the Lodhika-1 well. They identified thermogenic microseepages associated with iodine anomalies that were broadly coincident with dykes and lineaments in the Deccan Traps, and concluded that the latter probably formed the conduit for the hydrocarbon seepage from the underlying Mesozoic section and that the seepages were also coincident with where the Mesozoic sediments were thickest (most thermally mature). Thermogenic hydrocarbons are found in Deccan Basalts in the Cambay Basin (Kumar 2006).

Regional correlation

See Figure 21 for the regional correlation panel.

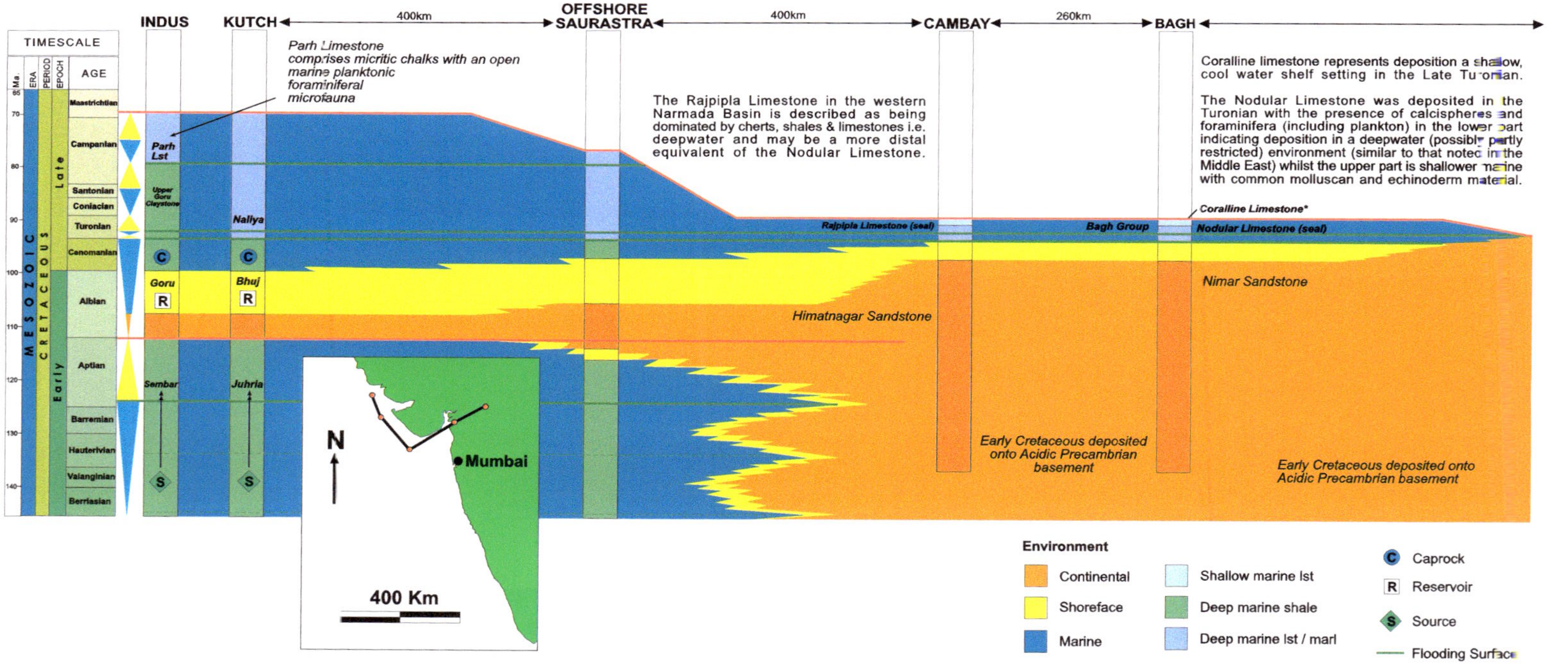

Fig. 21. Regional correlation panel showing the main ages and depositional ages of the studied interval together with the potential petroleum systems elements.

Lower Cretaceous: fluvial sandstones

The Jabalpur Group of the western Narmada is poorly dated on the basis of plant macrofossils (*Onychiopsis* and *Weichselia*) as Upper Jurassic–Lower Cretaceous and may range older than the Nimar Group. The Jabalpur Group has yielded the palynomorphs *Cicatricosisporites*, *Coptospora* and *Cooksonites*, which were assigned to the Berriasian–Aptian by Singh & Venkatachala (1988) and, consequently, its upper part may be broadly equivalent in age to the upper part of the Nimar Group. The Jabalpur Group also rests unconformably on acidic Precambrian basement and is unconformably overlain by non-marine carbonates and shales (mainly lacustrine) of the Lameta Formation, which are dated on the basis of palynology and dinosaur remains as Maastrichtian.

The older fluvial-dominated portion of the Nimar Sandstone Group probably correlates with the sandier Lower Goru Formation and the source-prone Sembar Formation of the Indus Basin; the lower part of the Bhuj Formation (i.e. the Ukra and Ghuneri members in the Kutch Basin) and, possibly, with the Dhrangadrha Group of Saurastra.

The Lower Cretaceous Himatnagar Sandstone Formation of the Cambay Basin is laterally equivalent to the Nimar Sandstone Group of the Narmada Basin to the east and the Dhrangadhra Group sandstones, which crop out to the west/SW in Saurastra.

Mid Cretaceous: onset of marine transgression: shallow-marine clastics

The presence of marine sandstones (Nimar Sandstone Group) some 400–500 km inland in the Narmada area indicates that the graben formed a shallow, possibly sometimes restricted (?) marine seaway from the Early Cretaceous onwards. The presence of marine sandstones, interfingering with fluvial deposits, also suggests that the fluvial systems operating in the Narmada Basin during the Early Cretaceous were externally draining.

Regionally, the upper marine portion of the Nimar Sandstone Group (Narmada Basin) may correlate with the mudstone-dominated Upper Goru Formation of the Indus Basin, which is equivalent to the upper part of the Umia/Bhuj Formation (Bhuj Member) of the Kutch Basin and possibly also with the Ranipat Formation of Saurastra. These shallow-marine sandstones are probably, in part, equivalent to the Jabalpur Group fluvial sandstones located some 460 km to the east further along the Narmada Graben. The Suredranagar Sandstone Member of the Wadwhan Formation is most likely equivalent to the upper shallow-marine portion of the Nimar Group sandstones of Narmada but it is more fossiliferous (? more marine) and may also be equivalent to the upper part of the Himatnagar Sandstone Formation of eastern Cambay some 150 km to the NE. The Dhrangdhra Group (Saurashtra) is believed to correlate with the Bhuj Member of the Kutch Basin. The Nimar Group of the Narmada area is broadly equivalent to the Bhuj Member of Kutch, the lower part of the Dhrangadrha Group (Than, Surajdeval and Ranipat formations) of Saurastra and the Lower Goru Formation of the Indus Basin.

Cenomanian–Turonian sea-level maximum

In the Narmada Basin, the transition from the shallow-marine sandstones of the upper part of the Nimar Sandstone Group to the deeper-water marls of the Nodular Limestone Formation records the continuation of a major marine transgression (Fig. 7). This lithofacies change to the marly limestones of the Nodular Limestone Formation broadly corresponds to a major global sea-level maximum in the Turonian (Hancock & Kauffman 1979) and indicates that a major marine transgression did advance at least as far east as Zeerabad in the Narmada Graben.

Regionally, the Nodular Limestone Formation correlates with the Upper Goru shales in the Indus Basin and the Wadhwan Formation of Saurastra (albeit, it is markedly thinner and less well developed). The Parh Limestone in the lower Indus Basin represents the younger Late Campanian flooding event.

It is likely that the Nimar Group to the middle part of the Nodular Limestone Formation represents an overall transgressive sequence (probably as two distinct transgressions separated by a sequence boundary) up to the level of the *Placenticeras* ammonite influx and is then regressive up to the top of the Coralline Limestone Formation. Firmgrounds and hardgrounds within the upper part of the Nodular Limestone and throughout the Coralline Limestone Formation, coupled with significant differences in lithology and thickness on a fairly local scale, are likely to represent higher-order sea-level fluctuations. The Coralline Limestone Formation (of the Narmada Basin) may be correlated with the Bhaduka Limestone Member of the Wadwhan Formation of Saurastra, which has a similar facies and macrofauna (Chiplonkar & Borkar 1971).

Upper Cretaceous unconformities

In Narmada and Cambay, the Coralline Limestone Formation is locally unconformably overlain by the Lameta Formation, which is dated as Maastrichtian (Tandon *et al.* 1995; Joshi & Ganapathi 2000). Consequently, the Coniacian–Campanian are absent (either not deposited or eroded). A similar break is noted in the Kutch Basin and in the Cambay Basin to the immediate west. The age of the Bagh Group (Nodular and Coralline Limestone formations) may fall entirely within the Turonian (to possibly basal Coniacian), indicating a significant unconformity between the top of this unit and the overlying Maastrichtian Lameta Formation.

In the Cambay and Kutch areas to the west and NW, a marked break in the Santonian–Campanian also appears to be present. This stratigraphic gap of around 15+ Ma occurred when global sea level was relatively high and, therefore, is more likely to represent a tectonic effect that would have a more marked significance in such a proximal location within the deposystem. Downdip, this time interval may actually be present/preserved.

Conclusions

This field campaign set out to constrain models for petroleum systems postulated to have developed beneath the Deccan volcanic sequence of offshore western India. Prior to the fieldwork, establishing the link between onshore outcrops in different basins and offshore data points was extremely problematic as the ages of many of the units were poorly constrained. The issue of correlation was further compounded by the plethora of lithostratigraphic schemes suggested within the published literature.

New biostratigraphic data collected during this study and a thorough revision of existing palaeontological data indicated that a revised correlation was required across the Cretaceous sequence of NW India (Fig. 21). Outcrop lithofacies and palaeontological analyses provided critical information on the environments of deposition and any lateral or vertical changes. By placing these observations in the correct chronostratigraphic context provided by the revised biostratigraphic framework, a basis for a sequence stratigraphic correlation from which meaningful predictions of offshore pre-Deccan petroleum systems can be obtained could be established.

The Early Cretaceous was a time of clastic deposition regionally, as evidenced by the thick sandstone packages observed in basins throughout the study area. The Lower Cretaceous sandstones are proven reservoirs in the Indus Basin in Pakistan and in the Kutch Basin in India. This study suggests that a similar reservoir potential may exist for the offshore portions to the south of Saurastra, in and around the present-day Mumbai Basin.

Petrographical studies have shown that the Lower Cretaceous sandstones in all of the studied basins have a similar provenance that is dominated by quartz, and which appears to mainly represent the recycling of Precambrian Aravalli Group metamorphic, metasediments and associated acidic plutonic rocks. The Lower Cretaceous sandstones cover a huge area and show remarkably little variation in their bulk provenance. Moreover, thin-section observations and laboratory measurements on plugs taken from the field samples indicate moderate to good reservoir quality. Based on their quartz-dominated composition, these sandstones may retain significant porosity during burial. Field observations of palaeocurrents in the basins studied show a prevalence of sediment transport directions towards the south and west (i.e. towards the offshore area of interest).

As discussed previously, there is increasing evidence of a marine influence towards the top of the Lower Cretaceous that marks the onset of a significant marine transgression, which reached a maximum in the Turonian. This event has been broadly correlated across the study area and is typically represented by a package of deep-shelf, marly limestones (e.g. the Nodular Limestone Formation, Narmada Basin). The deep-water marls and claystones deposited during this regional transgressive event are considered as a potential seal lithology to the underlying Lower Cretaceous sandstone-dominated package.

Prior to this work, no potential source rocks had been identified in the pre-Deccan. Data relevant to source potential across the area is summarized in Figure 20. Two source rocks, one poorly dated as Late Jurassic–Early Cretaceous (? older than Barremian) in the Cambay/Narmada Basin and one in the Lower Cretaceous (?Barremian–Aptian) of the Saurastra Basin, are now recognized based on the results of this fieldwork. Although both sources are dominated by terrestrial organic matter and are overmature at present, they do indicate the potential presence of better-quality, more marine-dominated source intervals downdip (i.e. towards the offshore). In addition, the samples analysed were only part of a widespread reconnaissance survey and, consequently, less thermally mature outcrops may be present elsewhere in the study area. Recent surface geochemical studies in Narmada and Saurastra over Deccan sequences also support the presence of a pre-Deccan source rock (Kumar 2006; Vishnu Vardhan *et al.* 2008; Mani *et al.* 2012).

This study demonstrates the critical contribution that fieldwork can make when working in frontier exploration areas where models for key elements that make up potential petroleum systems are poorly constrained. As technology marches on, the petroleum geoscientist is presented with increasingly more sophisticated ways to try to understand, image and predict the subsurface geology. This example serves as a reminder that fieldwork can provide an efficient and economical way of obtaining genuine geological data that can be used to inform commercial decisions regarding exploration strategy.

Special thanks are due to Drs Michael Oates and Simon Gozzard (BG Group) for their assistance in the field. Dave Nuttman is thanked for draughting many of the figures. Drs Liam Gallagher and Keith Gueinn (Network Stratigraphic Consulting) are thanked for the nannofossil and palynological analyses, whilst Drs Andrew Smith and Paul Taylor (Natural History Museum, London) are thanked for providing information on the echinoids and bryozoans, respectively. Special thanks are due to BG Group for supporting this work, which was undertaken while AR, JF and SKR were employees of this company.

References

ABDUL AZEEZ, K.K., SATISH KUMAR, T., BASAVA, S., HARINARAYANA, T. & DAYAL, L.M. 2011. Hydrocarbon prospects across Narmada–Tapti rift in Deccan Trap, central India: inferences from integrated interpretation of magnetotelluric and geochemical prospecting studies. *Marine and Petroleum Geology*, **28**, 1073–1082.

ADAMS, T.D., KHALILI, M. & SAID, A.K. 1967. Stratigraphic significance of some oligostegenid assemblages from the Lurestan province, northwest Iran. *Micropaleontology*, **13**, 55–87.

AHMAD, A.H.M. & AKHTAR, K. 1990. Clastic environments and facies of the Lower Cretaceous Narmada Basin, India. *Cretaceous Research*, **11**, 175–190.

AKHTAR, K. & AHMAD, A.H.M. 1991. Single cycle cratonic quartzarenites produced by tropical weathering: the Nimar Sandstone (Lower Cretaceous), Narmada Basin, India. *Sedimentary Geology*, **71**, 23–32.

AKHTAR, K. & AQUIL, M. 1984. Depositional environments of the Himatnagar Sandstone (Lower Cretaceous), Gujarat. *Petroleum Asia Journal*, **7**, 92–96.

AKHTAR, K., KHAN, M.M. & AHMAD, A.H.M. 1992. Diagenetic evolution of a Cretaceous quartz arenite, Narmada rift basin, India. *Sedimentary Geology*, **76**, 99–109.

ASLAM, M. 1991. Trace fossils from the Ranipat sediments (Early Cretaceous), Saurastra Basin, Gujarat, Western India. *Current Science*, **61**, 403.

ASLAM, M. 1992. Delta plain coal deposits from the Than Formation of the Early Cretaceous Saurastra basin, Gujarat, Western India. *Sedimentary Geology*, **81**, 181–193.

BABU, P.V.L.P. 1984. A structural synthesis of the Lower Narmada River basin and its bearing on the Mesozoic oil prospects. *Petroleum Asia Journal*, 7, 97–105.

BARDHAN, S., GANGOPADHYAY, T.K. & MANDAL, U. 2002. How far did India drift during the Late Cretaceous? *Placenticeras kaffrariium* Etheridge, 1904 (Ammonoidea) used as a measuring tape. *Sedimentary Geology*, **147**, 193–217.

BESSE, J. & COURTILLOT, V. 1988. Palaeogeographic maps of the continents bordering the Indian Ocean since the Early Jurassic. *Journal of Geophysical Research*, **93**, 11,791–11,808.

BHANDARI, A. 1999. Phanerozoic stratigraphy of western Rajasthan: a review. *In*: KATARIA, P. (ed.) *Geology of Rajasthan Status and Perspective. (A.B. Roy Felicitation Volume)*. M.L. Sukhadia University, Udaipur, 126–174.

BISWAS, S.K. 1982. Rift basins in western margin of India and their hydrocarbon prospects with special reference to Kutch Basin. *American Association Petroleum Geologists Bulletin*, **66**, 1497–1513.

BISWAS, S.K. 1987. Regional tectonic framework, structure and evolution of the western marginal basins of India. *Tectonophysics*, **135**, 307–327.

BISWAS, S.K. & DESHPANDE, S.V. 1983. Geology and hydrocarbon prospects of Kutch, Saurastra and Narmada basins. *In*: BHANDARI, L.L., VENATACHALA, B.S., KUMAR, R., SWAMY, S.N., GARGA, P. & SRINASTAVA, D.C. (eds) *Petroliferous Basins of India. Petroleum Asia Journal*, **6**, 111–126.

BLANDFORD, W.T. 1869. *Geology of the Area Between Tapti and Narbada Valley and the Adjoining Districts of Malwa and Gujarat*. Geological Survey of India, Memoirs, **6**.

BOSE, P.K. & DAS, N.G. 1986. A transgressive storm- and fair-weather wave dominated shelf sequence. Cretaceous Nimar Formation, Chakrud, Madhya Pradesh, India. *Sedimentary Geology*, **46**, 147–167.

BOSE, P.N. 1884. *Geology of Lower Narbada Valley between Nimavar and Kawant*. Geological Survey of India, Memoirs, **21**.

BOWN, P.R., RUTLEDGE, D.C., CRUX, J.A. & GALLAGHER, L.T. 1998. Lower Cretaceous. *In*: BOWN, P.R. (ed.) *Calcareous Nannofossil Biostratigraphy*. Chapman & Hall, London, 86–131.

BRAMWELL, N.P., CAILLET, G., MECIANI, L., JUDGE, N., GREEN, M. & ADAM, P. 1999. Chalk exploration, the search for a subtle trap. *In*: FLEET, A.J. & BOLDY, S.A.R. (eds) *Petroleum Geology of Northwest Europe: Proceedings of the 5th Conference*. Geological Society, London, 911–937, http://doi.org/10.1144/0050911

BURNETT, J.A., GALLAGHER, L.T. & HAMPTON, M.J. 1998. Upper Creaceous. *In*: BROWN, P.R. (ed.) *Calceareous Namofossil Biostiatigraphy*. Chapman & Hall, London, 132–199.

CASSHYAP, S.M. & ASLAM, M. 1992. Deltaic and shoreline sedimentation in Saurastra Basin, Western India: an example of infilling in an Early Cretaceous failed rift. *Journal of Sedimentary Petrology*, **62**, 972–991.

CHIPLONKAR, G.W. & BORKAR, V.D. 1971. A preliminary note on the Wadhwan Formation of Saurastra. *Current Science*, **40**, 433–434.

CHIPLONKAR, G.W. & BADVE, R.M. 1976*a*. Palaeontology of the Bagh Beds Part IV: inoceramidae. *Journal of the Palaeontological Society of India*, **48**, 11–12.

CHIPLONKAR, G.W. & BADVE, R.M. 1976*b*. On some inoceramid species occurring in the Bagh Beds of the Narmada Valley. *Journal of the University of Poona, Science and Technology*, **48**, 195–209.

CHIPLONKAR, G.W. & GHARE, M.A. 1976. Palaeontology of the Bagh Beds – Part VII: ammonoidea. *Bulletin of Earth Sciences*, **4 & 5**, 1–10.

CHIPLONKAR, G.W., BADVE, R.M. & GHARE, M.A. 1977. On the stratigraphy of Bagh Beds of the Lower Narmada valley. *In*: VENKATACHALA, B.S. & SASTRI, V.V. (eds) *Proceedings of Fourth Indian Colloquium Micropaleontology and Stratigraphy*. Institute of

Petroleum Exploration, Oil & Natural Gas Commission, Dehradun, 209–216.

DAS GUPTA, S.K. 1975. Revision of the Mesozoic–Tertiary stratigraphy of the Jaisalmer Basin, Rajasthan. *Indian Journal of Earth Science*, **2**, 77–94.

DASSARMA, D.C. & SINHA, N.K. 1975. *Marine Cretaceous formations of Narmada Valley (Bagh Beds), Madhya Pradesh and Gujarat*. Geological Survey of India, Palaeontologia Indica, **42**.

GANGOPADHYAY, T.K. & BARDHAN, S. 2000. Dimorphism and a new species of *Barroisiceras* de Grossouvre (Ammonoidea) from the Coniacian of Bagh, central India. *Canadian Journal of Earth Science*, **37**, 1377–1387.

HANCOCK, J.M. & KAUFFMAN, E.G. 1979. The great transgressions of the Late Cretaceous. *Journal of the Geological Society, London*, **136**, 175–186, http://doi.org/10.1144/gsjgs.136.2.0175

JAFAR, S.A. 1982. Nannoplankton evidence of Turonian transgression along Narmada Valley India, and Turonian–Coniacian boundary problem. *Journal of the Palaeontological Society of India*, **27**, 17–30.

JAITLY, A.V. & AJANE, R. 2013. Comments on *Placenticera mintoi* (Vredenburg, 1906) from the Bagh Beds (Late Cretaceous), Central India with special reference to Turonian Nodular Limestone Horizon. *Journal of the Geological Society of India*, **81**, 565–574.

JOHRI, D.N. & KANDPAL, L.D. 1983. Exploration for oil and natural gas in Saurastra plateau – a geophysical review. *In*: *Geophysics in Geological Mapping and Structural Studies, Petroleum Resources Exploration and Mining Geophysics*. Geological Survey of India, Special Publications, **2**, 89–100.

JOSHI, A.V. & GANAPATHI, S. 2000. Lameta Formation (Maastrichtian) of Narmada Rift Basin, Western India – a critical study of its depositional environments. *In*: GOVINDAN, A. (ed.) *Cretaceous Stratigraphy – An Update*. Geological Survey of India, Memoirs, **46**, 273–294.

KAILA, K.L., KRISHNA, V.G. & MALL, D.M. 1981. Crustal structure along Mehmadabad–Billimora profile in the Cambay Basin, India, from deep seismic soundings. *Tectonophysics*, **76**, 99–130.

KAILA, K.L., REDDY, P.R., DIXIT, M.M. & KOTESWARA ROE, P. 1985. Crustal structure along the Narmada–Son lineaments, central India, from deep seismic soundings. *Journal of the Geological Society of India*, **26**, 465–486.

KENNEDY, W.J., PHANSALKAR, V.G. & WALASZCZYK, I. 2003. *Prionocyclus germari* (Reuss, 1845), a Late Turonian marker fossil from the Bagh Beds of central India. *Cretaceous Research*, **24**, 433–438.

KRISHNA, J., SINGH, I.B., HOWARD, J.D. & JAFAR, S.A. 1983. Implications of new data on the Mesozoic rocks of Kachchh, western India. *Nature*, **305**, 790–792.

KUMAR, B., KALPANA, G., PATIL, D.J. & VARDHAN, V. 2009. Hydrocarbon prospects of Jamnagar Sub-basin, Saurastra: inferences from Surface Geochemical Surveys. *Petroview*, **III**, 21–26.

KUMAR, P. 1973. The spores dispersae of Jabalpur stage Upper Gondwana, India. *Palaeobotanist*, **20**, 91–126.

KUMAR, P. 1992. Palynology of Mesozoic sediments exposed near Ellichpur, Maharastra. *Palaeobotanist*, **39**, 381–386.

KUMAR, P. 1994. Palynology of carbonaceous clays of the Nimar formation, Jhabua district, Madhya Pradesh, India. *Journal of the Geological Society of India*, **44**, 671–674.

KUMAR, A. 2006. Formation evaluation of unconventional basaltic Deccan trap basement reservoirs of Gramij Field, Cambay Basin, India. *In*: *6th International Conference and Exposition on Petroleum Geophysics*, Kolkata, India, 534–538.

KUMAR, P. & KULSHRESHTHA, S.K. 1979. Palynostratigraphical studies of carbonaceous shales from Kotri, Narsingphur district, India. *Geophytology*, **9**, 52–61.

KUNDAL, P. & SANGANWAR, B.N. 1998. Stratigraphical, palaeogeographical and palaeoenvironmental significance of fossil calcareous algae from Nimar Sandstone Formation, Bagh Group (Cenomanian–Turonian) of Pipaldehla, Jhabua Dt, MP. *Current Science*, **75**, 702–708.

LUNDEGARD, P.D. 1992. Sandstone porosity loss – a 'big picture' view of the importance of compaction. *Journal of Sedimentary Petrology*, **62**, 250–260.

MANI, D., PATIL, D.J., KALPANA, M.S. & DAYAL, A.M. 2012. Evaluation of hydrocarbon prospects using surface geochemical data with constraints from geological and geophysical observations: Saurastra Basin, India. *Journal of Petroleum Geology*, **35**, 67–84.

MORGAN, W.J. 1981. Hot spot tracks and the opening of the Atlantic and Indian oceans. *In*: EMILIANI, (ed.) *The Sea Volume 7. The Oceanic Lithosphere*. Wiley, New York, 443–487.

MUDE, S.N., RAUT, S.D. & WAGH, R. 2012. Paleoenvironmental significance of Ichnofossils from the Bhuj Formation (Gondwana sediments) of the Kachch Region, North Western India. *World Journal of Environmental Biosciences*, **1**, 25–29.

MUKHERJEE, M.K. 1983. Petroleum prospects of Cretaceous sediments of the Cambay Basin, Gujarat, India. *Journal of Petroleum Geology*, **5**, 275–286.

MURTY, K.N., RAO, R.P., DHOKARIKAR, B.G. & VARMA, C.P. 1963. On the occurrence of plant fossils in Nimar Sandstones near Umrali District, Jhabua, Madhya Pradesh. *Current Science*, **32**, 21–23.

OGG, J.G., OGG, G. & GRADSTEIN, F.M. 2008. *The Concise Geological Time Scale*. Cambridge University Press, Cambridge.

OMAÑA, L., TORRES, J.J., LÓPEZ-DONCEL, R., ALENCÁSTER, G. & LÓPEZ-CABALLERO, I. 2014. A pithonellid bloom in the Cenomanian–Turonian boundary interval from Cerritos in the western Valles-San Luis Potosi platform, Mexico; Paleoenvironmental significance. *Revista Mexicana de Ciencias Geológicas*, **31**, 28–44.

PAL, A.K. 1971. A note on the biozonation of the Bagh Group of Madhya Pradesh. *In*: *Proceedings of the Fifty Eighth Session of the Indian Science Congress Association. Part III: Abstracts*. Indian Science Congress Association, Kolkata, 332.

PERCH-NIELSEN, K. 1985. Mesozoic Calcareous Nannofossils. *In*: BOLLI, H.M., SAUNDERS, J.B. & PERCH-NIELSEN, K. (eds) *Plankton Stratigraphy*. Cambridge University Press, Cambridge, 427–454.

PODDAR, M.C. 1964. Mesozoics of western India – their geology and oil possibilities. *In*: *Proceedings of the 22nd International Geological Congress, India, Part 1*, 126–143.

RAI, J. 2006. Discovery of nannofossils in a plant bed of the Bhuj Member, Kutch and its significance. *Current Science*, **91**, 519–526.

RAMASAMY, S. & MADHAVARAJU, J. 1993. Petrographic studies of the Bagh Beds along the Narmada Valley in Madhya Pradesh, India. *Gondwana Geological Magazine*, **4/5**, 65–79.

RATHORE, S.S., PRABHU, B.N., VIJAN, A.R., VIG, K.C. & MISRA, K.N. 1999. K–Ar age dating of glauconites from the Kutch Basin, India. *Proceedings of the Indian Academy of Science (Earth Planetary Science)*, **108**, 49–55.

RAVAL, U. 2003. Interaction of mantle plume with Indian continental lithosphere since Cretaceous. *In*: MAHADEVAN, T.M., ARORA, B.R. & GUPTA, K.R. (eds) *Indian Continental Lithosphere: Emerging Research Trends*. Geological Survey of India, Memoirs, **53**, 449–479.

RAVAL, U. & VEERASWAMY, K. 2003. India–Madagascar separation: breakup along a pre-existing mobile belt and chipping of the craton. *Gondwana Research*, **6**, 467–485.

RODE, K.P. & CHIPLONKAR, G.W. 1935. A contribution to the stratigraphy of the Bagh Beds. *Current Science*, **2**, 322–323.

ROY, A.B. & JAKHAR, S.R. 2002. *Geology of Rajasthan (Northwest India) Precambrian to Recent*. Scientific Publishers, Jodphur.

ROY CHOWDHARY, M.K. & SASTRI, V.V. 1962. On the revised classification of the Cretaceous and associated rocks of the Man River section of lower Narbada Valley. *Records of the Geological Survey of India*, **91**, 283–301.

SAHAY, B. 1984. A review of the geology and petroleum possibilities of the continental margins of India. *Paper OTC-4699 presented at the 16th Offshore Technology Conference*, 7–9 May 1984, Houston, Texas.

SAHNI, B. 1936. The occurrence of *Metonidium* and *Weichselia* in India. *Records of the Geological Survey of India*, **71**, 152–160.

SAHNI, M.R. & JAIN, S.P. 1966. Note on a revised classification of the Bagh Beds, Madhya Pradesh. *Journal of the Palaeontological Society of India*, **11**, 24–25.

SCOTESE, C.R. 2011. Paleogeographic and Paleoclimatic atlas. AAPG Search and Discovery Article 30192. *AAPG Annual Convention and Exhibition*, Houston, Texas, USA, 10–13 April 2011.

SINGH, D., ALAT, C.A., SINGH, R.N. & GUPTA, V.P. 1997. Source rock characteristics and hydrocarbon generating potential of Mesozoic sediments in Lodhika area, Saurastra Basin, Gujarat, India. *In*: *Proceedings of the Second International Petroleum Conference and Exhibition, PETROTECH-97*, New Delhi. B.R. Publishing Corporation, New Dehli, 205–220.

SINGH, H.P. & VENKATACHALA, B.S. 1988. Upper Jurassic–Lower Cretaceous spore–pollen assemblages in the peninsular India. *The Palaeobotanist*, **36**, 168–176.

SINGH, N.P. 1996. Mesozoic–Tertiary biostratigraphy and biogeochronological datum planes in Jaisalmer Basin, Rajasthan. *In*: PANDEY, J., AZMI, R.J., BHANDARI, A. & DAVE, A. (eds) *Contributions XV Indian Colloquium Micropalaeontology Stratigraphy*. DDMIPE and WIHG, Dehradun, 63–89.

SINGH, S.K. & SRIVASTAVA, H.K. 1981. Lithostratigraphy of Bagh Beds and its correlation with Lameta beds. *Journal of thePaleontological Society of India*, **26**, 77–85.

SMITH, A.B. 2010. The Cretaceous Bagh Formation, India: a Gondwanan window on Turonian shallow-water echinoid faunas. *Cretaceous Research*, **31**, 368–386.

TANDON, S.K., SOOD, A., ANDREWS, J.E. & DENNIS, P.F. 1995. Palaeoenvironments of the dinosaur bearing Lameta Beds (Maastrichtian), Narmada Valley, Central India. *Palaeogeography, Palaeoclimatology, Palaeoecology*, **117**, 153–184.

TAYLOR, P.D. & BADVE, R.M. 1995. A new cheilostome bryozoan from the Cretaceous of India and Europe: a cyclostome homeomorph. *Palaeontology*, **38**, 627–657.

TRIPATHI, S.C. 1995. Palaeontological and palaeoenvironmental studies of Bagh Group, Madhya Pradesh. *Records of the Geological Survey of India*, **128**, 104–105.

VERMA, R.G. & RAWAT, M.S. 1964. A note on the age of the Dhrangadhra Formation, western India in light of pollen and spores recovery. *Pollen et Spores*, **6**, (1).

VISHNU VARDHAN, C., KUMAR, B., KUMANAN, C.J., MANI, D. & PATIL, D.J. 2008. Hydrocarbon prospects in sub-Trappean Mesozoic Deccan Syneclise, India: evidence from surface geochemical prospecting. AAPG Search and Discovery Article 10143, http://www.searchand discovery.com/documents/2008/08007vardhan/

WILKINSON, I.P. 2011. Pithonellid blooms in the Chalk of the Isle of Wight and their biostratigraphical potential. *Proceedings of the Geological Association*, **122**, 862–867.

Reservoir architectures of interlava systems: a 3D photogrammetrical study of Eocene cliff sections, Faroe Islands

HENRIK VOSGERAU[1]*, SIMON R. PASSEY[2,4], KRISTIAN SVENNEVIG[1], MAX N. STRUNCK[1,5] & DAVID W. JOLLEY[3]

[1]*Geological Survey of Denmark and Greenland, Øster Voldgade 10, DK-1350 Copenhagen, Denmark*

[2]*Jarðfeingi (Faroese Earth and Energy Directorate), Brekkutún 1, PO Box 3059, FO-110 Tórshavn, Faroe Islands*

[3]*Geology & Petroleum Geology, University of Aberdeen, Meston Building, Kings College, Aberdeen AB24 3UE, UK*

[4]*Present address: CASP, West Building, 181A Huntingdon Road, Cambridge CB3 0DH, UK*

[5]*Present address: COWI, Parallelvej 2, DK-2800 Lyngby, Denmark*

**Corresponding author (e-mail: hv@geus.dk)*

Abstract: Following the intra-volcanic Rosebank discovery in the Faroe–Shetland Basin, NE Atlantic, there has been a need to find suitable analogues to characterize reservoir architectures, connectivities and compartmentalization of interlava clastic beds. The Faroe Islands, situated *c*. 160–190 km to the NW of this discovery, are an exposed remnant of the Palaeogene lava field host and the near-vertical cliff sections afford the opportunity to map lateral variations over many kilometres. This was achieved through 3D photogrammetry based on high-resolution digital photographs taken from a helicopter. The study focused on the Eocene Enni Formation, which is dominated by a mixture of simple and compound lava flows commonly separated by minor volcaniclastic beds, including the widespread Argir Beds. In general, the interlava beds are tabular shaped and <1 to *c*. 6 m thick. Locally they thicken in depressions formed by the wedging out or differential erosion of underlying lava flow lobes. Connectivity may be caused by the wedging out of successive lava flows leading to the lateral merging of interlava beds or more rarely be hydraulically connected via conglomeratic fills of major channels. Lateral compartmentalization may be caused by the intersection of interlava beds by dykes, lava tubes, lava-filled channels or reverse faults.

A major oil and gas discovery within a Paleocene–Eocene Flood Basalt Province on the NW European continental shelf has given rise to a new hydrocarbon play type. The Rosebank Field in the submerged Faroe–Shetland Basin consists of a series of fluvial reservoir siliciclastic sandstones interbedded with basalt lava flows, hyaloclastites and volcaniclastic sedimentary units (Duncan *et al.* 2009; Helland-Hansen 2009, Austin *et al.* 2014). This has subsequently been bolstered by the intra- and supra-volcanic Cambo discovery, *c*. 35–40 km along-strike to the SW (Austin *et al.* 2014; Fielding *et al.* 2014). These discoveries have led to a renewed interest in the interplay between volcanism and sedimentation, ranging from lava–sediment interactions (Hole *et al.* 2013) to the development of intra-volcanic drainage systems (Ellis *et al.* 2009; Williamson & Bell 2012; Schofield & Jolley 2013). The offshore discovery is hampered, however, by the absorption and scattering of acoustic signals by the encasing lava flows and, if imaged, by the seismic-scale resolution (tens of metres) of the data. These problems make it difficult to delineate and characterize the reservoir architectures (Austin *et al.* 2014). There is a need, therefore, to mitigate these risks and validate the interpretations (e.g. volume estimates) through well-exposed analogues. Through such analogues the reservoir architectures, such as the geometries, relationships, compartmentalizations and connectivities (migration pathways), of interlava clastic units can be studied.

The well exposed, near-vertical volcanic cliff sections of the Faroe Islands offer an excellent opportunity to study such analogues. The Faroe Islands are also situated *c*. 160–190 km to the NW

From: Bowman, M., Smyth, H. R., Good, T. R., Passey, S. R., Hirst, J. P. P. & Jordan, C. J. (eds) 2016. *The Value of Outcrop Studies in Reducing Subsurface Uncertainty and Risk in Hydrocarbon Exploration and Production*. Geological Society, London, Special Publications, **436**, 55–73.
First published online December 17, 2015, http://doi.org/10.1144/SP436.7

of the Rosebank and Cambo fields and form the only exposed remnant of the Faroe Islands Basalt Group (FIBG) that hosts the discoveries. This contribution presents the results of a 3D photogrammetric study of some of the seismic-scale cliff sections, focusing on the volcaniclastic (sedimentary) rocks interbedded between the basalt lava flows of the Eocene Enni Formation. The high-resolution photographs taken of the cliff sections allow for a detailed study of the formation, not just vertically, but also making it possible to map lateral variations over distances of many kilometres. Information about the geometries, thicknesses and lateral extents of the interlava beds and their relationships with the surrounding lava flows is presented. Additional observations concerning the potential compartmentalization of the interlava beds by igneous (intrusive and extrusive) rocks and faulting are also given. Although the reservoir properties of the interlava beds may be limited by low permeabilities, related to a high content of volcanic material (Ellingsgaard 2009), they are still considered a relevant analogue for high-quality, siliciclastic reservoir sandstones within lava-dominated successions. This is because the principal governing mechanisms (e.g. geometries, lateral extents, compartmentalization) in many respects are the same for both types of clastic (volcaniclastic v. siliciclastic) rocks.

All the mapped features are stored in a GIS database and the 3D data can be exported for 3D modelling using, for example, Petrel reservoir engineering software in order to constrain uncertainties in offshore interlava hydrocarbon discoveries (not undertaken in this study).

Geological setting and stratigraphy

The Faroe Islands are an exposed remnant of the FIBG that locally to the islands rarely exceeds 5 km in thickness (Rasmussen & Noe-Nygaard 1969, 1970; Waagstein 1988; Passey & Jolley 2009). The FIBG is predominantly composed of subaerial basalt lava flows commonly separated by minor volcaniclastic units of pyroclastic and/or sedimentary origin. This widespread volcanic activity, during the Paleocene and earliest Eocene (61–55 Ma; Storey *et al.* 2007), is associated with the arrival of the putative Iceland plume beneath Greenland and resulted in the eventual continental break-up between Greenland and Eurasia that culminated with the onset of seafloor spreading in magnetochron 24r, *c.* 56 Ma (Saunders *et al.* 1997; Storey *et al.* 2007). The FIBG extends from the Faroe Islands into the Faroe–Shetland Basin in an easterly to southeasterly direction for at least 200 km (Ritchie *et al.* 1999; Passey & Hitchen 2011). The FIBG covers an area of at least 120 000 km^2 around the islands, but stretches from offshore western Norway southwestwards to offshore western Ireland (Ritchie *et al.* 1999; Passey & Hitchen 2011).

The Faroe Islands were dissected by glacial action during the Quaternary, resulting in an archipelago composed of 18 main islands with a distinctive mountainous terrain (Jørgensen & Rasmussen 1986). The islands have an average height of *c.* 300 m above mean sea-level and no point is more than 5 km away from the coastline. The islands are typically elongated with a strong NW–SE trend and are separated by long, narrow fjords (Fig. 1). The islands are greatly affected by wave action and continually battered by the North Atlantic Current, resulting in the undercutting of primarily western coastlines leading to the formation of near-vertical sea cliffs hundreds of metres high. These inaccessible cliffs provide excellent vertical exposure through the lava field, but also laterally over many kilometres. The opposing sides of the cliffs are typically grass covered and continuous exposure is limited to stream sections. The FIBG is in general dipping slightly (*c.* 3°) towards the SE (Passey & Varming 2010).

Renewed mapping following the groundwork by Rasmussen & Noe-Nygaard (1969, 1970) has produced a revised lithostratigraphic subdivision of the islands (Table 1; Passey & Jolley 2009). Seven formations are recognized based on the presence of major disconformity surfaces and changes in lithology (Passey & Jolley 2009), but also on geochemistry (Rasmussen & Noe-Nygaard 1969, 1970; Waagstein 1988). Only the upper *c.* 3.4 km of the FIBG is exposed on the islands and is chiefly composed of the Beinisvørð, Malinstindur and Enni formations. The Beinisvørð Formation is characterized by laterally extensive simple lava flows decametres thick (Passey & Bell 2007). The Malinstindur Formation, however, conversely consists of compound lava flows of similar thicknesses, but composed of numerous overlapping and anastomosing lava flow lobes centimetres to several metres in thickness (Passey & Bell 2007). The Enni Formation is characterized by both simple and compound lava flows erupted from fissure systems and low angle shield volcanoes, respectively (Passey & Bell 2007).

The lava flow-dominated formations are generally separated by laterally extensive volcaniclastic sedimentary sequences, commonly composed of reworked pyroclastic material and deposited by various fluvial and mass flow processes (e.g. Hvannhagi and Sneis formations). Other minor pyroclastic, lacustrine, fluvial and mass flow deposits are found throughout the volcanic pile and some of these are also laterally extensive over many tens of kilometres (Passey & Jolley 2009). Coals formed in lakes and swamps are locally found, but relatively

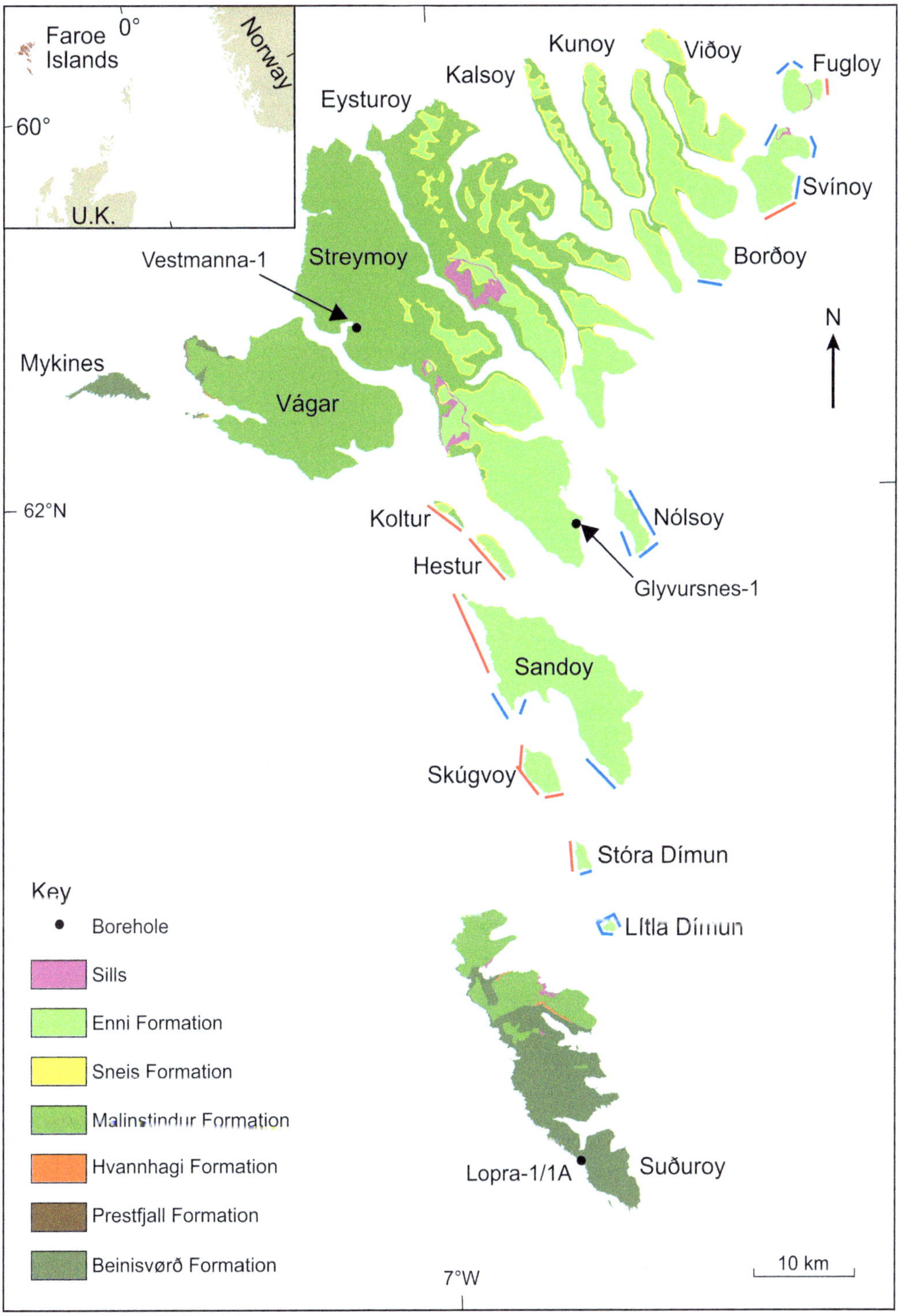

Fig. 1. Simplified geological map for the Faroe Islands Basalt Group, Faroe Islands. Modified after Passey & Jolley (2009) and references therein. The blue and red lines mark the positions of the photographed cliff sections. Those marked in red were registered for 3D photogrammetry.

Table 1. *Lithostratigraphic framework for the Faroe Islands Basalt Group*

Faroe Islands Basalt Group	Enni Formation
	Sneis Formation
	Malinstindur Formation
	Hvannhagi Formation
	Prestfjall Formation
	Beinisvørð Formation
	Lopra Formation

After Passey & Jolley (2009).

substantial seams are preserved in the <15 m-thick Prestfjall Formation (Rasmussen & Noe-Nygaard 1969, 1970; Lund 1989; Passey 2014).

This contribution primarily focuses on the Enni Formation, the youngest lava-flow-dominated formation exposed on the islands. The formation is at least 900 m thick, but at least 900 $\pm$ 200 m of strata has been removed from the top of the volcanic pile (Jørgensen 2006; Passey & Jolley 2009; Passey & Varming 2010). The lava flows vary petrographically, consisting broadly of plagioclase–phyric and olivine-bearing basalt irrespective of lava flow morphology (Noe-Nygaard & Rasmussen 1968; Waagstein 1988; Ellis *et al.* 2009; Passey & Jolley 2009). The heterogeneity in the morphology and petrology of the lava flows has helped identify key stratigraphic marker units within the formation that proved useful in the interpretation of the cliff sections.

The Høvdhamarin Flow is a 10–15 m-thick, whitish grey simple flow that forms a prominent bench in a generally poorly exposed sequence of compound lava flows of the basal *c.* 250 m of the Enni Formation (Passey 2009; Passey & Jolley 2009). The flow has been traced over a distance of at least 25–30 km from southern Streymoy southwards. The upper surface of the Høvdhamarin Flow is generally poorly exposed owing to weathering of a more vesicular upper crust, but the basal contact is generally sharp and planar. The upper part of the underlying compound lava flow(s) is also distinctly reddened.

The Argir Beds comprise a widespread volcaniclastic sequence that consistently crops out within the Enni Formation, *c.* 250 m above the base of the formation (Passey 2009; Passey & Jolley 2009; Passey & Varming 2010). The Argir Beds form the main subject of this study. The beds range in thickness from <0.5 to *c.* 6 m, but are typically about 2–3 m thick where present (Passey 2009; Passey & Jolley 2009). The beds are relatively easy to identify across the central islands from Stóra Dímun to Eysturoy (Fig. 1) owing to their pronounced thickness in a succession dominated by plagioclase–phyric compound and simple lava flows (Passey 2009; Passey & Jolley 2009). In contrast, on the northeastern islands the Argir Beds are either poorly developed or absent, but the surface they rest upon is still recognized by distinct changes in morphology and petrology of the sub- and superjacent lava flows (Ellis *et al.* 2009; Passey & Jolley 2009).

The Argir Beds can be crudely separated into a lower, dominantly greenish clay-rich sandstone succession, and an upper, reddish brown clay-poor sandstone succession (Passey 2009). The beds consist predominantly of fine- to medium-grained basaltic glass fragments, but the lower clay-rich succession additionally contains a matrix of secondary clays formed by the palagonitization (hydration) of basaltic glass fragments. The lower, clay-rich sandstones may show weak, discontinuous subhorizontal laminations, but are otherwise structureless. The lamination is picked out by thin, <1 cm-thick, laterally persistent, yellowish fine-grained sands, interpreted to represent fallout tuffs. The upper, clay-poor sandstones show scour and small-scale channel as well as broad, low-relief channel structures, several metres wide and a few decimetres deep. In places, the upper succession consists of cross-bedded, fining-upwards sandstones interpreted to represent point bar deposits formed by the lateral migration of channels (Fig. 2). Overall the Argir Beds are interpreted as deposited in a complex fluvial-floodplain environment (Passey 2009).

Initial reservoir data for the Argir Beds drilled in the Glyvursnes-1 borehole, onshore Streymoy, show that the clay-rich sandstones have lower average porosity and permeability values of 11.7% and 1.03 mD, respectively, compared with the clay-poor sandstones that have average porosities and permeabilities of 27.3% and 2.16 mD, respectively (Ellingsgaard 2009). Higher permeability values were obtained in a study on additional geotechnical and scientific cores which showed average porosity and permeability values for the Argir Beds of *c.* 24.8% and 13.3 mD, respectively (Ólavsdóttir 2009). The same study showed that all of the volcaniclastic beds analysed from the Enni Formation have average porosity and permeability values of *c.* 23.0% and 4.9 mD, respectively. Ellingsgaard (2009) measured slightly lower average porosity and permeability values of 18.1% and 1.1 mD, respectively for the interlava beds within the formation.

Detailed correlation and mapping of the key stratigraphic marker units of the upper part of the Malinstindur Formation, Sneis Formation and Enni Formation have shown thickness and facies variations across the fjords (Ellis *et al.* 2009). Ellis *et al.* (2009) related these variations and the pronounced NW–SE trend of the fjords (and islands) to similarly orientated lineaments inferred offshore in the Faroe–Shetland Basin to the SE

Fig. 2. Inclined sandstone beds with tangential lower contacts (some are emphasized with dashed lines) in the upper part of the Argir Beds, Norðastahorn horse stables, Streymoy. The inclined sandstone beds probably represent point bar deposits formed by the lateral migration of channels from right (east) towards the left (west) in the photo. The sedimentary structures within the upper, clay-poor sandstone succession are generally not very distinct in the field and are most easily observed in exposures orientated east–west, as in the present example, perpendicular to the presumed orientation of channels. The view is of the upper *c.* 2 m of a *c.* 6 m-thick section through the Argir Beds.

(Ritchie *et al.* 2011 and references therein). These lineaments, invariably referred to as transfer zones (Duindam & van Hoorn 1987; Rumph *et al.* 1993; Ellis *et al.* 2009), have been suggested to have influenced the distribution of Palaeogene sediments in the basin (e.g. Lamers & Carmichael 1999; Naylor *et al.* 1999) and may have played an important role transporting sediments from East Greenland, via the Faroe Islands, to the basin (Larsen *et al.* 1999; Whitham *et al.* 2004; Larsen & Whitham 2005; Jolley & Morton 2007; Ellis *et al.* 2009). The study of the Faroese cliff sections may help identify preferential trends, whether this be structural, drainage or otherwise.

3D photogrammetrical method

The 3D photogrammetrical method used in this study was developed at the Geological Survey of Denmark and Greenland and builds partly on earlier work by Dueholm *et al.* (1993) and Dueholm & Olsen (1993). The method allows the acquisition of geological data from vertical and oblique aerial photographs, with a 3D overview of the outcrops, by using a 3D stereo-plotter coupled with stereo-mirror technology.

Approximately 940 oblique photographs were collected around nine islands over 2 days, but only nine cliff sections were set up for 3D analysis, providing a cumulative distance of *c.* 20 km, with individual cliff sections ranging in length from 1.7 to 7 km. The photographs were taken out of the open door of a Bell 412 helicopter hired from Atlantic Airways in May 2010. The photographs were taken with GPS coordinates with an overlap of 60–80% using a hand-held Canon EOS-1Ds Mark III digital camera with a 36 × 24 mm CMOS sensor with 21 megapixels and a 35 mm Canon lens. The exposure time was set to a minimum of 1/250 s in order to reduce image blur owing to helicopter vibration and relative movement of the camera with respect to the cliff sections. During the photography the helicopter flew close to the cliff sections (<800 m) and at a constant altitude along straight lines approximately parallel to the cliffs.

The digitization of important geological features over several kilometres and the determination of

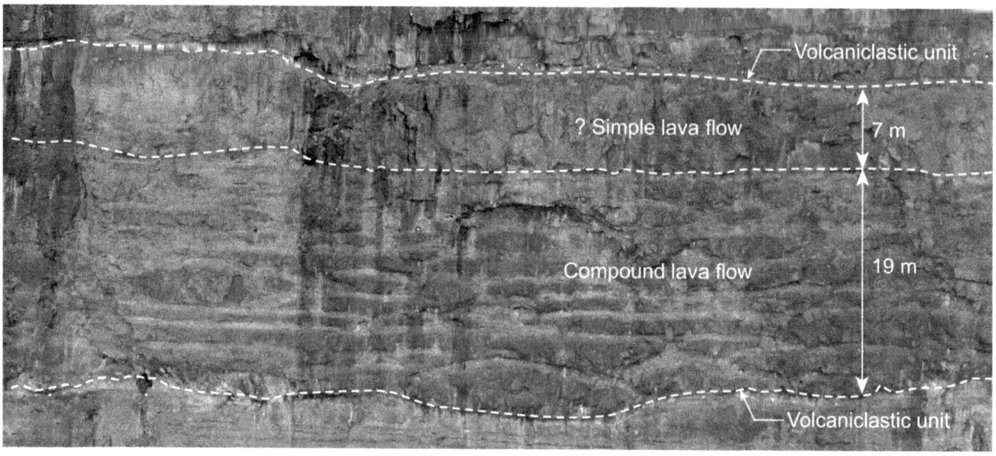

Fig. 3. Typical compound lava flow exposed in the cliff section on Fugloy (Fig. 1). The compound flow is made up of individual lava flow lobes with characteristic lens shapes and has light-coloured vesicular crusts that are prone to weathering. The palaeoflow direction for the compound lava flow is more or less perpendicular to the orientation of the cliff section. A possible simple lava flow is overlying the compound flow. At this distance, however, it cannot be ruled out that it is a relatively thick compound lava flow lobe, owing to the lack of an underlying weathered flow top or volcaniclastic unit.

strata thickness, strike direction and dip values were obtained from working on high-resolution visualization of the cliffs (Fig. 3). Although aerial photographs were not used, because of the prohibitive cost, the relative accuracy of mapping and measuring the geological features observed on the oblique photographs was down to the decimetre scale. All the mapped features are stored in a GIS database and 3D polylines can be exported as shape files suitable for 3D modelling using, for example, Petrel reservoir engineering software. Moreover, using 3D feature databases in ArcGIS, geological cross-sections can be generated automatically to obtain real representations of outcrops where the relative accuracy is as high as the resolution on the photographs, and then projected onto 2D profiles.

The study also had access to the original lithostratigraphic logs obtained from stream sections from many of the photographed islands, most of which have been presented elsewhere (Ellis *et al.* 2009; Passey 2009; Passey & Jolley 2009). These logs primarily helped with constraining the stratigraphy, but also partly validated measured thicknesses, etc.

The method allows rapid, airborne data collection and does not require any monitoring of equipment on the fly vessel. Furthermore, energy supply only includes batteries for camera and GPS equipment. The method is therefore very mobile and also suitable for collecting data from a boat or walking along cliff sections in the field. For a more comprehensive description of the method see Svennevig *et al.* (2015).

Terminology

Figure 3 shows the typical lava flow morphologies and intervening volcaniclastic units observed during the photogrammetrical study. The interlava units are regarded as being volcaniclastic based on the findings of other studies (Passey 2004, 2009; Passey & Jolley 2009). In those studies, volcaniclastic is used for a rock, irrespective of the mode of origin, that contains >60 vol% volcanic debris. The volcaniclastic units are invariably terrestrial, mainly consisting of (primary) fallout tuffs or as (secondary) reworked fluvial, lacustrine and mass flow deposits. The precise genetic interpretation is difficult to identify in the field and under the microscope as the strong alternation has typically obliterated the original rock texture.

The terminology of Self *et al.* (1997) has been used to describe the subaerial lava flows of the FIBG (Passey & Bell 2007; Passey 2009; Passey & Jolley 2009). The term ‘flow lobe’ is used to describe an individual package of lava surrounded by a chilled crust, and the term ‘lava flow’ is defined as the product of a single, more or less continuous outpouring of lava that can be composed of one or more flow lobes. The individual lava flows are usually separated from other flows by weathering surfaces and/or clastic beds. A lava flow consisting of a single flow lobe is referred to as a simple lava flow, whereas a compound lava flow is made up of two or more flow lobes of any geometry or size (Walker 1971). For both simple and compound lava flows each flow lobe can generally be separated into three parts: a lower vesicular crust, a massive core and an upper vesicular crust. The simple lava flows may especially show a relatively thick, reddened (oxidized), in places rubbly, upper crust which is the result of subaerial chemical weathering (Sabine 1971; Parra *et al.* 1987). Commonly this less resistant part of the lava flows has been removed by erosion implying that the overlying strata rests directly upon the solid core of the lava flow.

Description of the studied interlava beds

This study focuses on the photographed cliff sections of Sandoy, Skúgvoy, Stóra Dímun, Svínoy and Fugloy (Fig. 1), which contain distinctive interlava beds traceable over many kilometres. In contrast, interlava beds exposed in the photographed cliff sections of Koltur and partly Hestur are very thin, in general <1 m thick, implying that they are not very suitable for studying lateral variations in geometry and thickness of volcaniclastic sediments. These latter cliff sections are dominated by generally continuous compound lava flow sequences from the upper Malinstindur Formation to the base of the Høvdhamarin Flow. They do, however, reveal other relevant features included in the study, such as dykes intersecting the exposed volcanic succession and the presence of channels and lava tube structures intersecting interlava beds.

Overall the interlava beds are light yellowish brown or light reddish in colour and consist mainly of volcaniclastic sandstones and subordinate mudstones (Passey 2009; Passey & Jolley 2009). Observable sedimentary structures of the volcaniclastic sediments are limited to internal bedding marked by subtle colour changes between reddish and subordinate yellowish, dark grey and whitish beds. The bedding is generally sub-parallel to the surface the sediments rest upon except in a few places where internal beds onlap elevated parts of the underlying lava flows.

The reservoir architectures of the interlava beds are described by their component parts, namely their geometries, connectivities and compartmentalization. The numbering of interlava beds in the figures refers to the individual cliff sections as it has generally not been possible to correlate the beds between islands.

Geometries of the interlava beds

The lateral extent of the interlava beds generally exceeds the extent of the mapped cliff sections, which range in length from 1.7 to 7 km. The Argir Beds are known to be present across the central islands from Stóra Dímun to Eysturoy, a distance of more than 50 km, whereas they are either poorly developed or absent on the northeastern islands (Ellis *et al.* 2009; Passey & Jolley 2009). The Argir Beds were easily recognizable in the photographed cliff sections of Stóra Dímun, Skúgvoy and Sandoy. It is possible therefore, that some of the other less distinctive or poorly developed interlava beds may also be present across several islands, but this needs confirming with additional fieldwork. The thickness of the mapped interlava beds is generally <3 m with the exception of the Argir Beds, which locally attain thicknesses of *c.* 6 m.

Along the cliff sections the interlava beds rest on lava surfaces which are either planar to slightly undulating with a relief of around 1.5 m over distances of a few metres or show a more pronounced relief of around 10 m over a few decametres (Figs 4 & 5). The volcaniclastic sediments generally drape the relief with a fairly consistent thickness or locally double in thickness in depressions. The Argir Beds locally onlap and wedge out against a rising relief of the underlying lava flow, commonly formed by the local preservation of weathered and disintegrated lava, up to 7 m thick. These mounds of weathered rubbly lava are most likely remnants of the upper vesicular crust of the underlying lava flow, which elsewhere has been fully or partly removed by erosion (Fig. 6).

The relief of the upper boundary of the interlava beds is more or less parallel to the lower boundary of the beds, at least seen on a large scale. Where, however, the interlava beds contain intervening and laterally discontinuous lava flows, as described below, the upper boundary is typically sub-parallel to the top surface of the intervening lava flows. The large scale relief of the upper boundary of the interlava beds is levelled out by overlying lava flows, which thicken in depressions and thin over elevated ground (Fig. 4). On a finer scale the upper boundary may be irregular with a relief of *c.* 1.5 m over a distance of a few metres. In a few places the fine-scale relief includes marked concave-upwards structures into the overlying lava flow lobes (Fig. 7).

Intervening and laterally discontinuous lava flows locally separate the interlava beds into lower and upper parts. This applies for the Argir Beds in the cliff sections on Skúgvoy where intervening lava flows can be followed laterally over distances of 125–1250 m before they wedge out. The Argir Beds contain several wedging-out lava flows that typically occur at the same level above the base of the beds. The lava flows commonly fill up and level out depressions as well as forming topographic highs which are draped by the upper volcaniclastic sediments (Fig. 8).

The planar to subtle relief onto which the interlava beds rest may represent the original topography of the underlying lava flow after solidification, perhaps modified by minor erosion prior to the

Fig. 4. Two lava flow lobes, up to 12 m thick, thin abruptly towards each other and thereby form a depression between them, *c.* 6 m deep by *c.* 60 m wide. The volcaniclastic sediments of interlava Bed 1 drape the relief with a fairly consistent thickness, whereas the overlying simple lava flow overfills the depression and levels out the relief. In comparison, the relief of the lower boundary of interlava Beds 2/3/4 is considerably less and only slightly undulating along the cliff section. Cliff section on Svínoy (Fig. 1).

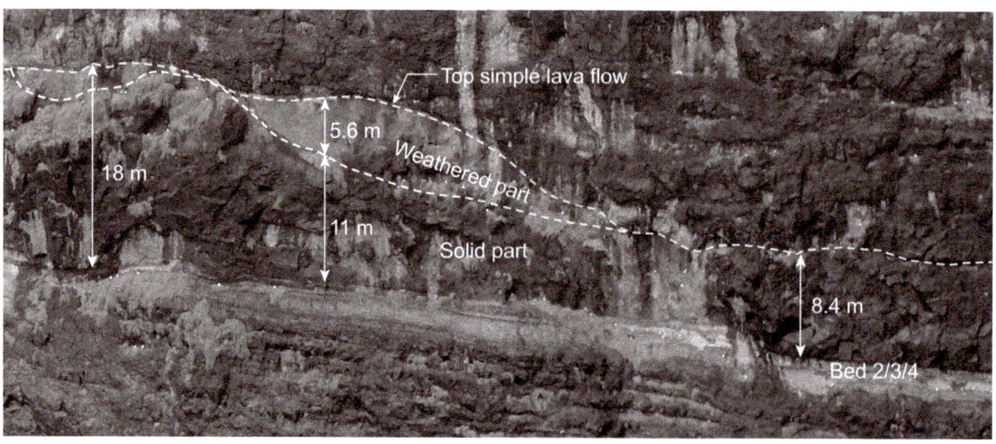

Fig. 5. The marked relief of the top surface of the simple lava flow overlying interlava Beds 2/3/4 reflects differential weathering and erosion of the upper vesicular crust. To the right (east), the deeply weathered part of the lava flow has been removed by erosion and only the solid core of the flow is preserved. To the left (west) the light grey weathered part of the lava flow is locally preserved. Further to the left, the weathering is less pronounced and the erosion therefore does not go as deep as seen to the right. The top surface of the simple lava flow is draped with a thin volcaniclastic bed and the relief is levelled out by the overlying compound lava flow. Cliff section on Svínoy (Fig. 1).

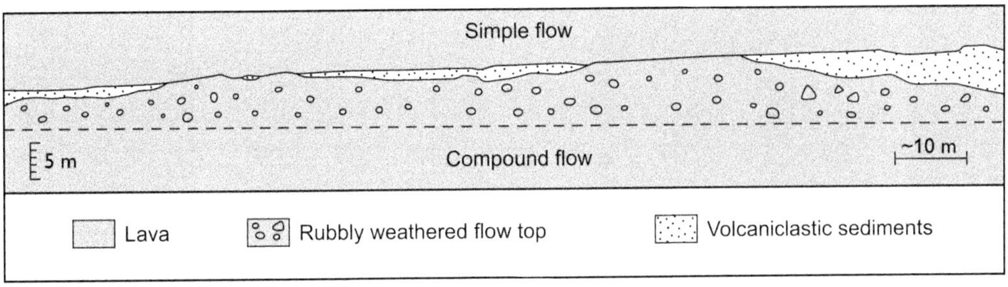

Fig. 6. The Argir Beds onlap an irregular relief of weathered and disintegrated lava formed by differential erosion of the upper crust of an underlying compound lava flow. The highs are directly overlain by a simple lava flow. The cross-section is constructed on the basis of 3D polylines of mapped boundaries in the cliff section on Skúgvoy (Fig. 1).

deposition of the volcaniclastic sediments. In places where the relief is pronounced, this reflects either a preserved marked relief of the underlying lava flows or marked differential erosion into these. The former occurs where thick lava lobes, forming the top of the lava flows, wedge out towards each other and leave depressions between them as observed in particular in the cliff section on Svínoy (Fig. 4). The terminating flow lobes are draped with volcaniclastic sediments showing angles of repose up to 40°. Differential weathering and erosion into the underlying lava flow may cause a relief of up to 10 m over a few tens of metres (Fig. 5). This supports the supposition that weathering and erosion of upper crusts may account for anomalously low normalized upper crust thicknesses for simple lava flows from Suðuroy and the presence of basalt lithoclasts in interlava beds that are presumably derived from lava flows (Passey 2004; Passey & Bell 2007).

The facts that the relief of the upper boundary of the interlava beds is more or less parallel to the lower boundary of the beds and that overlying lava flows fill depressions and level out the relief suggest that the overlying lava flows were non-erosive. This is supported by the fact that basal pockets of volcaniclastic sediment have not been observed within the lava flows. The finer-scale, irregular relief of the upper boundary may reflect loading into the volcaniclastic sediments by the overlying lava flows and small-scale variations in the topography of the

Fig. 7. The possible initiation of a lava flow apophysis (an invasive offshoot). The basal part of the lava flow has begun to invade the underlying volcaniclastic unit (Bed 1) and lift up the sediments in front of it. Cliff section on Fugloy (Fig. 1).

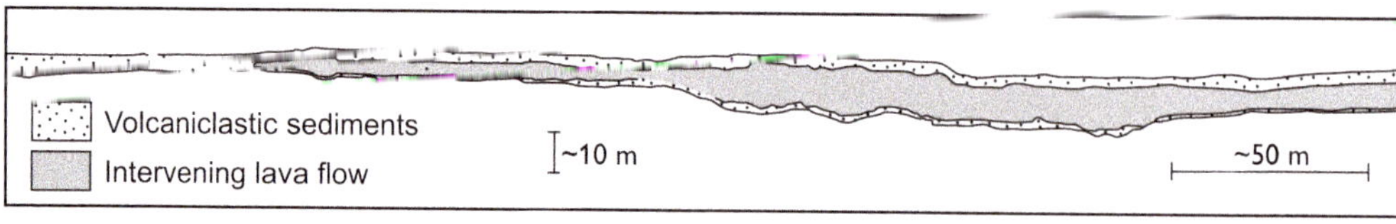

Fig. 8. A lava flow lobe within the Argir Beds which thins out towards the left (north) and to the right thickens in a depression, *c.* 7 m deep, in an underlying compound lava flow. The cross-section is constructed on the basis of 3D polylines of mapped boundaries in the cliff section on Skúgvoy (Fig. 1).

landscape the lava flows entered. The more peculiar concave-upwards structures into overlying lava most likely represent differential loading and the precursor for the development of apophyses (Fig. 7) or more commonly tree moulds (Fig. 9).

The fact that intervening lava flows occur at approximately the same level above the base of the interlava beds indicates that they represent the same volcanic event. The lava flows probably represent distal lava flow lobes connected to the same lava flow in a proximal direction as outlined for the Argir Beds on Skúgvoy in Figure 10. This interpretation is based on the mapped extent of the flow lobes and the positions of wedge-out points and the assumption that the flow lobes were formed by the same volcanic event. The radiating nature of the individual flow lobes suggests that the flow direction was towards the SW. Apart from Skúgvoy, the Argir Beds have also been mapped within the studied cliff sections on NW Sandoy and Stóra Dímun. In these places the Argir Beds contain no intervening lava flow lobes, showing that the interpreted lava flow did not reach these areas or was channelled between these islands.

Connectivity between interlava beds

Interlava beds are in several places seen to merge laterally owing to the wedging- out of intervening lava flows. This is especially well exposed along the 160 m-high, SE-facing cliff section of Svínoy,

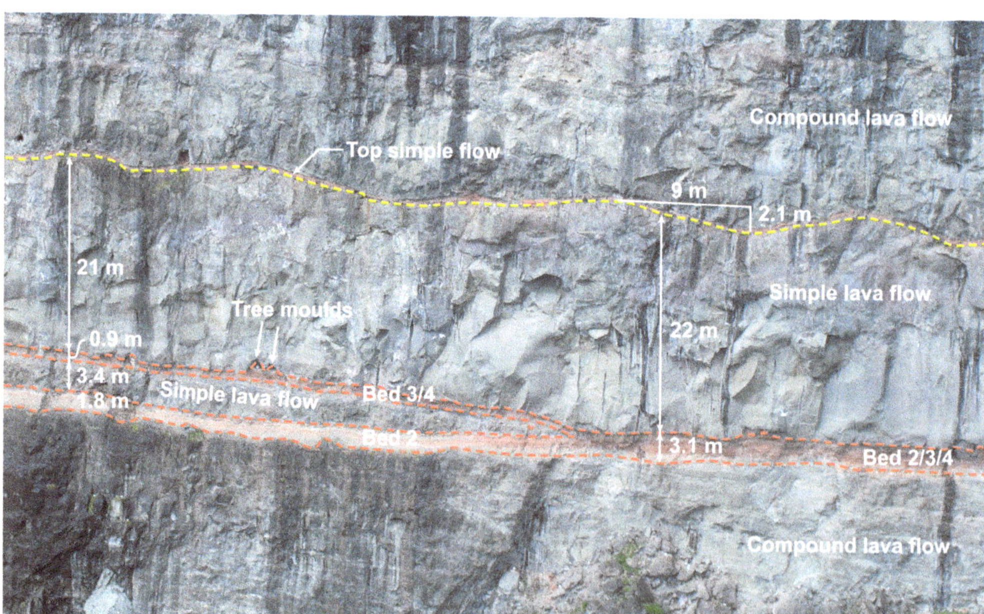

Fig. 9. The simple lava flow, separating interlava Bed 2 and Beds 3/4, is wedging out towards the right (NE) and terminates where there is minor rise in the underlying relief. The merged volcaniclastic sediments are slightly thicker than the individual interlava beds, Beds 2 and 3/4, they are composed of. Tree moulds at the base of the simple lava flow overlying Beds 3/4 imply a period of time past before the emplacement of the encasing lava flow. Note the undulating relief of the top surface of the thick simple lava flow that is draped by a thin volcaniclastic bed followed by a compound lava flow. Cliff section on Svínoy (Fig. 1).

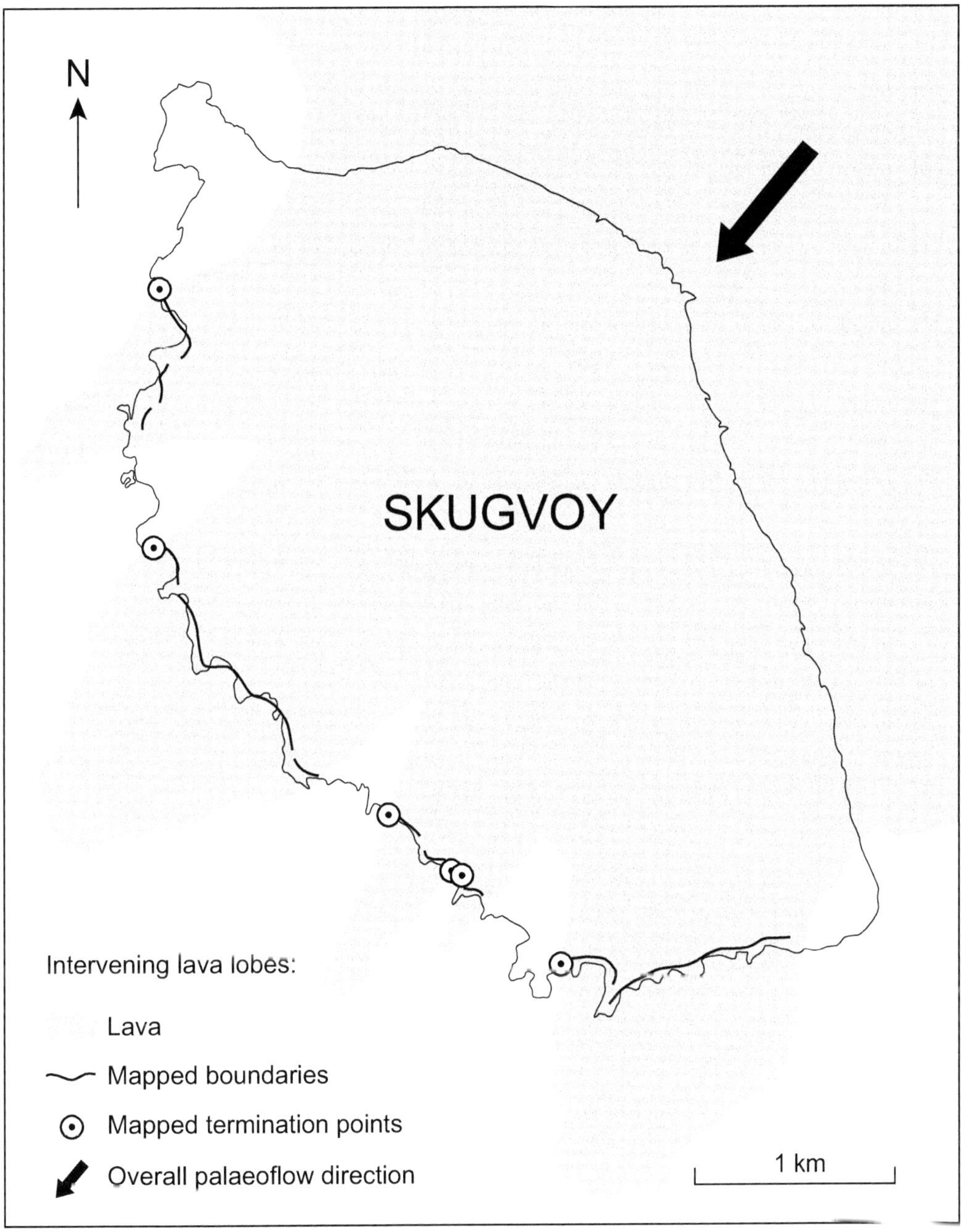

Fig. 10. Conceptual model suggesting that the lava flow lobes within the Argir Beds on Skúgvoy represent distal finger-like lobes emanating from the same lava flow in the NE (i.e. towards the proximal direction).

where simple lava flows, up 20 m thick, wedge out towards the NE, implying that four separate interlava beds merge into one interlava bed in this direction over a distance of *c.* 3 km (Fig. 11). The merged interlava bed is at least as thick as the sum of the individual interlava beds it is composed of (Fig. 9). There are twice as many interlava beds in the southwestern part of the cliff sections as opposed to the northeastern end. As a consequence, the *c.* 45 m thick subsection in the SW composed of

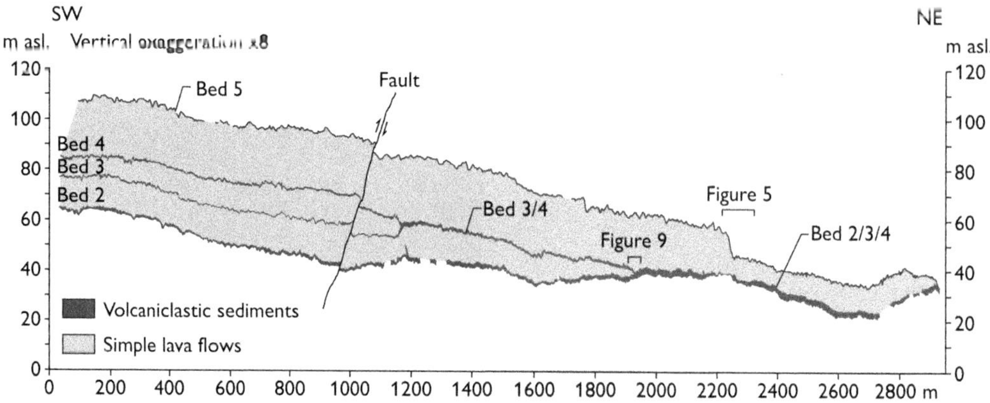

Fig. 11. Three simple lava flows, up to 20 m thick, wedge out from the SW towards the NE leading to the merging of the interlava beds in that direction. A close up of the merging point between interlava Bed 2 and Bed 3/4 can be seen in Figure 9. A reverse fault cuts the succession with a vertical displacement of *c.* 6 m. The cross-section is constructed on the basis of 3D polylines of mapped boundaries in the cliff section on Svínoy (Fig. 1).

four interlava beds and three lava flows is represented in the NE by a condensed section, consisting of a single interlava bed, <3 m thick. The thinning out of the simple lava flows towards the NE most likely reflects the original thinning out and termination of the flows in this direction, although an overprint of differential erosion of the uppermost vesicular and weathered part of the flows locally makes its mark on the relief (Fig. 5).

Interlava beds may be hydraulically connected via the conglomeratic fill of major channels, which are seen to cut through several interlava beds. Major channel fill structures, up 180 m wide and 45 m deep, occur within the volcanic successions in the studied cliff sections of Fugloy, Skúgvoy and Koltur. The channels on Skúgvoy and Koltur are essentially V-shaped with an irregular relief along their bottom and margins. In contrast the main and basal channel on Fugloy is 'box-shaped' with a relatively planar channel bottom, *c.* 60 m wide, with nearly vertical channel margins, and up to 27 m high (Fig. 12). The units that contain the channels show no mark relief away from the channels where they can be traced laterally over the entire length of the cliff sections (up to 1.7 km long). The channels show complex, multiple channel fill structures with fill consisting of volcaniclastic conglomerates and sandstone as well as lava flows. The lava flows tend to overfill the channels and extend beyond the channel margins, producing a relatively planar upper surface (Fig. 12). The sedimentary fill is best exposed within the channel on Skúgvoy, where it is dominated by dark-coloured conglomerates containing matrix-supported boulders, up to 0.8 m in size, and locally blocks of lava, up to 9 m long, the latter probably representing remnants of the collapsed channel margin. Channel axis orientations are estimated to be approximately NNW–SSE and WNW–ESE for the main channels on Fugloy and Skúgvoy, respectively.

The major channels observed on Fugloy, Skúgvoy and Koltur were probably formed during short, but intense erosional events, for example, by ephemeral streams triggered during heavy precipitation events. This is based on the very local occurrence of the channels and the fact that the units containing the channels show no marked erosional relief away from the channels as may be expected in places if formed over a significant period of time. Considering the coherent nature of solid lava compared with clastic sediments, the V-shape of some of the channels and the irregular relief of channel floors do not categorically reflect an immature nature and corresponding short period in which the channels were active. The conglomerates overlying the channel floors may have been deposited when erosion and bypass ceased, thereby giving way for the deposition of coarse-grained volcaniclastic sediments eroded and transported from further upstream. The fact that large boulders are matrix supported suggests that the conglomerates represent debris flows.

Lateral compartmentalization of interlava beds

Compartmentalization of individual interlava beds only occurs locally, but may have formed in several ways. Dykes, up to around 6 m wide, intersect the exposed volcanic successions including interlava beds in the cliff sections on Fugloy and Koltur (Fig. 13). There is a general paucity of dykes, however – only three dykes have been observed in the studied

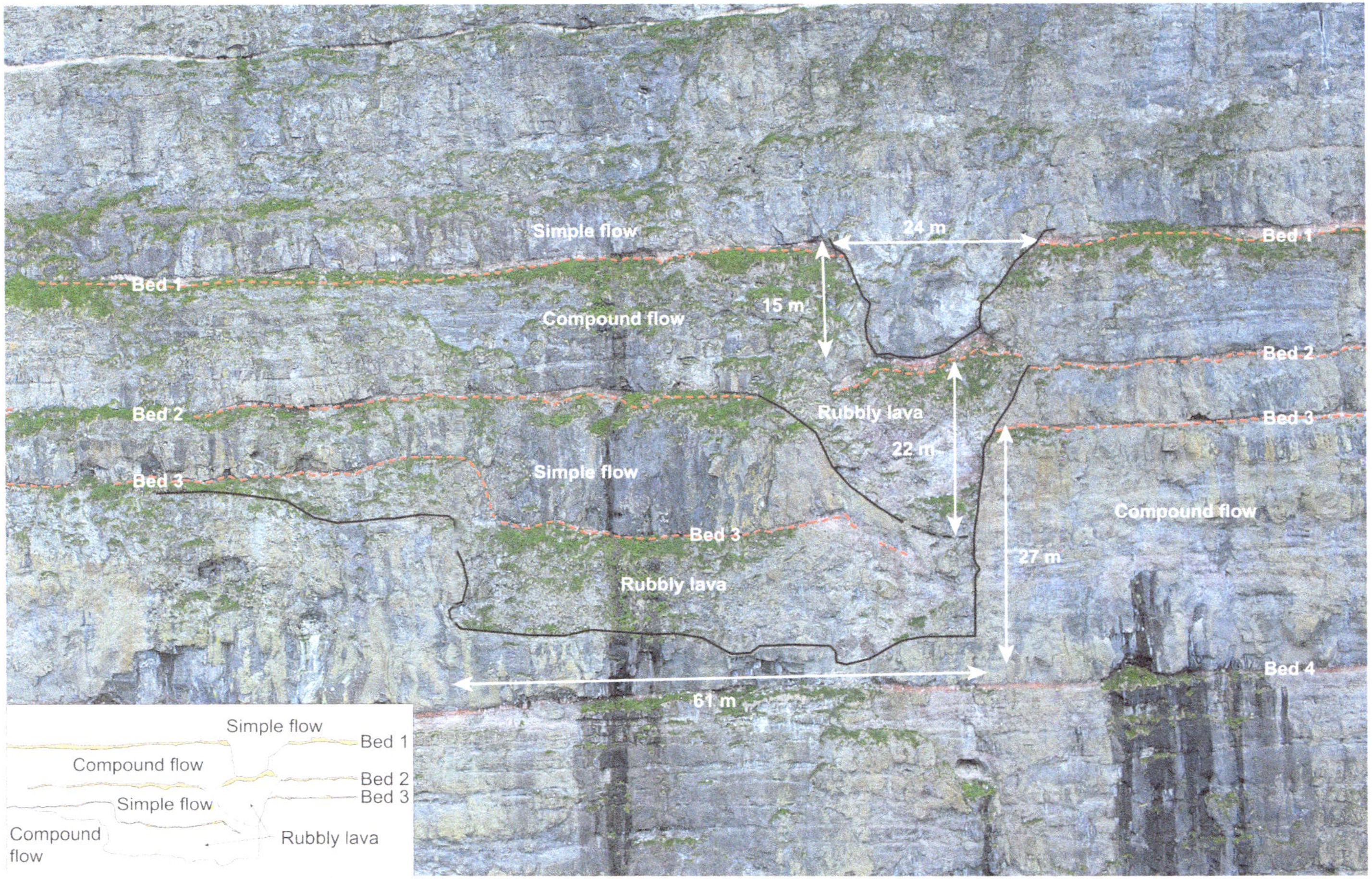

Fig. 12. Multiple channel fill structures (channel bottoms are marked with black lines) reflecting several erosional events. The main and basal channel is *c.* 61 m wide by *c.* 27 m deep. Channel fills consist of lava-dominated blocky deposits that were either formed by the brecciation of the lava flows that entered the channel or were more likely deposited by debris flows. The figure insert is a cross-section constructed on the basis of 3D polylines of mapped boundaries in the cliff section. Cliff section on Fugloy (Fig. 1).

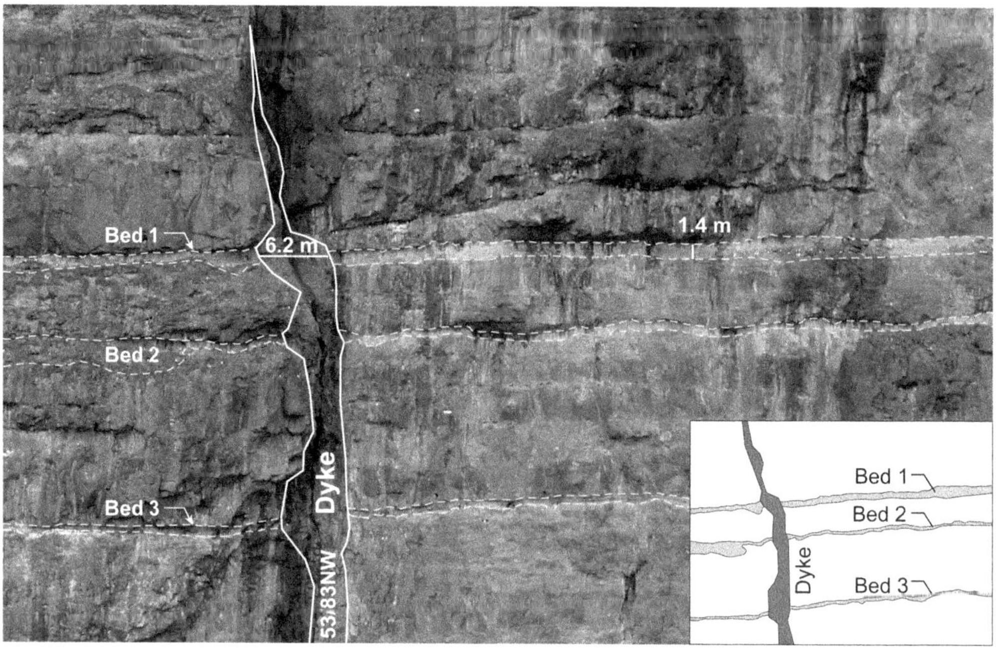

Fig. 13. A close-up of a rare dyke intersecting three interlava beds in the cliff section on Fugloy (Fig. 1). The dyke is up to *c.* 6 m wide and strikes NE–SW. Notice the lack of vertical displacement associated with the dyke, but the apparent widening where it intersects the volcaniclastic beds. The figure insert is a cross-section constructed on the basis of 3D polylines of mapped boundaries in the cliff section.

cliff sections. Two of these are from the cliff sections of Fugloy where they strike NE–SW and east–west, respectively.

Interlava beds are also truncated by channels and, where these are filled with lava flows, the lava may act as barriers to fluid flow. A channel, *c.* 15 m deep and 24 m wide, intersects an interlava bed in the cliff section on Fugloy, for example, and is filled with solid lava from an overlying and widespread simple lava flow (Fig. 12). In addition, a shallower channel intersects the Argir Beds on Sandoy and is filled with lava from an overlying compound lava flow. As mentioned above, however, if the channel fill is clastic in nature, then this may actually aid hydraulic connectivity between interlava beds at different levels.

A major lava tube, roughly circular in form, is exposed in the northern end of the cliff section at Fugloy. The lava tube has a diameter of nearly 50 m and the orientation of the long axis of the tube is estimated to be NNW–SSE. The tube intersects an interlava bed, whereas a vertical outflow of the lava tube, up to *c.* 6 m wide, intersects three overlying interlava beds (Fig. 14). The vertical outflow implies that the lava tube acted as an active conduit for flowing lava during several lava flow episodes or alternatively as a corridor for a succeeding intrusion. Either way, the filling of the lava tube may cause individual compartmentalization of four interlava beds and a lateral barrier to fluid flow.

A reverse fault intersects the volcanic succession exposed in the cliff section on Svínoy (Fig. 11). The fault strikes NNW–SSE and dips *c.* 20° towards the west. The vertical displacement is *c.* 6 m, enough to disconnect the individual interlava beds and juxtapose them against solid lava.

As previously described, volcaniclastic sediments locally wedge out against elevated topographies of weathered and disintegrated lava representing remnants of the upper vesicular crust of underlying lava flows (Fig. 6). It is doubtful that the weathering remnants of the upper rubbly lava crust form lateral barriers within the interlava beds as they will be more porous than the underlying solid lava cores. The weathered rubbly lava may, therefore, be considered as having high hydraulic conductivity unless reduced by subsequent diagenetic alternation.

Discussion

The 3D photogrammetrical study shows that interlava beds can be expected to have large lateral

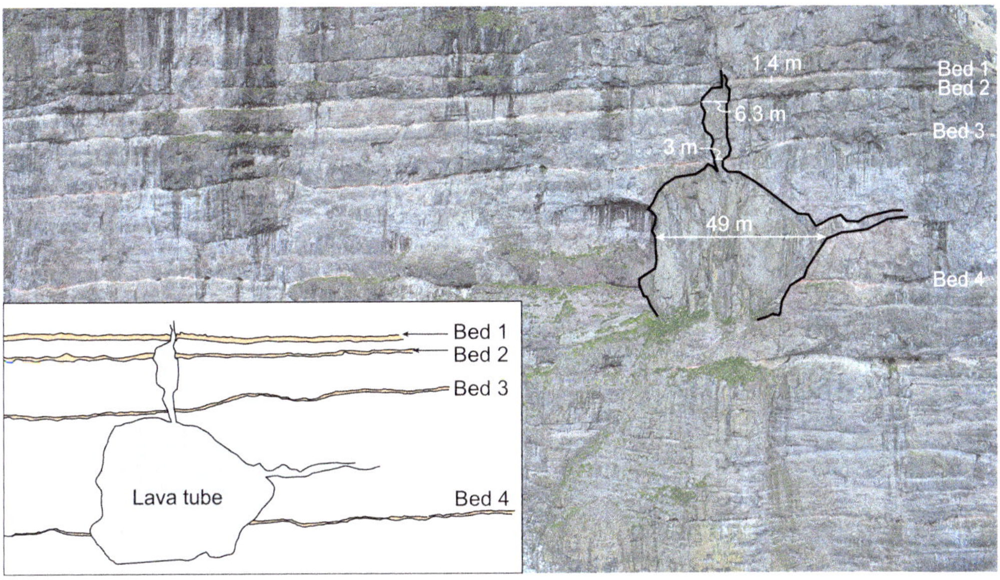

Fig. 14. Lava tube, up to 49 m wide, intersects interlava Bed 4 whereas a vertical apophysis from the tube, up to *c.* 6 m wide, intersects the three overlying interlava beds. The figure insert is a cross-section constructed on the basis of 3D polylines of mapped boundaries in the cliff section. Cliff section on Fugloy.

extents and merge with other interlava beds owing to the wedging out of intervening lava flows. Individual interlava beds are difficult to correlate from island to island except for the Argir Beds, which form a stratigraphic marker unit within the Enni Formation. In the present study, the Argir Beds have been mapped in cliff sections on Sandoy, Skúgvoy and Stóra Dímun, a lateral distance of *c.* 25 km. From other studies, however, the Argir Beds are known to have a lateral extent of >50 km (Passey & Jolley 2009; Passey & Varming 2010). It seems likely, therefore, that many of the other, less recognizable, interlava beds also extend beyond the studied cliff sections and are present on several islands. The difficulty in correlation arises, in part, from the merging/separation of interlava beds along-section and, therefore, they may be found at different stratigraphic levels between discontinuous cliff sections (i.e. between islands). This study has shown that, over a few kilometres, *c.* 45 m-thick sections can be condensed to <3 m. This highlights the difficulty of trying to correlate isolated drilled sections over several kilometres, if not tens of kilometres. This may be compounded at the distal edge of the lava field where the repose between lava flow emplacement is expected to be generally longer.

Owing to the presumed large areal extents of the interlava beds and their merging over significant distances, they could form widespread reservoirs under the right conditions in the sub-surface. Furthermore, major channels intersecting several interlava beds may contribute to the hydraulic connectivity between the interlava beds via their clastic, typically conglomeratic, fill. The conglomeratic fills may conversely drain the interlava beds of hydrocarbons and the channels may act as the main reservoir(s). The multiple channel fill structures observed in the cliff sections on Fugloy and Skúgvoy reflect several erosional events and that these areas, for some unknown reason, formed preferred corridors for the removal of deposits by channel erosion during significant periods of time. The volcanic material will have been transported down-system, where it may have mixed with and degraded the reservoir potential of non-volcanic, siliciclastic sandstones. The presence of the channels may have also allowed non-volcanic material to bypass the lava field and to be transported to the Faroe–Shetland Basin. This is supported by the presence of a diverse assemblage of Late Cretaceous dinoflagellate cysts in the Sneis Formation on Eysturoy (Passey & Jolley 2009).

The lateral extent and volume of potential intravolcanic reservoirs may be limited by the intersection of interlava beds by dykes, lava-filled channels and lava tubes. In addition, lateral compartmentalization of individual interlava beds may be caused by their subvertical displacement by reverse faulting. Although intersecting subvertical structures (faults, channels) may allow fluids and pressure to equilibrate over geological timescales, they act as seals over production timescales because they prevent cross-flow at normal production rates (Jolley *et al.* 2010). It is important, however, to keep in mind that the photogrammetry has essentially only revealed the sealing nature of the structures exposed obliquely or perpendicular to the cliff sections. The connectivity of individual interlava beds may, therefore, be re-established somewhere along-strike or on the longitudinal axis of the sealing structures or where the structures cease to exist. It is also worth noting the overall paucity of dykes, lava-filled channels and lava tubes observed across the many kilometres of studied cliff sections.

The large-scale thinning out of simple lava flows towards the NE observed in the cliff section on Svínoy fits well with the conceptual model of Ellis *et al.* (2009), who suggested that the inferred NW–SE orientated transfer zones, intersecting the Faroe Islands, acted as conduits of basaltic magma that erupted as lava and flowed out as lava perpendicular to the transfer zones. The Westray Transfer Zone is suggested to lie *c.* 15 km to the SW of the Svínoy cliff section. The NE–SW and west–east orientations of the two dykes in the cliff section of Fugloy is in accordance with the general orientations of dykes and faults measured on the Faroe Islands (Rasmussen & Noe-Nygaard 1969, 1970), which Ellis *et al.* (2009) suggest is caused by lateral oblique-slip (transpressional) movements along the transfer zones. However, structural components aligned more parallel with the orientation of inferred transfer zones have also been measured. This goes for the lava tube in the cliff section on Fugloy and the reverse fault in the cliff section on Svínoy, which are both orientated NNW–SSE. In addition, the long axis of the major channel in the cliff section on Skúgvoy is orientated WNW–ESE, a common fracture trend seen across the central islands and inferred for the Skopunarfjørður fault to the north of Sandoy (Passey 2009).

The general sub-parallel bedding to the lower boundary of the interlava beds and the fairly laterally consistent thickness of the beds points to deposition from suspension fallout. The majority of the units are most likely fallout tuffs, but as Suthren (1985) points out, it can be difficult to distinguish fallout tuffs from parallel-bedded reworked volcaniclastic sandstones without additional microscopic examination. Even then, original textures may be destroyed by alteration processes, as is the case for the Faroese units. The lack of pillow structures and brecciation along the bases of the overlying lava flows suggests that the volcaniclastic beds were not water saturated at the time of lava flow emplacement. The steep angles of repose (up to 40°), seen where interlava beds drape positive relief, indicate a general absence of remobilization after deposition. This might be due to the inherent cohesiveness (e.g. angularity) of the volcanic grains,

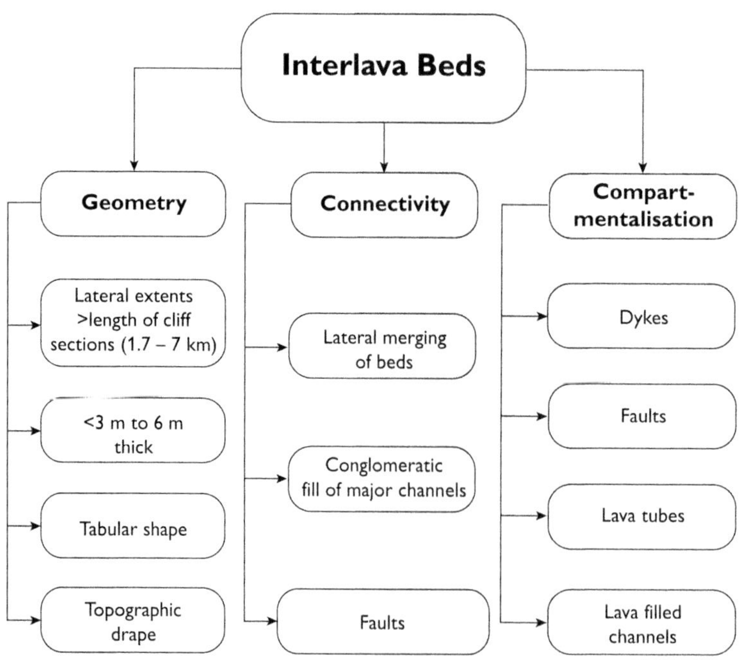

Fig. 15. Simplified overview of key findings of the study when it comes to the main subjects: geometry, connectivity and compartmentalization of interlava beds.

which would have increased their adherence to essentially roughened surfaces. Furthermore, dampness of the volcanic debris by (light) rainfall might have also increased the cohesiveness of the volcanic grains and hampered re-deposition by wind or water (cf. Hunter *et al.* 1983). Only under intense rainfall or water displacement (e.g. damming of rivers by lava flows) would saturation of the ash lead to its remobilization. Water can easily infiltrate what are initially very porous deposits, and therefore, are essentially dry/damp except during and shortly after intense rainfall (e.g. Rodolfo 1989). The water may also percolate down into porous weathered upper zones of underlying lava flows limiting the re-deposition of the volcaniclastic sediments by surface runoff.

The remobilization of the volcaniclastic sediment is, however, documented for the Argir Beds that exhibit a range of sedimentary structures and textures, which implies they have been deposited in a fluvial–floodplain environment (Passey 2009). The fluvial nature of the Argir Beds is reflected in the cliff sections by their ability to level out relief through lateral thickness variations. This includes onlap onto positive relief formed by the underlying lava flow(s). These features are clearly seen in the cliff sections on Skúgvoy and Stóra Dímun. The principle governing mechanisms for the geometries, lateral extents and compartmentalization of the partly fluvial Argir Beds and some of the other interlava beds are most likely the same for pure fluvial siliciclastic intra-volcanic sediments found offshore.

Conclusion

Three-dimensional photogrammetry of the volcanic cliff sections of the Faroe Islands has allowed for the correlation of lava flows and interlava units over many kilometres. The 3D aspect has made it possible to follow units around irregular coastlines, which was extremely difficult to do with confidence with normal 2D photographs. The orientation of many features (channels, lava tubes) and structures (faults) has also been calculated with a high degree of accuracy and the measurement of units is down to the decimetre scale. The study has highlighted the complexity of the volcanic sections that on first appearances are monotonous sequences. These complexities, including the merging of interlava beds, have been digitized and can be exported for 3D modelling using, for example, Petrel reservoir engineering software. Only nine of 27 photographed cliff sections were set up for detailed analysis and this study has focussed on one aspect that can be gleaned from the cliff sections; therefore, there is still a wealth of information that could be derived from this dataset. This includes the modelling of dip surfaces between islands, the construction of detailed cross-sections, the lateral study of lava flow architectures, fracture analysis and the production of videos and teaching resources for virtual field trips. The study has nevertheless documented features that are important in the exploration and/or appraisal of hydrocarbon plays in challenging intra-volcanic settings. The information may form important constraints for the estimation of reservoir size, geometry, connectivity and compartmentalization of interlava reservoirs. The main findings are summarized in Figure 15 and are as follows:

- The extent of the interlava beds generally exceeds the length of the studied cliff sections (1.7–7 km). The exception are local major channel fill complexes, up to 45 m deep and 180 m wide, which are filled in with volcaniclastic conglomerates.
- The mapped interlava beds are <1–6 m thick. Laterally they have a fairly consistent thickness and tabular shape owing to a non-erosive nature of overlying lava flows and a general draping nature of the volcaniclastic sediments.
- The relief of the lower boundary either reflects the original topography left behind by the solidified lava flows or was governed by differential weathering and erosion of the vesicular upper crusts prior to the deposition of the volcaniclastic sediments.
- Lava flows, up to 20 m thick, wedge out over distances of 1–3 km, leading to the merging of the interlava beds they separate. This results in condensed sections and would make correlations difficult between isolated locations (e.g. wells).
- Hydraulic connectivity may be present via the conglomeratic fill of major channels, which in a few places cut through several interlava beds.
- Lava-filled channels, up to 24 m wide and 15 m deep, may conversely form barriers to fluid flow between intersected interlava beds, as may lava tubes with diameters of up to 50 m.
- Dykes, up to 6 m wide, locally intersect volcanic successions including several interlava beds and may form barriers to lateral fluid flow. Dykes are very rare in the studied cliff sections.
- At one locality, interlava beds are laterally disconnected by a reverse fault with a vertical displacement of *c.* 6 m. This may have acted as a fluid pathway or acted as a seal.

The companies of the SINDRI Group are gratefully acknowledged for their generous support of the project. Asger Ken Pedersen is thanked for conducting the training in taking photos suitable for 3D studies and together with Lotte Melchior Larsen for valuable discussions concerning the geological features revealed in the photos. Erik V. Sørensen is thanked for ad hoc support in the photogrammetry laboratory. Thanks to Jette Halskov who produced the figures and Thomas Varming for participating in the fieldwork and for taking illustrative 'in action' photos of the process of photographing cliff sections from helicopter. Richard J. Brown and an anonymous reviewer are thanked for constructive reviews. The paper is published with the permission of the Geological Survey of Denmark and Greenland.

References

AUSTIN, J. A., CANNON, S. J. C. & ELLIS, D. 2014. Hydrocarbon exploration and exploitation West of Shetlands. *In*: CANNON, S. J. C. & ELLIS, D. (eds) *Hydrocarbon Exploration to Exploitation West of Shetlands*. Geological Society, London, Special Publications, **397**, 1–10.

DUEHOLM, K. S. & OLSEN, T. 1993. Reservoir analog studies using multimodel photogrammetry; a new tool for the petroleum industry. *American Association of Petroleum Geologists Bulletin*, **77**, 2023–2031.

DUEHOLM, K. S., GARDE, A. A. & PEDERSEN, A. K. 1993. Preparation of accurate geological and structural maps, cross-sections or block diagrams from colour slides, using multi-model photogrammetry. *Journal of Structural Geology*, **15**, 933–937.

DUINDAM, P. & VAN HOORN, B. 1987. Structural evolution of the West Shetland continental margin. *In*: BROOKS, J. & GLENNIE, K. W. (eds) *Petroleum Geology of North West Europe*. Graham & Trotman, London, 765–773.

DUNCAN, L., HELLAND-HANSEN, D. & DENNEHY, C. 2009. The Rosebank discovery: a new play type in intra basalt reservoirs of the North Atlantic volcanic province. *In*: *DEVEX 2009. 6th European Production and Development Conference – Crossing the Boundaries in Pursuit of Ultimate Recovery*, 12–13 May, Aberdeen Exhibition and Conference Centre.

ELLINGSGAARD, Ó. 2009. *Formation Evaluation of Interlava Volcaniclastic Rocks from the Faroe Islands and the Faroe–Shetland Basin*. MSc thesis, Lund University.

ELLIS, D., PASSEY, S. R., JOLLEY, D. W. & BELL, B. R. 2009. Transfer zones: the application of new geological information from the Faroe Islands applied to the offshore exploration of intra basalt and sub-basalt exploration. *In*: VARMING, T. & ZISKA, H. (eds) *Faroe Islands Exploration Conference: Proceedings of the 2nd Conference*. Annales Societatis Scientiarum Færoensis, Tórshavn, **50**, 205–226.

FIELDING, K. D., BURNETT, D., CRABTREE, N. J., LADEGAARD, H. & LAWTON, L. C. 2014. Exploration and appraisal of a 120 km^2 four-way dip closure: what could possibly go wrong? *In*: CANNON, S. J. C. & ELLIS, D. (eds) *Hydrocarbon Exploration to Exploitation West of Shetlands*. Geological Society, London, Special Publications, **397**, 145–162.

HELLAND-HANSEN, D. 2009. Rosebank – challenges to development from a subsurface perspective. *In*: VARMING, T. & ZISKA, H. (eds) *Faroe Islands Exploration Conference: Proceedings of the 2nd Conference*. Annales Societatis Scientiarum Færoensis, Tórshavn, **50**, 241–245.

HOLE, M., JOLLEY, D., HARTLEY, A., LELEU, S., JOHN, N. & BALL, M. 2013. Lava sediment interactions in an Old Red Sandstone basin, NE Scotland. *Journal of the Geological Society, London*, **170**, 641–655.

HUNTER, R. E., RICHMOND, B. M. & ALPHA, T. R. 1983. Storm-controlled oblique dunes of the Oregon coast *Geological Society of America Bulletin*, **94**, 1450–1465.

JOLLEY, D. W. & MORTON, A. C. 2007. Understanding basin sedimentary provenance: evidence from allied phytogeographic and heavy mineral analysis of the Palaeocene of the NE Atlantic. *Journal of the Geological Society, London*, **164**, 553–563.

JOLLEY, S. J., FISHER, Q. J. & AINSWORTH, R. B. 2010. Reservoir compartmentalization: an introduction. *In*: JOLLEY, S. J., FISHER, Q. J., AINSWORTH, R. B., VROLIJK, P. J. & DELISLE, S. (eds) *Reservoir Compartmentalization*. Geological Society, London, Special Publications, **347**, 1–8.

JØRGENSEN, G. & RASMUSSEN, J. 1986. *Glacial Striae, Roches Moutonnées and Ice Movements in the Faeroe Islands*. Geological Survey of Denmark, København, DGU Series, C.

JØRGENSEN, O. 2006. The regional distribution of zeolites in the basalts of the Faroe Islands and the significance of zeolites as palaeotemperature indicators. *In*: CHALMERS, J. A. & WAAGSTEIN, R. (eds) *Scientific Results from the Deepened Lopra-1 Borehole, Faroe Islands*. Geological Survey of Denmark and Greenland Bulletin, Copenhagen, **9**, 123–156.

LAMERS, E. & CARMICHAEL, S. M. M. 1999. The Paleocene deepwater sandstone play West of Shetland. *In*: FLEET, A. J. & BOLDY, S. A. R. (eds) *Petroleum Geology of Northwest Europe: Proceedings of the 5th Conference*. Geological Society, London, 645–659.

LARSEN, M. & WHITHAM, A. G. 2005. Evidence for a major sediment input point into the Faroe–Shetland Basin from the Kangerlussuaq region of southern East Greenland. *In*: DORÉ, A. G. & VINING, B. A. (eds) *Petroleum Geology: North-West Europe and Global Perspectives – Proceedings of the 6th Petroleum Geology Conference*. The Geological Society, London, 913–922.

LARSEN, M., HAMBERG, L., OLAUSSEN, S., NORGAARD-PEDERSEN, N. & STEMMERIK, L. 1999. Basin evolution in southern East Greenland; an outcrop analog for Cretaceous–Paleogene basins on the North Atlantic volcanic margins. *American Association of Petroleum Geologists Bulletin*, **83**, 1236–1261.

LUND, J. 1989. A late Paleocene non-marine microflora from the interbasaltic coals of the Faeroe Islands, North Atlantic. *Bulletin of the Geological Society of Denmark*, **37**, 181–203.

NAYLOR, P. H., BELL, B. R., JOLLEY, D. W., DURNALL, P. & FREDSTED, R. 1999. Palaeogene magmatism in the Faeroe–Shetland Basin: influences on uplift history and sedimentation. *In*: FLEET, A. J. & BOLDY, S. A. R. (eds) *Petroleum Geology of Northwest Europe: Proceedings of the 5th Conference*. Geological Society, London, 545–558.

NOE-NYGAARD, A. & RASMUSSEN, J. 1968. Petrology of a 3000 metre sequence of basaltic lavas in the Faeroe Islands. *Lithos*, **1**, 286–304.

ÓLAVSDÓTTIR, J. 2009. Reservoir quality of the intra-basalt volcaniclastic units onshore Faroe Islands as an analogy to the offshore intra- and post-basalt epiclastic sedimentary ricks. *In*: *3rd Faroe Islands Exploration Conference*, 15–16 September, Tórshavn, Faroe Islands.

PARRA, M., DELMONT, P., DUMON, J. C., FERRAGNE, A. & PONS, J. C. 1987. Mineralogy and origin of Tertiary interbasaltic clays from the Faeroe Islands, northeastern Atlantic. *Clay Minerals*, **22**, 63–82.

PASSEY, S. R. 2004. *The Volcanic and Sedimentary Evolution of the Faeroe Plateau Lava Group, Faeroe Islands and the Faeroe–Shetland Basin, NE Atlantic*. PhD thesis, University of Glasgow.

PASSEY, S. R. 2009. Recognition of a faulted basalt lava flow sequence through the correlation of stratigraphic marker units, Skopunarfjørður, Faroe Islands. *In*: VARMING, T. & ZISKA, H. (eds) *Faroe Islands Exploration Conference: Proceedings of the 2nd Conference*. Annales Societatis Scientiarum Færoensis, Tórshavn, **50**, 174–204.

PASSEY, S. R. 2014. The habit and origin of siderite spherules in the Eocene coal-bearing Prestfjall Formation, Faroe Islands. *International Journal of Coal Geology*, **122**, 76–90.

PASSEY, S. R. & BELL, B. R. 2007. Morphologies and emplacement mechanisms of the lava flows of the Faroe Islands Basalt Group, Faroe Islands, NE Atlantic Ocean. *Bulletin of Volcanology*, **70**, 139–156.

PASSEY, S. & HITCHEN, K. 2011. Cenozoic (igneous). *In*: RITCHIE, J. D., ZISKA, H., JOHNSON, H. & EVANS, D. (eds) *Geology of the Faroe–Shetland Basin and Adjacent Areas*, British Geological Survey & Jarðfeingi Research Report, **RR/11/01** 209–228.

PASSEY, S. R. & JOLLEY, D. W. 2009. A revised lithostratigraphic nomenclature for the Palaeogene Faroe Islands Basalt Group, NE Atlantic Ocean. *Earth and Environmental Science Transactions of the Royal Society of Edinburgh*, **99**, 127–158.

PASSEY, S. R. & VARMING, T. 2010. Surface interpolation within a continental flood basalt province: an example from the Palaeogene Faroe Islands Basalt Group. *Journal of Structural Geology*, **32**, 709–723.

RASMUSSEN, J. & NOE-NYGAARD, A. 1969. *Beskrivelse til Geologisk Kort over Færøerne i Målestok 1:50 000*. Danmarks Geologiske Undersøgelse, København, **1/24**, 370.

RASMUSSEN, J. & NOE-NYGAARD, A. 1970. *Geology of the Faeroe Islands (Pre-Quaternary)*. Geological Survey of Denmark, Copenhagen, **1/25**, 142.

RITCHIE, J. D., GATLIFF, R. W. & RICHARDS, P. C. 1999. Early Tertiary magmatism in the offshore NW UK margin and surrounds. *In*: FLEET, A. J. & BOLDY, S. A. R. (eds) *Petroleum Geology of Northwest Europe: Proceedings of the 5th Conference*. Geological Society, London, 573–584.

RITCHIE, J. D., ZISKA, H., JOHNSON, H. & EVANS, D. 2011. *Geology of the Faroe–Shetland Basin and Adjacent Areas*. British Geological Survey & Jarðfeingi Research Report, **RR/11/01**.

RODOLFO, K. S. 1989. Origin and early evolution of lahar channel at Mabinit, Mayon Volcano, Philippines. *Geological Society of America Bulletin*, **101**, 414–426.

RUMPH, B., REAVES, C. M., ORANGE, V. G. & ROBINSON, D. L. 1993. Structuring and transfer zones in the Faeroe Basin. *In*: PARKER, J. R. (ed.) *Petroleum Geology of Northwest Europe: Proceedings of the 4th Conference*. Geological Society, London, 999–1009.

SABINE, P. A. 1971. Bentonitic beidellite-mudstone from the Faeroe Islands. *Clay Minerals*, **9**, 97–106.

SAUNDERS, A. D., FITTON, J. G., KERR, A. C., NORRY, M. J. & KENT, R. W. 1997. The North Atlantic Igneous Province. *In*: MAHONEY, J. J. & COFFIN, M. L. (eds) *Large Igneous Provinces: Continental, Oceanic, and Planetary Flood Volcanism*. American Geophysical Union, Washington, D.C., Geophysical Monographs, **100**, 45–93.

SCHOFIELD, N. & JOLLEY, D. W. 2013. Development of intra-basaltic lava-field drainage systems within the Faroe–Shetland Basin. *Petroleum Geoscience*, **19**, 273–288.

SELF, S., THORDARSON, T. & KESZTHELYI, L. 1997. Emplacement of continental flood basalt lava flows. *In*: MAHONEY, J. J. & COFFIN, M. L. (eds) *Large Igneous Provinces: Continental, Oceanic, and Planetary Flood Volcanism*. American Geophysical Union, Washington, D.C., Geophysical Monographs, **100**, 381–410.

STOREY, M., DUNCAN, R. A. & TEGNER, C. 2007. Timing and duration of volcanism in the North Atlantic Igneous Province: implications for geodynamics and links to the Iceland hotspot. *Chemical Geology*, **241**, 264–281.

SUTHREN, R. J. 1985. Facies analysis of volcaniclastic sediments: a review. *In*: BRENCHLEY, P. & WILLIAMS, B. P. J. (eds) *Sedimentology: Recent Advances and Applied Aspects*. Geological Society, London, Special Publications, **18**, 123–146.

SVENNEVIG, K., GUARNIERI, P. & STEMMERIK, L. 2015. From oblique photogrammetry to a 3D model – Structural modeling of Kilen, eastern North Greenland. *Computers & Geosciences*, **83**, 120–126.

WAAGSTEIN, R. 1988. Structure, composition and age of the Faeroe basalt plateau. *In*: MORTON, A. C. & PARSON, L. M. (eds) *Early Tertiary Volcanism and the Opening of the NE Atlantic*. Geological Society, London, Special Publications, **39**, 225–238.

WALKER, G. P. L. 1971. Compound and simple lava flows and flood basalts. *Bulletin Volcanologique*, **35**, 579–590.

WHITHAM, A. G., MORTON, A. C. & FANNING, C. M. 2004. Insights into Cretaceous–Palaeogene sediment transport paths and basin evolution in the North Atlantic from a heavy mineral study of sandstones from southern East Greenland. *Petroleum Geoscience*, **10**, 61–72.

WILLIAMSON, I. T. & BELL, B. R. 2012. The Staffa Lava Formation: graben-related volcanism, associated sedimentation and landscape character during the early development of the Palaeogene Mull Lava Field, NW Scotland. *Scottish Journal of Geology*, **48**, 1–46.

Application of outcrop analogues in successful exploration of a sand injection complex, Volund Field, Norwegian North Sea

ANDREW HURST[1]*, MADS HUUSE[1,2], DAVIDE DURANTI[1,3], MARIO VIGORITO[1,4], ERIC JAMESON[5] & ANNE SCHWAB[5]

[1]*Department of Geology and Petroleum Geology, University Aberdeen, King's College, Aberdeen AB24 3UE, UK*

[2]*University of Manchester, Manchester, UK*

[3]*E.On, Stavanger, Norway*

[4]*Statoil ASA, Stavanger, Norway*

[5]*Marathon Oil Company*

**Corresponding author (e-mail: ahurst@abdn.ac.uk)*

Abstract: Identification and exploration drilling of the Volund Field as part of a sand injection complex is the first example of deliberate targeting of sandstone intrusions in oil exploration. Outcrop data were an important element in the process of constraining the uncertainty associated with reservoir presence and connectivity. A strong lobby against the relevance of sand injectites as exploration targets, and significant uncertainty associated with sub-surface sand injectite analogues associated with existing oil fields, combined to discourage and down-grade Volund's prospectivity. Few outcrop studies provided data of relevance to exploration of sandstone intrusions and original outcrop data were utilized in evaluation of Volund. Sills and saucer-shaped sandstone intrusions are the most common reservoir units observed at outcrop and similar features were identified in the 3D seismic across the Volund Field prospect. A large-scale sand injection origin rather than a depositional origin was proposed. Outcrops of sandstone intrusions demonstrated excellent reservoir quality in composite sandstone bodies that cross-cut depositional bedding. High-quality reservoirs with excellent vertical and lateral connectivity are observed and used to support the prediction of similar quality reservoirs in the Volund prospect.

This paper describes how outcrop data played a significant role in constraining exploration risk, in the pre-drill evaluation of part of a regionally developed sand injection complex. At the time, between 2000 and 2003, 'sand injectite plays' were undefined and to the best of our knowledge deliberate exploration of sandstone intrusions had not been undertaken. Prior to and concomitant with our evaluation, several significant Norwegian and UK oil fields, including Harding, Balder, Alba, Gryphon and Grane, all originally explored, appraised and developed as *in situ* deep-water clastic reservoirs, were known to have significant proportions of their reserves in sandstone intrusions (Hurst *et al.* 2005). Limited documentation of relevant outcrop data was available (largely summarized in Hurst & Cartwright 2007*a*; Hurst *et al.* 2011) and almost none of these data occurred on a scale relevant to mapping sub-surface relationships. Discovery of the Volund Field in 2004 by the Marathon Oil-operated exploration well N24/9-7 on the Norwegian continental shelf followed the interpretation of two 3D seismic surveys and well data from the area of interest (de Boer *et al.* 2007). Critically this interpretation included several wells with continuous core. The evaluation benefitted from independent work conducted by the Sand Injection Research Group (SIRG, based at the University of Aberdeen with current and former collaborative partners at the University of Manchester and the University of Wales, Cardiff, respectively) research that documented the regional occurrence and significance of sand injection complexes in the UK and Norwegian North Sea (Duranti *et al.* 2002; Huuse *et al.* 2001, 2002, 2003). Prior to drilling well N24/9-7, critical risks associated with exploration drilling were identified as the presence or otherwise of thick, good-quality, oil-charged sandstone up-dip

From: Bowman, M., Smyth, H. R., Good, T. R., Passey, S. R., Hirst, J. P. P. & Jordan, C. J. (eds) 2016. *The Value of Outcrop Studies in Reducing Subsurface Uncertainty and Risk in Hydrocarbon Exploration and Production*. Geological Society, London, Special Publications, **436**, 75–92.
First published online December 16, 2015, http://doi.org/10.1144/SP436.3

and in connectivity with a thin oil leg discovered in wells N24/9-5 and -6. Reservoir presence and quality were the background for engagement between Marathon and SIRG and are the focus of this paper.

Background geology

Volund overlies the NNE–SSW-trending axis of the South Viking Graben in Norwegian licence PL150 (Fig. 1; Jones *et al.* 2003; Huuse *et al.* 2004). In the area of interest depositional sandstones are known to occur in the Heimdal (Upper Palaeocene) and Hermod formations (Lower Eocene). The target reservoir, a saucer-shaped sandstone intrusion, is believed to emanate from underlying depositional Hermod sandstone; however, this cannot be confirmed from seismic or borehole data. The saucer-shaed intrusion is hosted within mudstone-dominated strata of the tuffaceous Balder Formation and the lower part of the Hordaland Group (Fig. 1b). Prior to drilling exploration well N24/9-7 the host strata were interpreted to be both top and lateral seals for the reservoir, thus defining an intrusive trap (Hurst *et al.* 2005).

Turbiditic slope- to basin-floor channel sandstones in the Hermod Formation that originated from the East Shetland Platform to the west are interpreted to be the parent units for the sandstone intrusions. By analogy with the Panoche Giant Injection Complex (PGIC; Vigorito *et al.* 2008), where turbiditic sandstones lying immediately below the sand injection complex sourced the sandstone intrusions, depositional sandstone in the Hermod Formation was the main parent unit for sandstone in the Volund Field. The thick, regionally developed Heimdal Formation is inferred to be a major source of aqueous fluid that supported sand injection (Ahmadi *et al.* 2003).

Pre-Volund exploration

Exploration wells drilled by Fina, N24/9-5 and N24/9-6 (Fig. 1a), discovered thin, 6 and 11 m respectively, oil legs in high-quality reservoir sandstone from which conventional core was taken. The discovery named Grieg Field was deemed sub-commercial and the operational licence was relinquished in 1994. In 2003 Marathon was awarded operatorship of Norwegian licence PL150. Part of

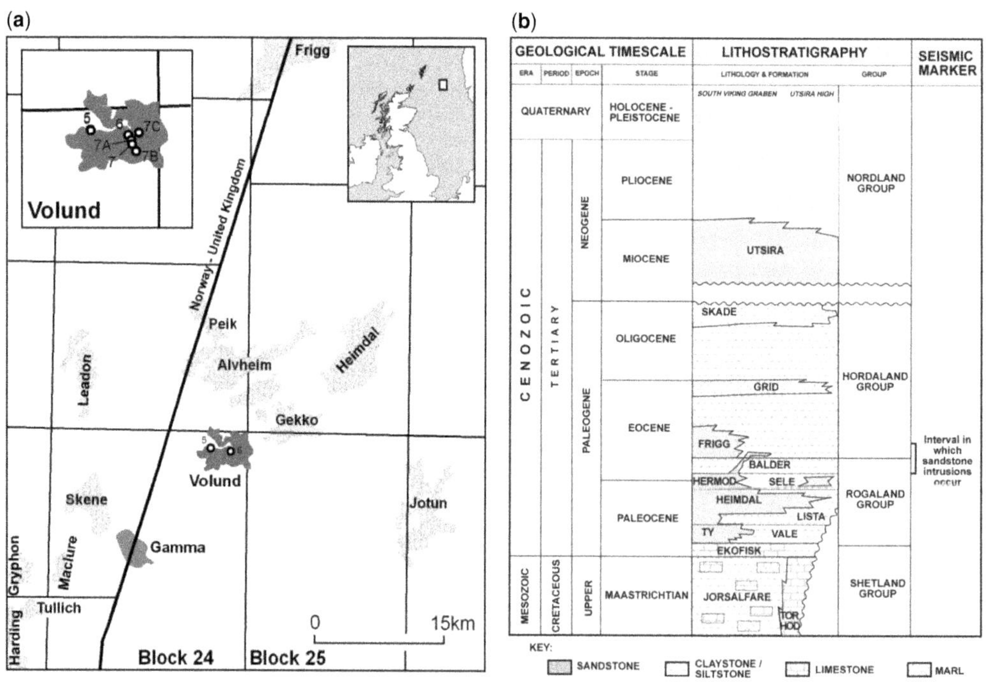

Fig. 1. (**a**) Location of the study area showing the location of Volund Field (licence PL150) and adjacent oil and gas fields in which significant sandstone intrusions are known to be present. Wells N24/9-5 (1993) and N24/9-6 (1994) discovered the sub-commercial Grieg Field, now known to be the lowest oil-bearing part of the Volund Field sand injection complex. The inset shows all exploration well locations on Volund Field. (**b**) Tertiary lithostratigraphy in the study area showing the interval in which the Volund sandstone intrusions occur.

a probable sand injection complex had been recognized independently in the licence area (Huuse *et al.* 2004; Fig. 2). The main targets for exploration were low-angle (*c.* 25° from bedding) dykes that emanated upward from the margins of the sandstone encountered by wells 24/9-5 and -6 (Lawrence *et al.* 1999). In 2003 the prospect was named Hamsun but was renamed Volund Field after drilling of the discovery well 24/9-7 in 2004 and approval of the Plan of Development in 2006.

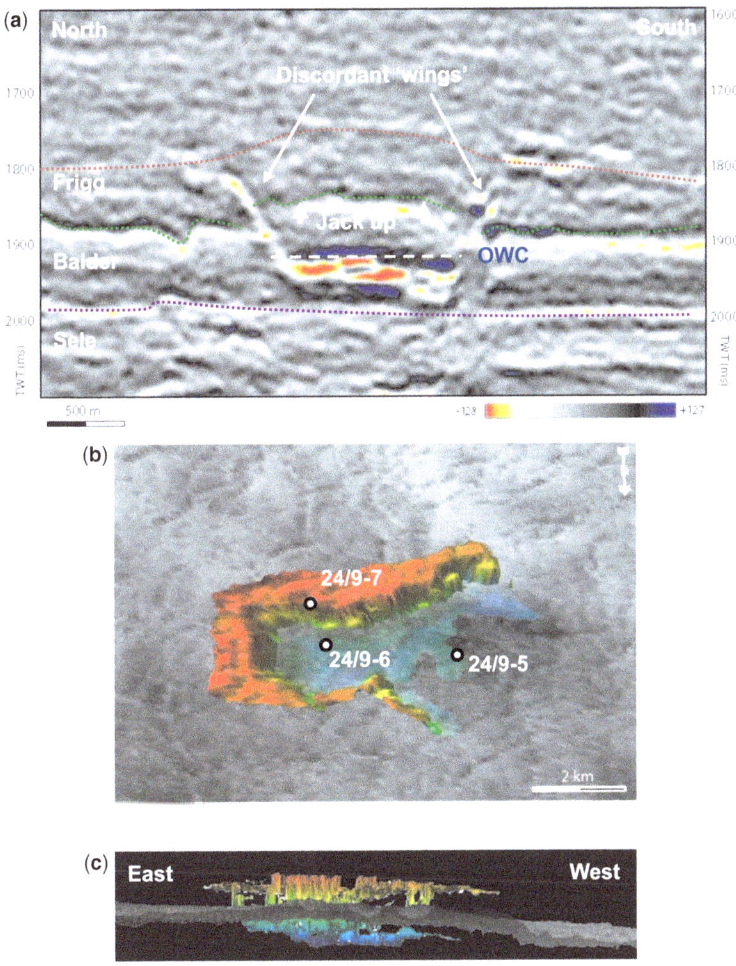

Fig. 2. (**a**) North–south-oriented seismic line across the Volund prospect. A basal sandstone (high-amplitude blue and red areas on the cross-section), which was cored in wells 24/9-5 and -6 contained a thin oil leg, 6 and 11 m, respectively. Bedding-discordant features ('wings') emanate from the basal sandstone, cut through approximately 200 m of stratigraphy and formed the target for exploration of the Volund prospect. The right-side (southern) wing has stronger amplitude than the left-side wing (northern) and was associated with a thicker development of sandstone. The upper tips of the wings develop close-to bedding-parallel reflectors and were originally interpreted as seafloor extrusions of sand but following exploration drilling proved to be of intrusive origin. Jack-up of the top Balder Formation is seen in the inter-wing area and corresponds to the thickness of sand within the Balder Formation. The wing tips become almost bedding parallel at the Top Frigg level. (**b**) Plan-view image of top massive sandstone surface of the Volund sandstone intrusion. Small-scale intrusions, unmappable using seismic data, exist above the picked surface, up to about the top Frigg level. The exploration target is the green–yellow section that represents the main reservoir volume. The blue area was penetrated by wells 24/9-5 and -6 (modified from Huuse *et al.* 2003). (**c**) Side view of the Volund prospect and the Top Balder surface (grey) highlighting the concordant lower sandstone body and the discordant wings cross-cutting more than 200 m of stratigraphy, including the entire Balder and Frigg formations.

Several sedimentological studies of core from wells 24/9-5 and -6 had interpreted the sandstones as turbidite channel facies, although one core description identified internal structures that were associated with sand fluidization. In independent unpublished studies in 2001 and 2003, SIRG examination of the core from both wells found no evidence for internal sedimentary structures diagnostic of depositional processes.

Although sand injectites within Palaeogene strata were known from nearby oil fields (Fig. 1a) these were not collectively associated with the definition of a 'sand injectite play'. Many sub-surface interpreters were reluctant to recognize large-scale sandstone intrusions as significant reservoir volumes within Palaeogene deep-water, basin-floor settings in which more conventional stratigraphic models for sandstone distribution within a sequence stratigraphic framework were well established (e.g. Den Hartog-Jaeger *et al.* 1993).

Sand injection complexes

Interpretation of extensive 3D seismic data acquired in the North Sea revealed that sand injection complexes appeared to be regionally developed in specific stratigraphic intervals within the Palaeogene (Huuse *et al.* 2001, 2003, 2009). Examination of continuous core from exploration and appraisal wells revealed that North Sea sand injection complexes were not restricted to the Palaeogene but occurred also in Lower Cretaceous and Upper Jurassic deep-water clastic successions (Fig. 3 in Hurst *et al.* 2007) although their significance as reservoir units appears to be or have been overlooked or underestimated.

Parts of possible large sand injection complexes were known from outcrops of the Santa Margarita Sandstone (Thompson *et al.* 1999), the Vocontian Basin (Parize & Fries 2003), the Nummidian Flysch (Parize & Beaudoin 1988) and the Hareelv Formation (Surlyk & Noe-Nygaard 2001). All these areas provide some insight into intra-injection-complex features but also demonstrate their limitations as sub-surface analogues because the relationships between parent units, sandstone intrusions and palaeo-seafloors are speculative or absent. With the exception of the Hareelv Formation, all the examples have limited lateral and vertical exposure, and all are difficult to place in a 3D stratigraphic context. These factors limit their value as sub-surface analogues.

Re-discovery of the Danian PGIC in the Panoche Hills by SIRG built on preliminary work by Smyers & Peterson (1971) and allowed study of the regional development of sandstone intrusions within a basinal context. Critically, metre- to decametre-thick parallel-to-bedding (sills), low-angle-to-bedding (saucer-shaped and conical intrusions) and deformed parent units were identified. For the first time it was possible to map the stratigraphic relationships between parent depositional sandstones, sandstone intrusions and sand extrudites (Vigorito *et al.* 2008), which allowed a much improved understanding of the processes responsible for sandstone intrusion (Vigorito & Hurst 2010). In 2002, however, SIRG's knowledge of the PGIC was limited to the characterization of several reservoir-scale sandstone units and identification of field relationships between probable parent depositional units, sandstone intrusions and sand extrudites (Hurst *et al.* 2005). More detail of these relationships and their relevance to sub-surface interpretation is presented later and was particularly significant as a pre-drill geological risk constraint at Volund.

Sub-surface analogues

At the time of the exploration evaluation of licence PL150 several oil fields on the UK and Norwegian continental shelves were recognized as having significant reservoir volumes in sandstone intrusions (*c.* 2450 × 10^6 BOE; Hurst *et al.* 2005). An extensive review of relevant seismic data was undertaken and many smaller, and some substantially larger, but similar, features to Volund were identified (Huuse *et al.* 2003, 2004). These were at least superficially similar to the features that constituted the sub-commercial Grieg Field (Lawrence *et al.* 1999). Although sub-surface analogues such as those defined using seismic images from the Alba Field (Fig. 3), where low-angle sandstone intrusions were proven (MacLeod *et al.* 1999; Duranti *et al.* 2002), were useful, comparison with Volund and other similar undrilled features was limited by the lack of compatibility between different seismic surveys shot over sand injection complexes and data quality. Sandstone distribution in Alba could not be imaged using 3D full-offset P-wave data (Newton & Flanagan 1993; MacLeod *et al.* 1999). Uncertainty associated with data from possible analogue oil fields decreased their efficacy as analogues for the Volund sandstone intrusions. Many interpreters remained secure with their interpretations of the Volund prospect as seismic artefacts or part of a deep-water channel despite the atypical (for a channel) planform geometry of the inferred reservoir units (Fig. 2b). In the face of such divergent views, outcrop analogue data clearly offered potential insight into the sub-surface interpretation. Despite provision of valid outcrop analogue data, many sub-surface interpreters remained deeply sceptical of the prospectivity of Volund as part of a sand injection complex.

Description of conventional core from the Harding Field revealed sedimentary structures that

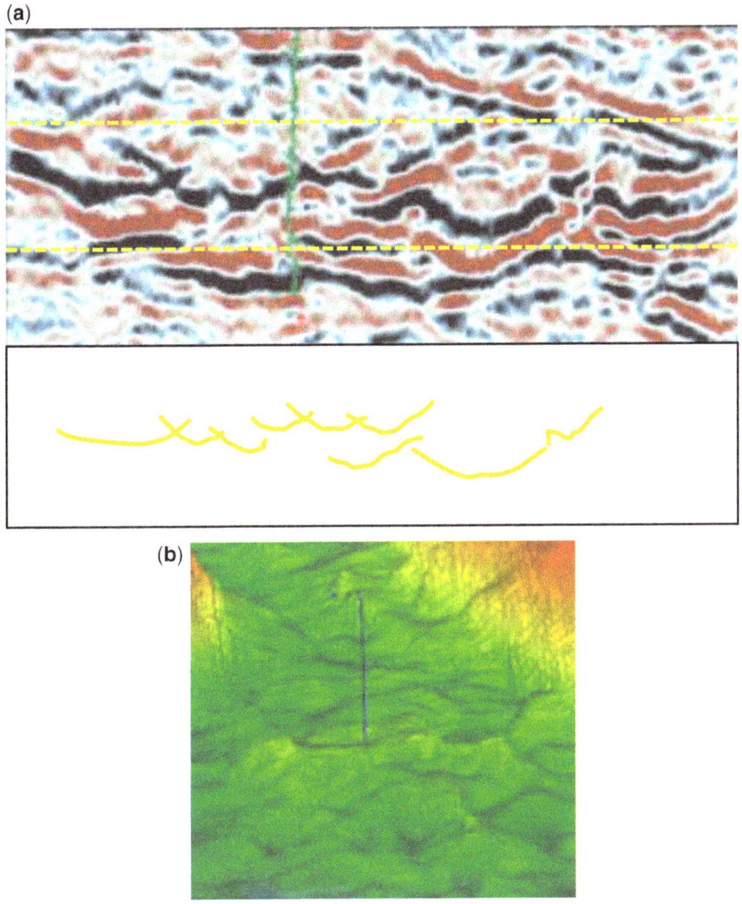

Fig. 3. (**a**) An arbitrary seismic line from the 1999 OBC converted wave (P-S) 3D seismic survey over the Alba Field (MacLeod *et al.* 1999) available prior to commencement of the Volund Field exploration drilling. The line of section cuts across the depositional slope-channel from which sandstone intrusions are likely to have emanated. Some saucer-shaped sandstone intrusions are identified between the dashed lines and displayed below the seismic line. Only the geometry of top of each sandstone intrusion is mapped and no intrusive thickness is inferred. Picking a bedding-parallel top reservoir in this interval is clearly inappropriate. (**b**) Irregular top reservoir from the basal sandstone intrusion, Volund Field (after Szarawarska *et al.* 2010).

were identified as the product of sand injection. The presence of larger-scale sandstone intrusions was considered as a possible sub-surface scenario but at the time only limited calibration with seismic data was undertaken (Dixon *et al.* 1995). Duranti *et al.* (2002) introduced a classification of internal structures observed from Alba Field core that were related to sand fluidization and placed them in context with 3D seismic data; the classification was further developed by Duranti & Hurst (2004). In general the paucity of documentation of internal structures from sandstone intrusions and the process by which they formed (see review in Hurst *et al.* 2011) limited their value in sub-surface evaluation.

In the period 1999–2003 SIRG regularly examined core, core descriptions and wireline log data from deep-water clastic reservoirs in which operators believed that sandstone intrusions might be present. Reasons for suspecting the presence of sandstone intrusions frequently included unexpectedly strong aquifer support in thin-bedded strata during field development, difficulty in correlating between wells, and ‘inconclusive’ biostratigraphic evaluations. (The quality of biostratigraphic data was rarely called into question but typically data did not fit readily with the inferred lithostratigraphy and sequence stratigraphy.) Frequently core descriptions recorded sandstone intervals where sedimentologists failed to recognize internal structures or interpreted unusual, irregular structures as 'possible slumping'. In the same intervals wireline logs often had small but quantifiable differences in response to adjacent sandstones (Fig. 8 in Duranti *et al.* 2002). When core from these intervals was examined, they invariably preserved evidence of sand injection (Duranti & Hurst 2004). When core was unavailable the inference of the presence of sandstone intrusions provided a reasonable explanation for enhanced vertical transmissibility and aquifer support.

Outcrop analogues and sub-surface risk constraint

Although some outcrop data were available in the public domain, all studied sandstone intrusions were smaller in scale and geometry than sub-surface features, a possible exception being a single large sandstone body within the Santa Cruz Mudstone (Miocene) of central California (Thompson *et al.* 1999). In most accounts dykes were the predominant sandstone intrusion described (Murchison 1827; Jenkins 1930; Peterson 1968; Taylor 1982; Surlyk & Noe-Nygaard 2001), possibly because identification of sills and their differentiation from depositional units was problematic. Sills were documented by Truswell (1972), Hiscott (1979) and Archer (1984); however all these are small-scale intrusions at spatially restricted localities with limited relevance as sub-surface analogues (fig. 6a–c in Hurst *et al.* 2011). It is now known that sand injection complexes form an organized architecture largely in response to the interaction between variations in pore-fluid pressure and the internal heterogeneity of the host strata (Vigorito & Hurst 2010; Hurst *et al.* 2011). Before 2003 this was less obvious; however, several important observations were made.

Sills

Typically sills form composite units that can be several tens of metres thick, and which in two dimensions are seen to interconnect vertically by upward-stepping and curving sill-segments and are cross-cut by dykes (Figs 4a & 5). The upper surfaces of sills are irregular to bedding because of both steps and erosion. Bedding is cut by sills, hence the margins of sills are diachronous; biostratigraphpic sampling adjacent to sills often records different ages along sill margins.

Logs of reservoir thickness show that 'top reservoir sandstone' does not correspond laterally to the same sandstone and those individual sandstone intrusions, and groups of sandstone intrusions, pinch out laterally (Fig. 4b). A thin sandstone present at the top of the right-hand log is part of three thin, laterally linked saucer-shaped intrusions that persist >100 m to the left, after which no sandstone sills occur at this level. In the central and right-hand field of view, the thin sandstones would, if hydrocarbon saturated, constitute top reservoir in the sill complex. Anomalously thick sandstone is recorded in the right-hand log where the section has crossed the confluence of two sills. The general architecture of this sill complex is a thinning of individual sills above and below the thickest sill with individual sills tending to have an overall curved external geometry in cross-section. Curvature decreases as the sills thin. In general these relationships are similar in other sandstone sill complexes.

Within composite sill-dominated units individual sills are typically composite and contain internal structures (Fig. 4c). Banding (Scott *et al.* 2009) is the most common internal structure and may form geometries that appear similar to cross-bedding and parallel lamination. On a micro-scale banding lacks the granular grading that typifies depositional structures with the margins of bands characterized by features that include concentrations of sand-sized mudstone clasts, micas, finely comminuted quartz grains and concentrations of elongate grains.

Saucer- and conical-shaped intrusions

Interpretation of seismic data typically identifies saucer- and conical-shaped intrusions (Huuse *et al.* 2003, 2004, 2005; Hurst *et al.* 2003*a*) and the Volund prospect appeared to comprise several partially complete saucer-shaped intrusions. Three features of these intrusions when viewed at seismic scale are particularly significant: saucer-shaped and conical external geometry predominate; typically they are composite bodies; and jack-up of host strata occurs above major sandstone intrusions. The presence of these bodies is particularly conspicuous in thickness maps of their host strata, most notably the Balder Formation in the South Viking Graben (Purvis *et al.* 2002; Huuse *et al.* 2004).

A seismic-scale exposure of a saucer-shaped intrusion shows a composite intrusion cross-cutting >170 m of host mudstone and jacking up of the overlying strata (Fig. 6). In the centre field of view multiple intrusions are stacked together and these become less evident toward the margins. Jack-up is well-developed at both margins but most easily visible at the right-hand (southern) margin. Jack-up (Fig. 6b) is a useful guide in sub-surface evaluation because it is indicative of reservoir presence and thickness. Jack-up of the top Balder Formation in Volund varies (Fig. 2) and exploration and development drilling demonstrated that this corresponded to variations in the thickness of sandstone present

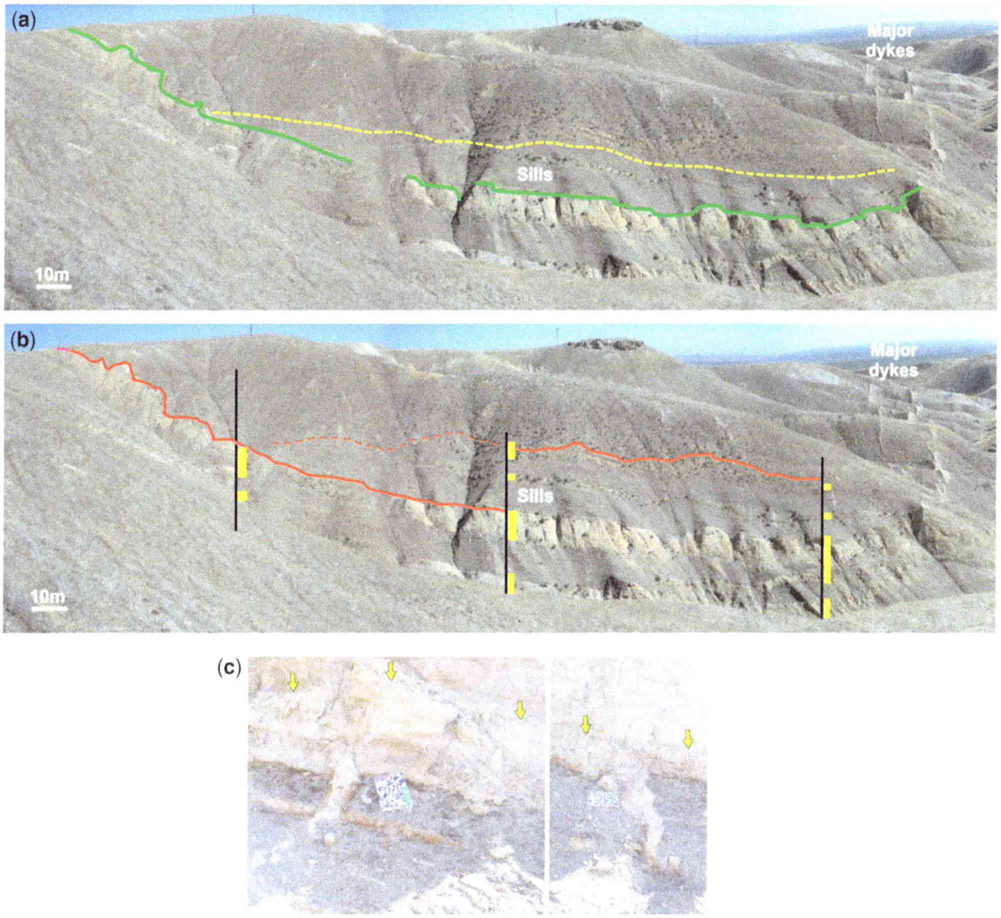

Fig. 4. A composite >30 m sill-rich unit. Strata dip into the field of view at approximately 30°. (**a**) Bedding (yellow-dashed line) is cut by the upper surface of the thick sill (solid green line) and other sills. (**b**) Fewer sills are present toward the left where the top of thick sill is the shallowest sill present in the composite sill-rich unit. By following bedding across the section different sandstones are observed to form the shallowest sandstone intrusion present and a >60 m horizontal section without sandstone constitutes the bedding-parallel 'top sandstone' (fine-dashed red line). The solid-red line is the actual top sandstone. Vertical logs of the sandstone units demonstrate the lateral variations of sandstone thickness and that top sandstone is represented by different units laterally. (**c**) Detail of internal structures within part of a composite stepping sill. The yellow arrows mark the boundary between two intrusions (after Hurst *et al.* 2007).

(Schwab *et al.* 2014). The significance of jack-up as a pre-drill indicator of sandstone reservoir presence and thickness is demonstrated by the post-Volund-exploration of a sand injectite prospect (Danica) with little evidence of jack-up; <10 m sandstone was present (Szarawarska *et al.* 2010). For direct comparison a seismic line is shown at true-vertical-scale and has almost identical geometry and dimensions to the saucer-shaped intrusion seen at outcrop (Fig. 6c).

Irregular and discordant upper surfaces to both sandstone intrusions and depositional sandstones modified by sand fluidization are characteristic of sand injectites. When seismic interpreters assume sedimentary conformity between a reservoir and its overlying seal, the mind-set is to focus on identifying and mapping a continuous, approximately bedding-parallel horizon. Where a break in the continuity of a bedding-parallel horizon occurs, faulting is often inferred. Because sandstone intrusions cross-cut bedding and may stop or thicken suddenly (Fig. 4a, b), mapping along a bedding-parallel horizon will inevitably transect a series of sandstone intrusions within mudstone at diverse angles to bedding, dependent on the slope of the individual intrusions. The net effect is that bedding-parallel

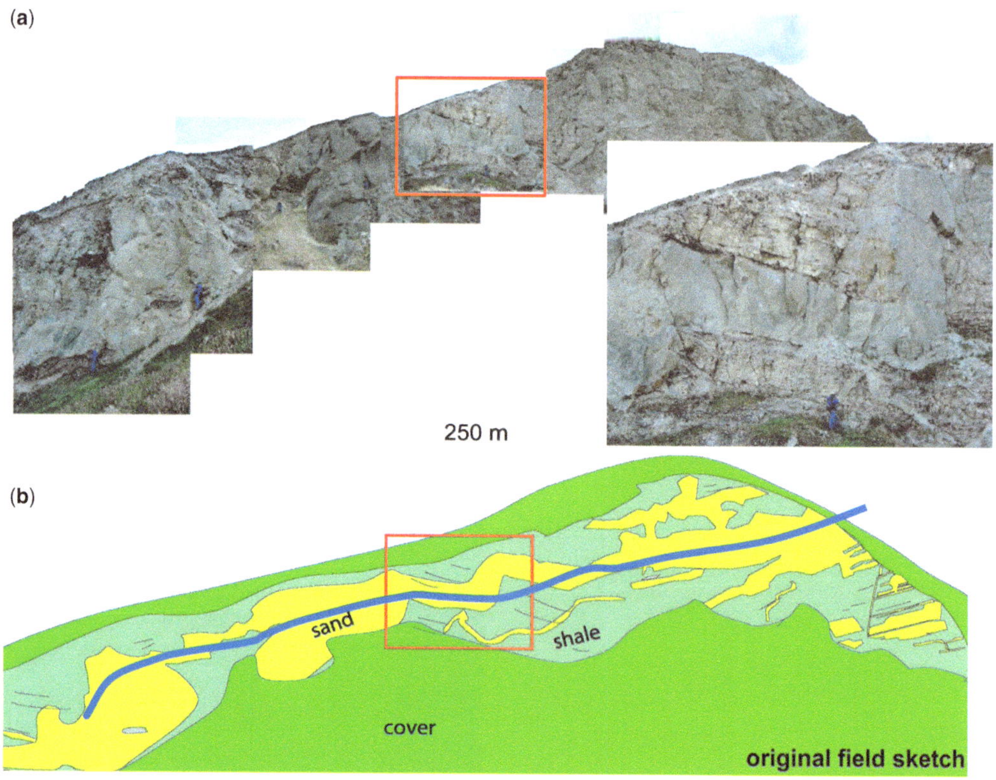

Fig. 5. (**a**) A stepping sill in highly fractured (post sand-injection) biosiliceous mudstone, which has a slight pink colouration. Steps are several metres high but in detail their margins are rounded by erosion caused by turbulent fluidized flow during sand intrusion. Orientation of steps indicates that the sill was emplaced in a plane approximately perpendicular to the rock-face. The red rectangle is shown in detail in the inset. (**b**) The blue line is an idealized well-track to demonstrate how easily mudstone can be encountered during drilling of sandstone intrusions despite high sandstone content and the high level of hydraulic continuity (modified from Huuse *et al.* 2007).

'top reservoir' picks are weak because in reality a laterally continuous, bedding-parallel sandstone is not present (Figs 3a & 4b). In Figure 3a a cross-section from Alba Field shows the lack of a continuous top reservoir horizon and the presence of numerous saucer-shaped intrusions. Because the upward terminations of saucer-shaped intrusions are below the resolution of seismic data (Huuse *et al.* 2007), a top reservoir horizon is typically a compromise between an interpreter's need to create a top reservoir map, the absence of a laterally continuous bedding parallel sandstone and the reality of sandstone intrusions that cross-cut stratigraphy. Consequently, top reservoir maps that are approximately bedding-parallel are typically rugose (Fig. 3b), although in reality they are likely to be even more rugose simply because thin, upward extensions of saucer-shaped intrusions are not resolved. Possible upward erosive surfaces (Fig. 7) may contribute to the rugose surfaces and are known from core in wells 24/9-5 and -6.

Erosive margins

Examination of core demonstrated that the fractures into which sand was injected are frequently modified by erosion (Hurst *et al.* 2011) and similar relationships are recorded from outcrops where features resembling turbidite sole-marks are observed on both upper and lower intrusion margins and on either side of vertical dykes (Keighley & Pickerill 1994; Fig. 11 in Hurst *et al.* 2007). The scale of erosion is best appreciated at outcrop. Modification by erosion along the top of sills is commonplace (Fig. 4a) and is typified by convex-upward geometry, and strongly discordant and often steep (>80° to bedding) truncation of bedding in the host strata (Fig. 7a). The 3D geometry

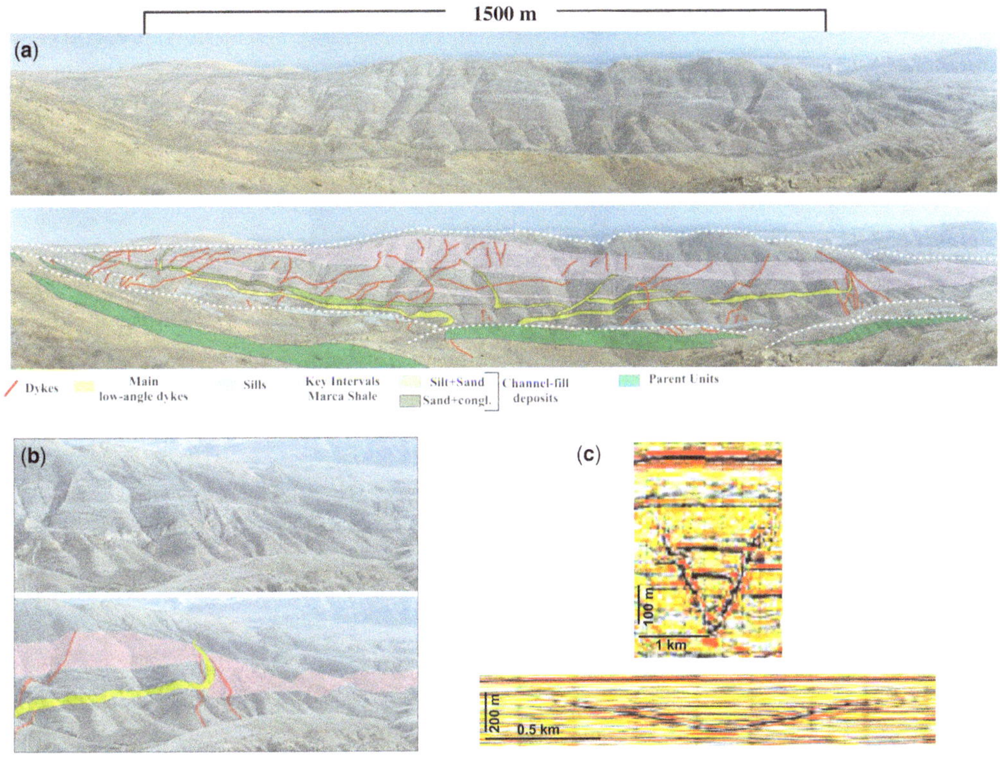

Fig. 6. (**a**) Exposure and an interpretation of a saucer-shaped sandstone intrusion (yellow). The intrusion is composite and cross-cut by dykes; only the main intrusions are shown. On the left side of the image the intrusion underlies, deviates from and cross-cuts a thin turbiditic very-fine-grained sandstone channel (pale green). The angle of discordance between the saucer-shaped intrusion and bedding varies between *c.* 10° and >80°. At the shallowest points the intrusion becomes approximately vertical and marks a clearly defined jack-up of the overlying Marca Shale Member (pink-shade) above where the sandstone occurs. (**b**) Detail of right-hand (southern) jack-up. The slope of the hillside and its curvature produce an exaggerated impression of the jack-up. (**c**) A conical intrusion with vertical exaggeration and the same section flattened to give true-vertical thickness, which is similar in profile to the outcrop.

of the Yellowbank Creek intrusion cannot be constrained by outcrop mapping (Thompson *et al.* 2007) and was described as having 'irregular geometry' (Hurst *et al.* 2011). However, the exposure is likely to be part of a large stepping sill, the apparent irregularities being a function of incomplete exposure and post-emplacement tectonic deformation.) Termed scallops (Hurst *et al.* 2005), these are characteristic of the upper surfaces of stepped sills and parent units that are modified by sand fluidization. In some cases upward erosion and truncation of bedding occur adjacent to the bases of dykes that emanate from parent units (Fig. 7b).

Erosive margins typically polish mudstone surfaces leaving a variety of sole markings, including flutes, all of which are indicative of turbulent flow during sand fluidzation. Erosion may be several metres into host mudstone (Fig. 7) and is likely to contribute to producing rugose top surfaces on sandstone intrusions (Fig. 3b).

Erosion of host strata has a locally significant effect on the mineralogy and reservoir quality of sandstone intrusions as granular mudstone particles are introduced into the sandstone intrusion, which on compaction are squeezed into adjacent pores, thus reducing reservoir quality. Lowering of reservoir quality occurs proximal to *in situ* mudstone, typically along the upper margins of intrusions (Scott *et al.* 2014) or adjacent to large mudstone intraclasts incorporated into sandstone intrusions.

Internal structures

Despite earlier descriptions of internal structures in dykes (Peterson 1968; Truswell 1972; Taylor 1982; Archer 1984) and fewer descriptions from sills

Fig. 7. Photo-pairs of sandstone intrusions without and with interpretations. (**a**) Upward erosion along the top surface of a sandstone intrusion. Individual convex-upward erosive features (margin marked by the red line) are termed scallops (after Hurst *et al.* 2005). In this example at least 5 m of mudstone is eroded (mudstone bedding marked by the stippled black lines). Note the presence of banding running approximately parallel to the right-hand erosive margin. (**b**) Erosive margins (red line) at the point of emanation of a dyke from a parent unit (bedding planes as yellow dashes).

(Kawakami & Kawamura 2002), 'geological folklore' in the early 2000s spoke of the structureless character of sandstone intrusions as a key differentiating feature relative to depositional sandstones (in sills, Truswell 1972; Hillier & Cosgrove 2002). Whatever the reason for these perceptions, the reality from outcrop study is that internal structures are common in sandstone intrusions (Fig. 8; Diggs 2007; Scott *et al.* 2009; Hurst *et al.* 2011).

Internal structures in sandstone intrusions, predominantly banding, have some superficial characteristics similar to depositional structures but are formed largely by combination of rapid settling during waning of fluidized flow, consolidation of grains during the collapse of meta-stable granular structure and shear along pre-existing laminae that occurred during compaction/consolidation. Margins of bands may be enriched in sand-sized mudstone particles and/or aligned elongate grains. Structureless sandstone occurs but in our experience rarely predominates.

Mudstone- (shale-)clast breccias were identified in core from Alba Field (Duranti *et al.* 2002; Duranti & Hurst 2004) and are common facies in sandstone intrusions (Fig. 9; Fig. 7 in Hurst & Cartwright 2007*b*). Mudstone-clast textures such as microfractures, jigsaw fabrics, locally eroded (rounded) mudstone-clast margins, laminae of small (typically large sand-size or coarser) angular mudstone clasts (Figs 4 & 5 in Duranti & Hurst 2004) are common, and all formed by instantaneous hydraulic fracture and subsequent rapid fill by fluidized sand. Many of the mudstone-clast textures are unlikely to form and preserve in depositional environments and allow the important distinction between linked (sandy) debrites (Haughton *et al.* 2003) and mudstone-clast breccias (Duranti & Hurst 2004). (In 2000 prior to publication of Haughton *et al.* (2003) and without knowledge of the content of that paper, some of the same core was described by SIRG and a sand injectite 'mudstone-clast breccia' origin ascribed – an example of the same data

Fig. 8. Internal structures in a sandstone sill. (**a**) Banding in a stepping sill with superficial similarity to depositional structures. Boundaries of individual bands may be finer-grained than within bands. Variations in grain packing and grain alignment are often the main defining characteristic in banding. Typically banding is sub-parallel to the margins of sandstone intrusions (see Fig. 7a); however, cross bedding (banding) occurs and has superficial similarity to depositional features. (**b**) Detailed view of (a) in which colour variations are caused by weathering and reflect slight differences in grain size and permeability of the different bands.

described with divergent interpretations.) Our early experience with core examination was that features associated with sand injectites were commonly attributed to sandy debrites or slumps, which lead to fundamentally different interpretations of reservoir geometry, connectivity and quality.

Hydraulically fractured mudstone

Intensely fractured mudstone is a common feature of sand injection complexes (Fig. 10) and formed by hydraulic fracture (pore-fluid pressure at least exceeding the fracture gradient), which led

Fig. 9. A mudstone-clast breccia (cf. Fig. 4, B4, Duranti & Hurst 2004). Mudstone clasts are generally elongate, microfractured and sometimes broken by <0.1 m, and often much narrower, aperture sand-propped fractures. Rounded clast-margins formed during sand injection by abrasion in the sand-charged turbulent fluidized flow. At a larger scale the unit is seen to be part of a depositional sandstone that is pervasively modified by hydraulic fracturing and sand injection. Pen is *c.* 0.15 m long.

to a high rate of fluid flow that eventually exceeded the minimum fluidization velocity of sand in adjacent parent, depositional units and thereby fluidized and injected sand into some fractures. Although the spatial significance of intensely hydraulically fractured mudstone within sand injection complexes was not established prior to the exploration of Volund Field, their association with sandstone intrusions was known. Often mudstone is so intensely and pervasively fractured that no trace of bedding or lamination remains (Fig. 10). Fractures are short (typically <0.1 m long), only locally planar and randomly oriented. Occasional fracture alignment and more planar geometry occur adjacent and sub-parallel to dykes. Hydraulically fractured mudstone has a general texture of disoriented angular clasts that are rarely >0.05 m across, which when buried form a compacted micro-fractured texture. Conchoidal fractures and polished (slickensided) surfaces are common. Fractures are frequently mineralized by gypsum or minor sulphide mineralization.

It is now known that intense hydraulic fracturing of mudstone is concentrated above a level termed the lithostatic equilibrium surface within which the Sill Zone occurs (Vigorito & Hurst 2010), where supra-lithostatic pore-fluid pressure prevailed during sand fluidization and injection.

Randomly oriented dykes

Earlier work in the Panoche Hills identified a crudely orthogonal orientation of dykes (Smyers & Peterson 1971); however, they appear only to have studied the upper part of the PGIC, effectively missing the Sill Zone in which many randomly oriented dykes as well as sills occur (Vigorito *et al.* 2008; Vigorito & Hurst 2010). Random orientation (Fig. 11) records intrusion under near-isotropic stress and is associated with supra-lithostatic pore-fluid pressure. In practical terms the random orientation creates an extremely high level of hydraulic continuity between all sandstone intrusions even when the net:gross (N/G) is low, thus enhancing drainage efficiency during production. Recognition in outcrop of randomly orientated dykes adjacent to sills and saucer-shaped intrusions, the main exploration targets, was critical to understanding how pressure continuity and reservoir connectivity even in low N/G might be expected at Volund.

Fig. 10. Hydraulically fractured mudstone with randomly oriented fractures. Crude vertical fracture alignment is present in the left field of view and is associated with an adjacent dyke margin. Gypsum crystallization (white/pale crystals) occurs along some fractures. No trace of primary lamination or bedding remains.

Discussion: reservoir risk constraint

From interpretation of the 3D seismic data Volund was mapped as a single sandstone unit of irregular thickness, in places >30 m thick (De Boer *et al.* 2007). No outcrops were known with similar geometry or thickness and it was suspected that the Volund interpretation was an oversimplification of reality, largely caused by the limitations of seismic resolution. Gross volume was inferred from seismic mapping but through comparison with outcrop data (Figs 4–6) it was recognized that sandstone intrusions are typically composite with a variable N/G distribution, a characteristic that could not be resolved from the seismic data. In this context the outcrop analogue data provided invaluable support to the seismic interpretation in terms of 'what may be encountered during drilling a saucer-shaped intrusion'.

Identifying jack-up and relating it to the likely presence and thickness of sandstone is a profoundly important risk constraint that is probably globally applicable. Comparison between outcrop and Volund data gave significant confidence to prediction of the presence of locally thick (>30 m) sandstone (compare Figs 2 & 6). Up to approximately 30 m of sandstone is present below the maximum jack-up (*c.* 35 m) of the top Balder surface in Volund. Jack-up in Volund is not uniform and development drilling validated a relationship between jack-up and reservoir thickness (Schwab *et al.* 2014). By comparison, very minor jack up identified on the Danica sand injectite prospect, when drilled, confirmed that minor jack-up correlated with thin (10 m) sandstone (Szarawarska *et al.* 2010).

Sandstone sills and saucer-shaped intrusions, and composites of these, form the most volumetrically significant parts of sand injection complexes (Fig. 3; Hurst *et al.* 2011). The opportunity to observe the composite character of sandstone intrusions at outcrop was particularly valuable because although seismic interpretation resolved thickness

Fig. 11. Randomly oriented sandstone dykes in hydraulically fractured mudstone, which are typical of the Sill Zone (Vigorito *et al.* 2008; Vigorito & Hurst 2010). Dykes are generally short (several metres), have narrow and variable aperture (typically <1 m) and irregular, erosive margins.

variations of intrusions (Fig. 2), their internal character is typically not resolved (Fig. 12a; Schwab *et al.* 2014). It was recognized that, when drilling along a saucer-shaped intrusion, encountering mudstone intervals should be expected, either because host mudstone may be included as large (tens of metres long) intra-clasts or because the intrusion margins may be irregular (Fig. 5). Prior to drilling Volund, bedding-discordant sandstone/mudstone contacts were known from other sand injection complexes (Dixon *et al.* 1995) but understanding their erosive character and relationship to larger-scale intrusive features became apparent more recently (Duranti *et al.* 2002; Hurst *et al.* 2003*b*; Duranti & Hurst 2004). Specifically the prevalence of randomly orientated dykes (Fig. 11) in association with sandstone sills and saucers justified the interpretation that even thin sandstone intrusions were likely to be in hydraulic continuity, which is independently verified in other fields that are influenced by the presence of sandstone intrusions (Briedis *et al.* 2007; Guargena *et al.* 2007; Satur & Hurst 2007).

Sequence stratigraphic and biostratigraphic frameworks are disrupted by sand injectites. A consequence of this is that reservoir units will disrupt the superposition and lateral continuity of strata (Figs 4, 6 & 12). A sandstone intrusion should be expected to cut across biostratigraphic zones and, in deep-water clastic settings, the reservoirs have no reason to coincide with low-stands of sea-level. Because sand injects along a fracture network in lower permeability host strata, material from the host strata becomes incorporated into the fluidized sand during fracturing and by erosion. Consequently some mudstone clasts may be carried into a younger part of the stratigraphy during injection. Well N24/9-7 proved that within a mudstone-clast breccia some of the mudstone clasts were reworked from the older Sele Formation more than 200 m below the Volund Field (De Boer *et al.* 2007).

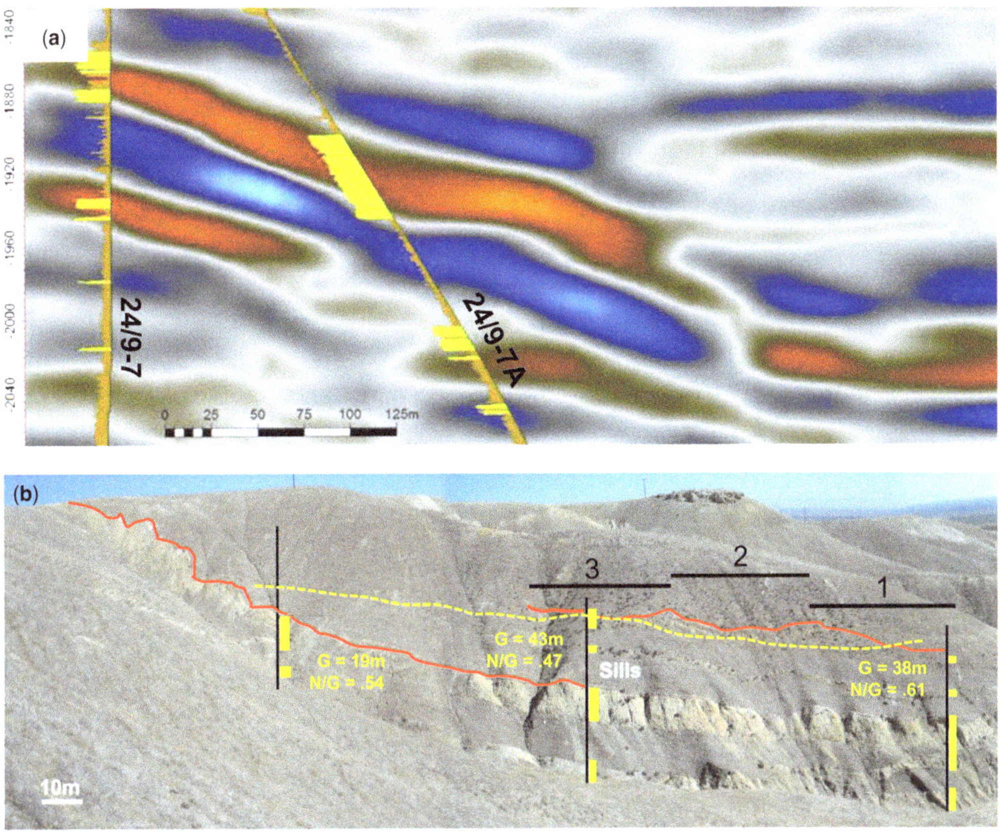

Fig. 12. (**a**) PAI_far offset seismic line through Volund Field displaying steps (see below and Figs 4 & 5) as part of an overall saucer-shaped geometry. From well penetrations the sandstone intrusion is known to comprise stacked sills with mudstone interbeds (adapted from Schwab *et al.* 2014). (**b**) Outcrop section from Figure 4 with the top sandstone intrusion (red line) within the composite intrusion and bedding (yellow dashes) shown. The top sill is defined as the shallowest saucer-shaped intrusion, which at the right-hand end of the outcrop is manifest as three small saucer-shaped intrusions (annotated 1–3) that overlie a *c.* 20 m-thick interval with low N/G in the composite intrusion. To the left of this the top of the sandstone intrusion is *c.* 20 m lower in the section and coincides with the top of the thickest sill present, which has a high N/G, and which steps up-section until it coincides with the bedding marker (yellow dashes). G, (known) gross thickness (m); N/G, net-to-gross ratio.

Conclusion

Outcrop data from sand injection complexes had not previously been applied to the evaluation of exploration targets: in the case of the Volund Field, outcrop data played a key role in measuring expectations and constraining the geological risk. Despite the distinctive geometry of the Volund Field, which comprises composite packages of saucer-shaped intrusions, a body of opinion favoured a sedimentary origin for the reservoir geometry and considered the intrusions unlikely to have good reservoir quality. Reservoir-scale outcrop data were used as analogues for the geometry and reservoir quality of the saucer-shaped intrusions recognized on seismic data, including mapping the distribution of mudstone-rich units and possible large intraclasts. Without the insight provided from the outcrop data exploration and appraisal drilling would have been less well prepared for the geology that was encountered.

Reservoir presence and thickness were demonstrated at outcrop by recognizing that stratal jack-up above the location of the sandstone intrusion is directly related to the thickness of sandstone present. Irregular discordant margins of saucer-shaped intrusions observed on seismic data were correlated with similar geometries that are caused by stepped and/or erosive margins that typify outcrops of similar intrusions. Discordance with bedding means that sandstone intrusions have diachronous margins and cross-cut biostratigraphic zones. High levels of hydraulic continuity are predicted in the Sill Zone, where most large volumes of sand are intruded, because of the lateral continuity of reservoir facies in sills and saucer-shaped intrusions and because these are inter-connected by a large number of generally short, low-aperture (<1 m), mostly randomly oriented dykes.

Internal structures are common in sandstone intrusions and form in response to sub-vertical fluid flow during emplacement of the sandstone intrusions or by consolidation processes immediately after their emplacement. Reservoir quality is generally very good but modified by the presence of mudstone clasts (mudstone-clast breccias) that reduce both pay volume and, where abundant, permeability. Structureless sandstone, although present, rarely predominates and is absolutely not a diagnostic characteristic of sandstone intrusions; banding is the predominant internal structure.

Drilling of exploration well N24/9-7, and sidetracks -7A, -7B and -7C proved the presence of high-quality reservoirs with similar facies to those observed at outcrop. Light oil with a gas cap is present and the Volund Field has been in successful commercial development since production began in 2010.

Peter Rawlinson and Wytze de Boer are thanked for their support and enthusiasm during the early stages of this work. Marathon Oil Company and their partners Lundin Norway AS and Det Norske AS are thanked for permission to publish.

Appendix

Details of outcrop locations shown in figures; all localities are in California, USA and their details are documented in the accompanying references.

Figures 4 and 12. Upper Cretaceous, Dosados Member, Moreno Formation, Moreno Gulch (Vigorito *et al.* 2008). (**4c**) Upper Cretaceous, Dosados Member, Moreno Formation, Escapardos Canyon (Hurst *et al.* 2007).

Figure 5. Eocene, Kreyenhagen Shale, Tumey Hill (Huuse *et al.* 2007).

Figure 6. Upper Cretaceous, Tierra Loma Member, Moreno Formation, Right Angle Canyon (36° 37′ 58″ N, 120° 41′ 56″ W).

Figure 7. (**a**) Miocene, Santa Cruz Mudstone Member, Panther Beach (Scott *et al.* 2009). (**b**) Miocene, Santa Cruz Mudstone Member, Baldwin Creek (Hurst *et al.* 2003*b*).

Figure 8. Upper Cretaceous, Dosados Member, Moreno Formation, Dosados Canyon (Vigorito *et al.* 2008).

Figure 9. Upper Cretaceous, Tierra Loma Member, Moreno Formation, Tumey Gulch.

Figures 10 and 11. Upper Cretaceous, Tierra Loma Member, Moreno Formation, West Tumey (36° 36′ 36″ N, 120° 39′ 21″ W).

References

Ahmadi, Z. M., Sawyers, M., Kenyon-Roberts, S., Stanworth, C. W., Kugler, K. A., Kristensen, J. & Fugelli, E. 2003. Paleocene. Chapter 13. *In*: Evans, D., Graham, C., Armour, A. & Bathurst, P. (eds) *The Millenium Atlas: Petroleum Geology of the Central and Northern North Sea*. Geological Society, London, 235–259.

Archer, J. B. 1984. Clastic intrusions in deep-sea fan deposits of the Rosroe formation, lower Ordovician, western Ireland. *Journal of Sedimentary Petrology*, **54**, 1197–1205.

Briedis, N. A., Bergslien, D., Hjellbakk, A., Hill, R. E. & Moir, G. J. 2007. Recognition criteria, significance to field performance, and reservoir modeling of sand injections in the Balder Field, North Sea. *In*: Hurst, A. & Cartwright, J. (eds) *Sand Injectites: Implications for Hydrocarbon Exploration and Production*. American Association of Petroleum Geologists, Tulsa, OK, Memoir, **87**, 91–102.

De Boer, W., Rawlinson, P. B. & Hurst, A. 2007. Successful exploration of a sand injectite complex: Hamsun Prospect, Norway Block 24/9. *In*: Hurst, A. & Cartwright, J. (eds) *Sand Injectites: Implications for Hydrocarbon Exploration and Production*. American Association of Petroleum Geologists, Tulsa, OK, Memoir, **87**, 65–68.

Den Hartog-Jaeger, D., Giles, M. R. & Griffiths, G. R. 1993. Evolution of Paleogene submarine fans of the North Sea in space and time. *In*: Parker, J. R. (ed.) *Petroleum Geology of Northwest Europe: Proceedings of the 4th Conference*. Geological Society, London, Petroleum Geology Conference Series, **4**, 59–71.

Diggs, T. N. 2007. Tesnus Formation, Marathon basin, Trans-Pecos Texas. *In*: Hurst, A. & Cartwright, J. (eds) *Sand Injectites: Implications for Hydrocarbon Exploration and Production*. American Association of Petroleum Geologists, Tulsa, OK, Memoir, **87**, 209–219.

Dixon, R. J., Schofield, K., Anderton, R., Reynolds, A. D., Alexander, R. W. S., Williams, M. C. & Davies, K. J. 1995. Sandstone diapirism and clastic intrusion in the Tertiary sub-marine fans of the Bruce-Beryl Embayment, Quadrant 9, UKCS. *In*: Hartley, A. J. & Prosser, D. J. (eds) *Characterisation of Deep Marine Clastic Systems*. Geological Society, London, Special Publications, **94**, 77–94.

Duranti, D. & Hurst, A. 2004. Fluidisation and injection in the deep-water sandstones of the Eocene Alba Formation (UK North Sea). *Sedimentology*, **51**, 503–531.

Duranti, D., Hurst, A., Bell, C. & Groves, S. 2002. Injected and remobilised sands of the Alba Field (UKCS): sedimentary facies characteristics and wireline log responses. *Petroleum Geoscience*, **8**, 99–107.

Guargena, C. G., Smith, G. B., Wardell, J., Nilsen, T. H. & Hegre, T. M. 2007. Sandstone injections at

Jotun oil field, Norwegian North Sea – modeling their possible effect on hydrocarbon recovery. *In*: Hurst, A. & Cartwright, J. (eds) *Sand Injectites: Implications for Hydrocarbon Exploration and Production*. American Association of Petroleum Geologists, Tulsa, OK, Memoir, **87**, 81–89.

Haughton, P. D. W., Barker, S. P. & McCaffery, W. D. 2003. 'Linked debrites in sand-rich turbidite systems – origin and significance. *Sedimentology*, **50**, 459–482.

Hillier, R. D. & Cosgrove, J. W. 2002. Core and seismic observations of overpressure-related deformation within Eocene sediments of the Outer Moray Firth, UKCS. *Petroleum Geoscience*, **8**, 141–149.

Hiscott, R. N. 1979. Clastic sills and dikes associated with deep-water sandstones, Tourelle formation, Ordovician, Quebec. *Journal of Sedimentary Petrology*, **49**, 1–10.

Hurst, A. & Cartwright, J. A. (eds) 2007*a*. *Sand Injectites: Implications for Hydrocarbon Exploration and Production*. American Association of Petroleum Geologists, Tulsa, OK, Memoir, **87**.

Hurst, A. & Cartwright, J. A. 2007*b*. Introduction, relevance of sand injectites to hydrocarbon exploration and production. *In*: Hurst, A. & Cartwright, J. (eds) *Sand Injectites: Implications for Hydrocarbon Exploration and Production*. American Association of Petroleum Geologists, Tulsa, OK, Memoir, **87**, 1–19.

Hurst, A., Cartwright, J. A., Huuse, M., Jonk, R., Schwab, D., Duranti, D. & Cronin, B. 2003*a*. Significance of large-scale sand injectites as long term fluid conduits: evidence from seismic data. *Geofluids*, **3**, 263–274.

Hurst, A., Cartwright, J. A. & Duranti, D. 2003*b*. Fluidisation structures in sandstone produced by upward injection through a sealing lithology. *In*: van Rensbergen, P., Hillis, R. R., Maltman, A. J. & Morley, C. K. (eds) *Subsurface Sediment Mobilization*. Geological Society, London, Special Publications, **216**, 123–137.

Hurst, A., Cartwright, J. A., Duranti, D., Huuse, M. & Nelson, M. 2005. Sand injectites: an emerging global play in deep-water clastic environments. *In*: Doré, A. & Vining, B. (eds) *Petroleum Geology: North-West Europe and Global Perspectives. Proceedings of the 6th Petroleum Geology Conference*. Geological Society, London, 133–144.

Hurst, A., Cartwright, J., Huuse, M. & Duranti, D. 2007. Sand injectites in deep-water clastic reservoirs: are they there and do they matter? Paper 120. *In*: Nilsen, T. H., Shaw, R. D., Steffans, G. S. & Studlick, J. R. J. (eds) *Atlas of Deepwater Outcrops*. American Association of Petroleum Geologists, Tulsa, OK, Studies in Geology, **56**, CD-Rom.

Hurst, A., Scott, A. & Vigorito, M. 2011. Physical characteristics of sand injectites. *Earth-Science Reviews*, **106**, 215–246.

Huuse, M., Duranti, D., Cartwright, J., Hurst, A. & Cronin, B. 2001. Seismic expression of large-scale sand remobilisation and injection in Paleogene reservoirs of the North Sea Basin and beyond. *In*: *63rd EAGE Conference and Technical Exhibition*. Amsterdam, 2001, Paper L-07.

Huuse, M., Duranti, D. *et al.* 2002. Detection, origin and significance of large-scale injected sand bodies in the North Sea Cenozoic. *In*: *64th EAGE Conference & Technical Exhibition*. Florence, 2002, Paper H-19.

Huuse, M., Duranti, D. *et al.* 2003. Sandstone intrusions: detection and significance for exploration and production. *First Break*, **21**, 33–42.

Huuse, M., Duranti, D. *et al.* 2004. Seismic characteristics of large-scale sandstone intrusions in the Paleogene of the South Viking Graben, UK and Norwegian North Sea. *In*: Davies, R. J., Cartwright, J. A., Stewart, S. A., Lappin, M. & Underhill, J. R. (eds) *3D Seismic Technology: Application to the Exploration of Sedimentary Basins*. Geological Society, London, Memoir, **29**, 263–277.

Huuse, M., Cartwright, J., Gras, R. & Hurst, A. 2005. Km-scale sandstone intrusions in the Eocene of the Outer Moray Firth (UK North Sea): migration paths, reservoirs, and potential drilling hazards. *In*: Doré, A. G. & Vining, B. (eds) *Petroleum Geology of NW Europe: Proceedings of the 6th Conference*. Geological Society, London, 1577–1594.

Huuse, M., Cartwright, J., Hurst, A. & Steinsland, N. 2007. Seismic characterization of large-scale sandstone intrusions. *In*: Hurst, A. & Cartwright, J. (eds) *Sand Injectites: Implications for Hydrocarbon Exploration and Production*. American Association of Petroleum Geologists, Tulsa, OK, Memoirs, **87**, 21–35.

Huuse, M., Jackson, C. A.-L., Cartwright, J. & Hurst, A. 2009. Large-scale sand injectites in the North Sea: Seismic and event stratigraphy and implications for hydrocarbon exploration. AAPG Search & Discovery Article 40481.

Jenkins, O. P. 1930. Sandstone dikes as conduits for oil migration through shales. *American Association of Petroleum Geologists Bulletin*, **14**, 411–421.

Jones, E., Jones, R. *et al.* 2003. Eocene. *In*: Evans, D., Graham, C., Armour, A. & Bathurst, P. (eds, coordinators) *The Millennium Atlas: Petroleum Geology of the Central and Northern North Sea*. Geological Society, London, 261–277.

Kawakami, G. & Kawamura, M. 2002. Sediment flow and deformation (SFD) layers: evidence for intrastratal flow in laminated muddy sediments of the Triassic Osawa Formation, northeast Japan. *Journal of Sedimentary Research*, **72**, 171–181.

Keighley, D. G. & Pickerill, R. K. 1994. Flute-like marks and associated structures from the Carboniferous Port Hood formation of eastern Canada: evidence of secondary origin in association with sediment intrusion. *Journal of Sedimentary Research*, **A64**, 253–263.

Lawrence, D. A., Sancar, B. & Molyneux, S. 1999. Large-scale clastic intrusion in the Tertiary of Block 29/4, Norwegian North Sea: origin, timing and implications for reservoir continuity. *American Association of Petroleum Geologists Bulletin*, **83**, 1324.

MacLeod, M. K., Hanson, R. A., Bell, C. R. & McHugo, S. 1999. The Alba field ocean bottom cable seismic survey: impact on development. *The Leading Edge*, **18**, 1306–1312.

Murchison, R. I. 1827. On the coal field of Bora in Sutherlandshire and some other stratified deposits in the

North of Scotland. *Geological Society of London*, **2**, 293–326.

NEWTON, S. K. & FLANAGAN, K. P. 1993. The Alba Field: evolution of the depositional model. *In*: PARKER, J. R. (ed.) *Petroleum Geology of Northwest Europe: Proceedings of the 4th Conference*. Geological Society, London, Petroleum Geology Conference Series, **4**, 161–171.

PARIZE, O. & BEAUDOIN, B. 1988. Clastic dykes and sills from Numidian Flysch (Sicily and Tunisia), sandy injection related to a high-density turbidity deposit. *AAPG Bulletin*, **72**, 1018.

PARIZE, O. & FRIES, G. 2003. The Vocontian clastic dykes and sills: a geometric model. *In*: VAN RENSBERGEN, P., MALTMAN, A. J. & MORELY, C. J. (eds) *Sub-surface Sediment Mobilization*. Geological Society, London, Special Publications, **216**, 51–72.

PETERSON, G. L. 1968. Flow structures in sandstone dykes. *Sedimentary Geology*, **2**, 177–190.

PURVIS, K., KAO, J., FLANAGAN, K., HENDERSON, J. & DURANTI, D. 2002. Complex reservoir geometries in a deep water clastic sequence, Gryphon Field, UKCS: injection structures, geological modelling and reservoir simulation. *Marine and Petroleum Geology*, **19**, 161–179.

SATUR, N. & HURST, A. 2007. Sand-injection structures in deep-water sandstones from the Ty Formation (Paleocene), Sleipner Øst Field, Norwegian North Sea. *In*: HURST, A. & CARTWRIGHT, J. (eds) *Sand Injectites: Implications for Hydrocarbon Exploration and Production*. American Association of Petroleum Geologists, Tulsa, OK, Memoir, **87**, 113–117.

SCHWAB, A. M., JAMESON, E. W. & TOWNSLEY, A. 2014. Volund Field: development of an Eocene sandstone injection complex, offshore Norway. *In*: MACKIE, T. (ed.) Geological Society, London, Special Publications, http://doi.org/10.1144/SP403.4

SCOTT, A. S. J., VIGORITO, M. & HURST, A. 2009. The process of sand injection: internal structures and relationships with host strata (Yellowbank Creek Injectite Complex, California, U.S.A.). *Journal of Sedimentary Research*, **79**, 1–18.

SCOTT, A., HURST, A. & VIGORITO, M. 2014. Outcrop-based reservoir characterization of a kilometer-scale sand injectite.complex. *AAPG Bulletin*, **97**, 309–343.

SMYERS, N. B. & PETERSON, G. L. 1971. Sandstone dikes and sills in the Moreno shale, Panoche Hills, California. *Geological Society of America Bulletin*, **82**, 3201–3207.

SURLYK, F. & NOE-NYGAARD, N. 2001. Sand remobilisation and intrusion in the Upper Jurassic Hareelv Formation of East Greenland. *Geological Society of Denmark Bulletin*, **48**, 169–188.

SZARAWARSKA, E., HUUSE, M., HURST, A., DE BOER, W., LU, L., MOLYNEUX, S. & RAWLINSON, P. B. 2010. Three-dimensional seismic characterisation of large-scale sandstone intrusions in the lower Palaeogene of the North Sea: completely injected v. in situ remobilised sandbodies. *Basin Studies*, **22**, 517–532.

TAYLOR, B. J. 1982. Sedimentary dykes, pipes and related structures in the Mesozoic sediments of south-eastern Alexander Island. *British Antarctic Survey Bulletin*, **51**, 1–42.

THOMPSON, B. J., GARRISON, R. E. & MOORE, C. J. 1999. A late Cenozoic sandstone intrusion west of S. Cruz, California. Fluidized flow of water and hydrocarbon-saturated sediments. *In*: GARRISON, R. E., AIELLO, I. W. & MOORE, J. C. (eds) *Late Cenozoic Fluid Seeps and Tectonics along the San Gregorio Fault Zone in the Monterey Bay Region, California GB-76. Annual Meeting of the Pacific Section*. American Association of Petroleum Geologists, Monterey, CA, 53–74.

THOMPSON, B. J., GARRISON, R. E. & MOORE, C. J. 2007. A reservoir-scale Miocene injectite near Santa Cruz, California. *In*: HURST, A. & CARTWRIGHT, J. (eds) *Sand Injectites: Implications for Hydrocarbon Exploration and Production*. American Association of Petroleum Geologists, Tulsa, OK, Memoir, **87**, 151–162.

TRUSWELL, J. F. 1972. Sandstone sheets and related intrusions from Coffee Bay, Transkei, South Africa. *Journal of Sedimentary Petrology*, **42**, 578–583.

VIGORITO, M. & HURST, A. 2010. Regional sand injectite architecture as a record of pore pressure evolution and sand redistribution in the shallow crust: insights from the Panoche Giant Injection Complex, California. *Journal of the Geological Society*, **167**, 889–904.

VIGORITO, M., HURST, A., CARTWRIGHT, J. & SCOTT, A. 2008. Regional-scale subsurface sand remobilization: geometry and architecture. *Journal of the Geological Society*, **165**, 609–612.

A new perspective on sequence stratigraphy of syn-orogenic basins: insights from the Tertiary Piedmont Basin (Italy) and implications for play concepts and reservoir heterogeneity

MASSIMO ROSSI* & JONATHAN CRAIG

Eni Upstream and Technical Services, Via Emilia 1, 20097 San Donato Milanese (Milano), Italy

**Corresponding author (e-mail: massimo.rossi@eni.com)*

Abstract: The Tertiary Piedmont Basin was reshaped as the Apennines progressively overrode the Alpine retroforeland. Nine basin-wide unconformities were followed by abrupt accommodation turnarounds or high-accommodation/high sediment supply intervals. Unusual stratigraphic discontinuities require a different perspective to reconcile allostratigraphy and sequence stratigraphy, highlighting a wider range of tectonically driven discontinuities (subaerial/subaqueous/regressive/transgressive), diachronism of accommodation and sediment–fairway turnarounds, unusual routing to deepwater, and relationships between fourth- and fifth-order sequences. This approach allows innovative play concepts and prediction of reservoir heterogeneity. Steep gradients and high sediment fluxes favour the development of coarse-grained deltas dominated by hyperpycnal processes and large-scale instability/sediment failures, often expressed by retrogressive slump scars, owing to over-steepening generated by drowning unconformities related to orogenic collapse or basin inversion. During increasing-accommodation intervals, subaqueous erosion leaves remnants of drowned deltas forming stratigraphic traps sealed by prodelta-slope muds. The sediments removed feed turbidites, generated during transgressive intervals and deposited at the toe of oversteepened slopes. These may eventually back-fill the slump scar system and show a fining-upward stacking-pattern owing to the re-establishment of the equilibrium profile. Another play concept relies on the combination of onlapping non-marine to marginal marine deposits against fault-bounded basement highs, sealed by marine muds deposited during the subsequent drowning.

The Alps–Apennines Junction: a different sequence stratigraphic perspective in structurally complex margins

Geological setting

The Tertiary Piedmont Basin (TPB) originated in the Alpine retroforeland after the mid-Eocene continental collision between the Adriatic Promontory and the European Plate. At present, it is developing at the Alpine–Apennine Junction (Figs 1–3), a complex 'knot' (Laubscher *et al.* 1992) where the Apennines progressively overrode the Alps retroforeland following the lateral substitution of the Alpine East-dipping subduction zone with the Apennine West-dipping subduction zone (Argnani 2012), causing: (1) the Oligocene orogenic collapse of Western Alps; and (2) the Late Oligocene–Miocene NW-ward propagation and oblique convergence of the Apennines (Ricci Lucchi 1986).

After the Late Rupelian, the Liguro-Provençal Basin opened (Seranne 1999). In the Early Miocene, rollback velocities significantly increased (Carminati *et al.* 2012), and the Apenninic slab retreat and consequent drifting of the Liguro-Provençal back-arc basin generated the counter-clockwise rotation of the Corse–Sardinia block and of the Apennines (Argnani 2012). After the Middle Miocene, the opening of the Tyrrhenian Basin generated a further counter-clockwise rotation of Apennines. In the TPB, rotations of approximately 50° took place (Gattacceca *et al.* 2007; Maffione *et al.* 2008). Muttoni *et al.* (2000) estimate a 24° rotation during Oligocene and Miocene for the Northern Apennines with the remaining rotation occurring during the Pliocene and Pleistocene.

The internal boundary of the Alpine axial sector lies within a broadly west–east-trending, 100 km-long left-lateral megashear belt (first documented by Rossi *et al.* 2009), accommodating the indentation of Adria and acting as a transfer zone between the Western Alps retrobelt and its southeastern prosecution, pivoting the counterclockwise rotation of Apennines and covered by the Upper Eocene to Miocene succession of the TPB. The present structural elevation of the different substrata

From: Bowman, M., Smyth, H. R., Good, T. R., Passey, S. R., Hirst, J. P. P. & Jordan, C. J. (eds) 2016. *The Value of Outcrop Studies in Reducing Subsurface Uncertainty and Risk in Hydrocarbon Exploration and Production*. Geological Society, London, Special Publications, **436**, 93–133.
First published online June 20, updated June 29, 2016, http://doi.org/10.1144/SP436.10

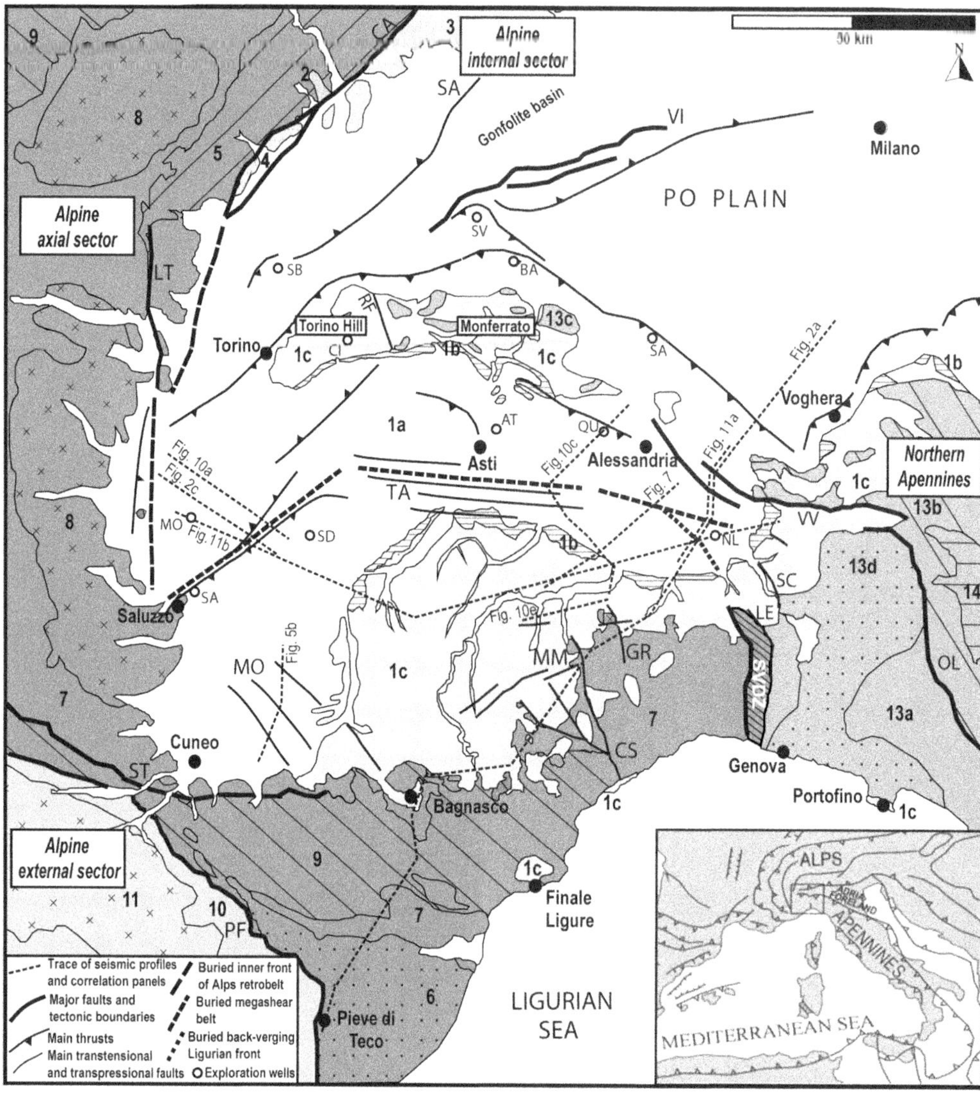

Fig. 1. Location map and structural framework of the Alps–Apennines Junction and of the Tertiary Piedmont Basin. Major faults and fault systems: CA, Canavese; SA, Salussola; VI, Villafortuna; LT, Lis-Trana; ST, Stura; OL, Ottone–Levanto; PF, Frontal Pennine Fault; SVDZ, Sestri–Voltaggio Deformation Zone; VV, Villavernia–Varzi; MO, Morozzo; CS, Celle–Sanda; GR, Grognardo; LE, Lemme; SC, Scrivia; MM, Merana–M. Acuto; TA, Tanaro; RF, Rio Freddo. Units: (1a) Pliocene to Holocene, (1b) Messinian, (1c) Upper Eocene to Tortonian, (2) Periadriatic Tertiary intrusives, (3) Southern Alpine units, (4) Canavese basement and cover units, (5) Sesia zone, (6) Helmintoid flysch units, (7) Piedmont–Ligurian ophiolite bearing nappes, (8) Internal crystalline massifs, (9) Gran San Bernardo nappe and relative Brianconnais sedimentary cover, (10) Dauphinois–Helvetic cover units, (11) Late-Hercynian deposits and basement units of the External crystalline massifs, (13) Ligurian units: (a) internal Ligurian units, (b) external Ligurian units, (c) Ligurian and chaotic complexes of Monferrato, (d) Antola unit, (14) Upper Eocene–Lower Miocene sub-Ligurian and Tuscan units (Bobbio window). Exploration wells: AT, Asti; BA, Balzola; CI, Cinzano; MO, Moretta; NL, Novi Ligure; QU, Quargnento; SA, Sartirana; SB, San Benigno; SV, Sali Vercellese; SA, Saluzzo; SD, Sommariva del Bosco.

decreases from the metamorphic Alpine axial zone to the non-metamorphic Cretaceous–Paleocene Ligurian flysch units in turn accreting over the Adria foreland (Figs 1 & 2). The Alpine axial belt corresponds in outcrop to the boundary between the Ligurian Alps and the Ligurian–Piedmont

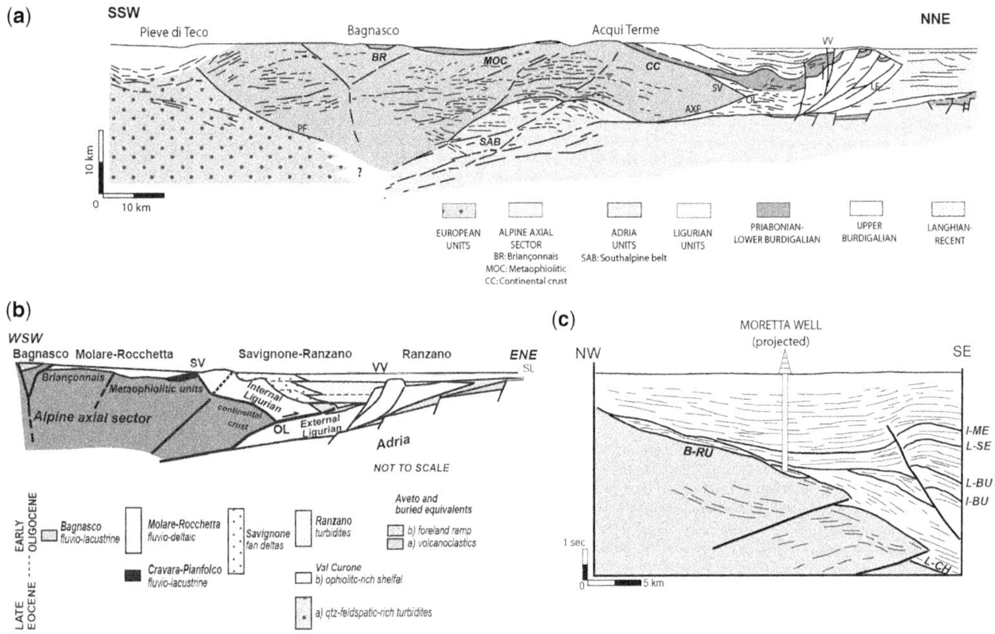

Fig. 2. Seismic line drawings and schematic cross-sections (from Rossi *et al.* 2009; Mosca *et al.* 2010) highlighting the main structural and stratigraphic features of Tertiary Piedmont Basin and surrounding areas of Western Alps and Northern Apennines (locations in Fig. 1). (**a**) Regional geological profile across the Alps–Apennines Junction (AXF Front of the Alpine axial zone, LF, Ligurian Front; OL, Ottone–Levanto Fault; PF, Penninic Front; SV, Sestri–Voltaggio Fault; VV, Villalvernia–Varzi Fault system); (**b**) Main Oligocene morphostructural and depositional features; (**c**) Seismic line drawing showing the timing and vergence of the Alpine and Apenninic thrust-fold belts (Bounding surfaces: B-RU, Base Rupelian; L-CH, Latest Chattian; I-BU, Intra-Burdigalian; L-BU, Late Burdigalian; L-SE, Late Serravallian; I-ME, Intra-Messinian).

Oceanic Nappes, locally recorded by west–east-oriented faults such as the Stura Fault (Capponi *et al.* 2009). The internal boundary of the Alps retrobelt south of the Insubric Fault lies mainly in the subsurface, where it is characterized by a broadly west–east-oriented high-angle transpressional fault (Fig. 1a–d), and then continuing in the Ottone–Levanto Fault in the easternmost TPB and its SE-wards prosecution (representing the boundary between Inner and Outer Ligurian Units).

During the Late Eocene and Early Oligocene (Fig. 3), the orogenic collapse following the collision between the Adria microcontinent and Europe generated a mosaic of partially interconnected sub-basins controlled by a transpressional–transtensional fault system mainly consisting of NW–SE to NNW–SSE-oriented transpressional faults (such as the Morozzo and Celle–Sanda fault systems, the Sestri–Voltaggio deformation zone, the Lemme, Scrivia and Grognardo faults) and WSW–ENE-oriented transtensional faults (such as the Merana– Monteacuto en-echelon fault system) associated with the west–east-oriented megashear belt (Mosca *et al.* 2010). At the boundary with the Adria foreland, the sinistral megashear is represented by the transtensional Villalvernia–Varzi fault system, that after Latest Oligocene was locally involved in the Apenninic frontal splays, so the constituent faults may have become transpressional (Fig. 2a, b).

After the Late Oligocene (Fig. 3), the NW–SE fault systems became predominant (Piana *et al.* 1997; Capponi *et al.* 2009) in relation to the development of incipient Apenninic compression. Around the Oligo-Miocene boundary, as the roll-back of Apennine subduction made the NE-verging Apenninic tectonics dominant, juxtaposed substrata accreted over the Adria foreland and the TPB developed as a wide and subsiding thrust-top basin, bounded to the north by frontal splays of the Monferrato that provided the separation from the adjacent Alpine retroforeland and to the northeast by the ensuing Apenninic foredeep.

The development of a deep-seated antiformal stack (Fig. 2a) took place during the Early Miocene (Fig. 3), although the western basin margin was still largely affected by reactivated structures of the Alps retrobelt while the Alpine subduction was

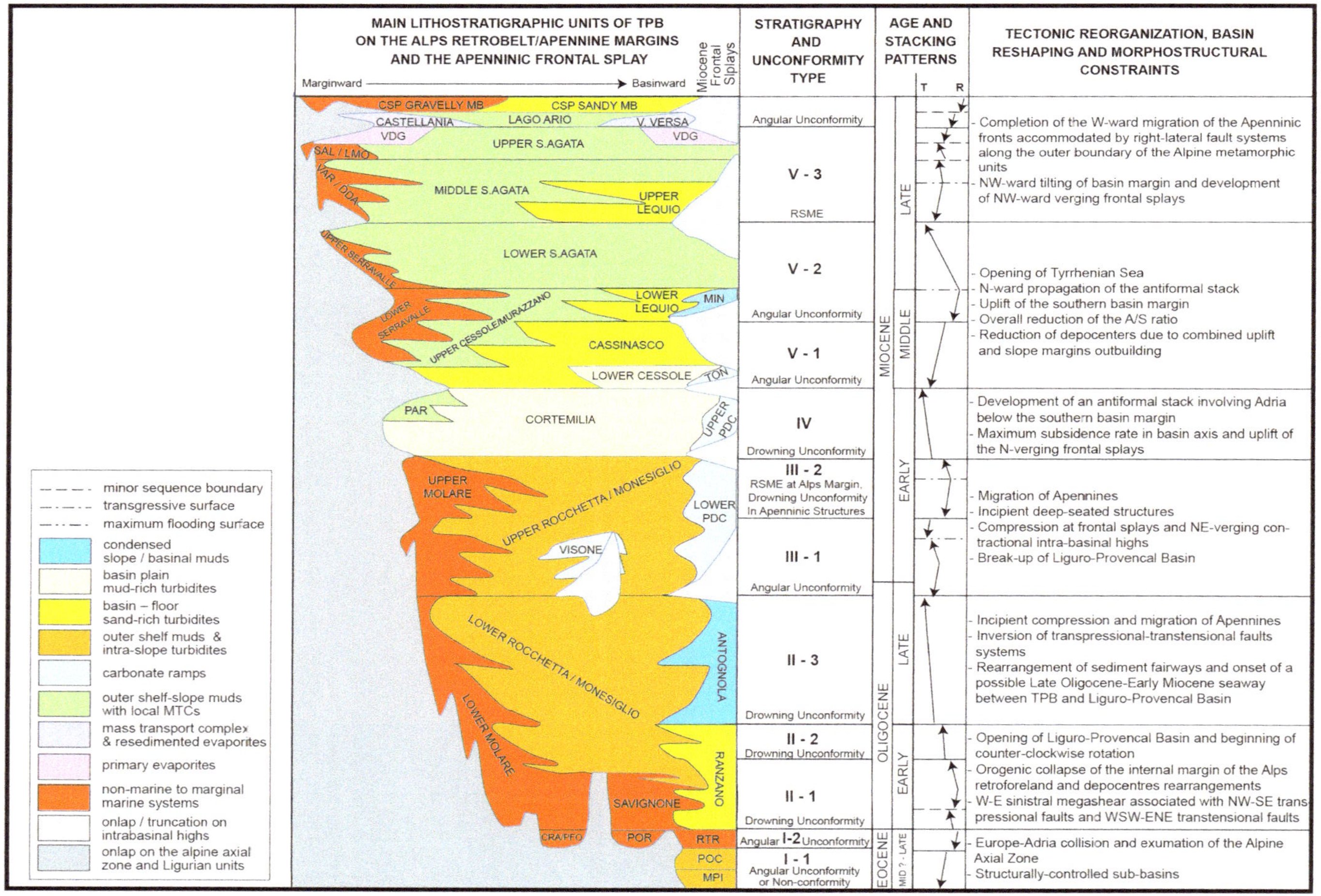

Fig. 3. Most representative lithostratigraphic units of Tertiary Piedmont Basin framed in the tectono-sedimentary evolution of the Alps–Apennines Junction; age, physical stratigraphy, T-R trends and tectonic reorganization are provided (for Upper Miocene lithostratigraphy, cf. Ghielmi *et al.* in prep.). Codes of formal and informal lithostratigraphic units: MPI, M. Piano; POC, Pizzo d'Oca; CRA, Costa Cravara; PFO, Pianfolco; POR, Portofino; RTR, Rio Trebbio; PDC, Pietra da Cantoni; PAR, Paroldo; TON, Tonengo; MIN, Mincengo; VAR, Vargo; DDA, Diano d'Alba; SAL, S.Alosio; LMO, La Morra; VDG, Vena del Gesso; CSP, Cassano Spinola.

progressively substituted by the Apenninic subduction. Since the Middle Miocene, the whole southern basin margin has been uplifted as the consequence of the northward propagation of the antiformal stack, locally associated with basin inversion driven by the northward tilting (Fig. 3). During the Late Miocene, the overriding of the Apennines on Western Alps was completed and NW-ward vergent Apenninic fronts of the Torino Hill accreted over the SE-vergent Alpine retroforeland (Figs 1, 2a–c & 3), pushed westwards and accommodated by the right-lateral reactivation of part of the Eo-Oligocene meagashear belt (Tanaro fault system, Ghielmi *et al.* in prep.), and associated with the NW-ward tilting of the southern basin margin.

Criteria for stratigraphic analysis in structurally complex margins

The tectono-sedimentary evolution of the TPB was controlled by changes in subsidence rates and shifts of depocentres in response to the combination of structural deformation and outbuilding of marginal clastic wedges. The Oligo-Miocene sedimentary succession defines a long-term broadly transgressive–regressive megacycle consisting of five unconformity-bounded stratigraphic units (UBSU *sensu* Salvador 1994) and nine allostratigraphic units (at the scale of allogroups, *sensu* NACSN 1983) that are the stratigraphic record of the Apennines overriding the Alps retroforeland. Within this framework, the bounding surfaces record changes in basin configuration (Figs 2–7). Alternating transtension, transpression and compression in both space and time has determined changes in the rate of tectonic subsidence leading to dramatic changes in accommodation space and to different types of unconformities and stacking patterns. Depositional systems and their internal architectures show considerable variation, mainly as a result of the interplay between high-frequency cyclicity and low-frequency sequences controlled by long-term changes in subsidence rates. This makes the Tertiary Piedmont Basin an ideal area in which to identify outcrop analogues according to variable tectonic and depositional styles in terms of play concepts and their internal reservoir architectures. Under these conditions, two main types of stratigraphic discontinuities can be identified, alternating in time and/or space as a consequence of the creation or elimination of accommodation space. When the rate of tectonic subsidence outpaced the rate of sedimentation, an overall transgression is recorded; differential subsidence and the downwarping of basin margins locally associated with faulting may have generated drowning unconformities characterized by the truncation of shallow marine deposits unconformably overlain by deeper-water facies. On the contrary, when the rate of tectonic subsidence decreased, an overall regression is recorded; uplift and inversion may have generated erosional truncation of deepwater sediments unconformably overlain by shallow-water deposits. The palaeobathymetric changes associated with these unconformities are often very abrupt and may exceed several hundreds of metres.

The sediment flux derived from the dismantling of the Alpine and Apennine orogenic belts was commonly in phase with uplift because of the greater erosion this produced. The timing of highest sediment flux could, however, be delayed until after the main uplift pulses, or climatic deterioration may have periodically resulted in a decrease of the accommodation/sediment supply ratio.

In addition, significant changes may be recorded at the same time on the different margins, particularly in relation to differences between areas adjacent to the Alps and intra-basinal highs generated by the Apenninic deformation. This includes the accumulation of shallow-water deposits bounded by subaerial unconformites and regressive surfaces of marine erosion along the southwestern inner basin margin that are coeval with unconformable boundaries, in places largely erosional, on the flanks of intrabasinal highs that can be interpreted as drowning unconformities.

Since very local correlative conformable surfaces occur, allostratigraphy is more suitable than sequence stratigraphy for describing the basin-fill architecture. In addition, classic sequence stratigraphic architecture cannot be satisfactorily used in these settings, where the vertical succession and internal organization of systems tracts may differ substantially from the usual models because the magnitude of accommodation changes is primarily driven by tectonic subsidence.

Published research on drowning unconformities has so far focused mainly on the drowning of carbonate platforms (Schlager 1999). On the contrary, little work has been done so far on drowning unconformities within clastic successions, despite the widespread occurrence of tectonically active geological settings characterized by abrupt and high-magnitude changes in the rate of tectonic subsidence (Vail *et al.* 1991), suggesting that these unusual stratigraphic discontinuities tend to be overlooked. In clastic systems, very different controlling factors are implied in terms of the mode and potential of erosion, depositional processes and sediment partitioning, which often result in the development of submarine erosional surfaces instead of merely producing condensation and omission surfaces. Contributions on this topic have been provided by Rossi *et al.* (2009), Mosca *et al.* (2010) and Salazar *et al.* (2011) for areas undergoing orogenic collapse and mega-shear zones at the junction between

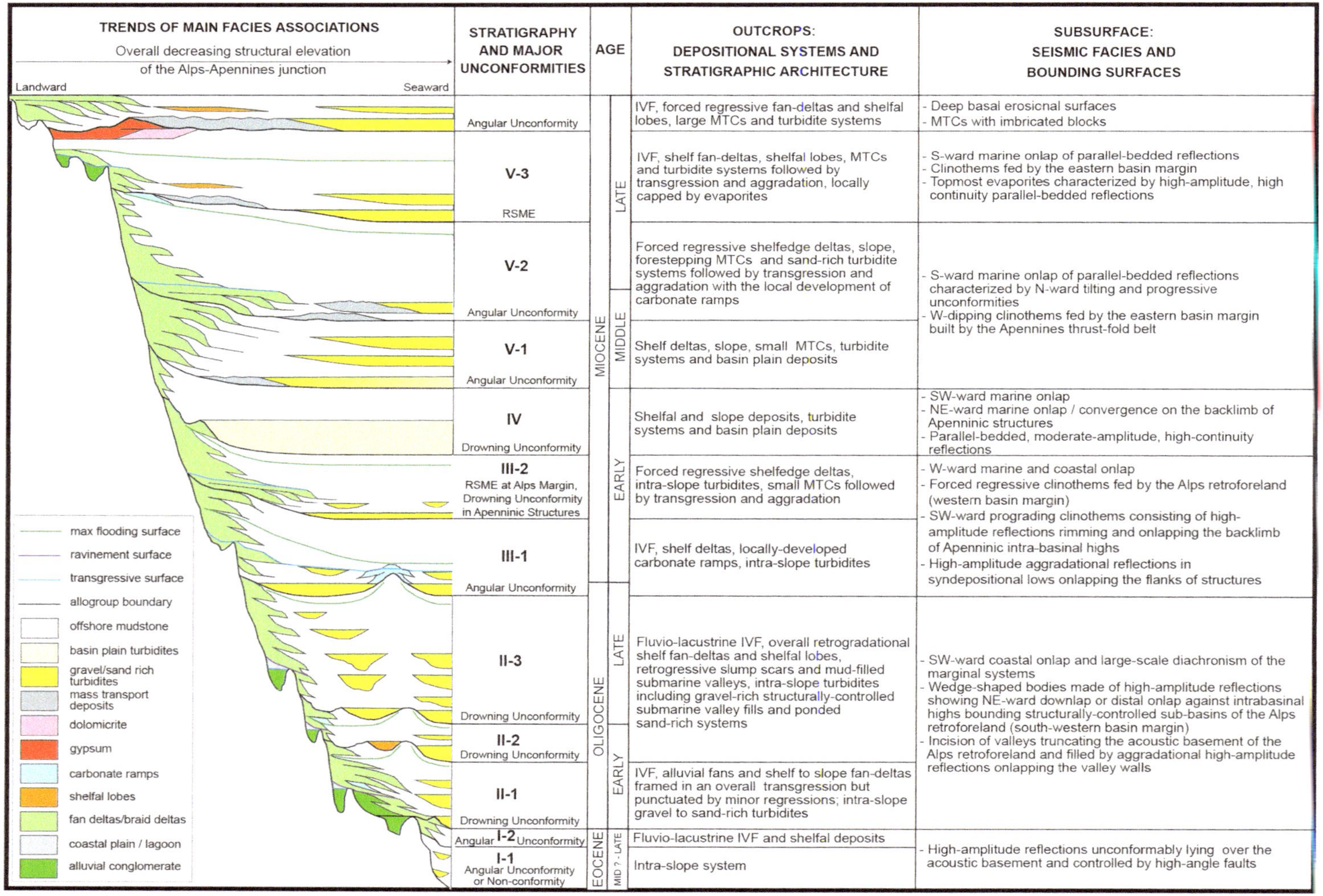

Fig. 4. Main stratigraphic and sedimentological features of the Tertiary Piedmont Basin. Trends of main facies associations, age, physical stratigraphy, outcropping depositional systems and buried seismic facies are provided.

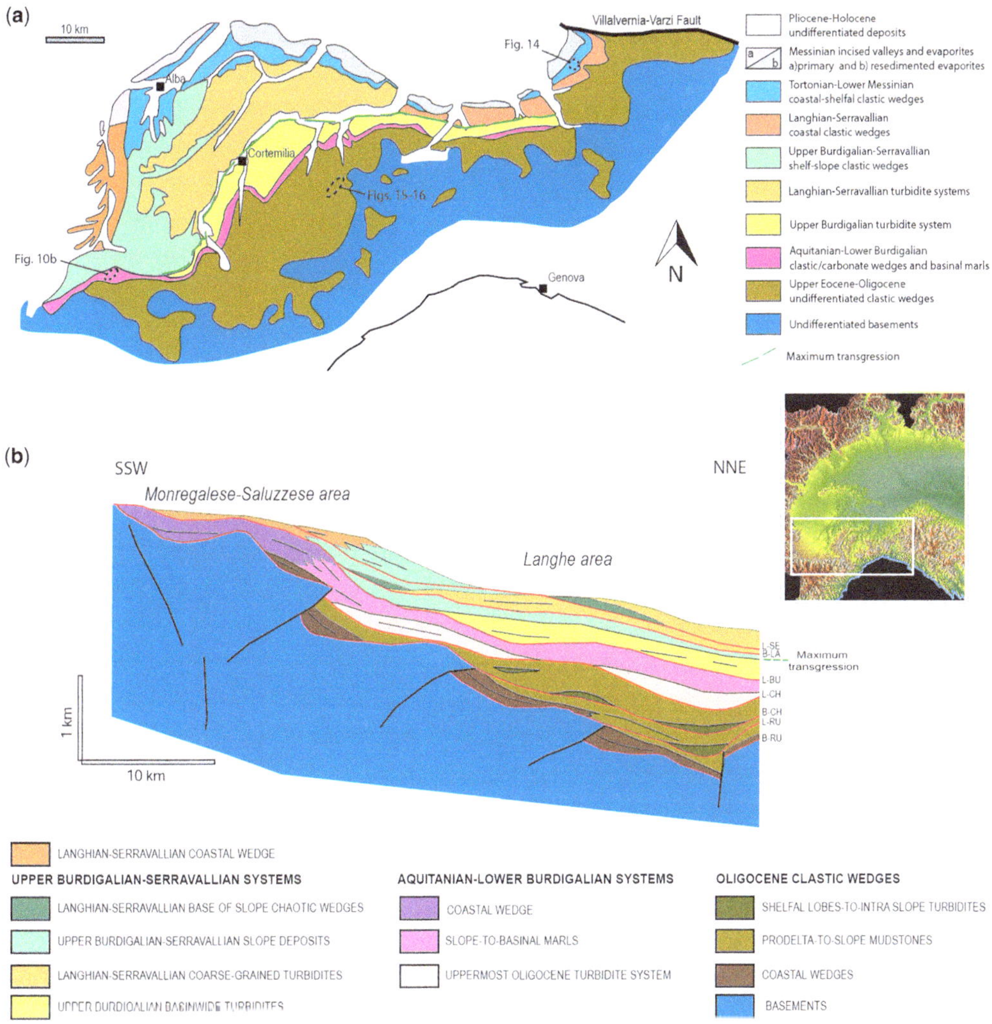

Fig. 5. Physical-stratigraphic framework (modified from Rossi *et al.* 2009). (**a**) Geological map of the Tertiary Piedmont Basin (TPB); dashed circled areas refer to the location of the outcrops shown in Figures 10 and 14–16. (**b**) Stratigraphic architecture of the Oligo-Miocene succession, highlighting key-surfaces and depositional geometries derived from a seismic line drawing (location in Fig. 1). Note: (1) the morphostructural elements controlling the sedimentation; (2) the diachronism of marginal depositional systems.

different orogenic belts or plate margins, by Mutti *et al.* (1994), Rossi *et al.* (2002), Covault & Graham (2010), Moscardelli *et al.* (2012), Ghielmi *et al.* (2013), Zecchin & Tosi (2014) and Rossi *et al.* (2015) for foreland basins and by Gawthorpe *et al.* (1994), Hansen *et al.* (2004), Martins-Neto & Catuneanu (2010) and Bell *et al.* (2014) for rift basins and epicontinental seas. In all these settings a different approach is required, integrating allo-stratigraphic and sequence stratigraphic concepts capable of framing high-resolution sequence stratigraphy within large-scale tectono-stratigraphic units.

This has important implications on all phases and disciplines of petroleum exploration and development, and at all scales from basin modelling to reservoir characterization. New criteria are required, with particular attention to the identification of: (1) a wider range of erosional bounding surfaces (subaerial v. subaqueous, and regressive v. transgressive); (2) the possible diachronous nature of changes in accommodation space and gateway turnarounds; and (3) unusual relationships between stratigraphy, stacking patterns, modes of sediment supply and sediment routing to deeper waters. This

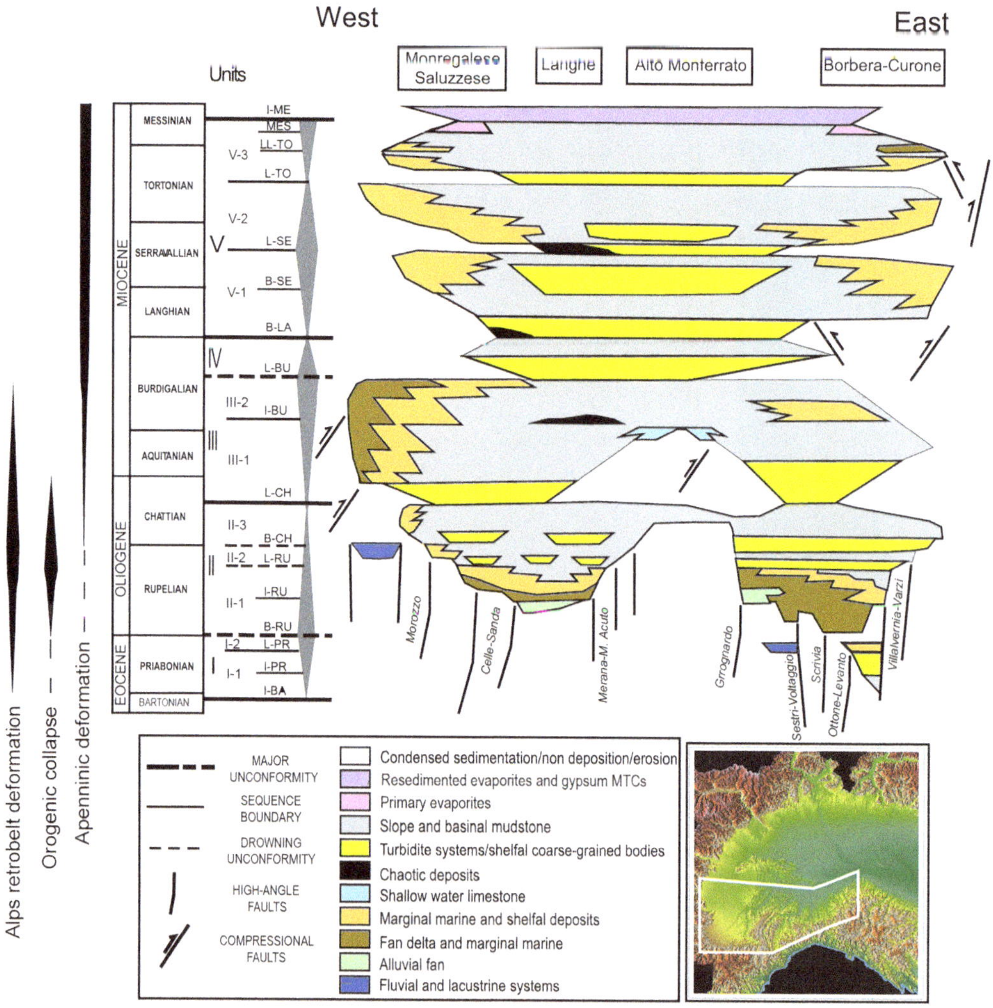

Fig. 6. Late Eocene–Miocene tectono-stratigraphic framework of Tertiary Piedmont Basin (modified from Rossi *et al.* 2009). Grey triangles define regressive v. transgressive stacking patterns. Thick solid lines mark the boundaries of major unconconformity-bounded stratigraphic units; intermediate lines mark the boundaries of allogroups (solid lines represent sequence bounderies; dashed lines represent drowning unconformities); thin lines mark minor sequence boundaries. I-BA, Intra-Bartonian unconformity; I-PR, Intra-Priabonian sequence boundary; L-PR, Late-Priabonian allogroup boundary; B-RU, Base Rupelian unconformity; I-RU, Intra-Rupelian sequence boundary; L-RU, Late Rupelian allogroup boundary; B-CH, Base Chattian allogroup boundary; L-CH, Latest Chattian unconformity; I-BU, Intra-Burdigalian allogroup boundary; L-BU, Late Burdigalian unconformity; B-LA, Base Langhian unconformity; B-SE, Base Serravallian sequence boundary; L-SE, Late Serravallian allogroup boundary; L-TO, Late Tortonian allogroup boundary; LL-TO, Latest Tortonian sequence boundary; MES, Messinian sequence boundary; I-ME, Intra-Messinian unconformity.

may contribute to delineate a more comprehensive approach potentially capable of reducing pitfalls generated by current sequence stratigraphic models biased by common assumptions and paradigms. In addition, little work has been done so far on sequence organization in relation to: (1) the symmetry v. asymmetry of a complete sea-level cycle controlled by greenhouse v. icehouse conditions, respectively; and (2) the gradual v. abrupt changes in a complete sea-level cycle controlled by eustatic v. tectonic mechanisms, respectively.

In this paper we describe the architecture of the sediment-fill of the TPB and its relation to changes in structural style through the integration of subsurface and outcrop data on the basis of seismic- and sequence-stratigraphy principles,

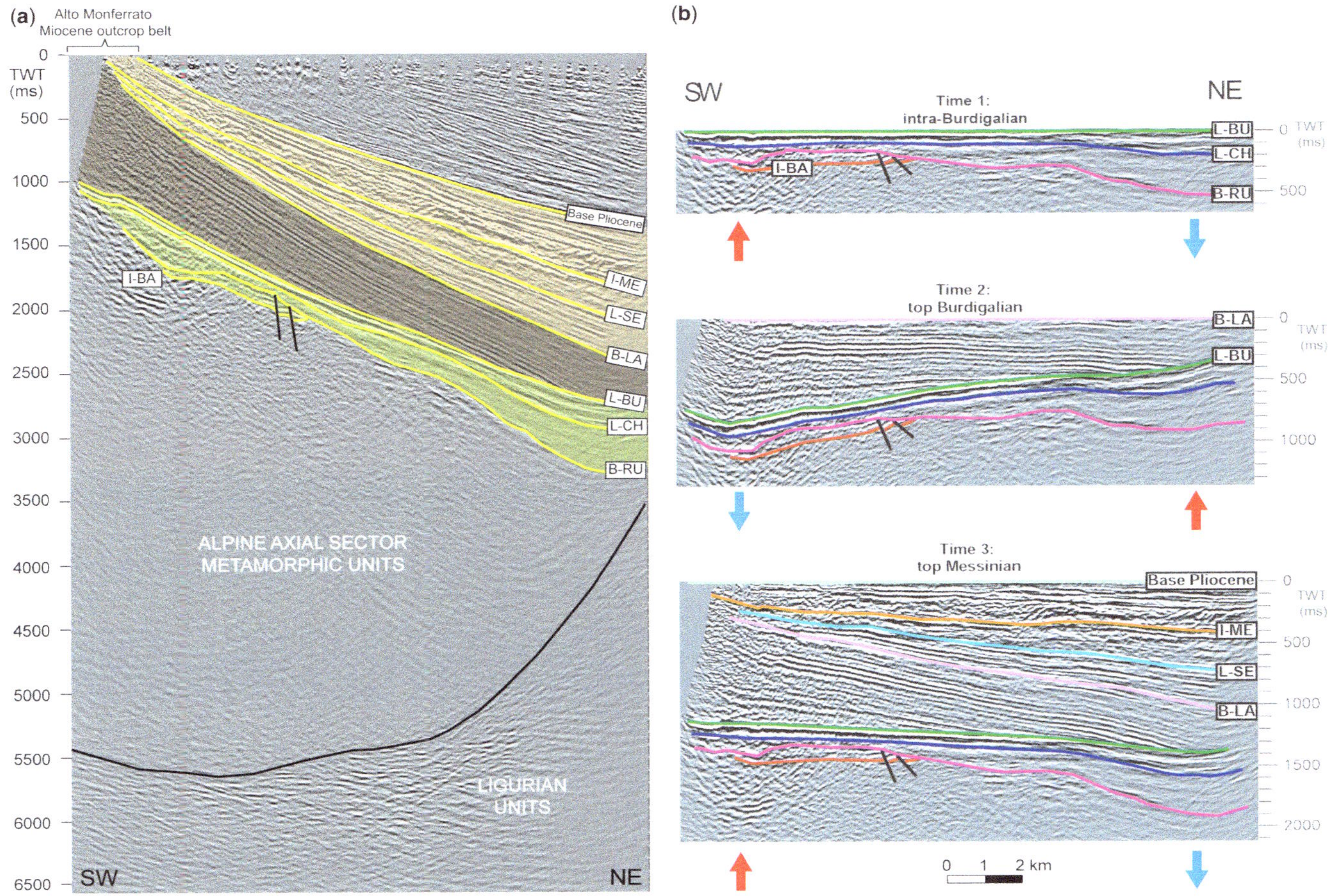

Fig. 7. (**a**) Seismic profile (location in Fig. 1) showing: (1) the buried internal margin of the Alps retroforeland, bounded by a high-angle transpressive fault separating the Alpine axial meta-ophiolitic units from the Ligurian units; (2) three Oligo-Miocene sedimentary wedges recording the collision and two subsequent major basin inversions affecting this margin (the outcrop-seismic tie is shown). (**b**) Three successive seismic flattenings on a portion of the seismic profile shown in (a), aimed at unravelling the tectono-sedimentary evolution of TPB. The initial basin configuration was controlled by a general accommodation increasing toward the less elevated morphostructural belt to the NE. The Late Burdigalian maximum accommodation interval records the first main basin inversion and generated an asymmetric basin wedging-out NE-wards. After Middle Miocene, the second main basin inversion took place, giving way to the uplift of the previous depocentre. Modified from Rossi *et al.* (2009).

respectively (Catuneanu *et al.* 2009 and references therein).

The characteristics of a variety of subaerial and subaqueous stratigraphic discontinuities as they appear in outcrop and in their buried prosecution are addressed, comparing their behaviour with the relevant tectonic constraints and discussing the implications for innovative play concepts and for reservoir architectures. In addition, a comparison is proposed with selected subsurface examples in prolific basins from different geological settings at both the exploration and reservoir scales.

Allostratigraphy, sequence stratigraphy and depositional systems

The major bounding surfaces are characterized by angular stratal relationships associated with erosional and/or non-depositional events, often marked by biostratigraphic gaps and/or hiatuses, respectively. These boundaries may pass laterally into paraconformable surfaces, while surfaces of correlative conformity rarely occur, these surfaces being generated by changes in basin configuration. They include both unconformities generated by relative sea-level falls and drowning unconformities generated by abrupt accommodation increases (Figs 2–7). The former are associated with periods of reduced regional accommodation space generally owing to low subsidence or uplift and are often expressed by forced regression wedges at the basin margins. The latter are associated with periods of major accommodation space-creation owing to increasing rates of regional subsidence as a result of downwarping of the basin margins. This led, respectively, to the generation of regressive and transgressive cycles of the order of 2 km thick in the depocentres and with an average duration of 5 myr. The overall trend in each unit is recorded by the stacking of the component depositional sequences bounded by unconformities towards the basin margins and correlative continuity surfaces towards the depocentres, the average time span of which normally does not exceed 1 myr.

The Late Eocene–Miocene succession in the Tertiary Piedmont Basin records the continental collision followed by the Alpine orogenic collapse and the progressive development of Apenninic deformation. The most evident stratigraphic surface that marks this transition is the Late Burdigalian maximum transgression, which coincides with the maximum rate of tectonically driven accommodation space creation, associated to the deposition of a 1 km-thick, highly efficient, basinwide turbidite system. This system separates the older SW-ward coastal onlap and aggradation of marginal marine systems related to the Alpine orogenic collapse (recorded by a predominantly transgressive succession), followed by the incipient uplift of the Apennines, from the younger outbuilding related to the Mid-Miocene uplift along both the southwestern and southeastern basin margins, which is affected by progressive unconformities and growth strata in the frontal splays and is characterized by a predominantly regressive succession (Figs 3–7). Second-order peak regressions, often associated with forced regressions, occurred during the Latest Priabonian, Early Burdigalian, Late Serravallian and Late Tortonian, whilst second-order peak transgressions occurred during the Late Chattian, Late Burdigalian and Early Tortonian.

The allostratigraphic units are briefly described, in ascending stratigraphic order. Their main features in both outcrop and subsurface, as well as the relationships with lithostratigraphic units and the morphotectonic evolution, are summarized in Figures 3 and 4. In Figure 3, the relationships between regional/local tectonic setting, stratigraphy/bounding surfaces, stacking patterns and diachronous lithostratigraphic units are shown in relation to the Alps retrobelt/Apenninic margin and Miocene intra-basinal highs. In Figure 4, the relationships between stratigraphy, transgressive v. regressive attitudes and related diachronism of depositional systems are shown in relation to the differential subsidence affecting the marginal hinge of the Alps–Apennines Junction referred to the whole Alps retroforeland–Apennine system.

Unit I (Upper Eocene) followed the continental collision and is mainly recorded in the eastern TPB (Borbera–Curone area in Figs 3, 4 & 6) by an overall forward-stepping succession resting unconformably (above the Intra-Bartonian Unconformity, I-BA) over Cretaceous–Paleocene flysch deposits (Ligurian Units) accreted in an area of low structural elevation east of the Alpine axial sector that bounds the south- and east-verging Alpine retroforeland. These Upper Bartonian (?) to Priabonian deposits consist of basinal mudstones and quartz-feldspathic turbidites (Allogroup I-1), unconformably overlain by Late Priabonian marginal marine ophiolitic-rich sandstones (Cavanna *et al.* 1989; Di Giulio 1989; Mutti *et al.* 1995, 2002) of the Allogroup I-2 (Figs 3 & 4). Towards the west, in a more marginal setting, the deposition was confined to small fault-bounded, extensional to strike-slip sub-basins (Fig. 7), where palaeovalleys and alluvial fan deposits overlain by fluvio-lacustrine facies were deposited locally nonconformably over the Alpine axial zone consisting of metaophiolitic rocks (Lorenz 1969).

Unit II (Oligocene) records most of the Alpine orogenic collapse. The lower boundary, representing the base of the lower allogroup, is marked by a drowning unconformity in the east, where a turbidite complex unconformably overlies (above the Base–Rupelian Unconformity, B-RU) Priabonian shallow marine deposits. Towards the west, marginal marine Oligocene deposits unconformably to disconformably overlie the Priabonian deposits, the Cretaceous–Paleocene flysch or the ophiolites of the Alpine axial zone (Turco *et al.* 1994).

The deepening-upward Oligocene succession was accommodated in fault-bounded depocentres driven by transpressional–transtensional tectonics. The succession, which progressively onlaps towards the south and west (i.e. toward the Alpine axial zone), records a marked SW-ward diachronism (Figs 3–6) of fluvio-deltaic deposition (first documented by Rossi *et al.* 2009) related to regional transgression from the NE (Figs 3 & 4). The overall transgression is punctuated by drowning unconformities generated by basinward tilting and high-angle synsedimentary faults. Deposition during Early Rupelian (Allogroup II-1) was dominated by alluvial fan-to-fan delta conglomerates overlain by paralic, marginal marine and shelfal facies driven by a higher-frequency cyclicity (Fig. 9). The sedimentation was controlled by the interplay of NW–SE- and WSW–ENE-oriented fault systems. The WSW–ENE system consists of transtensional en-echelon faults accommodating left-lateral displacements, while right-lateral transpression is recorded by NW–SE to NNW–SSE faults (Fig. 9). Sediment entry points are preferentially located at the intersection of these fault systems and mainly controlled the deposition of alluvial fans and fan deltas. Transfer faults acted as wave-dominated intra-basinal divides during transgressions (Rossi *et al.* 2009; Ghibaudo *et al.* 2014), separating open marine from paralic environments (Figs 8 & 9). During major periods of 'maximum flooding', these major highs became sites of condensed coral–algal carbonate sedimentation (Rossi *et al.* 2009). Syn-depositional lows were mainly sites of coarse-grained delta front deposits consisting of stacked coarse-grained graded beds with occasionally floating clasts and capped by bioturbated fine sandstone and siltstone.

The complex structural pattern resulted in an irregular shoreline, with dispersed palaeocurrents and fault-controlled embayments (Fig. 9) which locally deflected the sediment dispersion towards the west. The 'Intra-Rupelian sequence boundary' (I-RU) marks a minor regressive episode associated with fault-bounded incised valleys. NE-wards, fan deltas pass basinward into slope and turbidite systems closing northwards against the Adria foreland bounded by a high-angle extensional to transtensional fault system.

Allogroup II-2 (bounded at the base by the Late Rupelian Unconformity, L-RU) records a step-wise transgression associated with multiple drowning unconformities driven by flexural tilting associated with the orogenic collapse and locally enhanced by the reactivation of the WSW–ENE transfer faults within the left-lateral megashear belt (Fig. 8). This led to a marginward (SW-ward) shift of the fluvio-deltaic systems (Fig. 9), which pass basinward into shelfal gravity flow-dominated coarse-grained bodies (catastrophic systems *sensu* Mutti *et al.* 1996) encased in marine mudstones (Ghibaudo *et al.* 2014) and with palaeocurrents that may have been forced toward the East owing to the local inversion of the axial plunging of the Merana–Monteacuto transtensional fault system characterized by a 'scissor like' behaviour (Rossi *et al.* 2009).

This culminated in the deposition of Allogroup II-3, overlying the Base-Chattian Unconformity (B-CH), which records the onset of a period of maximum accommodation space, mostly expressed by a drowning unconformity often overlain by turbidite systems filling syn-depositional depressions and submarine valleys (Fig. 8). The main control on sedimentation at this time was the NNW–SSE fault system that controlled the basin orientation and affected sediment dispersion to turbidite bodies (Gelati & Gnaccolini 1998; Gelati *et al.* 2010*a*, *b*) and that represent the first, localized, evidence of a basin reorganization (Fig. 9) that was probably triggered by the incipient Apenninic deformation and the rotation owing to the opening of the Liguro-Provençal back-arc basin, whose NE apex was possibly recorded by shallow-water facies outcropping along the west Liguria coast. This was associated with the drowning of part of the southern basin margin and the deactivation of the sediment supply from the southeast. Palaeocurrents were mainly from the SW and NW (Cazzola & Fornaciari 1992; Bersezio *et al.* 2009), that is, from the Alpine axial zone, owing to the regional westward transgression.

Unit III (Uppermost Chattian–Lower Burdigalian) followed the onset of the NE-ward propagation of the Apenninic compressional deformation, which led to a reorganization in basin configuration and is marked by the L-CH Unconformity, which also represents the base of the lower allogroup (Allogroup III-1, Uppermost Chattian-Lowermost Burdigalian). This boundary is overlain by an abrupt shallowing event along the basin margins (Monregalese–Saluzzese area in Figs 5 & 6) and over major intra-basinal highs (Alto Monferrato in Fig. 6), where shallow-marine systems unconformably overlie the Upper Oligocene succession deposited during the period of maximum accommodation space at the top of the previous unit (Figs 3 & 4). The deep-water succession was dominated by turbidite systems fed mainly from the west and the NW (Gelati & Gnaccolini 1998; Ghibaudo *et al.* 2014). Along the western Alpine margin and the adjacent

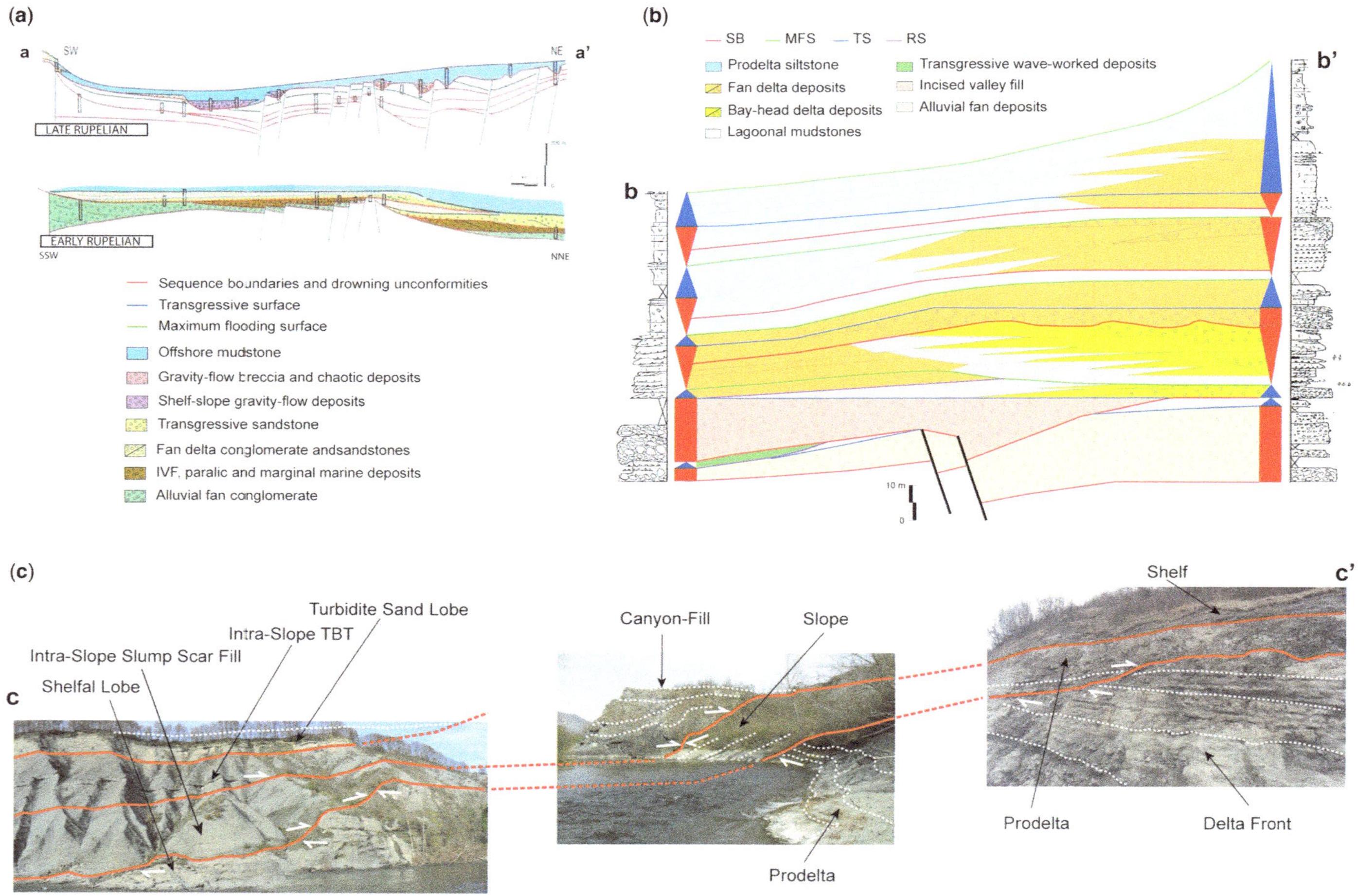

Fig. 8. Stratigraphic architecture of Oligocene deposits of TPB. (**a**) Early and Late Rupelian morphostructural and depositional features (location a–a′ in Fig. 9a). (**b**) Correlation panel showing the stratigraphic architecture of the Lower Rupelian sequences (location b–b′ in Fig. 9a). (**c**) Outcrop expression of the basinward changing character of the Late Rupelian Unconformity and of the underlying Lower Rupelian deposits. A Chattian turbidite sandstone lobe can be observed in the northern area (location c–c′ in Fig. 9a).

South Alpine fold-and-thrust belt, the deposits of Unit II are unconformably overlain by incised valleys (Gelati & Gnaccolini 1996) filled by fluvial underflows affected by wave reworking along the valley flanks, and by fan-deltas that prograded eastwards (Fig. 10a, b). The uplift of the frontal splays to the north provided an obstacle that deflected the palaeocurrents eastwards along the plunging axis of the basin. The occurrence of sand-rich deposits in the NW- and SE-ward directed palaeocurrents of many turbidite systems suggest that sediment was also supplied from the NW. This is also supported by the closure of the seaway that existed previously towards the Southern Alps foredeep.

The shape, facies and palaeocurrent directions (Felletti 2002) of the easternmost sector of the TPB suggest the presence of a transit zone towards the Bobbio area for sediment delivered to the Northern Apennines foredeep since the Early Miocene (Mutti 1975; Labaume & Rio 1994). The transtensional fault system along the southern edge of the Adria foreland was involved in the Apenninic tectonics and formed the transpressional southern boundary of the north-verging frontal splays (Figs 2 & 11). At a regional scale, a major sinistral displacement was accommodated along this fault system (e.g. also Laubscher *et al.* 1992), so that the megashear zone previously affecting the Alpine retroforeland partly changed its configuration (Ghibaudo *et al.* 2014). During the Early Miocene, wave-worked foramol ('temperate-type') carbonate platforms developed locally, unconformably overlying Oligocene slope mudstones, rimming the major syn-depositional highs and the Monferrato frontal splays (d'Atri 1990, 1995; Dela Pierre *et al.* 2002), and connected with the adjacent basinal areas by ramp deposits (Fig. 10d, e; Rossi *et al.* 2009).

Allogroup III-2, marked by the Intra-Burdigalian Unconformity (I-BU), records a forced regression in the marginal belts adjacent to the Alps (Figs 4, 6 & 10). In the eastern sectors affected by incipient Apenninic deformation, south- to SW-ward tilting of the backlimbs of several structures produced drowning unconformities recording the collapse of the previous intra-basinal highs, leading to their destabilization and partial cannibalization (Figs 6 & 10).

Unit IV (Upper Burdigalian) corresponds to a period of maximum accommodation space probably generated by the emplacement of a deep-seated antiformal stack below the southern margin (Figs 3 & 4). This major basin reorganization marks an acceleration of subsidence rates corresponding with the maximum transgressive interval at the scale of the whole basin-fill, recorded by a basin-wide, highly efficient turbidite system (Gelati & Gnaccolini 2003) that onlaps the flanks of the basin and the syn-depositional highs (Figs 3–7 & 10).

Lap-out terminations and lateral fringing toward intra-basinal highs occurs where basin plain turbidites onlap rapidly deepening-upward carbonate ramps. The lower boundary (L-BU surface) is a major drowning unconformity (Rossi *et al.* 2009), locally recording a basin inversion along the high-angle fault bounding the internal margin of the Alpine axial zone, the Scrivia fault and the Villalvernia–Varzi fault system (Figs 2, 6, 7 & 11).

With the deposition of this succession, a more regular physiography was established within a larger and more uniform basin, bounded to the north by the uplift of the frontal splays, where condensed successions were deposited. The marginal belts were located to the SW, where shallow-water sequences were deposited, defining during major transgressions a possible sea-way open to the northernmost part of the Liguro-Provençal Basin.

Unit V (Langhian–Messinian p.p.) records the uplift and inversion that started to affect the southeastern and southwestern parts of the basin owing to the northward propagation of the antiformal stack. Chaotic complexes developed locally to the SW. Subsequently, uplift extended to the whole southern basin margin, and a generalized northward tilting occurred, leading to progressive unconformities, and the narrowing and shifting of turbidite depocentres as a result of continuous tilting combined with the outbuilding of prograding clastic shelves (Figs 3–7 & 11). This unit is characterized by an overall regression, punctuated by phases of forced regression, indicating decreasing rates of subsidence and uplift, alternating with periods of relative sea-level rise (Rossi *et al.* 2009).

The initial uplift is marked by the Base–Langhian (B-LA) Unconformity (also representing the base of Allogroup V-1). In the south-central sector of the basin subsidence was continuing and decreased only later, marked by the development of a minor 'Intra-Langhian sequence boundary'. Along the southeastern margin of the basin, uplift associated with the innermost Apenninic arcs caused a dramatic truncation of the Upper Burdigalian basin plain turbidites, which are unconformably overlain by shelfal siltstones. Along the southwestern margin, the uplift is recorded by the onset of a large-scale progradation associated with a change in the toe of slope gravity flow systems from sand-rich to gravel-rich. The main depocentre shifted northwards (Fig. 7), where the turbidite systems changed from mud-rich and highly efficient to sand-rich and poorly efficient (Figs 3–6 & 11). A discrete forestepping pulse was associated with the relatively minor 'Base–Serravallian sequence boundary' (B-SE).

Allogroup V-2, unconformably overlying the Late-Serravallian Unconformity (L-SE), is characterized by progressive unconformities within

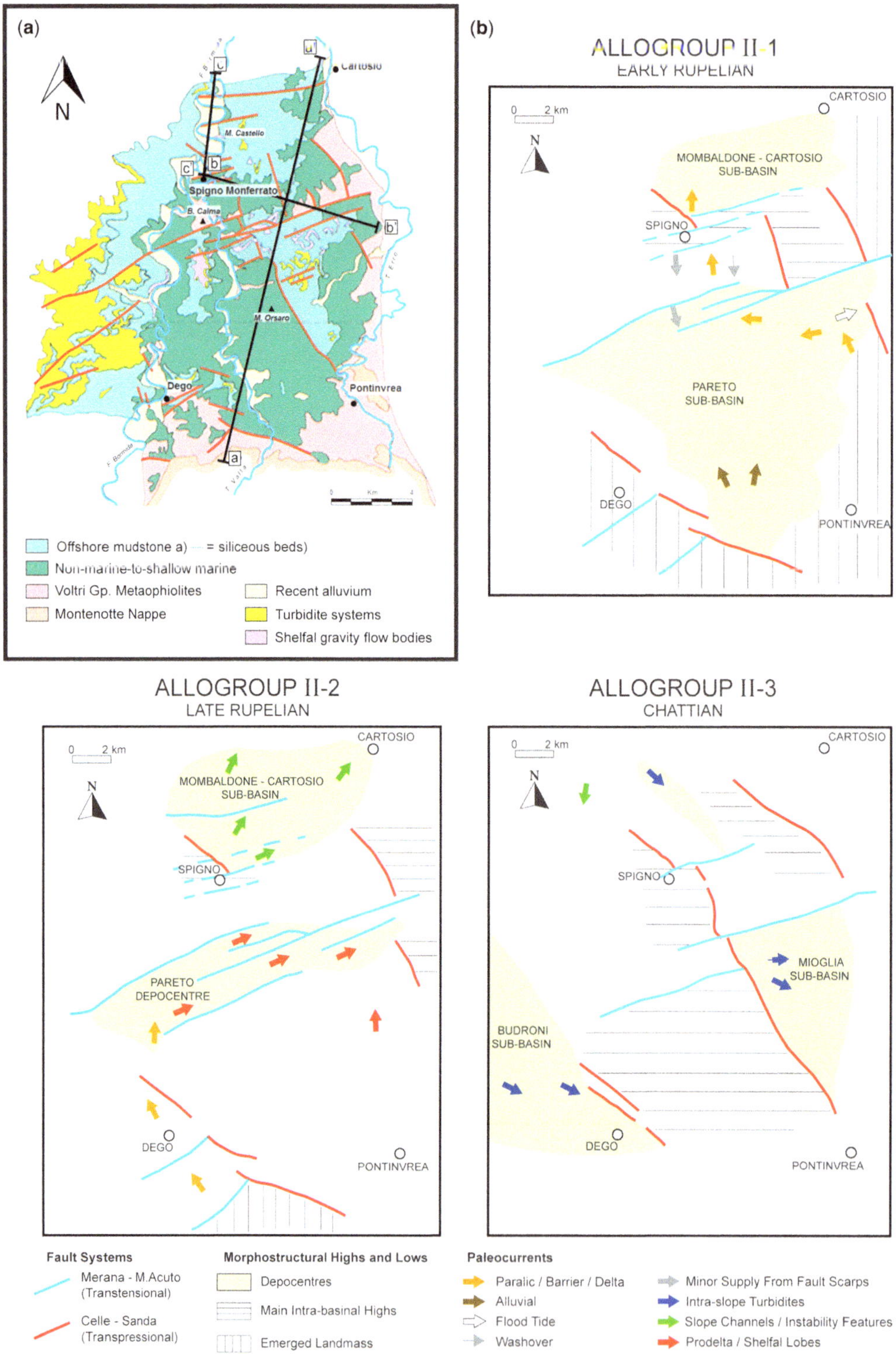

Fig. 9. (**a**) Geological map of the Oligocene succession between the Bormida and Erro Valleys. (**b**) Morphostructural reshaping and syn-depositional faults active during the sedimentation of Allogroups II-1 (Early Rupelian), II-2

fluvio-deltaic systems fed from the western, southern and eastern basin margins. The major depocentres for the turbidite systems shifted progressively northward owing to the combination of marginal uplift, basinward tilting and outbuilding of prograding wedges (Rossi *et al.* 2009). The L-SE surface marks the maximum rate of relative sea-level fall and the maximum development of lowstand forced regression dominated by catastrophic fluvial floods (*sensu* Mutti *et al.* 1996), capped by late lowstand aggradation and progradation coinciding with the maximum regression. Forestepping mass-transport complexes developed along the southwestern slope when the rate of progradation significantly increased. The Maximum Regression Surface (or Transgressive Surface, TS) coincides with the base Tortonian and is abruptly followed by a succession that records a deepening from shelfal siltstones to slope mudstones (Figs 3, 4, 6 & 11).

Allogroup V-3, bounded at the base by the Late Tortonian Unconformity (L-TO) that cuts into the underlying Lower Tortonian maximum transgression, is framed in the NW-ward migration of the Apenninic deformation, accommodated by right-lateral fault systems that reactivated part of the older sinistral megashear belt (Ghielmi *et al.* in prep.). This allogroup is characterized by lowstand fan deltas and incised valleys controlled by syn-depositional lows in synclines between the Apenninic thrusts, which pass basinward into shelfal lobes built by hyperpycnites fed by fluvial underflows or into shelf-edge collapsed delta front to pro-delta slumped facies (Rossi *et al.* 2009). Shoreface sandstones, lying unconformably above a regressive surface of marine erosion, rim the syn-depositional highs lateral to the main fairways and are overlain by transgressive surfaces of marine erosion consisting of slump scars that remove part of the shoreface sandstone, related to the northward propagation of deformation and the consequent collapse of the backlimb of the frontal splays (Figs 6 & 11).

A 'latest Tortonian sequence boundary' (LL-TO) is marked by the another phase of valley incision, while a 'Messinian sequence boundary' records the onset of the Mediterranean Salinity Crisis, whose initial stage is characterized in the TPB by the deposition of marginal primary gypsum diachronously passing basinward into dolomicrites and anoxic mudstones (Dela Pierre *et al.* 2011). The top of the allogroup is marked by the major Intra-Messinian Unconformity (Figs 3 & 4), overlain by mass-transport complexes, in turn unconformably overlain by fan-delta systems and the associated shelfal lobes (Ghielmi *et al.* in prep.).

Discussion

Bounding surfaces

The Alpine and Apennine mountain belts and their deformation history strongly influenced sedimentation in the Tertiary Piedmont Basin, by controlling: (1) the elevation and gradients of the basin margins; (2) the location of the main entry-points and sediment routes to deeper waters; and (3) the extent, shape and depth of the deep-water depocentres. Large-scale tectono-stratigraphic units are controlled by long-term changes (second to third order) of tectonic accommodation rates (Vail *et al.* 1991) and their boundaries are generated by stages of basin reorganization. Although deformation is usually considered quite continuous, in the Alps–Apennines Junction subsidence rates change according to the flip of subduction polarity and the consequent switch from the Alpine to the Apennines history (Argnani 2012), associated with east-dipping and west-dipping subduction, respectively (Doglioni 1991; Mariotti & Doglioni 2000). Stratigraphic events therefore mark intervals in which subsidence rates either increased or decreased. In addition, activation of detachment planes and deep-seated duplications further enhanced subsidence and uplift, respectively, at different location within the orogenic belt and the peripheral basins. Changes in elevation and basin morphology determine relative sea-level changes, the effects of which can be traced laterally. The relatively uniform plate motion and the gradual progression of deformation makes

Fig. 9. *(Continued)* (Late Rupelian) and II-3 (Chattian) in an area affected by the interplay of the Celle-Sanda (NW–SE-oriented) transpressional faults system and the Merana–Monteacuto (WSW–ENE-oriented) transtensional fault system. Sub-basins, local depocentres and syn-depositional highs are shown in combination with the average palaeocurrents measured in a variety of depositional settings from alluvial fan to intra-slope systems. The rearrangement of morphostructural elements and the consequent reorganization of depositional systems and their palaeocurrent patterns determine complex changes in basin infilling, play types and reservoir architectures. Note that: (1) an important morphostructural reshaping took place in this sector of TPB across the Early Rupelian–Late Rupelian boundary, with an inversion of the axial plunging of syn-depositional lows and highs driven by the scissor-like behaviour of the Merana–Monteacuto fault system (westward plunging during Early Rupelian and the present time, eastward plunging during Late Rupelian); (2) a stronger morphostructural reshaping determined a basin inversion across the Rupelian–Chattian boundary, with an overall northward basinward plunging during Rupelian and a SE-ward basinward plunging during Chattian when the marginal facies belt shifted towards the Western Alps owing to a regional transgression and local basin inversion took place in the Celle–Sanda fault system.

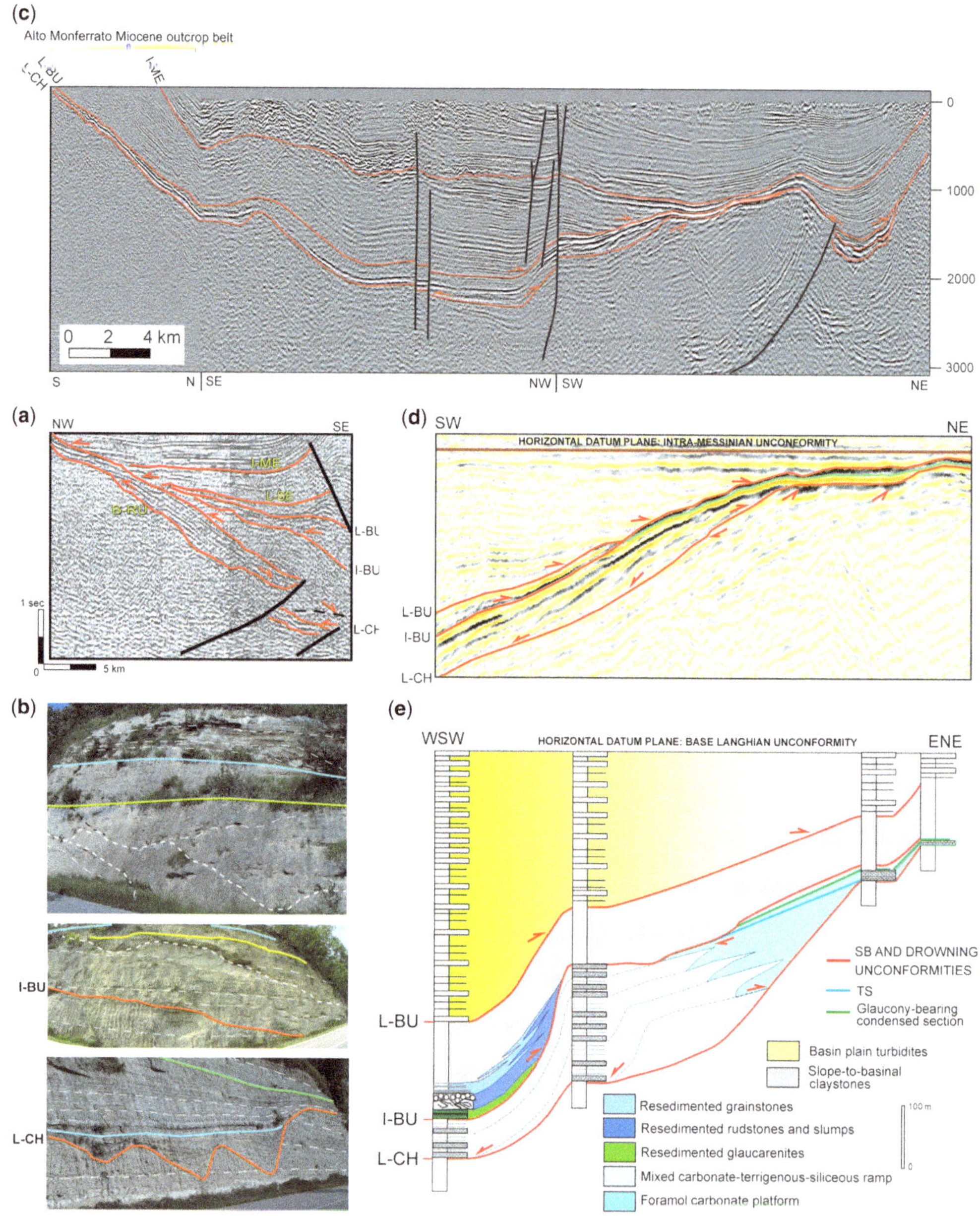

Fig. 10. Contrasting stratigraphic architectures of the Early Miocene Alpine v. Apenninic marginal basins. Valley incisions and regressive surfaces of marine erosion developed at the western margin of TPB adjacent to the Alps are coeval of erosional truncation and drowning unconformities on intra-basinal highs generated by the Apenninic tectonics. (**a**) Seismic profile at the western margin of the Alpine retroforeland (location in Fig. 1). (**b**) Outcrop expression of the Lower Miocene sequence set cropping out (location in Fig. 5) along the western margin of the Alpine retroforeland (codes of bounding surfaces: red, SB; blue, TS; green, MFS; yellow, increasing accommodation surface at the base of late Lowstand). (**c**) Seismic profile (location in Fig. 1) from the southern basin margin to the Monferrato Apenninic frontal splays (the outcrop-seismic tie is shown for the southern margin). (**d**) Flattening of a portion of the seismic profile shown in (c), showing the bounding surfaces and the geometry of depositional systems. (**e**) Correlation panel showing the stratigraphic architecture of the Lower Miocene sequence set on the western flank of a major intrabasinal high (location in Fig. 1).

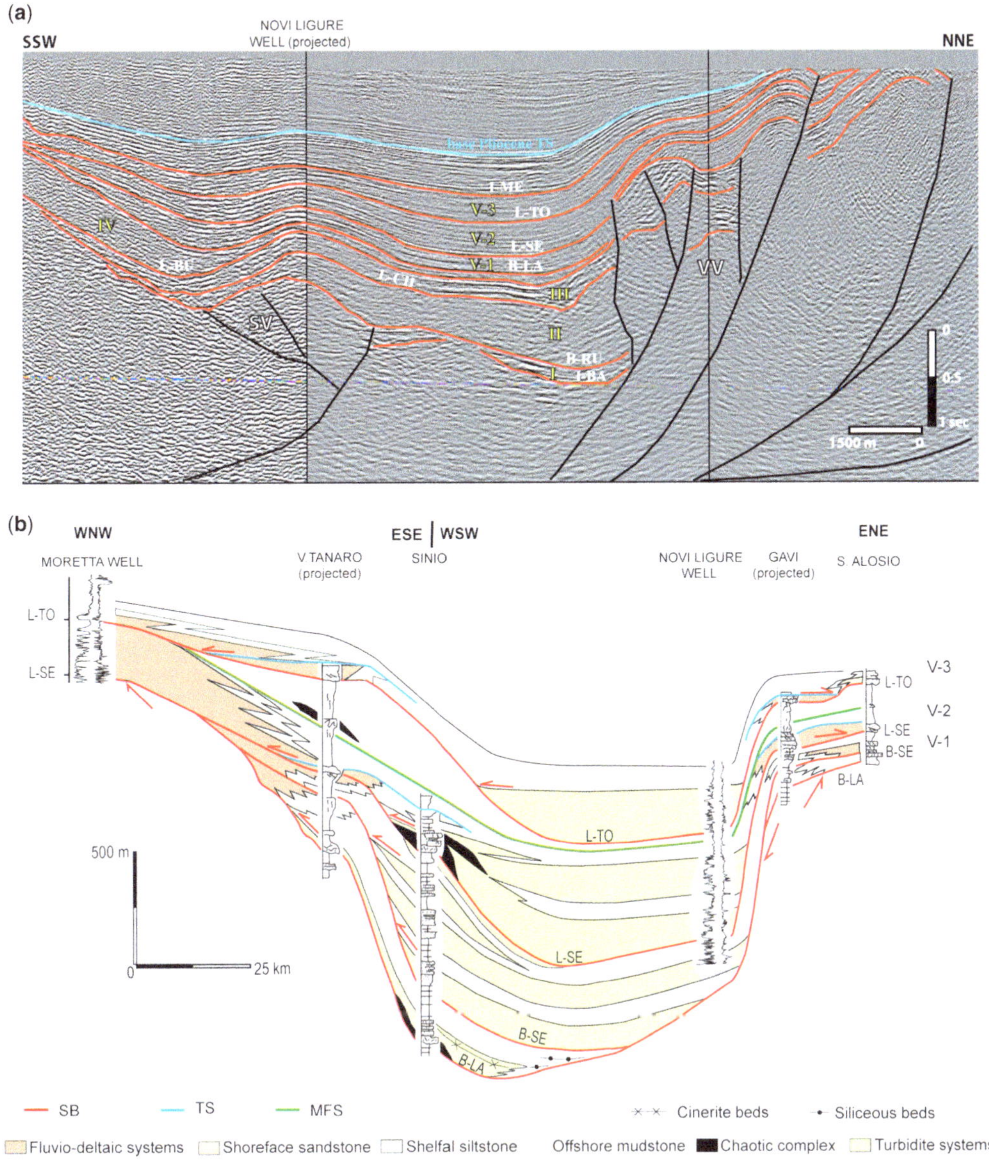

Fig. 11. Stratigraphic architecture of the Middle–Upper Miocene succession of TPB. The large-scale regression is primarily driven by the uplift of the southern basin margin. The interference with the Apenninic deformation active at the eastern margin may determine the occurrence of drowning unconformities generated by local tilting and abrupt increase of subsidence rates. The codes of bounding surfaces and units are the same used in the text. (**a**) Seismic profile showing structural and stratigraphic features of the southeastern basin margin affected by north-verging Apenninc frontal splays, and the relationships between the tilted basin margin and the Alessandria depocentre. Note the internal boundary of the Alps retroforeland, the buried expression of the back-verging Sestri–Voltaggio (SV) deformation zone (separating the Alpine axial meta-ophiolitic units from the Ligurian units) and the poly-history transtensional-transpressional evolution of the Villalvernia–Varzi (VV) fault system. (**b**) Correlation panel showing the stratigraphic architecture of the Middle–Upper Miocene sequence sets: note the basinward tilting of the western and southern basin margin.

it quite common to record in the stratigraphy an incipient change, marked by relatively minor coeval, but locally developed erosional surfaces, closely followed by a major and more widespread unconformity.

Where the bounding surfaces show erosional gaps, the morphology and the subaerial v. subacqueous nature of the erosional surfaces vary mainly according to: (1) the magnitude of morphostructural reshaping; (2) the location with respect to the basin margins and the tectonic hinge lines separating areas experiencing creation v. destruction of tectonic accommodation space; and (3) the structural elevation and uplift rates of the basin margins. In addition, their preservation potential may also be very different, mainly depending on the magnitude and type of the subsequent basin reshaping.

Changes in basin configuration according to different types of margin (i.e. influenced by tectonic accommodation space creation or destruction) determined accommodation and gateway turnarounds, deeply controlling stratigraphic architectures. Local diachronism in the change of accommodation rate also influenced sediment flux and routing to deeper waters in relation to sediment sourcing and processes.

The higher-frequency component units (fourth to fifth order) are bounded by minor (although locally prominent) unconformities that change basinward to correlative conformities because the shape of the basin only undergoes minor changes between the major basin reorganizations. This makes it possible to describe these units using a more conventional sequence stratigraphic approach. However, the succession and organization of systems tracts inside these sequences can show some differences from the classic architecture because of local tectonic deformation. In addition, it has been demonstrated that tectonic deformation may cause changes in accommodation space at rates, and of durations, similar to those generated by orbital forcing and glacio-eustatic variations (Mutti 1990; Burbank *et al.* 1992; Mellere *et al.* 2000).

Stacking patterns

Under these conditions, transgressive and regressive stacking patterns between unconformable surfaces are modulated by the interplay between longer- and shorter-term cycles, so that transgressive–regressive cycles cannot be used unambiguously. Allogroups (corresponding to sets of fourth-order sequences) may, in fact, exhibit transgressive (T), regressive (R), transgressive–regressive (T-R), regressive–transgressive (R-T) or regressive–transgressive–regressive (R-T-R) organization. Moreover, sometimes the same unit has opposite stacking patterns (regressive v. transgressive) on different margins or on marginal highs depending on locally different rates of subsidence and the type and amount of sediment supply, thereby precluding the use T-R cycles to define chronostratigraphically significant units.

T, R and T-R stacking patterns are more typical of allogroups where the change in the accommodation/sediment supply (A/S) ratio is mainly controlled by changes in the rate of tectonic subsidence. Decreasing rates of subsidence are often associated with forced regression; within overall regressive intervals, the shallowing is punctuated by unconformities and characterized by dramatic downward shifts of coastal onlap associated with relative sea-level falls of tectonic origin of the order of several hundred metres. Increasing rates of subsidence may not be sufficiently compensated by sedimentation rates (not exceeding 0.5 m/kyr in the Early Rupelian) so that an overall transgression is recorded, associated with a marked diachronism of marginal marine systems which progressively onlap SW-wards closer to the Alpine axial zone. In addition, the Apenninic slab retreat may be characterized by an increasing rate of tectonic subsidence through time so that the associated depositional units may be unconformably separated from previous transgressive units that are characterized by relatively lower rates of subsidence; a sudden deepening of the order of a few hundred to several hundred metres commonly occurs in this case. These peculiar units are defined here as 'Maximum Accommodation Intervals' and, differently from classic sequence stratigraphic models, are high accommodation/high sediment supply intervals associated with high sedimentation rates (e.g. 1 m/ky in Late Burdigalian) owing to the increasing flexural tilting that affected the basin margins as the Apenninic subduction increased, overriding the Alps retroforeland. In the stratigraphic record, such intervals are mainly represented by turbidite systems marking peak transgression and maximum deepening at the basin scale, such that stratigraphic expansion occurs instead of condensation from basin margins to basin centres. This behaviour is associated with a general enlargement of turbidite depocentres coupled with increasing in transport efficiency owing to higher mud availability in progressively more open marine source areas.

R-T-R and R-T stacking arrangements may correspond with the development of a complete depositional sequence consisting of lowstand (LST), transgressive (TST) and highstand (HST) systems tracts and of a truncated sequence framed within an overall regressive interval leading to limited preservation or non-preservation of the HST deposits, respectively. In such cases, lowstand deposits are commonly separated in an early LST forced regressive interval and a late LST aggradational and progradational interval. The generalized marginal uplift of the Alps–Apenines Junction and the constriction of depocentres owing to uplift and progradation from the margins resulted in the basin being oversupplied with a sand-rich turbidite complex (e.g. with low A/S ratio and sedimentation rate exceeding 0.5 m/kyr in the Serravallian depocentres).

Sequence sets and component sequences

The allogroups are third-order features that may be considered sequence sets consisting of the stacking and juxtaposition of smaller-scale component depositional sequences. During the orogenic collapse, the overall transgression is controlled by a high rate of subsidence and punctuated by transgressive surfaces of marine erosion corresponding to regional drowning unconformities. These features developed basinward of tectonic hinge-lines, which are sometimes associated with transtensional–transpressional faults, and mostly record abrupt deepening and erosion owing to tectonic oversteepening. Regressive surfaces of marine erosion (RSME) only develop landward of tectonic hinges, where a limited reduction of tectonically driven accommodation space may occur.

Following the onset of Apenninic deformation, erosional truncation, valley incision and RSME develop marginward, while drowning unconformities sometimes develop over the frontal splays and on the intra-basinal highs. The bounding surfaces are marked by a relative sea-level fall and forestepping of the coastal and marginal marine depositional systems to the SW, while NE-wards detailed seismic correlation supported by biostratigraphic analyses have shown that they correlate with major drowning unconformities marked by a high rate of relative sea-level rise and deepening-upward. Hence, such surfaces are often characterized by opposing relative sea-level variations in the highly elevated Alpine domain and in the less elevated Apenninic domain as a consequence of the different local tectonic regimes (uplift v. differential subsidence and tilting).

After the inversion and uplift of the southwestern basin margin, progressive unconformities, erosional truncation, fault-assisted valley incision and RSME typically developed marginward. This gave way to an overall decrease in the A/S ratio, which favoured the development of large-scale prograding wedges fed by fluvio-deltaic systems. The combination of marginal uplift and northward tilting of the southern basin margin with the development of prograding wedges produced a northward shift and a narrowing of the turbidite depocentres, with the result that the turbidite systems became progressively more confined, amalgamated and sand-prone.

The component depositional sequences are fourth-order features, spanning a time interval of 1 myr or less, and are more than 100 m thick in the depocentres. The thickness, stacking, internal architecture and preservation potential of component depositional sequences differs substantially depending on the A/S ratio of the low-frequency, higher-rank units to which they belong. They sometimes have clear relationships with morphostructural elements, suggesting a direct causal link with the deformation. However, in other cases there is no apparent relationship with the structural deformation, suggesting that the sequences are sometimes mainly controlled by other factors such as eustacy and/or climate.

Fifth-order elementary depositional sequences (EDSs) are the smaller sequentially arranged units, typically ranging from a few metres up to more than 10 m in thickness. At their most complete, they are normally R-T or R-T-R cycles consisting of shallowing and deepening-upward facies sequences.

These fifth-order sequences (EDSs) appear to be the true building blocks of the systems tracts of fourth-order sequences. It can be demonstrated that the concept of parasequence stacking, only considering shallowing-upward bedsets, does not represent the full spectrum of the high-frequency cycles, which are actually much more variable (Mutti 1990). The EDS concept, the complete expression of which is a decametric-thick R-T-R cycle composed of high frequency regressive (LST) and transgressive (TST) bedsets capped by a maximum transgression (possibly condensed) interval with or without a HST regressive interval at the top, mimics the behaviour of third- and fourth-order sequences, but at a smaller scale. Unlike parasequences, this concept takes into account a complete high-frequency cycle of relative base-level change and accounts for most marginal settings where LST deposits are not recorded and T-R cycles occur, which high-frequency sequence boundaries may coincide with transgressive surfaces (TS) or even with maximum flooding surfaces (MFS).

Application of outcrop analogues to petroleum geology

Regressive surfaces of marine erosion v. drowning unconformities

The marginal belts and the intrabasinal highs behaved differently during the Early Miocene propagation of the Apenninic deformation. Marginward, Oligocene valleys and satellite marginal basins adjacent to the internal boundary of the Alps retroforeland were unconformably overlain by Miocene deltas (Fig. 10a, b). Accommodation space was reduced along the uplifted marginal belts, as

Tertiary Piedmont Basin
Latest Chattian
Unconformity

Incised valley fill
SB generated by RSL fall cutting into Upper Oligocene TST-HST sandstones
Subaerial / shallow water erosional truncation below SB and coastal onlap above SB; abrupt shallowing above SB
Erosion and early LSTby-passing fluvial floods delivered sand and gravel basinward through sustained flows
Max erosional gap in valley axis; max non-depositional hiatus over interfluves
Late LST consists of fluvial flood deposits periodically reworked / redistributed by wave action
TST consists of distal delta front sandstone and siltstones; late TST passes over the interfluves
HST normal regression shows dark grey prodelta mudstones over the MFS overlain by a C-U delta front succession

---

Po Plain-Adriatic Basin
Intra-Messinian
Unconformity

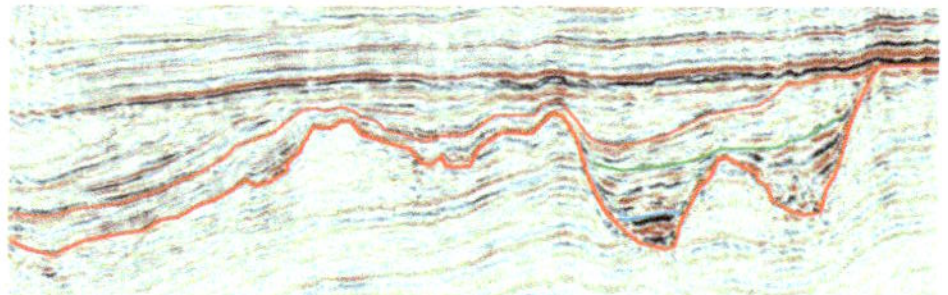

Incised valley fill
SB generated by RSL fall cutting into an Oligo-Miocene shelf-slope complex
Subaerial / shallow water truncation below SB and coastal onlap and abrupt shallowing above SB
Erosion and by-passing fluvial floods delivered sand and gravel basinward through sustained flows
Max erosional gap in valley axis; max non-depositional hiatus over interfluves
Late LST fluvial gravels (high-amplitude seismic facies) form a stratigraphic trap
TS due to base-Pliocene reflooding of Mediterranean Sea after the Messinian Salinity Crisis
TST / HST entirely comprised in the valley system and unconformably overlain by an intra-Pliocene valley reincision

---

Tertiary Piedmont Basin
Late Rupelian
Drowning Unconformity

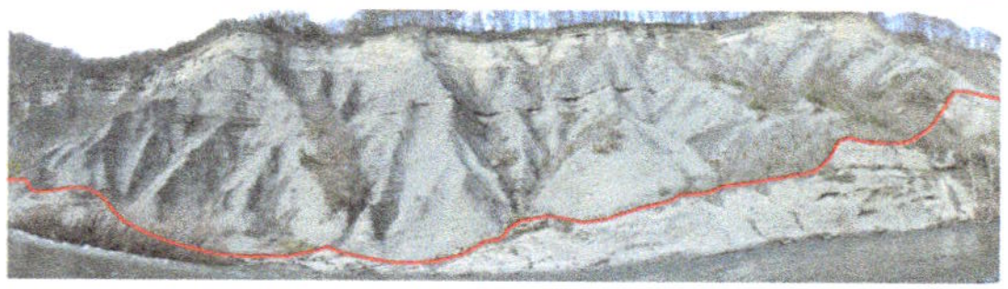

Retrogressive slump scars / submarine valley fill
Unconformity generated by RSL rise due to basinward tilting and oversteepening
The drowning unconformity shows erosional truncation below and marine onlap above
Max erosional gap in valley axis; max non-depositional hiatus over terraces
Abrupt deepening recorded in facies above the drowning unconformity
The subaqueous erosion cuts into shelfal sand lobes forming the reservoir of a remnant trap
The subaqueous erosion is filled by prodelta to slope muds forming top and lateral seal
Subaqueous erosion delivered sand and gravel basinward through surge flows

---

Po Plain-Adriatic Basin
Intra-Zanclean
Drowning Unconformity

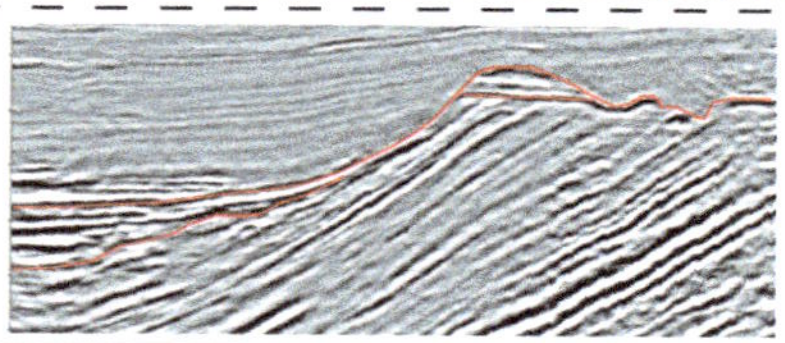

Retrogressive slump scars / submarine valley fill
Unconformity generated by RSL Rise due to increasing tectonic subsidence, tilting and oversteepening
The drowning unconformity shows erosional truncation below and marine onlap above
Max erosional gap in valley axis; max non-depositional hiatus over terraces
Abrupt deepening recorded in facies above the drowning unconformity
The subaqueous erosion cuts into fan deltas (unconformably overlying tilted slope deposits) that result in a remnant trap
The subaqueous erosional surface is filled by slope muds forming the top and lateral seal
Subaqueous erosion delivered sand and gravel basinward through surge flows

---

Tertiary Piedmont Basin
Base Rupelian
Non-conformity

Fluvial / coastal onlap against a metamorphic basement intrabasinal high
Non-conformity recording the basin-forming phase in a megashear belt
Fluvial / coastal (alluvial fan / fan delta) onlap above the non-conformity;
TS is a marine flooding surface coinciding with a ravinement surface
Non-depositional hiatus increases landward
Transgressive attitude and deepening-upward stacking pattern
Fluvial and marine deposits may form a stratigraphic trap topped by a MFS and laterally sealed by metamorphic rocks
In case of a fracture network, the metamorphic basement highs may also behave as reservoirs sealed by the MFS

Carupano Basin
Base Tertiary
Non-conformity

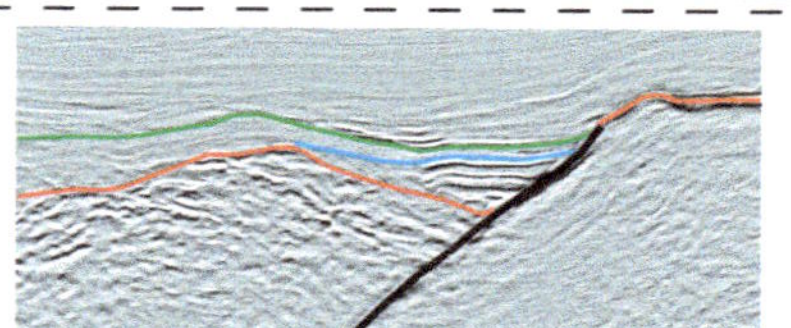

Fluvial / coastal onlap against metamorphic basement and localized intrabasinal basement highs
Non-conformity recording the basin-forming phase in a megashear belt
Fluvial / coastal onlap associated to a paleo-fault scarp or to a structurally-confined incised valley
The high-amplitude onlapping deposits may act as reservoirs of stratigraphic traps laterally sealed by metamorphic rocks
The lateral seal may be provided by the metamorphic basement
The top seal may be provided by TST-HST mudstones
In case of a fracture network, the lateral seal may fail but localized basement intrabasinal highs may behave as reservoirs
The top seal may be provided by the seaward change into prodelta muds (acoustically transparent seismic facies)

is recorded by Aquitanian incised valleys (Fig. 12) and Early Burdigalian forced regression in the fluvio-deltaic wedges that developed adjacent to the Alps. Above the Intra-Burdigalian unconformity, early Lowstand braid- and fan-deltas dominated in the delta front by coarse-grained hyperpycnal fluvial floods are overlain by late Lowstand normal mouth bars built by hypopycnal flows, suggesting a strong decrease in fluvial flood volumes related to the re-adjustment of the equilibrium profile. These deposits mark the maximum regression of the system. They are overlain by a transgressive interval which deepens rapidly upward to reach a maximum flooding surface which is, in turn, overlain by a highstand prograding wedge, with the re-establishment of progradation marked by instability features mainly triggered at the edge of tectonic hinge lines.

In more basinward locations in the eastern sectors, Oligocene turbidites and hemipelagites locally underwent erosional truncation over syn-depositional highs driven by incipient Apenninic thrusting. This was initially recorded by foramol–rhodalgal carbonate ramps rimming these highs.

Following the Intra-Burdigalian unconformity, in contrast to the Alpine sectors located in the west, the increasing Apenninic subsidence generated increasing accommodation rates and progressive unconformities along the flanks of intra-basinal highs. A drowning unconformity marks the collapse of the previous intra-basinal highs. The tectonic tilting and oversteepening generated destabilization and partial cannibalization of the older carbonate ramps. Resedimentation occurs at the toe of backlimbs and forelimbs, where onlapping or converging bedsets can be observed within an overall deepening-upward stacking pattern (Fig. 10c, d). This suggests that, while the Apennines progressively overrode the Alpine retroforeland, relative sea-level fall, testified by the presence of regressive surfaces of marine erosion, still occurred in the Alpine margin where the A/S ratio was very low or negative, while, at the same time, a dramatic relative sea-level rise produced drowning unconformities on the flanks of Apenninic intra-basinal highs where A/S ratio increased when these structures collapsed (Fig. 10).

This incipient Apenninic deformation foreshadows the major Late Burdigalian drowning unconformity related to the decrease of the Alpine subsidence and the westward propagation of the Apenninic subsidence, giving origin to the southward tilting of the whole basin. This resulted in the occurrence of progressive drowning unconformities framed in a step-wise deepening-upward succession leading to the maximum accommodation in the evolution of TPB.

This evolutionary outline is also indicated by the flattened seismic profiles in Figure 7, which show that the Late-Burdigalian sequence boundary is a major drowning unconformity driven by a major phase of accommodation space-creation and recorded by the deposition of basin plain turbidites. During the Oligocene, the tectonic accommodation space was generally greater to the north and the east (i.e. in the low elevation belts), while during the Latest Oligocene to Early Miocene interval the major depocentre abruptly shifted to the south.

In foreland basins, the rate of tectonic subsidence may increase or decrease quite abruptly in both time and space (Ghielmi *et al.* 2010). In the Po Plain-Adriatic Basin, the Intra-Zanclean Unconformity (Figs 12 & 13a) marks a major turning point related to a basin-reorganization mainly driven by the Apennine deformation (Rossi *et al.* 2015). The southern margin of the Alps remained relatively passive, so the northeastern margin of the Po Plain–Adriatic Basin was tilted towards the SW, producing a new pattern of accommodation space that deeply affected the type and the nature of bounding surfaces and basin-fill architecture, and determined an overall rearrangement of the sediment gateways.

At this time, the inner Apenninic margin records a downward shift of coastal onlap associated with an abrupt shallowing, while the outer ramp-foreland margin records an abrupt deepening owing to the tilting and the increasing subsidence rates associated with the outward shift of foredeep depocentres that may overlie previous marginal ramps, slopes and incised valleys (Fig. 13b). Consequently, the regressive surface of marine erosion developed on the inner more active margin to the south correlates with a drowning unconformity on the outer northern margin (Fig. 13).

The combination of onlapping non-marine to marginal marine deposits against the metamorphic basement, either in basin margins or around fault-bounded intra-basinal highs, and a marine flooding

Fig. 12. Summary of the different types of unconformity and their implications on the characteristics and potential of different play concepts; a brief description of the features of the stratigraphic discontinuities recognized in the Tertiary Piedmont Basin and their seismic expression in hydrocarbon-prolific basins is provided. The incised valley fill example crops out in the Corsaglia Valley near the village of San Michele di Mondovì, the slump scar/submarine valley fill example crops out in the Bormida Valley near the village of Mombaldone, the non-conformable onlap example crops out in the Bormida Valley near the village of Spigno Monferrato.

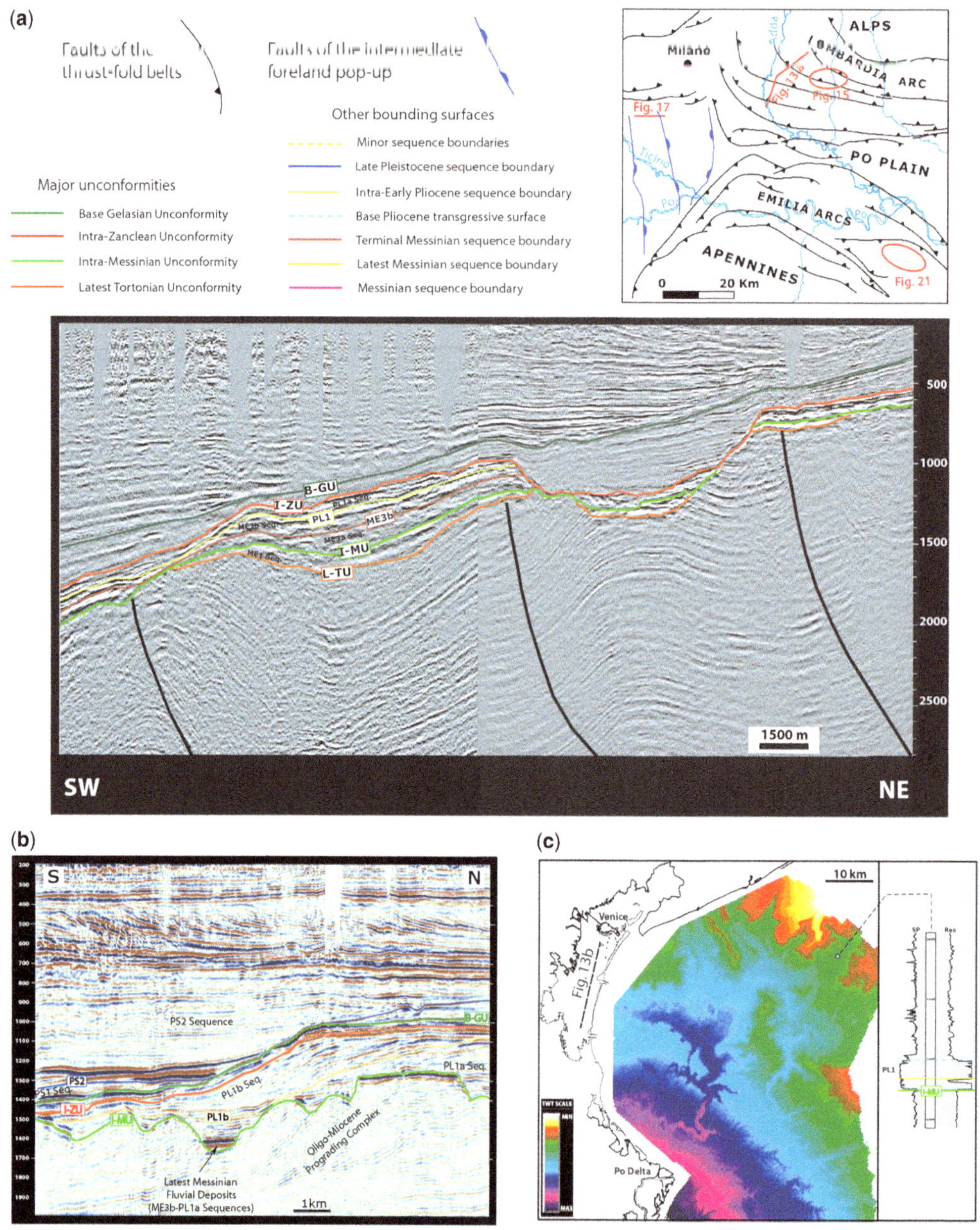

Fig. 13. Examples highlighting the differences between subaerial and subaqueous erosion in the Po Plain–Adriatic Basin (modified from Rossi *et al.* 2015). (**a**) Seismic profile across the Southern Alps thrust-fold belt, including location map and legend of bounding surfaces. Latest Tortonian subaerial valley incisions overlain by regressive surface of marine erosion are deeply truncated by the Intra-Zanclean large-scale submarine erosion regarded as a local expression of the drowning unconformity affecting the whole northern margin of Po Plain–Adriatic Basin. A further drowning linked to the outward migration of the Apenninic foredeep tilted the Pliocene slope that was onlapped by Gelasian turbidites. (**b**) Seismic profile showing a Latest Messinian incised valley cutting the Oligo-Miocene prograding complex. (**c**) Tributary network of incised valleys passing basinward to submarine canyons, shown on a two-way time (TWT) seismic map of the Adriatic foreland; in the inset, well-log expression of a Messinian incised valley fill in the northeasternmost basin margin.

surface or a drowning unconformity that provided their rapid transgression or their downwarping, respectively (Figs 5 & 12), may be a good analogue for stratigraphic traps along basin margins. This is especially the case in fault-bounded sub-basins or structurally controlled incised valleys (as seen for example in the landward onlap toward a palaeo-fault scarp that appears in the seismic profile shown in Fig. 12), provided that the basement rocks can guarantee an efficient lateral seal. In the TPB, the best examples are provided by Late Rupelian and Late Burdigalian coarse-grained deposits abruptly sealed by shelfal or slope mudstones deposited in response to increasing rates of accommodation. In case of a fractured metamorphic basement, only the fault-bounded intra-basinal highs located along the basinward plunging structurally controlled embayments may be considered favourable. In this case, the potential trap volume may consist of both the coarse-grained onlapping sediments and the metamorphic high; other interesting prospects may be associated with fault-bounded basement highs located basinward of marginal fan-deltas (as the isolated intrabasinal high developed basinward of the border fault in the example shown in Fig. 12), so that the highs may be directly sealed by prodelta mudstones, or associated with prominent highs where the basal non-conformity merges with a younger drowning unconformity or a marine flooding surface.

Drowning unconformities, retrogressive slump scars, the genesis of erosional remnant traps and marine onlap of gravity flow deposits in the Oligocene orogenic collapse sequence sets

In high A/S ratio settings, drowning unconformities are often characterized by the progressive tilting of shallow marine siliciclastic or carbonate systems, leading to oversteepening of basin margins or marginal highs and associated erosion and re-sedimentation, either in shelf or intra-slope settings. Drowning unconformities are a typical feature of the orogenic collapse of the Alpine retroforeland, punctuating the SW ward coastal onlap and transgression that progressively flooded the Alpine axial sector (Figs 3, 4, 6, 8 & 9), as shown by regional NE-ward directed palaeocurrents and SW-ward diachroneity of the marginal marine deposits, which become progressively younger towards the more structurally elevated belts.

The key factors that make these features (Fig. 12) attractive from the hydrocarbon exploration point of view are: (1) destabilization and erosion of recently deposited sands and gravels (consisting of either extrabasinal or intrabasinal grains), triggering gravity flow sedimentation during times of relative sea-level rise with onlap of the resulting turbidites at the toe of oversteepened slopes; (2) generation of retrogressive slumps scars and submarine valleys leaving erosional remnants in the underlying coarse-grained shallow water facies and so generating unusual stratigraphic or mixed play concepts; and (3) deposition of a thick and efficient prodelta to intra-slope mudstone seal (above either the onlapping gravity flow deposits or the remnant shallow marine deposits) owing to the rapidly increasing accommodation.

These features may occur at very different temporal and physical scales, as suggested by the recent and ancient examples shown in Figure 14. In the TPB, the collapse of shallow-water sands rimming morphostructural highs may occur in relation to the outward propagation of the Apenninic deformation characterized by a rising fold envelope (Fig. 14a). Also shelf edge deltas can be considered typically characterized by features related to collapse of the delta front slope, as shown in another ancient example from the Pliocene Palaeo-Orinoco delta (Fig. 14b). Smaller scales may be also implied, highlighted by recent geomorphic features (Fig. 14c, d) in backshore environments controlled by storm waves or the combined effects of major storms and fluvial entrenchment, which have virtually no preservation potential but are suggestive of a similar behaviour in relation to processes with seasonal to centennial periodicity.

Shallow-water deposits on moderately active margins undergoing flexural tilting may be remobilized in relation to drowning unconformities, through retrogressive slump scars, submarine valleys and gullies. In this case, the younger deeper-water deposits are separated from the shallow-water deposits by a major unconformity associated with a large stratigraphic gap. These surfaces can, therefore, be regarded as transgressive surfaces of marine erosion (TSME) showing locally prominent erosional surfaces changing laterally to paraconformable surfaces (Fig. 15a, b). This behaviour is the main difference between a drowning unconformity and a marine flooding surface because, although both of them are characterized by an abrupt relative base level rise, a flooding surface is a relatively conformable surface while a drowning unconformity is characterized by major erosion.

It should be also noted that very different behaviours and implications are associated with transgressive surfaces of marine erosion depending on whether the erosion takes place as a result of a relative sea-level rise associated with shoreface retreat and with a ravinement surface, or in relation to tectonic tilting and failure on oversteepened, above-grade ramps (cf. Prather 2003). In the former

(a)
Minor slump scar surfaces
Slump scar surface
Shelfedge deposits
Shoreface sand
RSME
TSME
Shelfal deposits
Shelfal deposits
(b)
RSME
Shelfedge delta deposits
TSME
Slope muds
Slope muds
Slump scar surface
(c)
N
25 m
(d)
20 m
N

case, the healing phase wedge that may develop basinward of the shallow water zone is normally mud-prone, but if the effect of tectonic oversteepening is added, then the underlying shallow-water sands and gravels can be affected by large-scale instability and large amounts of coarse-grained sediment can be delivered basinward to form high accommodation–high sediment supply intra-slope and basin-floor turbidites (Fig. 12) generated by the stacking of surge flows. In these situations, the retrogressive nature of slump scars, which provide a re-adjustment of the slope equilibrium profile, implies that progressively smaller volumes of sediments are removed during the retrogressive process. This, in turn, implies decreasing flow volumes of successive gravity flows so that a thinning-upward stacking pattern and a retrogradational organization is expected in the turbidite systems fed by retrogressive failures, as shown in Figure 16 (cf. Covault & Graham 2008). This may result in sequence architectures and the occurrence of turbidite systems that are very different from those classically established on passive margins, leading to play concepts that might have been overlooked. Gateways and sediment routing are also commonly reorganized. Although a sequence boundary and a drowning unconformity are conceptually very different, because they originated by relative sea-level fall v. relative sea-level rise conditions, respectively, it may be concluded that large volumes of turbidites may be equally delivered basinward in response to a drowning unconformity, that may generate as well a high-potential play concept, while a relatively conformable marine flooding surface is unable to convey sands into deeper waters.

A good example of these features is provided by the Late Rupelian Unconformity in the Tertiary Piedmont Basin that records a step-wise transgression, associated with multiple drowning unconformities, driven by the flexural tilting related to the orogenic collapse, locally enhanced by the reactivation of the WSW–ENE transfer faults related to the megashear zone. This led to a dramatic marginward (SW-ward) shift of the fluvio-deltaic systems that were replaced basinward by shelfal, gravity flow-dominated, coarse-grained bodies encased in marine mudstones (Rossi *et al.* 2009). A sudden deepening-upward is recorded in most of the basin, not developing a mere condensation but rather associated with erosional truncation and/or onlap relationships. The Upper Rupelian deposits infill areas where the accommodation rate was high and which experienced common instability along slopes undergoing tectonic oversteepening (Figs 5b, 8b, c & 9b). The sediment dispersion and the confinement of coarse-grained shelfal bodies were due to the interplay of NW–SE transpressional faults (Celle–Sanda fault system) and WSW–ENE transtensional faults (Merana–Monteacuto fault system). Local entry points were generated at the intersection between these fault systems. The palaeocurrent directions of the shelfal gravity flow deposits, which are mainly from WSW to ENE (evidencing an opposite dispersion with respect of the underlying Early Rupelian paralic deposits), suggest that the high-angle transfer faults underwent inversion of their plunging during the Late Rupelian, owing to their 'scissor-like' behaviour (Fig. 9).

The drowning continued also during Chattian, and is recorded by the deposition of intra-slope turbidites onlapping the underlying tilted succession (Figs 8c & 16a–c). This provides stratigraphic trapping potential related to marine onlap and lateral fringing of turbidites flowing perpendicularly (i.e. SE-wards) related to the Late Oligocene basin reorganization and associated rearrangement of sediment transport routes (Fig. 9), recording the collapse of the southern basin margin. New NW–SE transpressional faults developed, and the throw along some of the previously active transpressional fault planes was reversed, reflecting local basin inversion. The WSW–ENE transtensional faults were still partly active, controlling some entry points where the two fault systems intersect.

Fig. 14. Small- to very-small-scale ancient and modern examples of erosional remnants generated by high-frequency erosional surfaces in different bathymetric settings. (**a**) Late-Tortonian regressive surface of marine erosion in the TPB overlain by a thin shoreface bioturbated sandstone affected by slump scars owing to the sudden drowning as the deformation propagated and the structure collapsed (location in Fig. 5). (**b**) Intra-Pliocene regressive surface of marine erosion at Mayaro Beach (Trinidad east coast) overlain by shelfedge delta front-to slope sandstones affected by slump scars owing to the delta front collapse. (**c**) Modern erosional remnants (circled), up to a few m high and a few square metres in area, affecting backshore sands on the west coast of Flakstad (Norway); their development was determined by major storms on an elevated terrace shaped by the combined effects of storm waves and creek incision. (**d**) Modern erosional remnants (circled), up to a few decimetres high and a few square decimetres in area, affecting Pleistocene bioclastic grainstones on the NW coast of Sicily (Italy); their development was determined by wave erosion. Note, in yellow, the fractal geometry of these features shown in plan view on the satellite image. These relationships show striking similarities with the dip extraction on the 3D volume shown in Figure 16e. More in general, possible fractal relationships of approximately one order of magnitude may exist between the size of modern examples (c and d in this figure), instability affecting fourth/fifth order sequences (a and b in this figure), and major

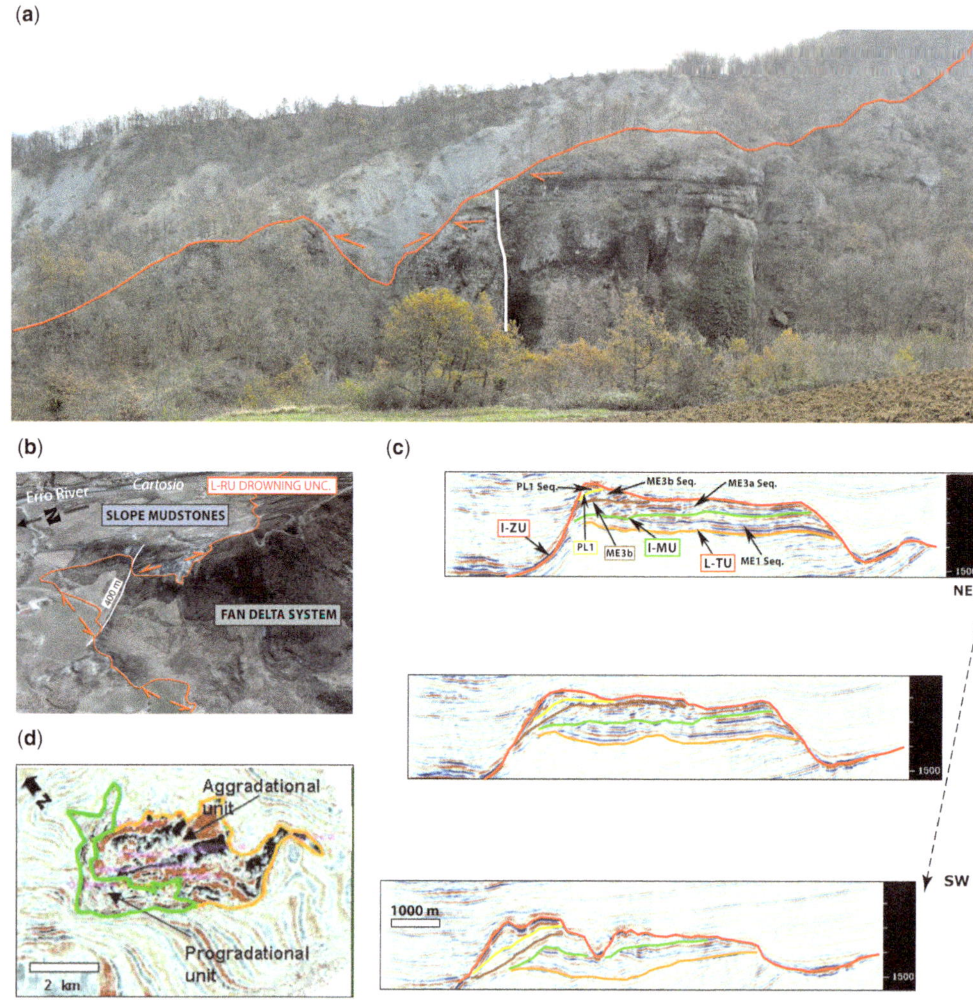

Fig. 15. Outcrop and subsurface expression of erosional remnant traps. An outcropping example of a lower Rupelian remnant cut by the Late Rupelian Unconformity is shown and compared with a similar situation occurring in the Po Plain–Adriatic Basin adjacent to the Southern Alps margin highlighted by seismic profiles and a time slice. Note: (1) the similar facies and truncation-onlap pattern; (2) the regressive surfaces of marine erosion (forced regression) taking place above Intra-Messinian Unconformity (ME3a ME3b, PL1surfaces); (3) the genesis of the remnant resulting from the basin reorganization recorded by the Intra-Zanclean Unconformity, expressed in this location by a transgressive surface of submarine erosion (drowning unconformity) generated by the coalescence of retrogressive slump scars and submarine canyons (modified from Rossi *et al.* 2015). (**a**) Outcrop expression of an erosional remnant shaped by the Late Rupelian submarine erosion and sealed by prodelta to slope mudstones. (**b**) Satellite image of the erosional remnant, showing the unconformity and the location of the outcrop (see also Fig. 5). (**c**) Seismic expression of an erosional remnant shaped by submarine erosion and sealed by prodelta to slope mudstones (Po Plain–Adriatic Basin; location and legend of bounding surfaces in Fig. 13). (**d**) Time slice of the four-way remnant trap shown in (c), with the Intra-Messinian Unconformity separating aggradational from progradational units. Location and legend of major unconformities in Figure 13.

An example documenting further drowning in an overall context of increasing tectonic accommodation driven by major reshaping of the Po Plain-Adriatic Basin, is shown in Figure 13a, where Gelasian foredeep turbidites onlap tilted Pliocene slope deposits that in turn overlaid the Intra-Zanclean drowning unconformity.

There are also several well-documented outcrop analogue examples where the reservoir consists of truncated Lower Rupelian shallow water deposits sealed by Upper Rupelian prodelta and intra-slope mudstones onlapping slump scars and submarine valleys (Fig. 12).

In Figure 15, Lower Rupelian deposits, consisting of fan delta front conglomerates overlain by transgressive alternating sandstones and conglomerate bodies, are affected by high-angle faults. The canyon fill overlying the drowning unconformity consists predominantly of mudstone, but multiple reshaping of the submarine valley has generated internal discontinuities marked by pebbly mudstones which onlap the flanks. This suggests that the erosional process was retrogressive and that the network of slope scars and valleys later acted as transport routes for gravels eroded from more landward or more lateral locations. The erosional remnants of Lower Rupelian conglomerates and sandstones preserved between the incision axes remain as prominent features and form potential stratigraphic traps sealed by mudstones during the continuing transgression.

In Figure 8c, Lower Rupelian deposits consisting of partly collapsed delta front sandstones and mudstones are unconformably overlain by prodelta mudstones. Approximately 1 km basinward, the same drowning unconformity is colonized by suspension-feeders marking an abrupt change from the Cruziana to the Zoophycos ichnofacies and is overlain by distal prodelta mudstones containing hyperpycnites consisting of very fine sandstone and siltstone. The multiple, step-wise nature of Late Rupelian drowning unconformities suggests progressive collapse and downwarping. This process, and the related overall upward-deepening character of the succession, is expressed by a younger, relatively minor and more localized erosional unconformity which is onlapped by a coarse breccia filling the canyon. A few kilometres to the north, in a more basinward location (Figs 8c & 16), the Late Rupelian drowning unconformity shows retrogressive slump scars cutting Lower Rupelian shelfal lobes that are coeval with the marginal delta front-prodelta facies described earlier. The slump scars are overlain by prodelta to slope mudstones. Multiple reshaping is indicated by internal minor discontinuities onlapped by a few sandstone–mudstone beds, confirming that the erosional process was retrogressive and the erosional depression was periodically receiving sands eroded from a more landward location (Figs 15a, b & 16a, b). Similar situations may occur in different tectonic settings which are characterized by abrupt increase in the rate of tectonic subsidence.

In foreland basins, tectonic subsidence may accelerate or decelerate. Low-subsidence margins recording the effects of relative sea-level falls may later experience an abrupt increase in subsidence rate, commonly associated with the outward shift of the foredeep depocentres and the associated turbidite systems that may overlie previous marginal ramps (Rossi *et al.* 2015). For example, in segments of the Apenninic foreland affected by low subsidence rates in relation to lateral variations in thickness and stratigraphy of the pre-collision successions, incised valleys formed during the Latest Messinian major base level fall cut down into the Oligocene–Middle Miocene prograding complex. The valley fill includes late LST Uppermost Messinian fluvial conglomerates and TST–early HST Lower Pliocene marine mudstones (Fig. 12). Such a major sequence boundary generated by a relative sea-level fall contrasts strongly with the dramatic increase in accommodation space owing to the increase of subsidence rate related to the NE-ward propagation of the Apennines thrust-fold belt, so that the incised valley fill was later affected by slump scars, which are sometimes abruptly overlain by Pleistocene turbidites (Fig. 13b, c).

In areas undergoing abruptly increasing subsidence rates as a result of the retreat and steepening of the Apenninic ramp that nucleated along the previously slowly subsiding Alpine retroforeland, subaqueous erosional surfaces were generated by drowning unconformities (Fig. 12). One of the best examples is the Intra-Zanclean Unconformity, which is expressed by multiple, retrogressive slump scars in the lower slope, which merge upslope into submarine valleys and canyons and eventually into shelf-edge gullies (Fig. 16d–f). These acted as sediment by-passing routes, which were subsequently filled with slope mud and, in a later stage, by the progressive aggradation of successive turbidite sandstone lobes that passively filled the available accommodation space (Fig. 13a). In areas where South-Alpine fronts of Messinian age undergo such a high rate of subsidence, retrogressive slump scars typically leave erosional remnants over the previous structural highs, which constitute four-way stratigraphic traps sealed by slope mudstones (Fig. 12). The basinward (southern and western) side of these features corresponds to the drowned Latest Messinian shoreline break, while the eastern and northern sides are affected by larger erosion (Fig. 15c, d). These features can be regarded as a peculiar, relatively deep-water type of transgressive surface of marine erosion driven by sediment failures generated by tectonic oversteepening.

The slopes shaped by the Intra-Messinian Unconformity, separating older primary evaporites from mass-transport complexes and gypsum-turbidites (first documented by Rossi & Rogledi 1988), provide seismic-scale examples of flow transformation processes (Artoni *et al.* 2010; Rossi *et al.* 2015). The stratigraphic relationships across the unconformity are well preserved, especially along the flanks of the intermediate foreland. The erosional surface is characterized by a stepped morphology extending

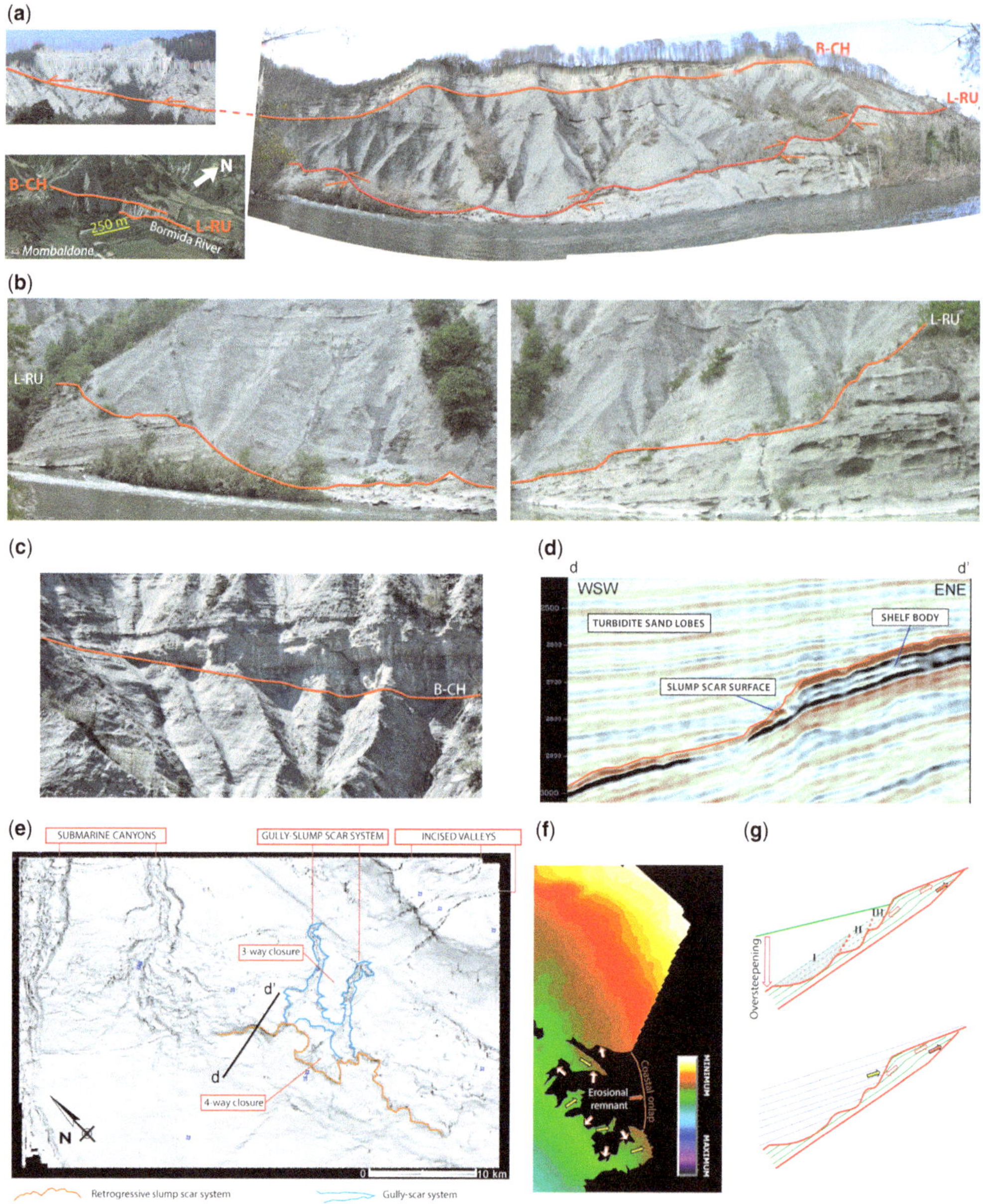

Fig. 16. Genesis of retrogressive slump scars produced by submarine erosion. An outcropping example of the Late Rupelian Unconformity, expressed in this location by retrogressive slump scars cutting shelfal lobes, is shown and compared with similar cases occurring in the Po Plain–Adriatic Basin. Further tilting coupled with basin reshaping and increasing rate of tectonic subsidence determined the marine onlap of Chattian turbidite sandstone lobes over the scar-fill mudstones. (**a**) Outcrop expression of retrogressive slump scars generated by the Late Rupelian drowning unconformity; the scar-fill is then unconformably overlain by a Chattian sandstone lobe. In the inset, oblique view from a satellite image showing the unconformities and the location of the outcrop (see also Fig. 5). (**b**) Detail of the western and eastern flanks of the Late Rupelian incision. (**c**) Detail of the marine onlap and lateral fringing of the Chattian sandstone lobe. (**d**) Seismic expression of multiple slump scars affecting shallow-water Messinian gypsum in relation to a drowning unconformity on the tilted Adriatic foreland ramp (modified from Casnedi *et al.* 2006). Location in (e). (**e**) Illuminated dip attribute of the Adriatic foreland ramp affected by retrogressive slump scars, and submarine canyons (modified from Rossi *et al.* 2015). Note: (1) the post-evaporitic Messinian incised valley-gully system (locally modified and re-oriented as NE–SW deep-water canyons during Pliocene);

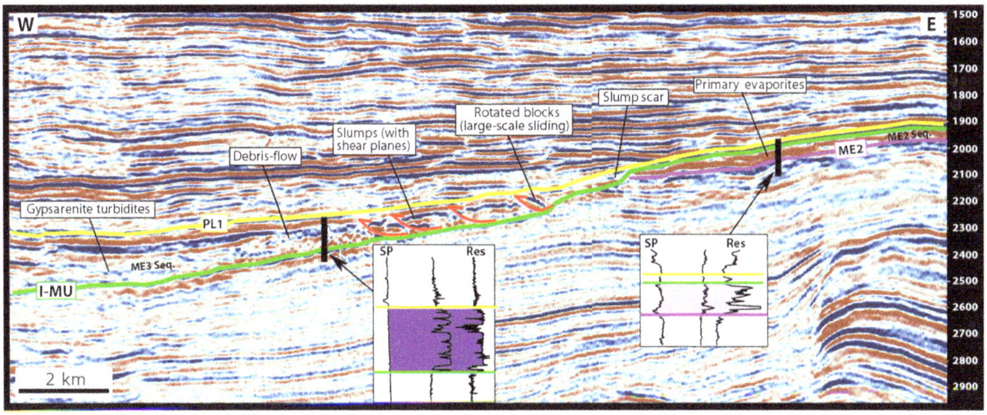

Fig. 17. Stratigraphic relationships between the primary and resedimented evaporites, separated by the Intra-Messinian Unconformity affecting the intermediate foreland between the Northern Apennines and Southern Alps thrust fold belts (location and legend of bounding surfaces in Fig. 13). The truncation and onlap terminations respectively below and above the unconformity, as well as the different stacking pattern of primary v. resedimented evaporites are shown. Note the high-resolution seismic imaging of downcurrent flow transformations in resedimented evaporites (from Rossi *et al.* 2015).

retrogressively towards the marginal high and passing into a flat paraconformable surface over the foreland, where primary evaporites are preserved. This is the location in which the minimum erosional gap and the maximum non-depositional hiatus coexist, while the opposite occurs basinward. While primary evaporites show a thickening-upward organization, the resedimented evaporites are arranged in a thinning- and fining-upward stacking pattern possibly related to the generation of an inverted stratigraphy in the re-sedimented deposits as a result of the progressive erosion of deeper levels. This interval consists of chaotic deposits passing basinward into gypsum–turbidites. A well-defined seismic-scale facies tract (Fig. 17) can be envisaged, showing a basinward transition of different seismic facies suggestive of downcurrent flow transformation, from slump scars, rotated blocks and imbricated blocks to slumps, then into a low-amplitude and disorganized seismic facies interpreted as debris-flows that pass into high-amplitude parallel-bedded seismic facies related to interbedded gypso-rudites and gyps-arenites. This eventually changes further basinward to a moderate-amplitude bedded seismic facies interpreted as alternating gypsum–arenites and basinal mudstones in the terminal depositional zone. The fining-upward stacking pattern is expressed laterally by the back-stepping of bedded turbidites with respect to more proximal chaotic facies associations. The back-stepping is interpreted to originate from the retrogressive slump scars associated with the destabilization of primary evaporites and related to the gradual re-establishment of the equilibrium profile along the slope after the Intra-Messinian structural reshaping and oversteepening, leading to a progressive decrease in the flow volumes.

The Tertiary Piedmont Basin megashear margin is also a good outcrop analogue for the Carupano Basin that has formed mainly since the Eocene in the right-lateral megashear zone at the junction between the Caribbean and South America plates (Escalona & Mann 2011; Garciacaro *et al.* 2011; Soto *et al.* 2011). The structural elements that control the stratigraphic architecture and the distribution of depositional systems in the Carupano Basin include extensional faults and high-angle strike-slip faults which extend onshore in Trinidad where they bound the Central Range and form the boundary with the northern offshore. The long-term cyclicity in the basin is expressed by a major T-R cycle, consisting of an Eocene–Miocene overall transgressive sequence, overlain by a Plio-Pleistocene overall

Fig. 16. (*Continued*) (2) a large slump scar system produced in the foreland ramp by submarine slides. Note how the slump scar-gully system may leave erosional remnants forming three- and four-way stratigraphic plays. (**f**) TWT map of the northern Adriatic Sea showing a submarine erosional network leaving remnants defining a potential three-way stratigraphic trap play concept. (**g**) Conceptual sketch of the development of the retrogressive slump scar; basinward, a turbidite body may be predicted, where a back-stepping organization is implied by the reducing volume of successive sediments removed by this process.

regressive sequence, separated by the base-Pliocene maximum transgression. During the deposition of the Eocene–Miocene transgressive interval, sedimentation rate did not keep pace with the high subsidence rate of the strike-slip basin. On the contrary, during the Pliocene uplift associated with the eastward-propagating Caribbean thrust-and-fold belt resulted in a major change in accommodation space. The combination of the high sediment supply from the dismantling orogenic belt and the sediment flux associated with the Paleo-Orinoco delta (at least in recent times) triggered a major phase of regression.

Seismic profiles show a marked diachronism of depobelts in the transgressive sequence, punctuated by large-scale erosional surfaces generated by drowning unconformities (Fig. 18). While syn-depositional lows may provide sufficient tectonic accommodation space to prevent erosional truncation generated by drowning phases, high blocks and oversteepened margins commonly exhibit retrogressive slump scars and submarine valley incisions. Continental and shallow marine (both siliclastic and carbonatic) systems commonly underwent large-scale erosion that appears to have been filled initially by prodelta and slope muds and then, as the long-term transgression progressed, were onlapped by turbidite systems. As in the Tertiary Piedmont Basin, three- to four-way remnant stratigraphic traps may occur, characterized by alluvial fan and fan delta reservoirs sealed by slope mudstones (compare with Fig. 15a, b) which, in addition, may also provide efficient lateral seals for marine onlap traps related to intra-slope turbidites (compare with Fig. 16a–c) or to shingled turbidites deposited at the toe of prograding slopes (Fig. 12). Other challenging play concepts that have interesting similarities with TPB outcrop analogues are related to fault-bounded fan-deltas abruptly closing against the metamorphic basement and, a few kilometres basinward, to drowned basement highs sealed by prodelta mudstones (Fig. 12).

In rift basins, the increase in the rate of tectonic subsidence passing from pre-rift phase to syn-rift phase, implying a relatively rapid increase of tectonic subsidence, generates an overall transgressive regime characterized by a deepening-upward succession. It is likely that this process is punctuated by the development of drowning unconformities, especially if half-graben development and/or salt evacuation causes tilting and oversteepening of the underlying strata. There are examples in the Upper Jurassic sequences of Central Graben of the Central North Sea (Howell & Flint 1996; Jeremiah & Nicholson 1999; Hampson *et al.* 2009*a*, *b*), where sequences affected by growth strata deposited in coastal plain and marginal marine settings and related to fault activation and reactivation commonly have unconformable lower boundaries (Fig. 19a, b). Oxfordian–Lower Kimmeridgian coastal

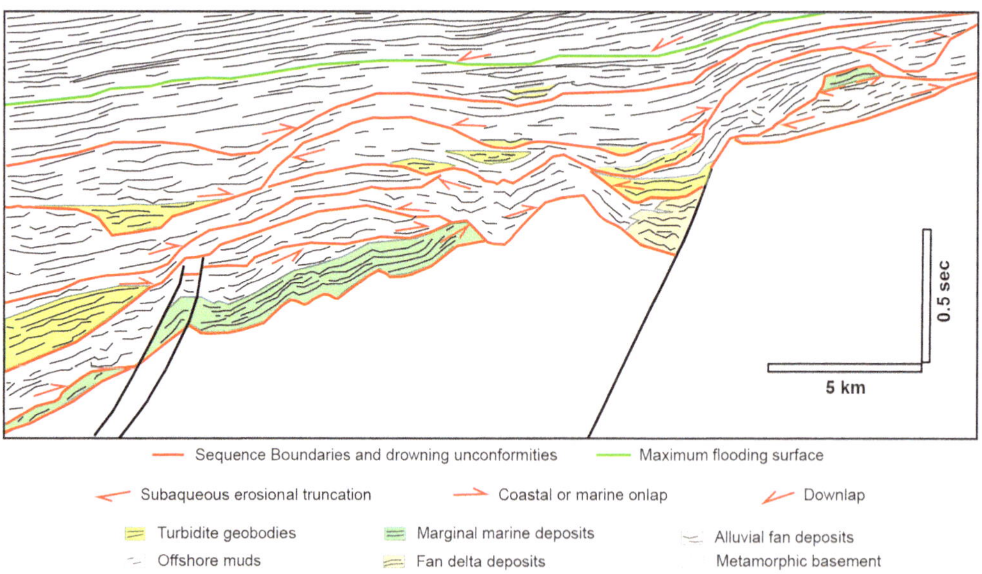

Fig. 18. Seismic line drawing in the Carupano Basin, showing striking similarities with the stratigraphic architecture recorded in the Tertiary Piedmont Basin. Both basins are affected by a major megashear zone determining changes in the accommodation/sediment supply ratio, and by the common development of drowning unconformities associated with large-scale subaqueous truncation and related erosional remnants.

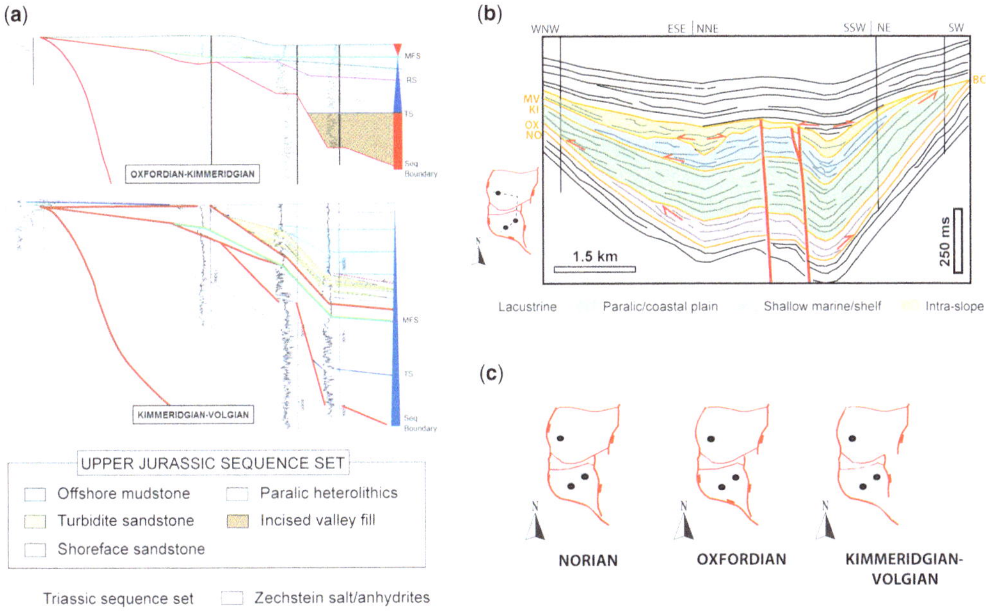

Fig. 19. Jurassic sequences in the Central Graben (CNS), UK. Blue and red triangles indicate transgressive and regressive stacking patterns, respectively. (**a**) Correlation panels of the Oxfordian–Volgian sequences, showing the effects of a major drowning unconformity; the gravity flow deposits onlapping the drowning unconformity derived by the oversteepening of previous HST deposits probably triggered by the diapir growth. (**b**) Seismic line-drawing showing local inversions of basin margin driven by scissor faults, controlling changes in space creation v. destruction and sediment fairways during an overall transgressive and deepening-upward succession. (**c**) Maps showing the uplifted belts (grey shade) owing to the tilting and associated with onlap and/or truncation relationships driven by the changing position of tectonic hinge-lines in relation to the scissor fault mechanism.

plain growth strata and incised valleys within the Pentland Formation are incised into Triassic non-marine deposits. Upper Kimmeridgian–Volgian shallow marine (Fulmar Formation) to deeper-water depositional systems generated by gravity flow processes (Jacqui Sandstone) may also fill accommodation space created by the reactivation of growth faults or accumulate at the toe of oversteepened slopes generated on the flanks of uplifting diapirs, respectively. In the first case, the lower boundary may coincide with a transgressive surface of marine erosion that can be a tectonically enhanced ravinement surface reworking material supplied by the shoreface retreat process. In the second case, gravity flow deposits are bounded (and fed) by the collapse of underlying sandy-muddy highstand wedges, interposed between the Oxfordian–Lower Kimmeridgian incised valleys and the drowning event (Fig. 19a, b).

Another interesting similarity with the Rupelian deposits outcropping in TPB is shown in Figure 19b, c, where the scissor-like behaviour of the transfer faults that locally segmented the basin margin determined the inversion of space creation v. space destruction depo-belts, and so controlled the local sediment routing. The Late Jurassic differential accommodation occurred during an overall marine transgression, while that in the Late Triassic occurred in a continental setting and was associated with a lake-expansion event.

High-resolution sequence stratigraphy in high A/S ratio settings and implications for reservoir heterogeneity: interplay between third- and fourth-order sequences

The stacking pattern and internal architecture of shallow-marine Lower Rupelian fourth-order sequences and their component systems tracts in the Tertiary Piedmont Basin (Fig. 8b) are a good analogue for predicting reservoir partitioning as a result of the common complex interplay of different orders of cyclicity. The palaeogeography of the area (Rossi *et al.* 2009) was controlled by an embayment formed by the morphological relief of a previous alluvial fan to the south, the westward plunge of uplifted basement rocks to the east, and to the north by an intrabasinal high controlled by a WSW–ENE-oriented, high-angle en-echelon transfer fault system that acted as a wave-dominated intrabasinal divide separating marine areas to the north from paralic protected environments to the south (Fig. 9b). An ENE- to WSW-oriented incised valley controlled by fault reactivation records the late LST overlying a fourth-order sequence boundary.

During high frequency maximum flooding, lignite beds occurred in embayed settings, while in open marine areas firm ground colonization developed in syn-depositional lows and condensed sections, consisting of thin coral-algal limestones, were deposited over syn-depositional highs. The complex structural pattern initially produced a very irregular shoreline, with dispersed palaeocurrents owing to the presence of fault-controlled embayments that were characterized by the local tidal modulation of bay-head deltas during transgressive and highstand systems tracts, recorded by bidirectional palaeocurrents and by the reshaping of fluvial floods giving origin to sigmoidal bedding. Toward the end of Early Rupelian, increasing accommodation space resulted in both a gradual change toward marine conditions and a more regular shoreline.

The correlation panel in Figure 8b shows the high-resolution internal architecture of a complex system partly confined in a structurally controlled embayment. The WSW-directed palaeocurrents record a local effect driven by an irregular shoreline produced by the various structural elements that define a WSW–ENE-oriented bay separated from the open sea to the north by a high-angle fault system that during the Early Rupelian was plunging westwards. It is also interesting to note that the further deepening occurred during the Late Rupelian was related to the reconfiguration of the basin which resulted in a local inversion of the plunge of the fault-bounded basin owing to the scissor fault mechanism. As the drowning event and the increasing rate of accommodation made the fault throw greater, shelfal gravity flows fed from the deltas, which shifted to the south in relation to the transgression associated with the drowning, were confined by the reactivated syn-depositional high and flowed eastwards along the inverted axial plunge (Fig. 9b).

A similar situation occurs at the transition between the thermal doming/pre-rift and rift stages in the North Viking Graben in the Northern North Sea. The Tarbert Formation of the Brent Group consists of a diachronous and internally very complex ensemble of shallow-marine transgressive deposits (Helland-Hansen *et al.* 1992; Rattey & Hayward 1993; Hampson *et al.* 2004) overlying and replacing basinward, at the scale of third-order sequences, the lower delta plain deposits of the Ness Formation (Fig. 20a). The overall transgressive nature of the Tarbert Formation deposits is due to an increasing rate of tectonic subsidence foreshadowing the onset of the rifting, which became fully established during the Callovian. In this small area of the Brent complex, the Tarbert Formation is mainly of Late Bajocian–Early Bathonian age and is overlain and replaced basinward by the offshore deposits of the Heather Formation (Fig. 20a).

When viewed from the perspective of the fourth-order sequences, the simple retrogradational stacking of parasequences predicted by the classical parasequence model does not adequately represent the actually much more complex interplay of high-frequency sequences (Hampson *et al.* 2009*a*, *b*) and the variability of the processes that produce rapidly vertically and laterally changing fluvial, tidal and wave signals (Løseth *et al.* 2009). To unravel the actual reservoir architecture, it is necessary to identify both the shallowing- and deepening-upward systems tracts of the fourth-order sequences. Owing to the overall transgression (third-order TST), three fourth-order sequences are stacked landward (Fig. 20b, c). In ascending stratigraphic order, the first of the fourth-order sequences in a basinward position consists of shallowing-upward intervals capped by thin condensed intervals, mainly recording LST and TST, while landward the LST wedge onlap and the sequence boundary is directly overlain by estuarine TST valley fill deposits, eventually capped by a regressive HST interval. The LST of the second fourth-order sequence marks a basinward shift with respect to the underlying HST, but a landward shift if compared with the LST of the first sequence and the same situation also occurs in the third, younger sequence. The trend is transgressive and aggradational overall, but the shoreline trajectory (*sensu* Helland-Hansen & Martinsen 1996) is punctuated by regressive surfaces of marine erosion corresponding with the fourth-order sequence boundaries when a convex-up trajectory would develop at the transition from HST to LST (Fig. 20c). The rate of relative sea-level rise increased with time, so that subaerial exposure and valley incision is predicted only in the older sequence when the rate of accommodation space creation was not so great. The stratigraphic architecture shows a striking similarity to the fourth-order Lower Rupelian sequences of the Tertiary Piedmont Basin (Fig. 20d).

Regressive surfaces of marine erosion and associated turbidite systems of the Mid–Late Miocene uplift and basin inversion sequence set

Hydrocarbon play concepts in the low A/S ratio settings of the Tertiary Piedmont Basin involve the progressive unconformities associated with the

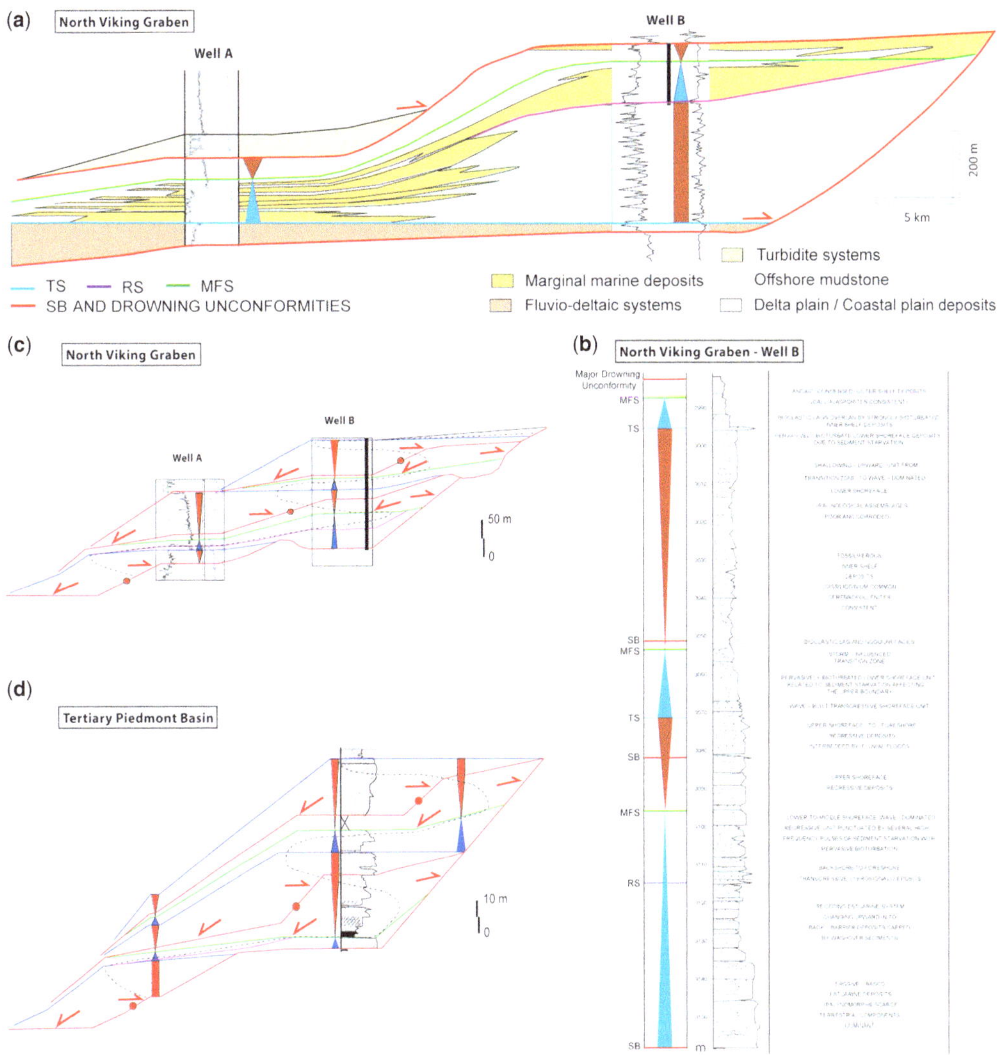

Fig. 20. Comparison between the overall transgressive stratigraphic architecture of the Early Rupelian sequence set in the Tertiary Piedmont Basin and the Bajocian–Bathonian sequence set in the North Viking Graben. Note, in both cases, how the stacking of component sequences and systems tracts permit to determine that the actual reservoir architecture is more complex than the one predicted considering only a relatively conformable transgressive interval. Blue and red triangles indicate transgressive and regressive stacking patterns, respectively; red dots show the position of the equilibrium point; the black dashed lines show the shoreline trajectories. (**a**) Low-resolution sequence stratigraphy; (**b**) high-resolution sequence stratigraphy stacking pattern in the cored section (**c**) high-resolution sequence stratigraphy (well-log correlation panel in the late TST-HST interval); (**d**) high-resolution sequence stratigraphy stacking pattern and predicted lateral change (outcrop analogue of Fig. 8b).

basin inversion that has occurred since the Middle Miocene. An overall progradation occurred along the western, southern and eastern basin margins, driven by the progressive uplift and local basin inversion (Fig. 7) that began during the Langhian. Angular unconformities to the east, where the proximity to the Apenninic frontal fault splays superposed shelfal deposits over tilted and truncated basin plain turbidites, suggest a relative sea-level fall of tectonic origin in the order of at least several hundred metres. Along the southwestern margin of the basin, the presence of shelfal-slope deposits that prograded over lowermost Langhian toe-of slope, channelized and locally chaotic deposits suggests that the Langhian relative sea-level fall was at least 100 m (see also Fig. 11b).

The Base-Serravallian and Late-Serravallian unconformities record further downward shifts of coastal onlap associated with forced regression, owing to the continuing uplift of basin margins. The forced regression wedge was mainly built by catastrophic fluvial underflows (Mutti *et al.* 2002), as suggested by the dominance of gravity-flow deposits consisting of coarse-grained graded sandstone beds, sometimes with water escape structures, capped by bioturbated fine sandstone and siltstone. The falling relative sea-level would have favoured this process, enhancing the by-pass of the shelf and thereby increasing the volume of sediments delivered to the basin. The local occurrence of fore-stepping chaotic complexes records the increasing tectonic destabilization of basin margins. The narrowing of the turbidite depocentre (Fig. 11) was due to the combination of the basinward tectonic tilting and the deposition of the off-lapping prograding wedges. In this setting, shelf and shelf-edge deltas may originate from sustained flows evolving from fluvial floods. Slope remobilization is also common, favoured by the very high sediment flux owing to the general uplift which occurred at this time. Fore-stepping turbidite systems related to regressive surfaces of marine erosion developed, especially during the early Lowstand (*sensu* Posamentier & Allen 1999 or falling stage systems tract *sensu* Hunt & Tucker 1992) on basin margins, dominated by sustained flows related to fluvial floods fed by delta systems and characterized by a coarsening- and thickening-upward stacking pattern.

A remarkable exception to this behaviour is related to the growth of thrust-propagation folds and their collapse in relation to the outward propagation of deformation in the frontal splays of TPB, where the structural elevation produces dip towards the hinterland, as shown by the southward-dipping envelope of fold crests, because the rate of fold uplift is lower than the rate of subsidence of the basin (cf. Doglioni & Prosser 1997).

Shoreface sandstones, unconformably overlying a regressive surface of marine erosion, rim syn-depositional highs away from the main fairways along which fan delta systems built by gravitational catastrophic fluvial floods were deposited preferentially. Later, shoreface sandstones may have been locally affected by transgressive surfaces of marine erosion in the form of slump scars which removed some of the sand in relation to the progressive northward propagation of deformation leading to the abrupt deepening of the backlimb. These surfaces can, therefore, be considered as the local record of small-scale drowning unconformities of tectonic origin, overlain basinward by slumped slope facies (Fig. 14a).

High-resolution sequence stratigraphy in low A/S ratio settings and implications on reservoir heterogeneity: interplay between fourth- and fifth-order sequences

The Western Alps retroforeland was affected by an overall regression during the Early Miocene. In the subsurface, gravelly fan delta systems are arranged in a forced regression mainly driven by the uplift of the basin margin, while in a lateral outcropping zone a late LST–TST valley fill is overlain by a HST normal regression passing upward into an early LST forced regression. The forced regression is characterized by a succession of seaward-dipping fifth-order erosional surfaces overlain by high-frequency regressive cycles deposited in a braid-delta front environment by hyperpycnal fluvial floods, producing graded beds with outsize mudstone clasts and parallel laminations. The last cycle, overlain by the coarsest and most channelized facies, is rapidly fining-and thinning-upward, and is, in turn, overlain by clinostratified mouth bars suggesting the re-establishment of hypopycnal flows during the third-order late LST–TST (Fig. 10b).

In the eastern margin of TPB, along the uplifting Apenninic margin, Serravallian braid-deltas developed changing-upward from normal to forced regression (Fig. 11b). The rate of accommodation space was decreasing rapidly at the time so the expression of the fourth- and fifth-order sequences changed accordingly. The fifth-order coarsening- and shallowing-upward prodelta-to-delta front facies sequences deposited by hyperpycnal flows and capped by condensed mudstones mark a high-frequency LST–TST cyclicity within a fourth-order regression. Above the Late-Serravallian fourth-order sequence boundary, the beds become thicker and contain large-scale water escape structures that suggest more rapid deposition. The component fifth-order sequences become sharp-based and overlain by fining- and deepening-upward proximal delta front facies, suggesting a more landward setting where the boundaries of the EDSs coincide with the TS. The overall shallowing resulted in the LST of the older EDSs being overlain by TST of younger EDSs, implying that the fifth-order LST deposits should occur further basinward. At the top, normal mouth bars exhibiting cross bedsets, periodically interrupted by bioturbated abandonment surfaces and with evidences of frequent autogenic lateral shifts, suggest the transition from the fourth-order early LST to late LST.

A similar situation is typical of Messinian deltas recorded on both the South-Alpine (Fig. 15c) and Apenninic (Fig. 21) margins of the Po Plain–Adriatic Basin (Fantoni *et al.* 2001; Rossi *et al.* 2002) and belonging to an intermediate category between

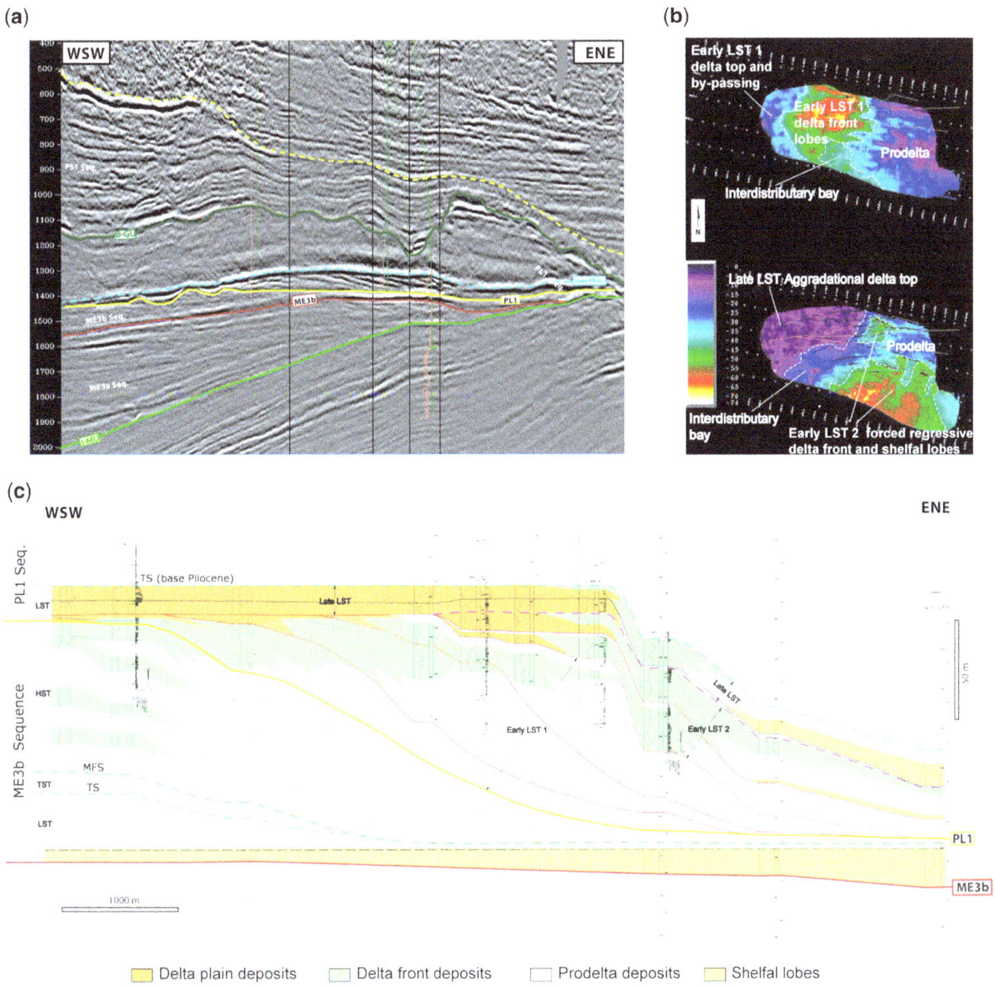

Fig. 21. Latest Messinian braid-delta system in an Apennine thrust-top basin at the southern margin of the Po Plain–Adriatic Basin (from Rossi *et al.* 2015). Location and legend of bounding surfaces in Figure 13. (**a**) Random profile from a 3D seismic survey, showing the sequence boundaries, the HST normal regression of the ME3b sequence, the PL1 valley incision cutting the ME3b topsets, and the early LST forced regression of PL1 sequence capped by the aggradational/progradational late LST deposits. (**b**) Isochronopach maps of successive stages of growth of the fluvio-deltaic system, showing the early LST delta front lobes, the resumption of the late LST aggradational component to the West, and the autogenic lateral shifts of delta front lobes interplaying with the forced regression. (**c**) Correlation panel, showing the fourth-order systems tracts and the component Elementary depositional sequences; note the normal regression of the ME3b sequence, the forced regression during the early LST of the PL1 sequence and the resumption of significant aggradation during the late LST.

the accommodation-dominated and supply dominated end-members of Porebski & Steel (2006). On the Apenninic margin, uppermost Messinian shelf-edge deltas occur (Fig. 21a, b). These systems show a convex-up shoreline trajectory, passing to a flat and then to a descending trajectory recording the maximum rate of relative base-level fall (around the fourth-order inflection point), eventually capped by an aggradational–progradational package with a concave-up shoreline trajectory marking a decreasing rate of relative base level fall (Rossi *et al.* 2015). This final wedge, immediately predating the Pliocene transgression, marks the late Lowstand maximum regression. These deltas are mainly built by graded and parallel-laminated sandstone beds that resulted from catastrophic fluvial floods triggering sustained, hyperpycnal flows associated with periodic by-passing of the delta front clinothemes during the maximum rates of regression. The deposition of aggradational sandstone lobes commonly occurs on prodelta-shelf bottomsets, onlapping the prodelta toe marginward and in turn downlapped by delta front prodelta prograding lobes with abundant soft-sediment deformation (Fig. 21c), possibly testifying to an autoacceleration mechanism (*sensu* Leva López *et al.* 2014).

The actual internal architecture of these systems is seen to be even more complex where high-resolution 3D seismic volumes with vertical resolution of approximately 10 m enable the architecture of fifth-order EDSs and their vertical and lateral stacking patterns to be defined. The transition from latest HST to early LST, when viewed within the relationships between fourth-order sequences and EDSs in high-resolution seismic volumes, reveals a more complex pattern (Fig. 21). The PL1 fourth-order downward shift marks the onset of a seismically recognizable accommodation space reduction, conceptually referable to the transition from rising to falling limb (SB *sensu* Posamentier & Allen 1999). The basinward shift is quite large and can be estimated to be more than 10 km, suggesting that this event actually records a fourth-order sequence boundary. The shoreline trajectory can be split into repeated, distinct rising, flat and descending tracts. High-frequency sequence boundaries record successively basinward-stepping wedges. These surfaces fit fifth-order sequence boundaries, because their downward shifts are in the order of 1 km (one order of magnitude lower than those of the fourth-order sequences). It appears that in the early Lowstand systems tracts the older EDSs show aggradation and progradation turning to progradation, while the younger EDSs are dominated by pure progradation.

Conclusions

In structurally complex margins, where changes in the tectonic accommodation/sediment supply ratio play a major role, reconciling allostratigraphy and sequence stratigraphy through framing high-resolution sequence stratigraphy within large-scale tectono-stratigraphic units may provide a model-independent approach (Catuneanu *et al.* 2009) that incorporate together allostratigraphic units (NACSN 1983), unconformity-bounded stratigraphic units (Salvador 1994), T-R sequences (Embry & Johannessen 1992), depositional sequences (Hunt & Tucker 1992; Posamentier & Allen 1999) and shoreline trajectories (Helland-Hansen & Martinsen 1996). Such an approach may better explain situations deviating from current sequence stratigraphic models and may therefore predict more precisely the basin-fill architectures and their relationships with tectonic deformation (impacting on petroleum system), and the three-dimensional distribution of depositional systems and their component elements (impacting on reservoir heterogeneity). Applying this approach to hydrocarbon exploration (Fig. 12) may permit: (1) new types of play concepts to be identified, for example stratigraphic or mixed remnant traps generated by drowning unconformities in siliciclastic systems; (2) challenging play concepts, for example stratigraphic traps against basement highs, to be assessed; (and 3) the occurrence and characteristics of turbidites delivered in deeper waters during relative sea-level rises to be better constrained.

Allostratigraphy v. sequence stratigraphy

Sequence boundaries developed on orogenic margins or landward of major tectonic hinge lines may correlate with drowning unconformities on intra-basinal highs or basinward of major tectonic hinge lines. No correlative conformable surfaces occur, so that allostratigraphy is more suitable than sequence stratigraphy for the description of the basin-fill architecture. The classic sequence stratigraphic architecture cannot be satisfactorily used in these settings, not least because the vertical succession and the internal organization of systems tracts may differ significantly from the usual models.

In such situations, accommodation turnarounds can vary in both time and space, related to locally varying changes in subsidence rate generated by changes in basin configuration. Such changes in basin shape can have a significant impact on basin margin gradients, and lead to a rearrangement of sediment gateways, sediment flux and flow types (e.g. hypopycnal v. hyperpycnal, sustained flows v. surge flows).

Drowning unconformities and erosional remnant play concept

Shallow-water deposits in regions affected by flexural or local tilting may be remobilized owing to drowning unconformities through retrogressive slump scars, submarine valleys and gullies. Resedimented deposits are separated from shallow-water deposits by a major unconformity corresponding to an allogroup boundary (associated to large stratigraphic gaps) marked in large areas by an abrupt deepening. These surfaces can be therefore regarded as transgressive surfaces of marine erosion showing locally prominent erosional surfaces changing laterally to paraconformable surfaces.

New play concepts generated by drowning unconformities affecting shallow marine siliciclastic systems are related to large-scale instability on oversteepened, above-grade slopes (cf. Prather 2003). This process gives way to deep but localized erosion of marginal wedges through retrogressive slump scars and submarine canyon incisions, leaving reservoir remnants. Prodelta to slope muds usually fill the erosional surfaces, eventually covering the morphologically prominent remnants and thus sealing these three- to four-way traps.

In this setting, another play concept consists of the combination of onlapping non-marine to marginal marine deposits against a magmatic or metamorphic basement and a drowning unconformity that provided their downwarping, or by fault-bounded fractured intrabasinal basement highs.

Sediment routing and turbidite play concepts in transgressive settings

When transgressive surfaces of marine erosion relate to drowning unconformities, different implications occur with respect to the erosion generated by a ravinement surface associated with shoreface retreat. While in this case the healing phase wedge occurring basinward of the shallow water zone is normally mud-prone, when a tectonic oversteepening is added the underlying shallow-water sands and gravels can be affected by large-scale instability processes and delivered basinward to form shelfal, intra-slope or basin-floor turbidites generated by the stacking of surge flows. The retrogressive nature of slump scars that re-adjust the slope equilibrium profile implies that progressively smaller volumes of sediments are removed. This in turn implies decreasing flow volumes of successive gravity flows resulting in a thinning-upward stacking pattern and a retrogradational organization of turbidite systems (cf. Covault & Graham 2008). Predictions on sequence architecture and on the occurrence of turbidite systems may thus differ from those classically established on passive margins, leading to play concepts that might have been overlooked so far.

Progressive collapse and downwarping of basin margins result in the multiple, step-wise nature of drowning unconformities. Initial erosion is filled by slope facies but the continuation of this process and the overall deepening-upward is expressed by further tilting of slope muds, unconformably overlain by deep-water turbidites onlapping gently inclined muds, commonly developed on the basinward side of underlying remnants. The overall deepening-upward is usually also associated with an increasing efficiency of turbidite systems with time because they are initially supplied by the erosion of coarse grained shallow marine systems and afterwards by the erosion of shelfal, more mud-rich sediments.

High-resolution sequence stratigraphy and implications for reservoir architectures

It appears that fifth-order EDSs can be considered the building blocks of fourth-order systems tracts. The EDS concept, whose complete expression may comprise a decametric R-T cycle consisting of high-frequency LST and TST bedsets capped by a maximum transgression (possibly condensed) interval eventually preserving at the top a HST regressive interval, mimics at small scale the third- and fourth-order sequences. Differently from parasequences, this concept relies upon a complete high-frequency cycle of relative base-level change and also accounts for the landward sectors where LST deposits are missing and high-frequency sequence boundaries may coincide with TS or eventually with MFS.

The internal reservoir architecture of shallow marine systems may be therefore much more complex than expected. The stacking of EDSs offers a more precise analysis of shoreline trajectories, helping in the identification of the relative magnitude of downward shifts of coastal onlap and reducing the uncertainties in the recognition of the maximum rates of relative base-level fall and rise.

We are indebted to two anonymous reviewers, whose constructive criticism permitted us to greatly improve the paper, and to M. Bowman for his careful editorial review. We also thank S. Rogledi, M. Ghielmi and B. Vigna for data and discussions that contributed to improve parts of this paper. M. Pedini of the Eni Graphics Department is acknowledged for preparation of the figures. Cooperation through the years with several colleagues on either the Tertiary Piedmont Basin or the other basins chosen to show petroleum geology applications is also appreciated: a list of all their names would be too long but let us mention U. Biffi, M. Minervini, R. Polino, P. Mosca, R. Gelati, M. Gnaccolini, M. Orlando, H. Brinkhuis, F. Lottaroli, S. Torricelli, G. Knezaurek, M. Bertamoni, F. Dela Pierre, A. Irace, F. Piana, A d'Atri, C. Bertok, M. Mantovani, M. Lazzerini Denchi, I. Marini, E. Spadafora, E. Capoferri, C. Monesi, P. Flores, J. Leutscher, E. Trincianti, A. Ciculo, M. Meda, G. Bagnoli. Last but not least, we thank Eni Upstream & Technical Services for the permission to publish this paper.

References

Argnani, A. 2012. Plate motion and the evolution of Alpine Corsica and Northern Apennines. *Tectonophysics*, **579**, 207–219.

Artoni, A., Bernini, M. *et al.* 2010. Mass-transport deposits in confined wedge-top basins: surficial processes shaping the Messinian orogenic wedge of Northern Apennine of Italy. *Italian Journal of Geosciences*, **129**, 101–118.

Bell, R.E., Jackson, C.A.-L., Elliott, G.M., Gawthorpe, R.L., Sharp, I.R. & Michelsen, L. 2014. Insights into the development of major rift-related unconformities from geologically constrained subsidence modelling: HaltenTerrace, offshore mid Norway. *Basin Research*, **26**, 203–224.

Bersezio, R., Felletti, F., Riva, R. & Micucci, L. 2009. Bed thickness and facies trends of turbiditic sandstone bodies: Unravelling the effects of basin confinement, depositional processes and modes of sediment supply. *In*: KNELLER, B., MARTINSEN, O.J. & MCCAFFREY, B. (eds) *External Controls on Deep-Water Depositional Systems*. SEPM, Tulsa, Oklahoma, Special Publications, **92**, 303–324.

BURBANK, D.W., PUIGDEFABREGAS, C. & MUNOZ, J.A. 1992. The chronology of the Eocene tectonic and stratigraphic development of the eastern Pyrenean foreland basin, northeast Spain. *Geological Society of America Bulletin*, **104**, 1101–1120.

CAPPONI, G., CRISPINI, L., FEDERICO, L., PIAZZA, M. & FABBRI, B. 2009. Late Alpine tectonics in the Ligurian Alps: constraints from the Tertiary Piedmont Basin conglomerates. *Geological Journal*, **44**, 211–224.

CARMINATI, E., LUSTRINO, M. & DOGLIONI, C. 2012. Geodynamic evolution of the central and western Mediterranean: tectonics v. igneous petrology constraints. *Tectonophysics*, **579**, 173–192.

CASNEDI, R., GHIELMI, M., ROSSI, M., CAZZOLA, L. & SERAFINI, G. 2006. Geometrical analysis and seismic modelling of an outer foredeep margin in the lower Messinian Laga turbidite complex (Central Apennines, Abruzzo, Italy). *Bollettino della Società Geologica Italiana*, **125**, 203–220.

CATUNEANU, O., ABREU, V. *ET AL*. 2009. Towards the standardization of sequence stratigraphy. *Earth-Science Reviews*, **92**, 1–33.

CAVANNA, F., DI GIULIO, A., GALBIATI, B., MOSNA, S., PEROTTI, C.R. & PIERI, M. 1989. Carta geologica dell'estremità orientale del Bacino Terziario Ligure Piemontese. *Atti Ticinesi di Scienze della Terra*, **32**.

CAZZOLA, C. & FORNACIARI, M. 1992. Geometria e facies dei sistemi torbiditici di Budroni e Noceto. *Atti Ticinensi di Scienze della Terra*, **33**, 177–190.

COVAULT, J.A. & GRAHAM, S.A. 2008. Turbidite architecture in proximal foreland basin-system deep-water depocenters: insights from the Cenozoic of Western Europe. *Austrian Journal of Earth Sciences*, **101**, 26–51.

COVAULT, J.A. & GRAHAM, S.A. 2010. Submarine fans at all sea-level stands: tectono-morphologic and climatic controls on terrigenous sediment delivery to the deep sea. *Geology*, **38**, 939–942.

D'ATRI, A. 1990. Facies e sequenze deposizionali nella Formazione di Visone (Miocene inferiore, Bacino terziario ligure piemontese). *Memorie della Società Geologica Italiana*, **45**, 723–729.

D'ATRI, A. 1995. Biostratigrafia della Formazione di Visone (Miocene inferiore, Bacino terziario ligure-piemontese). *Bollettino del Museo Regionale di Scienze Naturali di Torino*, **13**, 345–375.

DELA PIERRE, F., PIANA, F. & CLARI, P. 2002. Il Monferrato. *In*: POLINO, R. (coord.) *Il sistema Alpino-Appenninico nel Cenozoico: Torino (Italy)*. Società Geologica Italiana, 81° Riunione estiva. Guida alle escursioni, 116–128.

DELA PIERRE, F., BERNARDI, E. *ET AL*. 2011. The record of the Messinian salinity crisis in the Tertiary Piedmont Basin (NW Italy): the Alba section revisited. *Palaeogeography, Palaeoclimatology, Palaeoecology*, **310**, 238–255.

DI GIULIO, A. 1989. Litostratigrafia e petrografia della successione eo-oligocenica del Bacino Terziario Ligure Piemontese, nell'area compresa tra le valli Grue e Curone (Provincia di Alessandria, Italia settentrionale). *Bollettino della Società Geologica Italiana*, **109**, 279–298.

DOGLIONI, C. 1991. A proposal of kinematic modelling for W-dipping subductions and possible applications to the Tyrrhenian–Apennines system. *Terra Nova*, **3**, 423–434.

DOGLIONI, C. & PROSSER, G. 1997. Fold uplift v. regional subsidence and sedimentation rate. *Marine and Petroleum Geology*, **14**, 179–190.

EMBRY, A.F. & JOHANNESSEN, E.P. 1992. T–R sequence stratigraphy, facies analysis and reservoir distribution in the uppermost Triassic–Lower Jurassic succession,-Western Sverdrup Basin, Arctic Canada. *In*: VORREN, T.O., BERGSAGER, E., DAHL-STAMNES, O.A., HOLTER, E., JOHANSEN, B., LIE, E. & LUND, T.B. (eds) *Arctic Geology and Petroleum Potential*. Norwegian Petroleum Society, Special Publications, Elsevier, Amsterdam, **2**, 121–146.

ESCALONA, A. & MANN, P. 2011. Tectonics, basin subsidence mechanisms, and paleogeography of the Caribbean-South American plate boundary zone. *Marine and Petroleum Geology*, **28**, 8–39.

FANTONI, R., MASSARI, F., MINERVINI, M., ROGLEDI, S. & ROSSI, M. 2001. Il Messiniano del margine subalpino-padano: relazioni tra contesto strutturale e stratigrafico-deposizionale. *Geologia Insubrica*, **6**, 95–108.

FELLETTI, F. 2002. Complex bedding geometries and facies associations of the turbiditic fill of a confined basin in a transpressive setting (Castagnola Fm., Tertiary Piedmont basin – NW Italy). *Sedimentology*, **49**, 645–667.

GARCIACARO, E., MANN, P. & ESCALONA, A. 2011. Regional structure and tectonic history of the obliquely colliding Columbus foreland basin, offshore Trinidad and Venezuela. *Marine and Petroleum Geology*, **28**, 126–148.

GATTACCECA, J., DEINO, A., RIZZO, R., JONES, D.S., HENRY, B., BEAUDOIN, B. & VADEBOIN, F. 2007. Miocene rotation of Sardinia: new paleomagnetic and geochronological constraints and geodynamic implications. *Earth and Planetary Science Letters*, **258**, 359–377.

GAWTHORPE, R.L., FRASER, A.J. & COLLIER, R.E.L. 1994. Sequence stratigraphy in active extensional basins: implications for the interpretation of ancient basin-fills. *Marine and Petroleum Geology*, **11**, 642–658.

GELATI, R. & GNACCOLINI, M. 1996. The stratigraphic record of the Oligocene–Early Miocene events at the south-western end of the Piedmont Tertiary Basin. *Rivista Italiana di Paleontologia e Stratigrafia*, **102**, 65–76.

GELATI, R. & GNACCOLINI, M. 1998. Synsedimentary tectonics and sedimentation in the Tertiary Piedmont Basin, Northwestern Italy. *Rivista Italiana di Paleontologia e Stratigrafia*, **104**, 193–214.

GELATI, R. & GNACCOLINI, M. 2003. Genesis and evolution of the Langhe basin, with emphasis on the latest Oligocene-Earliest Miocene and Serravallian. *Atti Ticinensi di Scienze della Terra*, **44**, 3–18.

GELATI, R., GNACCOLINI, M. *ET AL*. 2010*a*. *Note illustrative della Carta Geologica d'Italia alla scala*

1:50 000, Foglio 211 Dego. 1-123. ISPRA – Istituto Superiore per la Protezione e la Ricerca Ambientale.

Gelati, R., Gnaccolini, M. et al. 2010*b*. *Carta Geologica d'Italia alla scala 1:50 000, Foglio 211 Dego*. ISPRA – Istituto Superiore per la Protezione e la Ricerca Ambientale.

Ghibaudo, G., Massari, F. & Chiambretti, I. 2014. Oligo-Miocene tectono-sedimentary evolution of the Langhe Sub-basin: fromcontinental to basinal setting (Tertiary Piedmont Basin – Northwestern Italy). *Journal of Mediterranean Earth Sciences*, **6**, 53–144.

Ghielmi, M., Minervini, M., Nini, C., Rogledi, S., Rossi, M. & Vignolo, A. 2010. Sedimentary and tectonic evolution in the eastern Po Plain and northern Adriatic Sea area from Messinian to Middle Pleistocene (Italy). *In*: Sassi, F.P. (ed.) *Nature and Geodynamics of the Lithosphere in Northern Adriatic. Rendiconti Lincei Scienze Fisiche e Naturali*, **21**(Suppl. 1), 131–166.

Ghielmi, M., Minervini, M., Nini, C., Rogledi, S. & Rossi, M. 2013. Late Miocene–Middle Pleistocene sequences in the Po Plain–Northern Adriatic Sea (Italy): the stratigraphic record of modification phases affecting a complex foreland basin. *Marine and Petroleum Geology*, **42**, 50–81.

Ghielmi, M., Rogledi, S., Vigna, B., Violanti, D. in prep. La successione Messiniana e Plio-Pleistocenica del Bacino di Savigliano (parte occidentale del Bacino Terziario Piemontese).

Hampson, G.J., Sixsmith, P.J. & Johnson, H.D. 2004. A sedimentological approach to refining reservoir architecture in a mature hydrocarbon province: the Brent Province, UK North Sea. *Marine and Petroleum Geology*, **21**, 457–484.

Hampson, G., Sixsmith, P.J. & Johnson, H.D. 2009*a*. A sedimentological approach to refining reservoir architecture in a mature hydrocarbon province: the Brent Province, UK North Sea. *Marine and Petroleum Geology*, **21**, 457–484.

Hampson, G., Sixsmith, P.J., Kieft, R.L., Jackson, C.A.L. & Johnson, H.D. 2009*b*. Quantitative analysis of net transgressive shoreline trajectories and stratigraphic architectures: mid-to-late Jurassic of the North Sea basin. *Basin Research*, **21**, 528–558.

Hansen, J.P.V., Clausen, O.R. & Huuse, M. 2004. 3D seismic analysis reveals the origin of ambiguous erosional feature at a major sequence boundary in the eastern North Sea: near top Oligocene. *In*: Davies, R.J., Cartwright, J.A., Stewart, S.A., Lappin, M. & Underhill, J.R. (eds) *3D Seismic Technology: Application to the Exploration of Sedimentary Basins*. The Geological Society, London, Memoirs, **29**, 83–89.

Helland-Hansen, W. & Martinsen, O.J. 1996. Shoreline trajectories and sequences: description of variable depositional-dip scenarios. *Journal of Sedimentary Research*, **B66**, 670–688.

Helland-Hansen, W., Ashton, M., Lømo, L. & Steel, R. 1992. *Advance and Retreat of the Brent Delta: Recent Contributions to the Depositional Model*. Geological Society, London, Special Publications, **61**, 109–127, http://doi.org/10.1144/GSL.SP.1992.061.01.07

Howell, J.A. & Flint, S.S. 1996. A model for high resolution sequence stratigraphy within extensional basins. *In*: Howell, J.A. & Aitken, J.F. (eds) *High Resolution Sequence Stratigraphy: Innovations and Applications*. Geological Society, London, Special Publications, **104**, 129–137.

Hunt, D. & Tucker, M.E. 1992. Stranded parasequences and the forced regressive wedge systems tract:deposition during base-level fall. *Sedimentary Geology*, **81**, 1–9.

Jeremiah, J.M. & Nicholson, P.H. 1999. Middle Oxfordian to Volgian sequence stratigraphy of the Greater Shearwater area. *In*: Fleet, A.J. & Boldy, S.A.R. (eds) *Petroleum Geology of Northwest Europe: Proceedings of the 5th Conference*. Geological Society, London, 153–170.

Labaume, P. & Rio, D. 1994. Relationships between the subligurian allochton and the Tuscan foredeep turbidites in the Bobbio window (NW Apennines). *Memorie Società Geologica Italiana*, **48**, 309–315.

Laubscher, H.P., Biella, G.C., Cassinis, R., Gelati, R., Lozej, A., Scarascia, S. & Tabacco, I. 1992. The collisional knot in Liguria. *Geologische Rundschau*, **81**, 275–289.

Leva López, J., Kim, W. & Steel, R.J. 2014. Autoacceleration of clinoform progradation in foreland basins: theory and experiments. *Basin Research*, **26**, 489–504.

Lorenz, C. 1969. Contribution a l'ètude stratigraphique de l'Oligocene inferieur des confins Liguro-Pièmontais (Italie). *Atti dell'Istituto di Geologia dell'Università Genova*, **6**, 253–888.

Løseth, T.M., Ryseth, A.E. & Young, M. 2009. Sedimentology and sequence stratigraphy of the middle Jurassic Tarbert Formation, Oseberg South area (northern North Sea). *Basin Research*, **21**, 597–619.

Maffione, M., Speranza, F., Faccenna, C., Cascella, A., Vignaroli, G. & Sagnotti, L. 2008. A synchronous Alpine and Corsica–Sardinia rotation. *Journal of Geophysics Research*, **113**, B03104, http://doi.org/10.1029/2007JB005214

Mariotti, G. & Doglioni, C. 2000. The dip of the foreland monocline in the Alps and Apennines. *Earth and Planetary Science Letters*, **181**, 191–202.

Martins-Neto, M.A. & Catuneanu, O. 2010. Rift sequence stratigraphy. *Marine and Petroleum Geology*, **27**, 247–253.

Mellere, D., Stefani, C. & Angevine, C. 2000. Polyphase tectonics through subsidence analysis: the Oligo-Miocene Venetian and Friuli Basin, north-east Italy. *Basin Research*, **12**, 159–182.

Mosca, P., Rogledi, S., Polino, R. & Rossi, M. 2010. New data for the kinematic interpretation of the Alps–Apennines junction (Northwestern Italy). *International Journal of Earth Sciences*, **99**, 833–849.

Moscardelli, L., Wood, L.J. & Dunlap, D.B. 2012. Shelf-edge deltas along structurally complex margins: a case study from eastern offshore Trinidad. *AAPG Bulletin*, **96**, 1483–1522.

Mutti, E. 1975. Remarks on the Miocene Bobbio Formation. *In*: Mutti, E., Parea, G.C. et al. (eds) *Example of Turbidite Facies and Facies Associations from Selected Formations of the Northern Apennines, Nice, International Association of Sedimentology*, IX International Congress of Sedimentology, Field trip a11, **24**, 107–131.

MUTTI, E. 1990. Relazioni tra stratigrafia sequenziale e tettonica. *Memorie Società Geologica Italiana*, **45**, 627–655.

MUTTI, E., DAVOLI, G., FIGONI, M. & SGAVETTI, M. 1994. Part 1: conceptual stratigraphic framework. *In*: MUTTI, E., DAVOLI, G., MORA, S. & SGAVETTI, M. (eds) *The Eastern Sector of the South-central Folded Pyrenean Foreland: Criteria for Stratigraphic Analysis and Excursion Notes*. 2nd High-resolution Sequence Stratigraphy Conference, 20–26 June 1994, Tremp, Spain, Istituto di Geologia, Parma University, Parma, 3–16.

MUTTI, E., PAPANI, L., DI BIASE, D., DAVOLI, G., MORA, S., SEGADELLI, S. & TINTERRI, R. 1995. Il Bacino Terziario Epimesoalpino e le sue implicazioni sui rapporti tra Alpi ed Appennino. *Memorie di Scienze Geologiche*, **47**, 217–244.

MUTTI, E., DAVOLI, G., TINTERRI, R. & ZAVALA, C. 1996. The importance of ancient fluvio-deltaic systems dominated by catastrophic flooding in tectonically active basins. *Memorie di Scienze Geologiche, Padova (Italy)*, **48**, 233–291.

MUTTI, E., DI BIASE, D., FAVA, L., MAVILLA, N., SGAVETTI, M. & TINTERRI, R. 2002. The Tertiary Piedmont Basin. *In*: MUTTI, E., RICCI LUCCHI, F. ET AL. (eds) *Revisiting Turbidites of the Marnoso-Arenacea Formation and their Basin-Margin Equivalents: Problems with Classic Models: Florence (Italy), Turbidites Workshop, Excursion Guidebook*. Part II, 1–25, Dipartimento di Scienze della Terra (Università di Parma) and Eni-Divisione Agip, Parma.

MUTTONI, G., LANCI, L., ARGNANI, A., HIRT, A.M., CIBIN, U., ABRAHAMSEN, N. & LOWRIE, W. 2000. Paleomagnetic evidence for a Neogene two-phase counterclockwise tectonic rotation in the Northern Apennines (Italy). *Tectonophysics*, **326**, 241–253.

NACSN – NORTH AMERICAN COMMISSION ON STRATIGRAPHIC NOMENCLATURE 1983. North American stratigraphic code. *AAPG Bulletin*, **67**, 841–875.

PIANA, F., D'ATRI, A. & ORIONE, P. 1997. The Visone Formation: a marker for the Early Miocene tectonics in the Alto Monferrato domain (Tertiary Piemonte Basin, NW, Italy). *Memorie di Scienze Geologiche*, **49**, 145–162, Padova.

POREBSKI, S.J. & STEEL, R.J. 2006. Deltas and sea-level change. *Journal of Sedimentary Research*, **76**, 390–403.

POSAMENTIER, H.W. & ALLEN, G.P. 1999. Siliciclastic sequence stratigraphy: concepts and applications. Society of Economic Paleontologists and Mineralogists (SEPM). *Concepts in Sedimentology and Paleontology*, **7**, 210pp.

PRATHER, B.E. 2003. Controls on reservoir distribution, architecture and stratigraphic trapping in slope settings. *Marine and Petroleum Geology*, **20**, 529–545.

RATTEY, R.P. & HAYWARD, A.B. 1993. Sequence stratigraphy of a failed rift system: the Middle Jurassic to Early Cretaceous basin evolution of the Central and Northern North Sea. *In*: PARKER, J.R. (ed.) *Petroleum Geology of Northwest Europe: Proceedings of the 4th Conference*. The Geological Society, London, 215–249.

RICCI LUCCHI, F. 1986. The Oligocene to recent foreland basins of the northern Apennines. *In*: ALLEN, P.A. & HOMEWOOD, P. (eds) *Foreland Basins*. IAS, Special Publications, Blackwell Scientific, Oxford, **8**, 105–139.

ROSSI, M. & ROGLEDI, S. 1988. Relative sea-level changes, local tectonic setting and basin margin sedimentation in the interference zone between two orogenic belts: seismic stratigraphic examples from Padan foreland basin, northern Italy. *In*: NEMEC, W. & STEEL, R.J. (eds) *Fan Deltas: Sedimentology and Tectonic Settings*. Blackie & Son, Glasgow, 368–384.

ROSSI, M., ROGLEDI, S., BARBACINI, G., CASADEI, D., IACCARINO, S. & PAPANI, G. 2002. Tectono-stratigraphic architecture of Messinian piggyback basins of Northern Apennines: the Emilia folds in the Reggio-Modena area and comparison with the Lombardia and Romagna sectors. *Bollettino della Società Geologica Italiana*, special issue, 437–447.

ROSSI, M., MOSCA, P., ROGLEDI, S., POLINO, R. & BIFFI, U. 2009. New outcrop and subsurface data in the Tertiary Piedmont Basin (NW-Italy): unconformity-bounded stratigraphic units and their relationships with basin-modification phases. *Rivista Italiana di Paleontologia e Stratigrafia*, **115**, 305–335.

ROSSI, M., MINERVINI, M., GHIELMI, M. & ROGLEDI, S. 2015. Messinian and Pliocene erosional surfaces in the Po Plain-Adriatic Basin: insights from allostratigraphy and sequence stratigraphy in assessing play concepts related to accommodation and gateway turnarounds in tectonically active margins. *Marine and Petroleum Geology*, **66**, 192–216, http://doi.org/10.1016/j.marpetgeo.2014.12.012

SALAZAR, M., MOSCARDELLI, L., FISHER, W. & LORENTE, M.A. 2011. Tectonostratigraphic evolution of the Morichito piggyback basin, Eastern Venezuelan Basin. *Marine and Petroleum Geology*, **28**, 109–125.

SALVADOR, A. 1994. *International Stratigraphic Guide: A Guide to Stratigraphic Classification, Terminology and Procedure*. The International Union of Geological Science and the Geological Society of America, Boulder, CO.

SCHLAGER, W. 1999. Type 3 sequence boundaries. *In*: HARRIS, P.M., SALLER, A.H. & SIMO, J.A. (eds) *Advances in Carbonate Sequence Stratigraphy: Application to Reservoirs,Outcrops, and Models*. Society of Economic Paleontologists and Mineralogists, Special Publications, Tulsa, Oklahoma, **63**, 35–46.

SERANNE, M. 1999. The Gulf of Lion continental margin (NW Mediterranean) revisited by IBS: an overview. *In*: DURAND, B., JOLIVET, L., HORVATH, F. & SERANNE, M. (eds) *The Mediterranean Basins: Tertiary extension within the Alpine Orogen*. Geological Society, London, Special Publications, **156**, 15–36.

SOTO, D., MANN, P. & ESCALONA, A. 2011. Miocene-to-recent structure and basinal architecture along the Central Range strike-slip fault zone, eastern offshore Trinidad. *Marine and Petroleum Geology*, **28**, 212–234.

TURCO, E., DURANTI, D., IACCARINO, S. & VILLA, G. 1994. Relationships between foraminiferal biofacies and lithofacies in the Oligocene Molare Formation and Rigoroso Marl: preliminary results from the Piota River section (Tertiary Piedmont Basin, NW Italy). *Giornale di Geologia*, **56**, 101–117.

VAIL, P.R., AUDEMARD, F., BOWMAN, S.A., EISNER, P.N. & PEREZ-CRUZ, G. 1991. The stratigraphic signatures of tectonics, eustasy and sedimentation. *In*: EINSELE, G., RICKEN, W. & SEILACHER, A. (eds) *Cycles and Events in Stratigraphy*, Springer-Verlag, Berlin, 617–659.

ZECCHIN, M. & TOSI, L. 2014. Multi-sourced depositional sequences in the Neogene to Quaternary succession of the Venice area (northern Italy). *Marine and Petroleum Geology*, **56**, 1–15, http://doi.org/10.1016/j.marpetgeo.2014.03.015

Ordovician shallow-marine tidal sandwaves in Algeria – the application of coeval outcrops to constrain the geometry and facies of a discontinuous, high-quality gas reservoir

J. PHILIP P. HIRST

BP Reservoir Development, Sunbury-on-Thames, Middlesex TW16 7LN, UK

Present address: Sedimentology Consultant, Church Lane, Warfield, Berkshire RG42 6EG, UK (e-mail: philiphirst2@btinternet.com)

Abstract: The Tiguentourine field in the Illizi Basin, Algeria, is a multi trillion cubic feet (TCF) gas accumulation, with production primarily from Late Ordovician proglacial density flow deposits with permeabilities typically in the range 100–0.1 mD. The proglacial succession is truncated by an erosive surface, above which a thin interval of shallow-marine sandstones with higher permeabilites is present in a few wells. These well-sorted, high-energy sandstones were, in turn, succeeded by Silurian-age graptolitic shales that form both the seal and the source rock. Although the shallow-marine sandstone is a high-quality reservoir interval, it is thin (<12 m), discontinuous and not resolvable on seismic; it therefore represents a challenging target to predict.

Coeval equivalents of the reservoir interval are exposed in the Tassili N'Ajjer, over 200 km south of the field. An erosive surface at the top of the proglacial sequence is succeeded by sandwaves with wavelengths of >100 m. The geometries and facies relationships have been documented from Landsat and outcrops; they were then compared to modern marine settings. The varied facies and the localized absence of the shallow-marine unit at outcrop serve to explain why the better quality sandstone is variably developed in the subsurface and remains a high risk target.

The In Amenas gas complex is situated in the south of the Illizi Basin, Algeria (Fig. 1a); hydrocarbons have been produced from the Lower Palaeozoic succession since 2006. A simplified chronostratigraphy over the main culmination, the Tiguentourine field, is presented in Figure 2 (Bennacef *et al.* 1971; Beuf *et al.* 1971; Whiteman 1971; Eschard *et al.* 2005; Le Heron & Craig 2008; Eschard *et al.* 2010). The primary reservoir interval comprises the complex Late Ordovician glacial deposits of the Tamadjert Formation. The post-glacial, organic-rich 'hot shales' at the base of the succeeding Imirhou Formation form an effective top-seal, as well as being a prolific regional source rock (Lüning *et al.* 2000). The Tiguentourine field wells have been extensively cored, and the synglacial sediments include proglacial turbidite outwash sandstones and interbedded debrites (Hirst 2012). Conventional core analyses (porosity and permeability data) and petrographical data from thin sections indicate that the high-density turbidites are important reservoirs (Hirst *et al.* 2002). In addition, it is noted that in some wells there is a sandstone immediately above the erosively truncated glacial succession with much better reservoir quality; this sandstone is the subject of this study, but it is only locally developed and has thus proved a challenge to predict.

The Ordovician succession is extensively exposed in the Tassili N'Ajjer region, which is located more than 200 km to the south of In Amenas (Fig. 1b) (Beuf *et al.* 1971), where an erosional cap to the synglacial sediments is also noted. Sedimentological interpretations of the cores and outcrops indicate the well and outcrop facies above the erosional event are comparable with a shallow-marine environment being interpreted. The exposed coeval interval is therefore considered to be a useful analogue to help in interpreting the subsurface. The outcrops have been studied to evaluate their potential to assist in understanding the depositional history of the sandstone overlying the transgressive marine erosion surface, and further, to determine whether it is possible to better predict the thickness and lateral extent of relatively thin and patchy shallow-marine, tidally influenced sandstone units.

Geological setting of SE Algeria

During Palaeozoic times, the Saharan platform was part of the northern passive margin of the Gondwana supercontinent (Smith 1997; Scotese *et al.* 1999; Ghienne *et al.* 2007). This landmass consisted of several terranes that accreted during the late

From: Bowman, M., Smyth, H. R., Good, T. R., Passey, S. R., Hirst, J. P. P. & Jordan, C. J. (eds) 2016. *The Value of Outcrop Studies in Reducing Subsurface Uncertainty and Risk in Hydrocarbon Exploration and Production*. Geological Society, London, Special Publications, **436**, 135–150.
First published online May 26, 2016, http://doi.org/10.1144/SP436.11

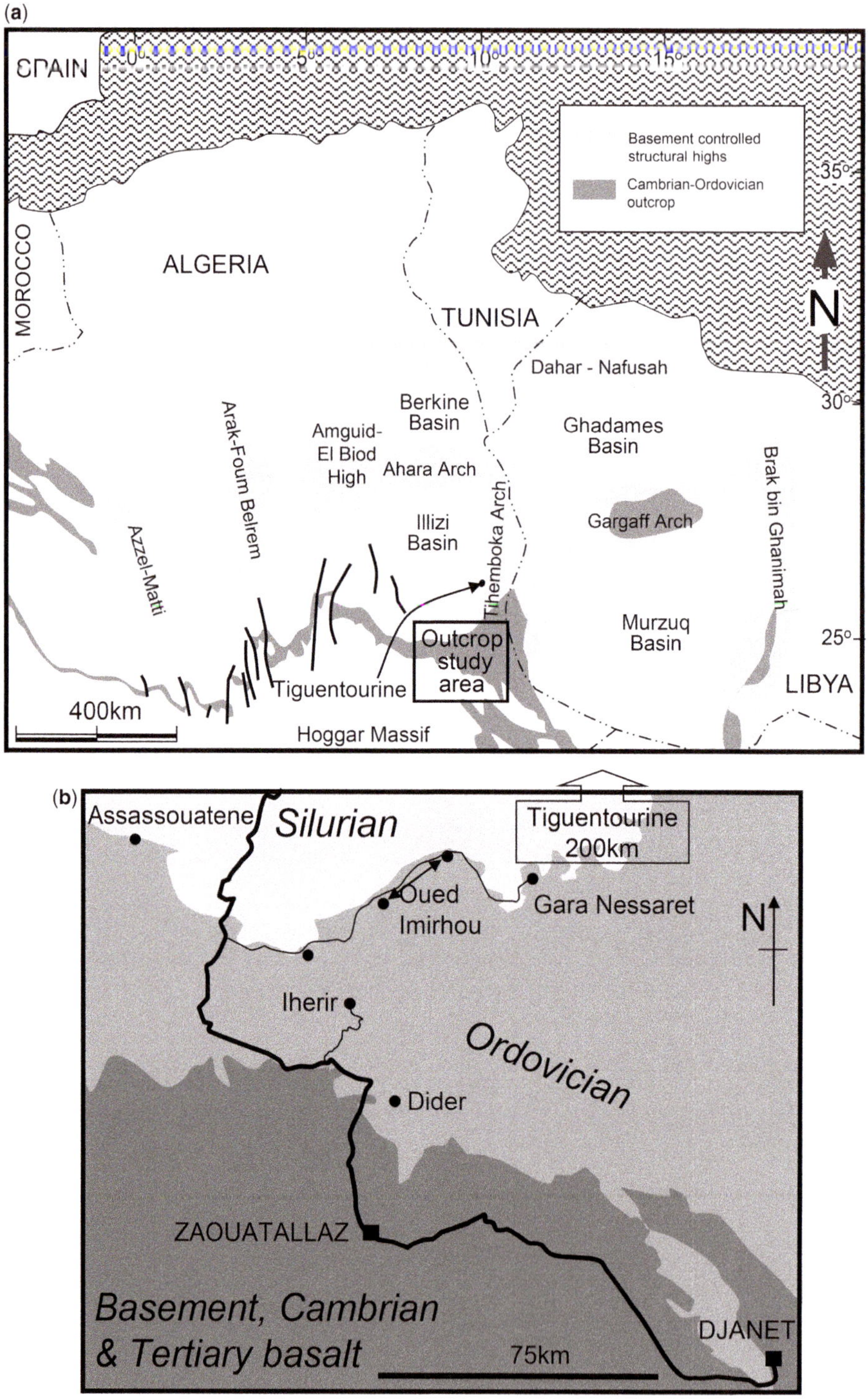

Fig. 1. (**a**) Map of central North Africa illustrating the main structural features and Cambrian–Ordovician outcrop. The Tiguentourine field and the Tassili N'Ajjer outcrop area are located in the south of the Illizi Basin (indicated by a rectangle in b). (**b**) Detailed map of the southern margin of the Illizi Basin, with the locations of the studied outcrops, the main road (thick line) and the access tracks (thin line) shown.

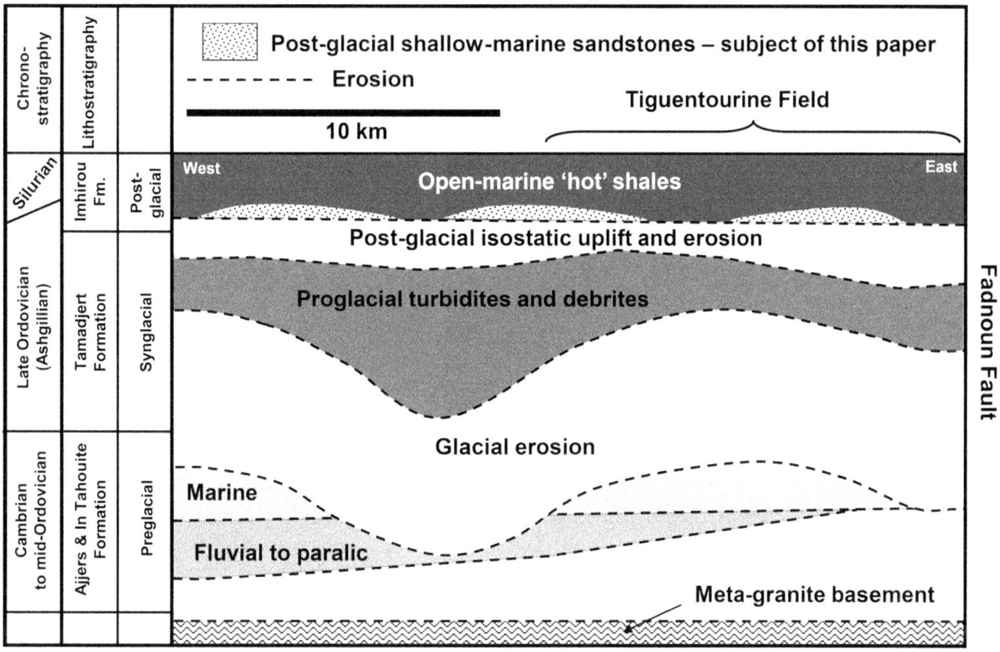

Fig. 2. Schematic stratigraphy of the Cambrian–Lower Silurian succession across the Tiguentourine field, southern Illizi Basin. Lithostratigraphic terms are from Bennacef *et al.* (1971) and Beuf *et al.* (1971, p. 263).

Proterozoic Pan-African Orogeny (Craig *et al.* 2010; Eschard *et al.* 2010; Galeazzi *et al.* 2010). Major north–south faults resulting from this accretion were reactivated several times during geological history, although, away from these faults, the landmass was relatively rigid and stable. A consequence of this periodic tectonic activity was the development of structural highs above several of these Pan-African lineaments; these include the Amguid-el Biod High and the Tihemboka Arch along the west and east sides, respectively, of the Palaeozoic depocentre that forms the Illizi Basin (Fig. 1) (Boote *et al.* 1998; Eschard *et al.* 2010; Galeazzi *et al.* 2010). By latest Ordovician times, when the glacial–post-glacial sediments were deposited, there was little or no tectonic activity affecting the sediment thicknesses across the Tassili N'Ajjer outcrops. Palaeoflows also appear not to have been influenced by the Tihemboka Arch, as the latest Ordovician proglacial palaeoflows have a relatively uniform NW orientation (Beuf *et al.* 1971; Girard *et al.* 2012).

The Late Ordovician glaciation

Late Ordovician glacial deposits have been recognized across North and South Africa, South America and Arabia, and several separate ice centres have been interpreted (Beuf *et al.* 1971; Hambrey & Harland 1981; Moreau *et al.* 2005; Saltzman & Young 2005; Ghienne *et al.* 2007; Le Heron & Dowdeswell 2009). Algeria was situated towards the margin of the North African ice sheet, with the Tiguentourine field and the Tassili N'Ajjer outcrops studied here entirely beneath ice at the glacial maxima (Hirst *et al.* 2002, Roussé *et al.* 2009, Lang *et al.* 2012). During the Late Ordovician Hirnantian Stage, at least two glacio-eustatic sea-level falls are interpreted in the *N. extraordinarius* Zone, followed by a glacio-eustatic sea-level rise in the *N. persculptus* Zone through to early Silurian times (Brenchley *et al.* 1994, 1995; Paris *et al.* 1998; Underwood *et al.* 1998; Sutcliffe *et al.* 2000; Le Heron & Craig 2008; Le Heron & Dowdeswell 2009). Post-glacial sedimentation is marked initially by a shallowing event, possibly the product of isostatic uplift as the ice front retreated, resulting in shallow-marine sediments cropping out in the Tassili N'Ajjer (Hirst *et al.* 2002; Dixon *et al.* 2008). With further melting of the ice masses across Africa, global sea-level rise outpaced any further isostatic rebound, resulting in the deposition of a thick succession of offshore graptolitic shale; included near the base are high-gamma 'hot shales', which have been recorded across the Saharan region (Legrand 1988; Lüning *et al.* 2000).

Subsurface depositional context and reservoir quality

The synglacial interval in the Tiguentourine field has been studied using core from numerous wells in the field; a bulk of the Upper Ordovician reservoir interval has been interpreted as the deposits of proglacial outwash turbidites and debrites (Hirst 2012). Proglacial, high-density turbidites with a low mud content comprise the primary reservoir in the succession, whilst the more mud-prone turbidite sandstones are poor reservoirs (Fig. 3) (Hirst *et al.* 2002). However, the highest recorded porosities and permeabilities are from a 11 m-thick blocky sandstone unit at the top of the reservoir interval, the lowest 1 m of which was cored in well Tg-315 (Figs 3 & 4). The base of this blocky sandstone is an abrupt, probable erosive surface, whilst the unit is capped by the Silurian, high-gamma, graptolitic marine shales. The interval comprises coarse to very coarse sandstone, in contrast to the mainly fine-grained sandstone in the turbidites and debrites of the underlying succession. Sedimentary features include poorly defined tractional bedforms, and sparse subcentimetre-diameter horizontal and vertical burrows. The grain size, bedforms and burrows are interpreted as the deposits of a shallow-marine, moderate- to high-energy setting above the erosional surface, indicating a significant change in depositional environment from the preceding synglacial outwash turbidites.

The erosional surface is recognized in many of the Tiguentourine field wells, although interpretations rely mostly on log character calibrated to the few cores available over this interval. The succession below and above the erosional event is variable in terms of both thickness and facies across the field. This variation is illustrated by the five wells in Figure 5 and the map indicating the facies encountered immediately below the Silurian high-gamma mudstones (Fig. 6). Towards the top of the synglacial succession, the sediments tend to be finer grained, with interbedded mudstone and siltstone layers succeeding the density flow deposits; this is seen in the higher gamma response in Figure 5 and the mudstone layers in Tg-315 (Fig. 4). Rare graptolites are recorded in the mudstones in some wells, including TRN-2. In several wells, a coarsening- and cleaning-upwards is noted immediately below the erosive surface. This coarsening-up interval, sometimes with ripple cross-laminae, tends to be reddened or colour mottled with a spectrum of small horizontal and vertical burrows including *Chondrites*, *Arenicolites* and *Palaeophycus*. It is interpreted to be a burrowed firmground formed during a shallowing event, which was, in turn,

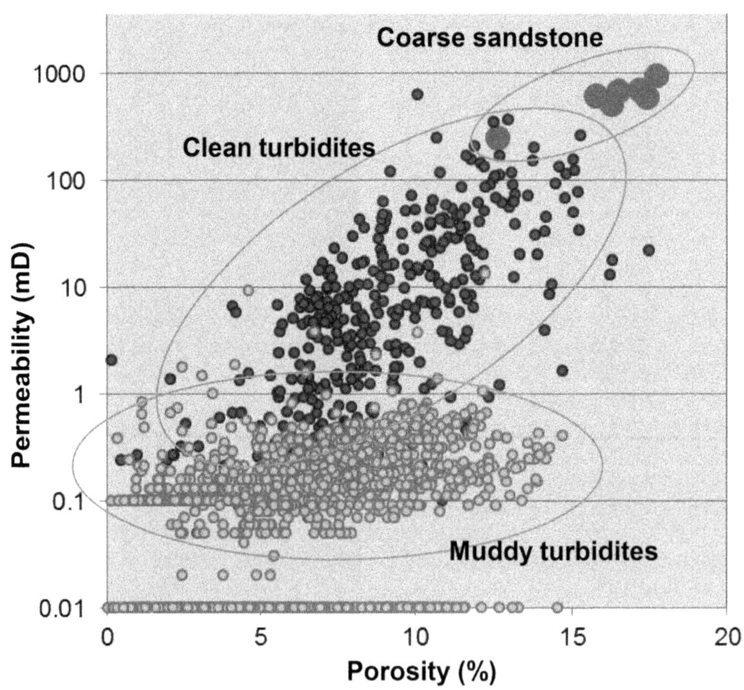

Fig. 3. Cross-plot of core-derived porosity and permeability analyses for the clean and muddy high-density synglacial turbidites and the post-glacial coarse sandstone after Hirst *et al.* (2002).

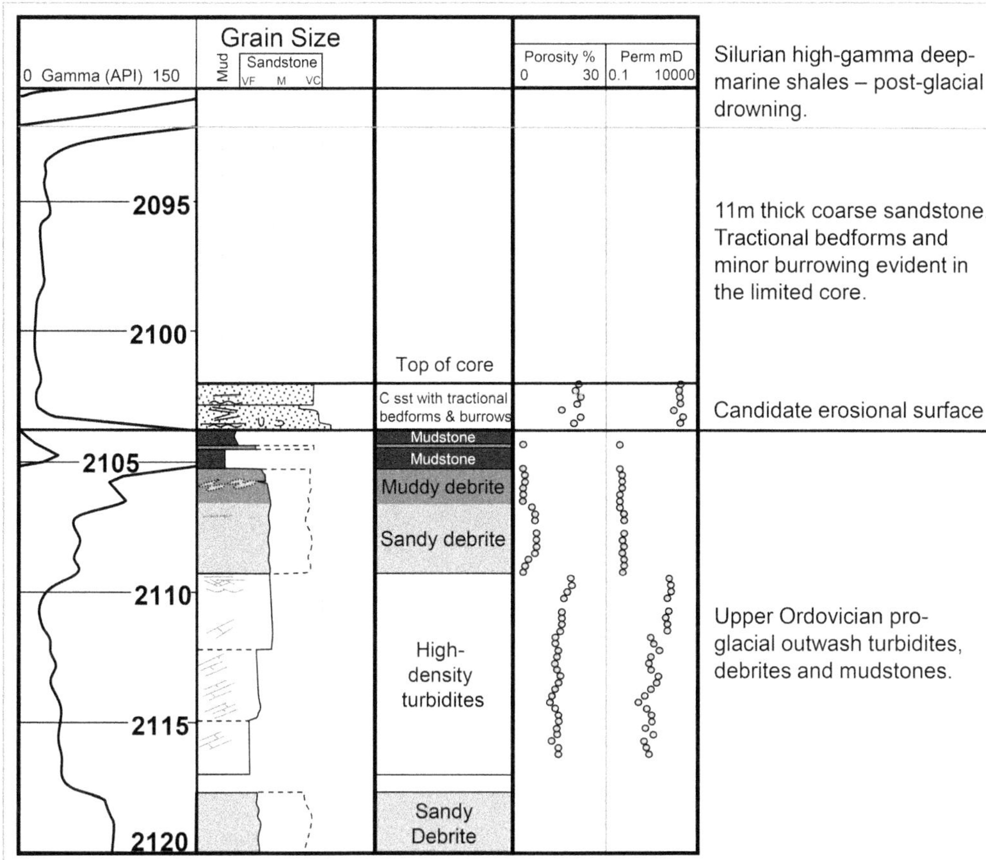

Fig. 4. Gamma log, core description and porosity–permeability data for the upper part of the reservoir in Tiguentourine well Tg-315. sst, sandstone.

truncated by the erosive event (Tg-7, Tg-311 and TRN-2: Fig. 5); this firmground has been interpreted over a wide area (Fig. 6).

In many of the wells, either with or without the firmground, the sediments above the erosive event are succeeded directly by high-gamma graptolitic shales (e.g. Tg-312 and Tg-7: Figs 5 & 6). In only a few wells were sandstones deposited above the erosional event; the sandstone is either the coarse and massively bedded deposit described in Tg-315 or may be a finer-grained, thin-bedded, heterolithic sediment as in TRN-2 and Tg-311 (Fig. 5). These finer sandstones may be wavy bedded with some ripple cross-lamination, mud drapes and sparse burrows. They are also interpreted to be marine in origin but not such high energy as the coarse sandstones, whilst the erosive surface without sandstone is likely to have been an area of the seafloor swept clean of sandstone. From the map (Fig. 6), it is evident that the thick sand with excellent reservoir quality recovered in Tg-315 is sparsely developed; equally problematic from a reservoir perspective are the shape and orientation of each accumulation.

Ordovician synglacial deposits at outcrop

To gain a better understanding as to why the Tiguentourine field post-glacial deposits vary in both thickness and facies, the coeval succession has been studied where it crops out at several localities some 200 km south of the field (Fig. 1b) (Beuf *et al.* 1971).

Oued Imhirou

Oued Imhirou is a major wadi immediately to the south of the east–west-orientated falaise (escarpment) in which Silurian–Devonian aged rocks are exposed. Along Oued Imhirou, the uppermost late Ordovician proglacial deposits are observed to have been erosively truncated; at the selected location, a

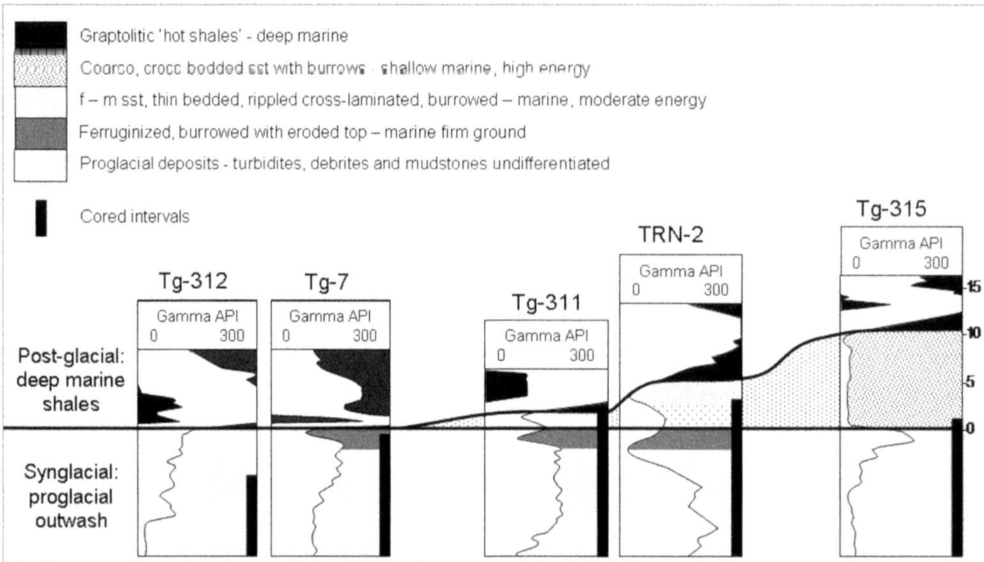

Fig. 5. Gamma log and sedimentological interpretation for five Tiguentourine field wells covering the interval from the upper part of the proglacial sediments up to the base of the Silurian high-gamma shales. The datum is the erosional surface. f–m sst, fine–medium sandstone.

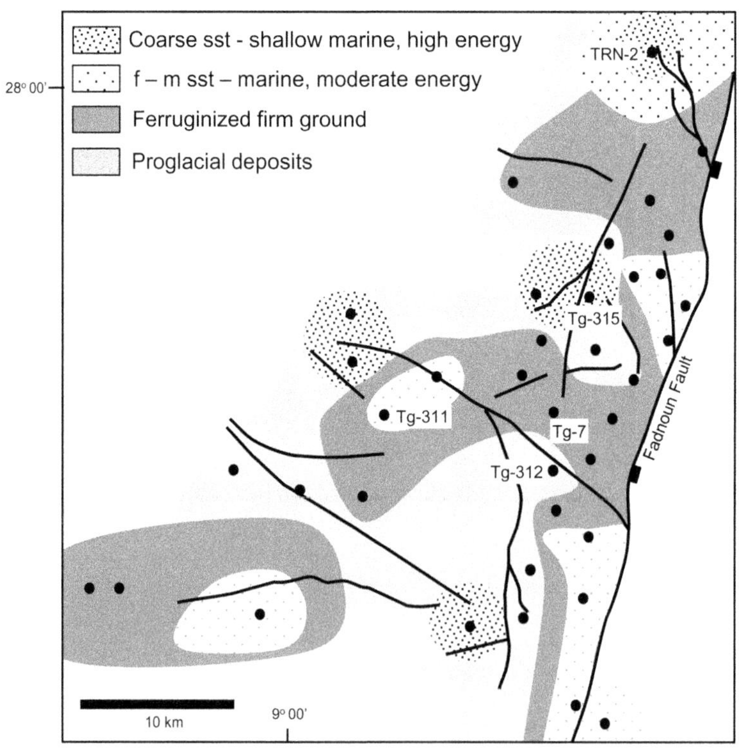

Fig. 6. The sediments subcropping the Silurian shales in the crestal part of the Tiguentourine field to to the west of the Fadnoun Fault (mapped from well data, including wells noted in the figure). f–m sst, fine–medium sandstone.

sedimentary log of the interval immediately above and below this event is presented (Fig. 7a, b). The underlying proglacial sediments comprise predominantly fine sandstone deposited as 1 m or longer wavelength, low-amplitude bedforms (Fig. 7c), which have been termed climbing dune cross-stratification (Ghienne *et al.* 2010). In this area, they have been interpreted as antidune bedforms deposited by high-density turbidity currents that flowed across a proglacial outwash fan (Hirst 2012). The antidunes are truncated by a sharp, planar erosion surface (Fig. 7a), above which the fine sand was reworked into thin, tabular beds. These become ferruginized upwards, particularly the upper 60–80 cm (Fig. 7d). Sedimentary structures were not discernable in this unit, although *Arenicolites* and *Skolithos* burrows were evident. Coarse sand grains and granules are sparsely distributed across the upper surface (Fig. 7e), along with poorly preserved elongate serrated depressions that may be graptolite impressions. The sandstone is succeeded by the Silurian graptolitic shales that crop out in the falaise behind the logged interval (Fig. 7a).

The infauna suggest a shallow-marine environment, with marine energy the likely cause of the underlying sharp, erosive surface; the succeeding thin bedded sandstones are interpreted to be reworked from the proglacial succession before being blanketed by the 'hot shales'.

Gara Nessaret West

Two further localities have been studied, both in the Gara Nessaret area (Fig. 8) some 60 km east of the Oued Imhirou outcrop. The more westerly of these two localities is adjacent to a wadi in which approximately 35 m of stacked sheet and channelized, fine- to sometimes medium-grained sandstones are exposed (Fig. 8 area A; Fig. 9). Long-wavelength bedforms are evident and they are interpreted, as at Oued Imhirou, to be antidune bedforms deposited by high-density turbidity currents in a proglacial setting (Hirst 2012). They are truncated by a planar erosive surface above which an approximately 7 m sheet of coarse-grained sandstone is exposed over a wide area (Fig. 8 area B; Fig. 10a). Other than being coarser grained than the proglacial sediments, this sandstone is friable and porous with little detrital clay, in contrast to the well-cemented turbidite interval (Fig. 10b). Internally, the lower part of the unit is variably cross-bedded, with bedsets up to 0.5 m high. The palaeocurrents were towards the east and burrows, mainly *Skolithos*, are preserved in some of the beds

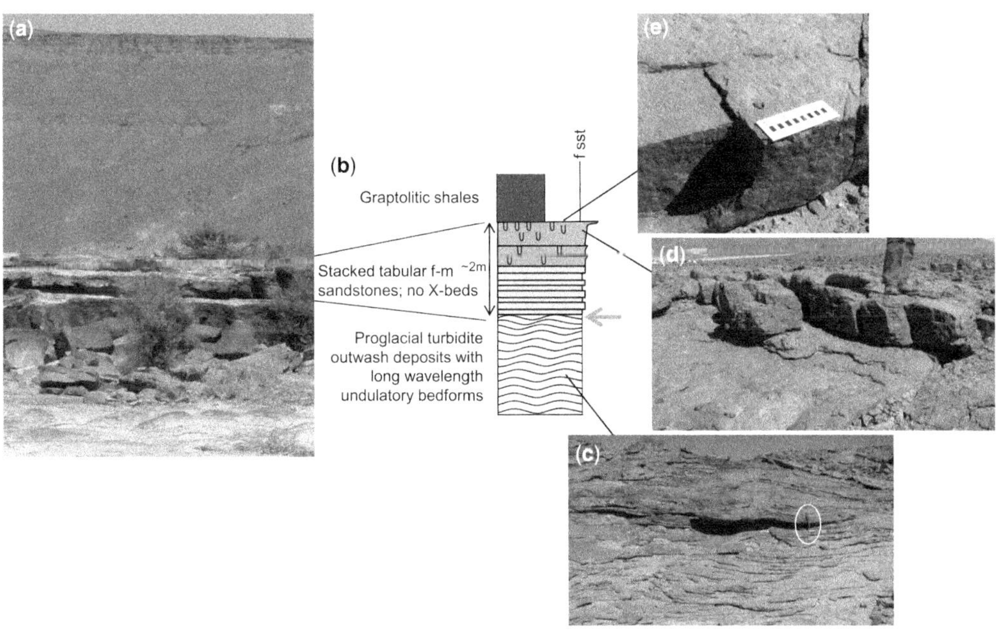

Fig. 7. Oued Imhirou outcrop. (**a**) General view of the uppermost Ordovician sandstones, with the Silurian shales up to the Devonian sandstones exposed in the escarpment in the background. (**b**) Summary log of the uppermost Ordovician sandstones with a planar erosion surface (arrow). (**c**) Long wavelength antidune bedforms within a proglacial high-density turbidite. (**d**) Well-cemented topmost sandstone bed with a burrowed and slightly coarser upper surface (**e**).

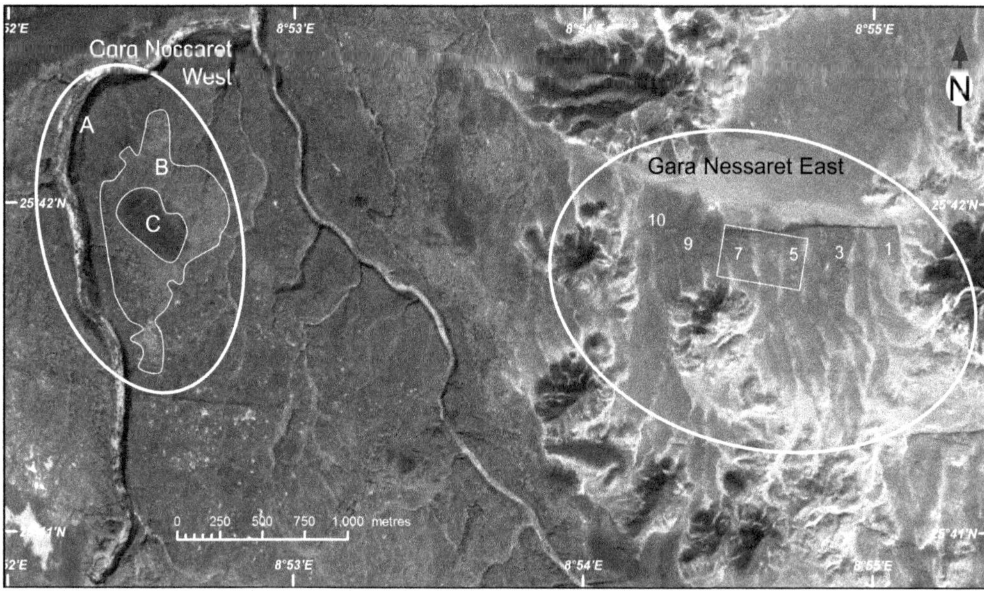

Fig. 8. Landsat image of the Gara Nessaret study areas. The Gara Nessaret West area is adjacent to a major wadi, with an outlier of Silurian shales (area C) above an approximately 7 m coarse sheet-like sandstone (area B), which sits on the eroded top of the proglacial turbidites (area A). Gara Nessaret East comprises a series of north–south-orientated curvilinear ridges, with pale wind-blown sand in the troughs: the ridges, which terminate abruptly northwards, are numbered 1–10 from east to west. The box across ridges 5, 6 and 7 is the area of detailed study in Figure 12. (Image purchased from Spot Image SA.)

(Fig. 10c). The upper part of the coarse unit is more consistently cross-bedded, with stacked low-angle planar sets dipping between 10° and 15° consistently eastwards (Fig. 10d); individual sets can be traced laterally for several tens of metres. The uppermost surface of the sandstone is more intensely burrowed than the body of the sand unit with small *Arenicolites* and *Skolithos*, and is succeeded abruptly by the lowermost Silurian graptolitic shales (Figs 8 & 10).

The grain size, paucity of detrital clay and large scale of some of the cross-beds are indicative of a high-energy setting. The widespread and non-channelized geometry of the sandstone blanketing the erosional event, along with the infauna, suggest a shallow-marine environment in which the dominant sediment transport direction was to the east.

Gara Nessaret East outcrop

The coarse-grained sandstone unit described above is exposed 3.5 km further east at Gara Nessaret East, where it crops out as north–south-orientated curvilinear ridges; these can be picked out on the satellite image because of the paler wind-blown sand that has accumulated in the troughs (Fig. 8). These ridges have been studied previously by Beuf *et al.* (1971, fig. 71) and from satellite imagery by Dixon *et al.* (2008). In this study, the ridges have been numbered 1–10 (Fig. 8) and, looking east towards ridge 5 from ridge 6 (Fig. 11), they can be seen to be low-amplitude features approximately 3.5 m high. Measuring across their axes, the 10 ridges on Figure 8 extend approximately 1460 m (trough to trough), giving an average width of 146 m. They are, however, variable, ranging from a maximum of up to 190 m down to less than 100 m wide. Individual ridges also vary along their axes both in terms of width and the orientation of their axis, with gentle bends of up to 20° noted (Fig. 8).

Internal stratification is dominated by planar surfaces that dip mainly between 10° and 12° towards the east, with a maximum inclination of 16° (Fig. 12b). Some cross-laminae indicate eastwards migration of small dunes or large ripples down the lee slopes of the larger sets (Fig. 12c). Whilst the ridges are relatively straight, with some bends of up to 20° noted above, the component sets, which record successive accretionary events, locally exhibit a higher degree of curvature in plan, indicating that the lee side periodically was more of a three-dimensional (3D) form rather than a simple planar slip-face (Fig. 12a). The ridges are mainly composed of medium sandstone; the top of some sets comprises

Fig. 9. General view NW across the wadi at Gara Nessaret West. The wadi floor and wall comprise stacked high-density turbidites that terminate upwards abruptly at the flat surface outside the wadi. In the near distance, an approximately 7 m-thick sandstone sheet rests on this surface and this sheet is, in turn, succeeded by the Silurian hot shales in lower slopes of the escarpment.

coarse sandstone and there are rounded mud clasts up to 3 cm in diameter on some of the eastwards-dipping surfaces. Burrows are sparse, and include small *Arenicolites* and *Skolithos*.

The basal contact with the underlying proglacial sandstone is not exposed immediately beneath the ridges but regionally it is an eroded, planar surface, as seen at the previous locality 3.5 km to the west and elsewhere in the general area. Although the sandstone ridges thin from crest to trough, sandstone is continuous across the trough, so that it comprises an extensive sheet (Fig. 13a). Within the trough, the sandstone consists of stacked, erosionally truncated, eastwards-dipping cross-beds and these are likely to be the remnants of large-scale migrating bedforms that were mostly removed by currents ahead of the following ridge. The geometry of the ridges is schematically represented in Figure 13b. The final elements noted at the outcrop are thin units of thin-bedded, fine-grained, rippled sandstone, which sit in the base of the troughs (Fig. 13c). The northwards palaeoflow of these ripples is orthogonal to the dominant eastwards palaeotransport direction evident in the ridges. The finer nature of the sand and the different current direction suggests these thin beds were deposited within the troughs after the ridges had stabilized.

The overall geometry noted in this study (Fig. 13b) is in agreement with an earlier representation by Beuf *et al.* (1971) (Fig. 13d), in that the major bedforms dip eastwards and the thin-bedded sandstones ('plaquettes' of Beuf *et al.* 1971) in the troughs are depositionally separate from the ridges. There is, however, no indication of the undulose base to the unit that was indicated by Beuf *et al.* (1971), and the regionally developed planar erosive surface is considered more plausible. Also, it is clear from this study that the coarser sandstone that forms each ridge is in continuity with the next ridge, although it is much thinner across the troughs.

Depositional setting

At each of the studied localities, the late Ordovician proglacial sediments deposited by turbiditic outflows (Hirst 2012) were truncated at a relatively planar surface interpreted as a significant erosional

Fig. 10. Gara Nessaret West. (**a**) Coarse 7 m-thick sand sheet. (**b**) Petrographical thin section of the coarse sandstone. (**c**) Easterly directed cross-beds in the lower part of the sheet with *Skolithos* burrows (inset). (**d**) Eastwards-dipping planar cross-sets in the upper part of the sheet.

event. Above this surface, a 1–7 m-thick interval of sandstone, somewhat coarser than the fine-grained proglacial sand, has been described. The grain size, paucity of detrital clay and large scale of some of the cross-beds at the Gara Nessaret locations are indicative of a high-energy setting. The widespread development and non-channelized geometry of the sandstone blanketing the erosional event, along

Fig. 11. View eastwards from the crest of sandstone ridge 6 towards ridge 5 illustrating that the ridges are low amplitude relative to their width; note the sharp termination northwards (left of the number 5), probably due to a fault.

Fig. 12. (**a**) Higher-resolution Landsat image of ridges 5, 6 and 7 (Quickbird data from Dixon *et al.* 2008); the white lines pick out two surface traces of the eastwards-dipping planar sets that are evident in the end view (**b**). These sets exhibit a higher degree of curvature in plan view than the gross sand ridge does. (**c**) Detail of small cross-sets resulting from bedforms that migrated eastwards down the main sets. Small clasts are also evident.

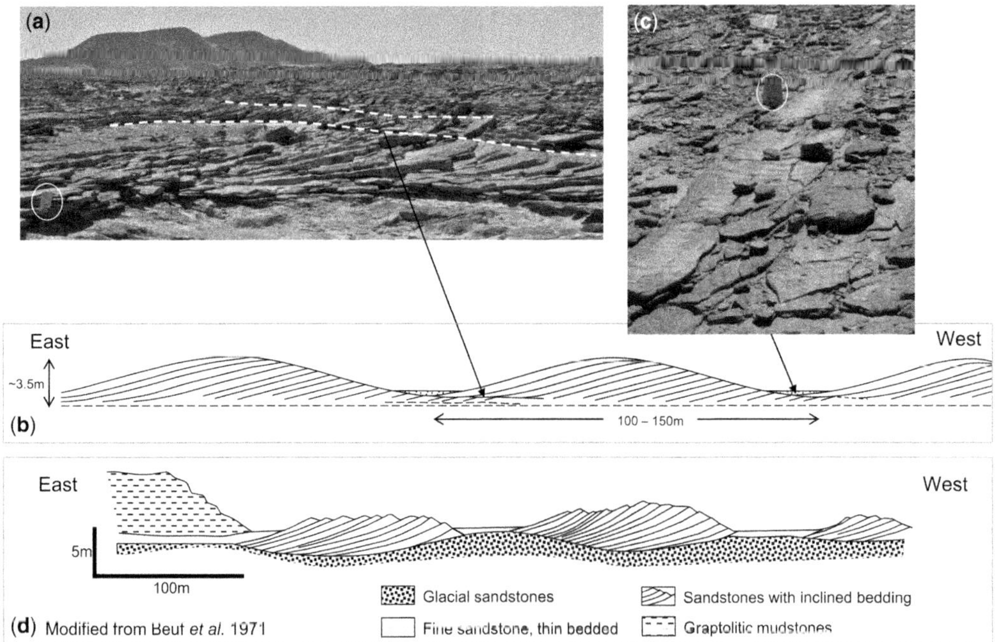

Fig. 13. (**a**) View south along a trough between two sandstone ridges. Cross-bedded sandstone is evident in the trough, indicating that the sandstone comprises stacked eastwards-migrating bedforms and that the sandstone is contiguous from one ridge to the next (20 cm-high notebook for scale). (**b**) Simplified sketch to illustrate the internal structure of the sandstone ridges (vertical exaggeration *c.* ×4). (**c**) Straight crested ripples formed of fine sand, within an inter-ridge trough indicating northwards palaeoflow along the trough (the notebook is for scale). (**d**) Modified sketch from Beuf *et al.* (1971) for comparison with the sketch constructed in (b).

with the infauna, suggest a shallow-marine environment in which the dominant palaeocurrent direction in the Gara Nessaret area was towards the east. Whilst it has been suggested these units may have been raised beaches (Beuf *et al.* 1971), the distinctive sand ridges at Gara Nessaret East are reminiscent of sand ridges described from many tidally influenced shallow-marine shelves (Stride *et al.* 1982; Reading 1996; Leeder 2011). The features documented here are much smaller than the prominent sand ridges described from the southern North Sea, which were largely formed longitudinally to peak tidal current directions (Stride *et al.* 1982) and are termed tidal bars by Olariu *et al.* (2012); some of these ridges are more than 30 m high. Smaller tidal bars formed by similar ridge parallel currents have been described by Berné *et al.* (2002) from the East China Sea. Unlike these modern examples, the somewhat smaller Algerian features do not coarsen-up. Furthermore, with most of the Ordovician palaeoflow indicators towards the east across the ridges, they more closely resemble modern compound tidal dunes or sandwaves formed transversely to peak tidal current directions (Olariu *et al.* 2012). They are similar in size and geometry to the compound tidal sandwaves described from the mouth of the San Francisco Bay entrance (Barnard *et al.* 2006).

The regional erosional event was a product of widespread shallowing of the marine shelf some time following the deposition of proglacial sediments. The shallowing is interpreted to be the product of isostatic rebound following retreat of the ice front southwards away from the immediate area, resulting in the establishment of a high-energy shallow-marine environment. No indication of subaerial exposure has been recognized, although this cannot be discounted. The grain size of the shallow-marine sediments is considerably coarser than that of the underlying proglacial sandstone, indicating that the unit was not derived simply through winnowing of the facies, as seen in the subcropping interval. It is suggested that a succession of shallower-water marine, or even fluvial deposits may have introduced coarser material into the area and was subsequently reworked.

The tidally influenced sand unit was, in turn, overlain by Silurian high-gamma graptolitic shales (Lüning *et al.* 2000), indicating an overall deepening of the marine environment that would be consistent with the final demise of the major ice cap (Le Heron & Craig 2008).

Discussion

Comparison with modern shallow-marine tidal shelves

It is evident that the post-glacial interval cropping out in the Tassili N'Ajjer varies in both facies and thickness over relatively short distances. In order to gain a better understanding as to how variable a marine substrate can be, it is instructive to compare these Ordovician deposits with modern shallow-marine sediments. There have been numerous studies of active tidally influenced environments; however, often relatively small areas have been described and typically they lack the spectrum of deposits described from Algeria (e.g. Dalrymple 1984; Berné *et al.* 2002; Barnard *et al.* 2006). In contrast, the shallow coastal waters of NW Europe, inundated following a sea-level rise since the last glacial maximum at approximately 18 ka BP, provide an extensive and well-studied region in which tidal activity has been a major control on sediment movement and deposition; significant sand banks are noted in some areas (e.g. Stride *et al.* 1982). Large areas of the modern seabed have been mapped in detail in order to understand variations in the lithology, bedforms, fauna and mineral resources; the synthesis by James *et al.* (2011) covers 12 755 km^2 of the northern part of the English Channel. This study provides an indication of the potential variability of the seabed sediments across a shelf with strong tidal currents. It is not proposed that this area is an exact analogue for the Ordovician of Algeria; however, the area includes sandwave deposits of comparable size to those in Algeria, along with swept seabed surfaces in close juxtaposition. Within a subset of the studied area (*c.* 30 km square SW of Dungeness: Fig. 14), a range of depositional morphologies has been mapped. These include rocky substrate, thin sediment and rock, coarse gravelly sediment, and areas that are predominantly sandy. Some of these sandy areas have been sculpted by tidal currents into megaripples, sandwaves with megaripples, sand banks and sediment ridges (James *et al.* 2011). The variation in deposits is the product of multiple factors including water depth, substrate resistance, current regime, sediment supply and sediment calibre. The current regime is a combination of current direction, flow duration, and the relationship between ebb and flood currents, as well as the peak flow strength; these aspects are, in part, influenced by seabed topography and local coastal morphology. This complex set of factors will affect sediment transport direction and create bedload partings and convergences, to form swept seabed surfaces free of sediment and sediment sinks.

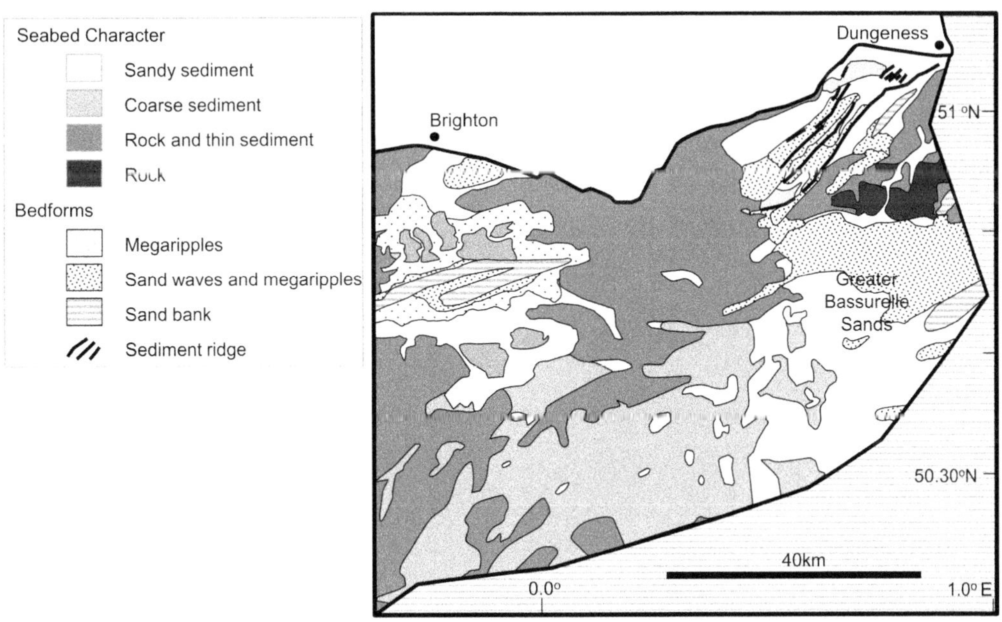

Fig. 14. Simplified sketch of the seabed character in the English Channel offshore Brighton–Dungeness (based on the detailed work of James *et al.* 2011).

The result is the proximity of very different seabed morphologies, including tidal bars and sediment ridges that formed longitudinally to peak tidal current directions, and megaripples and compound tidal sandwaves that formed transversely to peak tidal current directions (James *et al.* 2011). It is probable that a similarly complex range of depositional controls were in operation across the late Ordovician shelf, although elucidating the causes and effects is not possible. By analogy, it would be expected that the tidally influenced, post-glacial deposits cropping out in the Tassili N'Ajjer would exhibit a wide range of shallow-marine facies over short distances. Equally, the similar succession cored over the Tiguentourine High, whilst not demonstrably in continuum with the outcrops 200 km to the south, may also be expected to exhibit a significant degree of facies variability.

With the understanding gained from the modern example that the Tassili N'Ajjer outcrops are part of a contiguous but varied seabed, the contrasting well profiles presented in Figure 5 can now be reviewed in terms of the spectrum of shallow-marine lithologies and morphologies. The erosional event, interpreted to be the product of isostatic rebound, would have cut down into the proglacial deposits, as seen in well Tg-312 – this well would be similar to areas of the modern English Channel seabed where, locally, the bedrock is exposed and with any loose material being removed by the currents. In other wells, a subtle coarsening- and cleaning-upwards profile is noted below the erosional event, and often conditions were appropriate for the formation of a firmground in the immediate substrate, as seen in wells Tg-7, Tg-311 and TRN-2. Above the firmground in well Tg-311, the seabed accumulated a heterolithic thin-bedded and burrowed sediment similar to the outcrop at Oued Imhirou. Coarser sandstone is developed in TRN-2 and this is interpreted as part of a compound tidal sandwave similar to those exposed at Gara Nessaret East, whilst the well-developed sandstone in Tg-315 is slightly thicker, but similar in profile, grain size and facies to the outcrop in Gara Nessaret West. In addition, petrographical analysis from the outcrop indicates a markedly better reservoir quality of this unit relative to the preceding proglacial sediments.

Have outcrop studies helped to reduce uncertainty?

The late Ordovician–early Silurian stratigraphy and facies are similar in the subsurface and at the outcrops; the succession comprises proglacial outwash sediments, an erosional event overlain by shallow-marine sandstones, which, in turn, were blanketed by deeper marine 'hot shales'. The outcrops are therefore considered to be an appropriate analogue for the Ordovician high-porosity reservoir encountered locally at the Tiguentourine High. In turn, these outcrops have been compared with modern tidally influenced seabeds to develop a depositional model. The reservoir unit that is the subject of this study is the high-energy coarse-grained shallow-marine sandstone initially identified as a separate reservoir entity in well Tg-315. The Tassili N'Ajjer outcrops and the modern English Channel seabed would suggest this interval would likely to be laterally discontinuous over kilometres, with adjacent areas of non-deposition; nearby wells indicate this to be the case. Thus, the outcrop studies have helped to explain the variations in facies and thickness, and also why the sandstone is devoid of detrital mud relative to the proglacial succession. Uncertainty concerning the expected spectrum of facies within this interval and the lateral relationships has therefore been reduced through outcrop studies.

Have outcrop studies helped to reduce risk?

A depositional model has been formulated to explain why the post-glacial sandstone interval is variable in thickness and facies, and that there will be areas devoid of any reservoir because the seabed was swept clean of sand due to local conditions (areas of current divergence). Predicting this sub-seismic interval remains an exploration and development challenge – it continues to be a high risk target because of the innate variation in the depositional system and the apparent small areal extent of the better quality sandstones relative to well spacing. Determining bedform dip direction may constrain the long orientation of the sandwaves, but this may not significantly impact well placement. Equally, it is evident that there may be more positive outcomes – elsewhere in the region, a thicker, more sheet-like unit may be developed.

Conclusions

- Gas production from the Tiguentourine field is primarily from the thick succession of proglacial high-density, fine-grained turbidite sandstones deposited during the late Ordovician glacial event. However, the best-quality reservoirs are part of the post-glacial succession, but these are volumetrically minor, as they are thin and patchily developed.
- The late Ordovician–early Silurian stratigraphy and facies are similar in the subsurface and at the coeval outcrops 200 km to the south; the outcrops are considered to be useful analogues to aid in the interpretation of the Tiguentourine field sandstone geometries.

- The studied succession comprises proglacial outwash sediments truncated by an erosional event, which is interpreted to be the result of isostatic rebound following the southwards retreat of the ice front. This erosional surface is variably overlain by shallow-marine heterolithics and coarse sandstones, which constitute the best potential reservoir. These shallow-marine sediments, in turn, were blanketed by deeper marine 'hot shales' as the sea level rose in response to the demise of the ice cap; these shales form both the seal and the source rock.
- The post-glacial high-quality arenites were deposited as shallow-marine tidally influenced sandstones that are variably developed at outcrop, with the thickness of the unit ranging from <1 to >7 m. In some localities, they were formed into low-amplitude ridges approximately 3.5 m high; these are interpreted to be compound tidal sandwaves constructed by dominant currents with an easterly palaeoflow, which was transverse to the north–south axes of the ridges. The average wavelength of these sandwaves was of the order of 150 m.
- Comparison with modern tidally influenced settings and, in particular areas, of the English Channel seabed, supports the idea that sandstone depositional morphology is varied in such settings, with areas devoid of sandstone adjacent to significant sandstone ridges. This varied seabed morphology suggests that reservoir potential will be inconsistent and prediction of the sandstones challenging.
- A depositional model has been developed using both coeval outcrop and modern analogous systems. These have helped to improve understanding and reduced uncertainty regarding the range of facies that may be encountered over short distances across a tidal shelf. However, with the innate variability associated with tidal shelf environments, there remains significant risk in attempting to exploit these reservoirs. There is, however, an upside outcome that the shallow-marine deposit elsewhere on the shelf could comprise a thicker and more contiguous sheet sandstone than has been encountered so far.

Gratitude is extended to Sonatrach for permission to publish this work; the interpretations have benefited from discussions with numerous colleagues at BP including Toby Harold, Rachel Kieft and Martin Wells. I am indebted to Ceri James and Howard Johnson for their constructive reviews.

References

BARNARD, P.L., HANES, D.M., RUBIN, D.M. & KVITEK, R.G. 2006. Giant sand waves at the mouth of San Francisco Bay. *Eos, Transactions of the American Geophysical Union*, **87**, 285–289.

BENNACEF, A., BEUF, S., BIJU-DUVAL, B., DE CHARPAL, O., GARIEL, O. & ROGNON, P. 1971. Example of cratonic sedimentation: Lower Paleozoic of Algerian Sahara. *American Association of Petroleum Geologists Bulletin*, **55**, 2225–2245.

BERNÉ, S., VAGNER, P. *ET AL.* 2002. Pleistocene forced regressions and tidal sand ridges in the East China Sea. *Marine Geology*, **188**, 293–315.

BEUF, S., BIJU-DUVAL, B., DE CHARPAL, O., ROGNON, P., GARIEL, O. & BENNACEF, A. 1971. *Les Grès du Paléozoïque Inférieur au Sahara*. Editions Technip, Paris.

BOOTE, D.R.D., CLARK-LOWES, D.D. & TRAUT, M.W. 1998. Palaeozoic petroleum systems of North Africa. *In*: MACGREGOR, D.S., MOODY, R.T.J. & CLARK-LOWES, D.D. (eds) *Petroleum Geology of North Africa*. Geological Society, London, Special Publications, **132**, 7–68.

BRENCHLEY, P.J., MARSHALL, J.D. *ET AL.* 1994. Bathymetric and isotopic evidence for a short lived Ordovician glaciation in a greenhouse period. *Geology*, **22**, 295–298.

BRENCHLEY, P.J., CARDEN, G.A.F. & MARSHALL, J.D. 1995. Environmental changes associated with "first strike" of the late Ordovician mass extinction. *Modern Geology*, **20**, 69–82.

CRAIG, J., GRIGO, D. & REBORA, A., SERAFINI, G. & TEBALDI, E. 2010. From Neoproterozoic to Early Cenozoic: exploring the potential of older and deeper hydrocarbon plays across North Africa and the Middle East. *In*: VINING, B.A. & PICKERING, S.C. (eds) *Petroleum Geology: From Mature Basins to New Frontiers*. Proceedings of the 7th Petroleum Geology Conference. Geological Society, London, 673–705, http://doi.org/10.1144/0070673

DALRYMPLE, R.W. 1984. Morphology and internal structure of sandwaves in the Bay of Fundy. *Sedimentology*, **31**, 365–382.

DIXON, R.J., PATTON, T.L. & HIRST, J.P.P. 2008. Giant sandwaves from the Late Ordovician of the Tassili N'Ager, Algeria. *In*: *Extended Abstracts from Marine and River Dune Dynamics*, 1–3 April 2008, Leeds, United Kingdom, 75–78.

ESCHARD, R., ABDALLAH, H., BRAIK, F. & DESAUBLIAUX, G. 2005. The Lower Paleozoic succession in the Tasilli outcrops, Algeria: sedimentology and sequence stratigraphy. *First Break*, **23**, 27–36.

ESCHARD, R., BRAIK, F., BEKKOUCHE, D., BEN RAHUMA, M., DESAUBLIAUX, G., DESHCAMPS, R. & PROUST, J.N. 2010. Palaeohighs: their influence on the North African Paleozoic petroleum systems. *In*: VINING, B.A. & PICKERING, S.C. (eds) *Petroleum Geology: From Mature Basins to New Frontiers*. Proceedings of the 7th Petroleum Geology Conference. Geological Society, London, 707–724, http://doi.org/10.1144/0070707

GALEAZZI, S., POINT, O.N., HADDADI, N., MATHER, J. & DRUESNE, D. 2010. Regional geology and petroleum systems of the Illizi–Berkine area of the Algerian Saharan Platform: an overview. *Marine and Petroleum Geology*, **27**, 143–178.

GHIENNE, J.-F., LE HERON, D.P., MOREAU, J. & DEYNOUX, M. 2007. The late Ordovician glacial sedimentary system of the West Gondwana platform. *In*: HAMBREY, M.J., CHRISTOFFERSON, P., GLASSER, N.F. & HUBBARD, B. (eds) *Glacial Sedimentary Processes and Products*. International Association of Sedimentologists, Special Publications, **39**, 295–319.

GHIENNE, J.-F., GIRARD, F., MOREAU, J. & RUBINO, J.-L. 2010. Late Ordovician climbing dune cross-stratification; a signature of outburst floods in proglacial outwash environments. *Sedimentology*, **57**, 1175–1198, http://doi.org/10.1111/j.1365-3091.2009.01142.x

GIRARD, F., GHIENNE, J.-F. & RUBINO, J.-L. 2012. Channelized sandstone bodies ('cordons') in the Tassili N'Ajjer (Algeria & Libya): snapshots of a late Ordovician proglacial outwash plain. *In*: HUUSE, M., REDFERN, J., LE HERON, D.P., DIXON, R.J., MOSCARIELLO, A. & CRAIG, J. (eds) *Glaciogenic Reservoirs and Hydrocarbon Systems*. Geological Society, London, Special Publications, **368**, 355–379, http://doi.org/10.1144/SP368.3

HAMBREY, M.J. & HARLAND, W.B. 1981. *Earth's Pre-Pleistocene Glacial Record*. Cambridge University Press, Cambridge.

HIRST, J.P.P. 2012. Ordovician proglacial sediments in Algeria: insights into the controls on hydrocarbon reservoirs in the In Amenas field, Illizi Basin. *In*: HUUSE, M., REDFERN, J., LE HERON, D.P., DIXON, R.J., MOSCARIELLO, A. & CRAIG, J. (eds) *Glaciogenic Reservoirs and Hydrocarbon Systems*. Geological Society, London, Special Publications, **368**, 319–353, http://doi.org/10.1144/SP368.17

HIRST, J.P.P., BENBAKIR, A., PAYNE, D.F. & WESTLAKE, I.R. 2002. Tunnel valleys and density flow processes in the Upper Ordovician glacial succession, Illizi Basin, Algeria: influence on reservoir quality. *Journal of Petroleum Geology*, **25**, 297–324.

JAMES, J.W.C., PEARCE, B. ET AL. 2011. *The Marine Aggregate Levy Sustainability Fund Synthesis Study in the Central and Eastern English Channel*. British Geological Survey Open Report OR/11/01.

LANG, J., DIXON, R.J., LE HERON, D.P. & WINSEMANN, J. 2012. The Cambrian to Silurian succession of the Illizi Basin, Algeria: depositional model and sequence stratigraphic correlation of Upper Ordovician glaciogenic deposits, Illizi Basin, Algeria. *In*: HUUSE, M., REDFERN, J., LE HERON, D.P., DIXON, R.J., MOSCARIELLO, A. & CRAIG, J. (eds) *Glaciogenic Reservoirs and Hydrocarbon Systems*. Geological Society, London, Special Publications, **368**. 393–317, http://doi.org/10.1144/SP368.1

LEEDER, M.R. 2011. *Sedimentology and Sedimentary Basins: from Turbulence to Tectonics*. 2nd edn. Wiley-Blackwell, Hoboken, NJ.

LEGRAND, P. 1988. The Ordovician–Silurian boundary in the Algerian Sahara. *Bulletin of the British Museum of Natural History (Geology)*, **43**, 171–176.

LE HERON, D.P. & CRAIG, J. 2008. First order reconstructions of a Late Ordovician Saharan ice sheet. *Journal of the Geological Society, London*, **165**, 19–29. http://doi.org/10.1144/0016-76492007-002

LE HERON, D.P. & DOWDESWELL, J.A. 2009. Calculating ice volumes and ice flux to constrain the dimensions of a 440 Ma North African ice sheet. *Journal of the Geological Society, London*, **166**, 277–281, http://doi.org/10.1144/0016-76492008-087

LÜNING, S., CRAIG, J., LOYDELL, D.K., STORCH, P. & FITCHES, B. 2000. Lower Silurian hot shales in North Africa and Arabia: regional distribution and depositional model. *Earth-Sciences Reviews*, **49**, 121–200.

MOREAU, J., GHIENNE, J.-F., LE HERON, D., RUBINO, J.-L. & DEYNOUX, M. 2005. A 440 Ma old ice stream in North Africa. *Geology*, **33**, 753–756.

OLARIU, C., STEEL, R.J., DALRYMPLE, R.W. & GINGRAS, M. 2012. Tidal dunes v. tidal bars: the sedimentological and architectural characteristics of compound dunes in a tidal seaway, the lower Baronia Sandstone (Lower Eocene), Ager Basin, Spain. *Sedimentary Geology*, **279**, 134–155. http://doi.org/10.1016/j.sedgeo.2012.07.018

PARIS, P., DEYNOUX, M. & GHIENNE, J.-F. 1998. Découverte de Chitinozoaires à la limite Ordovicien–Silurien en Mauritanie; implications paléogéographiques. *Comptes Rendus de l'Académie des Sciences*, Paris, **326**, 499–504.

READING, H.G. 1996. *Sedimentary Environments: Processes, Facies and Stratigraphy*. 3rd edn. Wiley-Blackwell, Hoboken, NJ.

ROUSSÉ, S., SANDVIK, S.E., MURAT, B., HUTCHINSON, A., SAADI, K. & LE GUERROUÉ, E. 2009. Depositional model and allostratigraphic architecture of Late Ordovician syn-glacial strata from the Tiguentourine Field (Illizi Basin, Algeria). Extended abstract presented at the *8th PESGB/HGS Conference on African E&P*, 9–10 September 2009, London.

SALTZMAN, M.R. & YOUNG, S.A. 2005. Long-lived glaciation in the Late Ordovician? Isotopic and sequence stratigraphic evidence from western Laurentia. *Geology*, **33**, 109–112.

SCOTESE, C.R., BOUCOT, A.J. & MCKERROW, W.S. 1999. Gondwana palaeogeography and palaeoclimatology. *Journal of African Earth Sciences*, **28**, 99–114.

SMITH, A.G. 1997. Estimates of the Earth's spin (geographic) axis relative to Gondwana from glacial sediments and palaeomagnetism. *Earth-Science Reviews*, **42**, 161–179.

STRIDE, A.H., BELDERSON, R.H., KENYON, N.H. & JOHNSON, M.A. 1982. Offshore tidal deposits: sand sheet and sand bank facies. *In*: STRIDE, A.H. (ed.) *Offshore Tidal Sands: Processes and Deposits*. Chapman & Hall, London, 95–125.

SUTCLIFFE, O.E., DOWDESWELL, J.A., WHITTINGTON, R.J., THERON, J.N. & CRAIG, J. 2000. Calibrating the Late Ordovician glaciation and mass extinction by the eccentricity cycles of Earth's orbit. *Geology*, **28**, 967–970.

UNDERWOOD, C.J., DEYNOUX, M. & GHIENNE, J.-F. 1998. High palaeolatitude (Hodh, Mauritania) recovery of graptolite fauna after the Hirnantian (end Ordovician) extinction event. *Palaeogeography, Palaeoclimatology, Palaeoecology*, **142**, 91–105.

WHITEMAN, A.J. 1971. Cambro–Ordovician rocks of Al Jazaïr (Algeria) – a review. *American Association of Petroleum Geologists Bulletin*, **55**, 1295–1335.

Outcrop analogues for hydrocarbon reservoirs in the Columbus Basin, offshore east Trinidad

ANDREW BOWMAN

BP Trinidad & Tobago LLC, 5–5a Queens Park West, Port of Spain, Trinidad & Tobago, West Indies

Present address: BP Middle East Region, BP International Centre for Business & Technology, Chertsey Road, Sunbury on Thames, Middlesex TW16 7LN, UK (e-mail: andrew.bowman@uk.bp.com)

Abstract: The Mayaro Formation, which is exposed in SE Trinidad, is the key outcrop analogue for prolific hydrocarbon reservoirs in the Columbus Basin, offshore east Trinidad, because it was deposited in the same basin setting by the same depositional system (the palaeo-Orinoco) as the offshore reservoirs. Sequence and parasequence stacking patterns affect the distribution of different types of reservoir sand bodies and heterogeneities, with the potential to influence fluid flow in subsurface hydrocarbon reservoirs. Spectacular outcrops of successions of storm-wave-influenced sand bodies, tens to hundreds of metres thick, deposited in axial and medial delta-lobe settings are characterized by high sand–shale ratios (0.7–>0.95), and are analogues for many producing reservoirs in the basin. Fluvial- and tide-influenced sand bodies deposited in distal distributary settings are also recognized in the outcrops, but may be difficult to distinguish from medial delta-lobe deposits in subsurface data. More complex, thinly bedded distal delta-lobe deposits, with subsurface reservoir potential, exhibit lower sand–shale ratios (0.3–0.6). Complex depositional architectures, with a high proportion of thinly bedded sandstones, also occur in association with delta-front deposits. The distal delta-lobe and delta-front deposits are analogous to some of the most challenging to develop reservoirs in the Columbus Basin.

The Pliocene-age Mayaro Formation is exposed in SE Trinidad (Fig. 1) and comprises stacked shelf-edge delta deposits (Bowman & Johnson 2014) that are inferred, through the integration of outcrop and subsurface data, to record the same depositional system (the palaeo-Orinoco Delta) developed within the same basin setting as producing reservoirs in the Columbus Basin. The Columbus Basin, which is located offshore SE Trinidad (Leonard 1983; Gibson *et al.* 2012), is a prolific hydrocarbon province from which approximately 1 Bbbl (billion barrels) of oil and 25 TCF (trillion cubic feet) of gas have been produced (GoRTT MEEA 2013). Earlier studies in the basin have documented the sequence stratigraphy of the subsurface (Wood 2000; Sydow *et al.* 2003), and the detailed facies associations and sequence stratigraphy of the outcrops (Bowman 2003; Bowman & Johnson 2014). This study builds on this literature to illustrate how knowledge of the outcrops can help to understand and reduce subsurface uncertainty as the Columbus Basin matures and increasingly complex reservoirs are exploited.

The Columbus Basin is characterized by very high rates of sediment accumulation that have impacted reservoir distribution and architecture; fields commonly comprise stacked 100–300 m-thick progradational to aggradational deltaic reservoir successions (Hallam & Lumkin 2003; Lumsden *et al.* 2002; Sydow *et al.* 2003). The basin's most prolific reservoirs comprise laterally extensive, fine-grained, high net-to-gross sandstones (>0.9), which are tens to hundreds of metres thick, and preserve evidence of storm- and wave-influenced delta-front deposition. These high-quality reservoirs are characterized by excellent porosities and permeabilities (*c.* 25–30% porosity, and permeabilities of hundreds of mD up to >1000 mD), and, as a consequence, they have typically been developed with limited numbers of high-angle or horizontal wells characterized by high rates of production (*c.* 150–200 MMSCFD (million standard cubic feet/day)). These highly productive reservoirs have accounted for much of the historic gas and oil production in the basin (Lumsden *et al.* 2002; Pucknall *et al.* 2003; Sydow *et al.* 2003). Some of these highly productive reservoirs comprise sand bodies deposited in fluvial- and tide-influenced lower delta-plain to proximal delta-front settings, and, as a result, they have a wider range of more complex reservoir architectures (Bally *et al.* 1999). Production also comes from heterolithic reservoirs, deposited in distal deltaic environments, that are characterized by a high proportion of thin beds, more variable rock quality

From: Bowman, M., Smyth, H. R., Good, T. R., Passey, S. R., Hirst, J. P. P. & Jordan, C. J. (eds) 2016. *The Value of Outcrop Studies in Reducing Subsurface Uncertainty and Risk in Hydrocarbon Exploration and Production*. Geological Society, London, Special Publications, **436**, 151–192.
First published online February 24, 2016, http://doi.org/10.1144/SP436.8

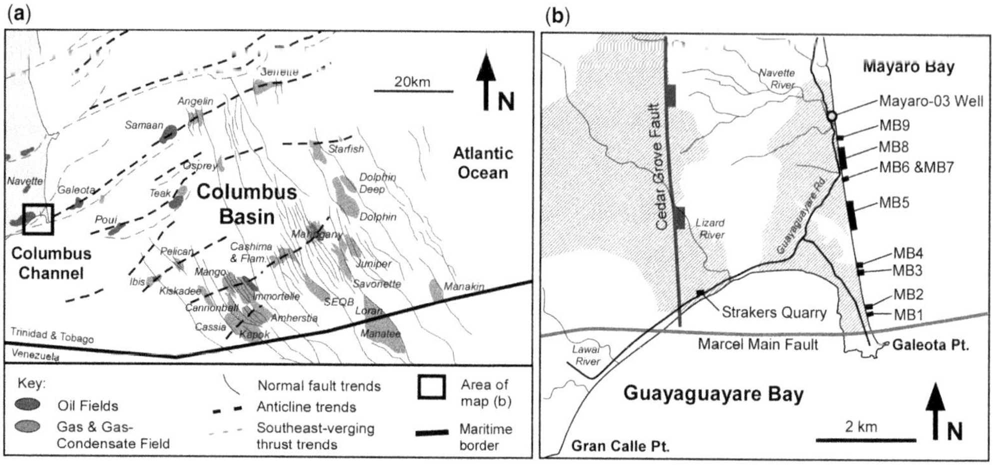

Fig. 1. (**a**) Map showing the location of major structural features and principal hydrocarbon fields in the Columbus Basin, offshore east Trinidad. The location of the outcrops relative to the offshore basin is shown by the black box, which locates map (b). (**b**) Map showing the location of the Mayaro Formation outcrops, which are exposed in SE Trinidad along the shore of Mayaro Bay (after Bowman 2003). The hashed area shows the distribution of the Mayaro Formation (simplified after Kugler 1961). The MB1 (stratigraphically oldest) to MB9 (stratigraphically youngest) outcrops are located in the hanging wall of the Cedar Grove growth fault.

(*c.* 10–20% porosity and 1–100 mD permeability) (Henderson *et al.* 2012; Taylor *et al.* 2012), and wells with lower (<80 MMSCFD) and less sustained rates of production (Baillie & James-Romano 2012). Rapid burial of sediment in the Columbus Basin means that the basin is relatively cool and diagenesis is limited at most reservoir depths (*c.* 3000–5000 m). As a result, reservoir rocks are relatively unconsolidated and sand production is a significant issue (Lumsden *et al.* 2002; Samsundar & Chung 2006); the friable Mayaro Formation outcrops amply demonstrate this risk to field development and production risk.

Recognizing and understanding different depositional elements and reservoir architectures in the subsurface of the Columbus Basin can be challenging. Seismic imaging is often hindered by shallow gas zones (Rampersad *et al.* 2014) and abundant normal faulting (Muller *et al.* 1999; Wood 2000). Core data are relatively limited, in part due to early difficulties in recovering and preserving core from young, friable, fine-grained sandstones and economic constraints in coring fields with stacked pay; whilst well-logging tools have inherent limitations in terms of their depth of investigation and vertical resolution (Passey *et al.* 2006). Finally, in the Columbus Basin there is often limited offset dynamic data. This reflects the preponderance of highly productive, fault-bounded, reservoir segments that can be effectively drained by a few high-rate wells (Hallam & Lumkin 2003; Pucknall *et al.* 2003).

The Mayaro Formation exposures can assist in reducing reservoir description uncertainty by helping to improve the understanding of those geological factors that can influence exploration, appraisal and production. Because the outcrops provide a wider and more detailed view of reservoir type and heterogeneity than is often enabled by limited subsurface data, a good understanding of the outcrops provides reservoir geologists with the knowledge to better appreciate subsurface uncertainty and risk, and help contribute towards more informed subsurface decision-making. However, the Mayaro Formation outcrops do have some shortcomings. Their dip means that they can only provide limited insight into lateral facies trends and lateral changes in reservoir architecture.

Since the outcrops provide information about the distribution, thickness and quality of sand bodies in the basin, they can provide essential input into subsurface play fairway definition and the risking of segments within play fairways. Similarly at the prospect scale, knowledge of the outcrops can help to inform uncertainty associated with activities defining segment closures, and the distribution and quality of the reservoir, and therefore help to inform exploration and appraisal well programmes. During field appraisal and early development, when decisions underpinning significant financial commitments are being made and implemented (e.g. development scheme, well type, well count, well completions), it is critical that geological features with the potential to influence fluid flow in the reservoir are identified and understood (Jackson & Muggeridge 2000; Ainsworth 2005, 2010). Importantly, insights from outcrop analogues can support predictions of sand-body extent beyond field closures: critical for trying to estimate aquifer volumes, connectivity and energy. In the appraisal and early development phase, dynamic data are often limited, so the insights provided by a high-quality outcrop analogue like the Mayaro Formation can help in understanding the character of features such as parasequences and their bounding surfaces that may define reservoir flow units (Weber & van Geuns 1990; Jackson & Muggeridge 2000; Ainsworth 2005, 2010), and provide hints of subtle geological features that may impact fluid flow later in the life of the field. As reservoirs are produced and fields mature, factors such as declining reservoir pressure and the interaction of aquifer waters with variable reservoir permeability means that subtle geological features, such as high-permeability streaks, dipping mudstones and bioturbated intervals, which are often difficult to identify in the subsurface, can begin to impact well rates, water-cut and the identification of late field life infill-drilling opportunities. It is also important to think about the potential range of fluid–rock interactions because hydrocarbon fields in the Columbus Basin may contain stacked pay segments of oil, gas condensate and dry gas. In these circumstances, the relative impact of heterogeneities may vary from one reservoir segment to another within a single hydrocarbon field. For example, a laterally extensive lower-permeability bioturbated sandstone interval may have a more limited impact on recovery in a dry gas-bearing segment than in an oil-bearing segment because of the higher mobility of gas. The outcrops can also provide analogue data for subsurface structural studies (Gibson 1994) and, although this aspect of the outcrops is not addressed in this study, it is important that in the subsurface the overprint of structure onto depositional elements is considered. For example, faults developed in lower net-to-gross layered reservoirs with higher clay content are more likely to have clay-smeared fault planes than faults developed in high net-to-gross homogenous reservoirs. Hence, the risk of fault compartmentalization is likely to be greater in lower net-to-gross layered reservoirs.

Finally, the unusually high-accommodation-space setting of the Columbus Basin means that the Mayaro Formation outcrops can aid in the understanding of hydrocarbon reservoirs in other high-accommodation settings such as the Baram Delta (Hiscott 2001; Hodgetts *et al.* 2001; Back *et al.* 2001, 2005; Saller & Blake 2003), and may also provide insights applicable to hydrocarbon reservoirs developed in shelf-margin settings such as the Gulf of Mexico (Mayall *et al.* 1992) and the Niger Delta (Damuth 1994; Saugy & Eyer 2003).

Geological setting of the Mayaro Formation

The Columbus Basin lies in the plate boundary zone between the Caribbean Plate and the South American Plate, close to where both plates meet the Atlantic Plate at the Lesser Antilles Trench accretionary prism and the South American passive margin, respectively (Babb & Mann 1999; Pindell & Kennan 2009). The basin is part of the youngest of a series of foreland basins that developed in northern South America during Eocene–recent times in response to flexural loading as the Caribbean Plate moved from west to east relative to the South American Plate (Di Croce *et al.* 1999). During the Late Eocene–middle Oligocene, the evolution of these foreland basins caused the palaeo-Orinoco drainage system to divert away from its original northerly flow to the easterly drainage system observed today (Hoorn *et al.* 1995; Diaz de Gamero 1996). The basin is structurally complex and characterized by large-scale growth faulting, superimposed on an original foreland basin fabric, and partially modified by transpression (Gibson *et al.* 2012). The large-scale growth faults define diachronous growth-fault-bounded depocentres (Fig. 2), which have successively developed further to the east through time in response to regional tectonics and sediment loading (Pocknall *et al.* 1999; Wood 2000; Sydow *et al.* 2003; Gibson *et al.* 2012). Each of the growth-fault-bound depocentres is filled with up to 12 km of Pliocene–recent upper-slope to delta-top sediments, supplied by deltas fed by the palaeo-Orinoco drainage system. As a result, the Columbus Basin records some of the highest rates of Neogene sedimentation and accommodation space creation in the world, 5–10 m ka^{-1} (Wood 2000; Sydow *et al.* 2003; Carvajal & Steel 2009).

The Mayaro Formation is exposed along the southern end of Mayaro Bay in SE Trinidad (Fig. 1), and records storm-wave-influenced shelf and shelf-edge delta deposition (Wood 2000; Bowman 2003; Sydow *et al.* 2003; Bowman & Johnson 2014). It forms the upper part of a >5 km-thick, partially exhumed, regressive deltaic package that records the Pliocene–recent infill of Trinidad's Southern and Columbus basins by the palaeo-Orinoco Delta (Leonard 1983; Rohr 1991; Carr-Brown 1998; Pocknall *et al.* 1999). Onshore outcrop mapping and review of well penetrations in nearby onshore oil fields by Kugler (1961, 1996, 2001) characterized the relationship between the Mayaro Formation and the slightly older Gros Morne Formation as being unconformable. The Gros Morne Formation records delta progradation into the Southern Basin, immediately to the west of the Columbus Basin (Leonard 1983; Rohr 1991; Carr-Brown 1998; Di Croce *et al.* 1999). This unconformable contact is not exposed in Mayaro Bay. Similarly,

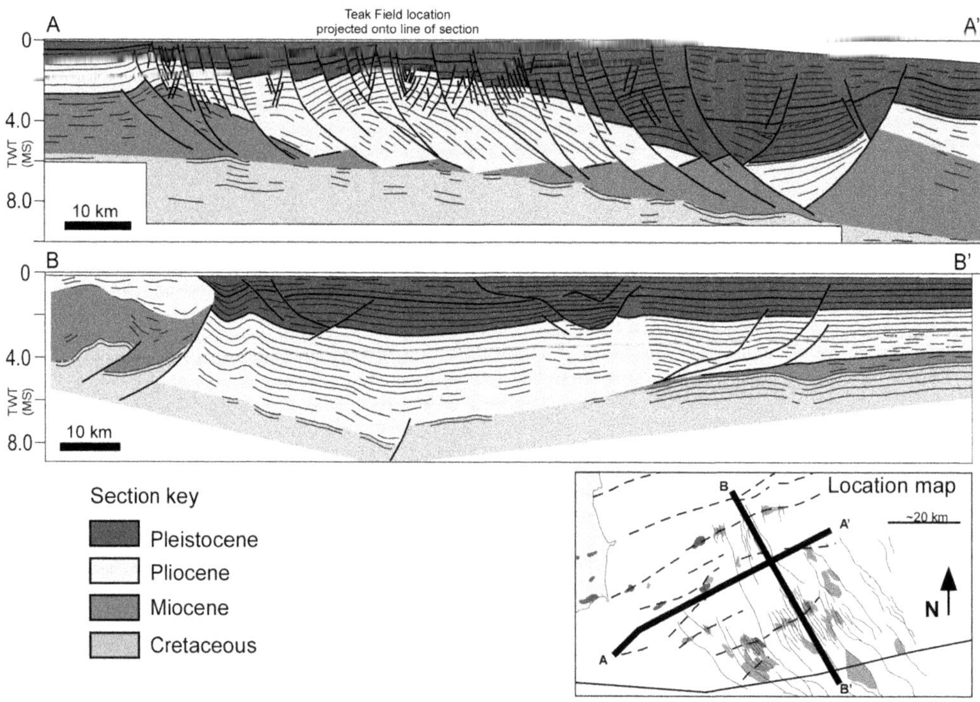

Fig. 2. Two interpreted geoseismic cross-sections illustrating the structure and stratigraphy of the Columbus Basin (modified from Gibson *et al.* 2012). Line A–A' shows a SW–NE-orientated section that highlights the large-scale growth faults which define the diachronous depocentres, which are characteristic of the basin. Line B–B' illustrates the underlying foreland basin fabric of the basin, with thrusting and loading to the north and thinning towards the south. The inset map (simplified from Fig. 2) locates the geoseismic lines in the Columbus Basin. TWT, two-way time.

Kugler mapped the Mayaro Formation as being unconformably overlain by the younger Palmiste Formation, which records lower delta-plain and shallow-water deltaic deposition (Kugler 1961, 1996, 2001; Carr-Brown 1998; Kugler 2001). Importantly, the Mayaro Formation was deposited in the hanging wall of the Cedar Grove Fault – a large 'basin-scale' growth fault that was mapped at the surface in a series of small river valleys approximately 3–5 km inland and west of Mayaro Bay by Kugler (1961, 1996, 2001) – and is imaged by onshore 2D seismic data (H. Vincent pers. comm.). It is interpreted to be one of the earliest growth faults developed in the Columbus Basin (Bowman 2003; Sydow *et al.* 2003; Dixon 2005).

Chronostratigraphically significant floras and faunas are scarce in the Plio-Pleistocene onshore and offshore Trinidad (Jones 1998; Wood 2000; Pocknall *et al.* 2001). Biostratigraphic data from the Mayaro Formation is consistent with, but not diagnostic of, the lower Pliocene N19 planktonic foraminiferal zone (Frampton & Wilson 2002; Zippi 2007), whilst biostratigraphic data from the overlying Palmiste Formation suggests an age of approximately 3.6–3.4 Ma (Pocknall *et al.* 2001). At the eastern end of the Southern Basin, thick progradational deltaic successions, which subcrop off Trinidad's southern coast, have been mapped on seismic and well-log data, and dated to 5.0–4.2 Ma: the upper packages in these successions are correlatable to the Gros Morne Formation (Dixon 2005; Winter 2006). Hence, Mayaro Formation deposition is interpreted to have occurred between approximately 4.0 and 3.6 Ma, in the latter part of the Early Pliocene.

Importantly, this means that the Mayaro Formation is time-equivalent to deep stratigraphy within the most westerly and youngest growth-fault-controlled Columbus Basin depocentre. Therefore, it is stratigraphically the closest analogue for the thick dominantly shelf-edge delta successions that form reservoirs offshore (Wood 2000; Bowman 2003; Sydow *et al.* 2003; Dixon 2005) (Fig. 3). Despite recognition of this relationship at an early stage of exploration and development activity in the basin (Leonard 1983), there have been few published outcrop studies. The exposures have previously been interpreted to record deposition on

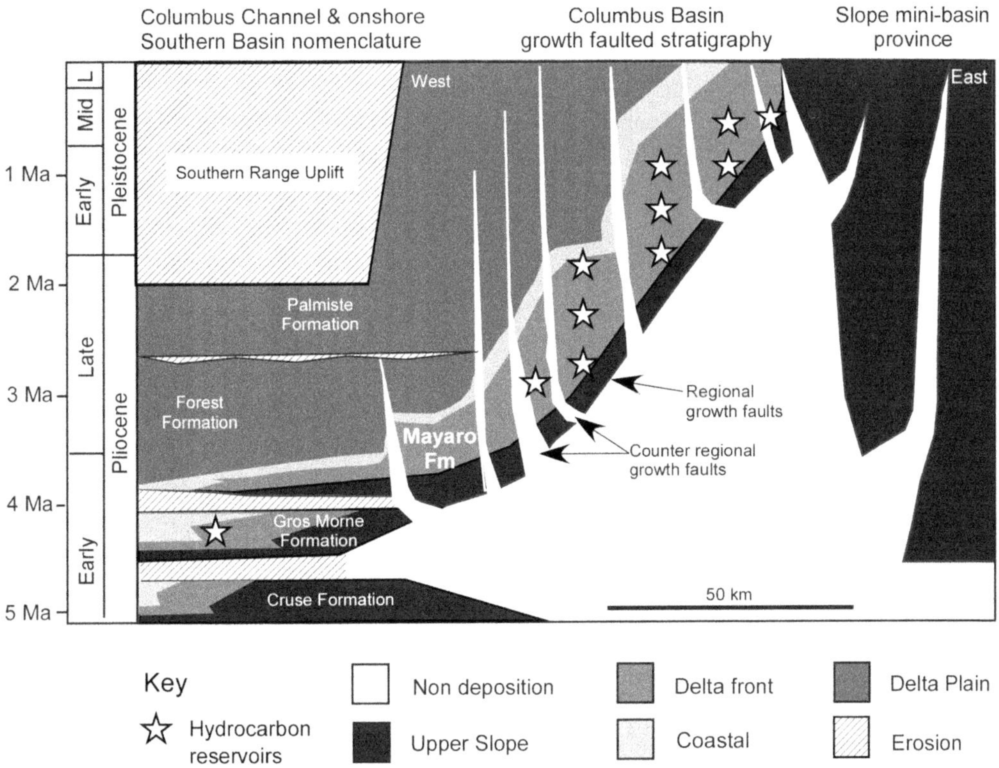

Fig. 3. A SW–NE-orientated schematic chronostratigraphic cross-section through the Plio-Pleistocene section of the Columbus Basin (modified from Bowman & Johnson 2014). It highlights the gross depositional setting of the basin-fill, the northeasterly progradation of the palaeo-Orinoco Delta, and the approximate stratigraphic and spatial location of the major reservoir intervals.

mixed-influence deltas (Farrelly 1987), fluvially influenced deltas (Algar 1993), wave-dominated shorefaces and wave-dominated deltas (Wood 2000). The interpretation of dominantly storm-wave-influenced shelf and shelf-edge delta deposits is preferred (Bowman & Johnson 2014), which, when coupled with the unusually thick aggradational sand-prone successions, has distinctive implications for reservoir distribution and architecture.

Methodology

Traditional sedimentary field data derived from detailed measured sections at a variety of scales, and supported by field sketches, interpretations of outcrop photographs and measurement of palaeocurrent data, have been critical in identifying the range of facies associations, depositional elements and potential reservoir architectures that are likely to exist in the offshore reservoirs. In order to understand better the limited resolution of subsurface data, when compared to the higher-resolution outcrops, gamma-ray log and probe-permeameter datasets were also acquired over key exposures.

Gamma-ray log data were collected with a hand-held scintillometer at a vertical spacing of 25 cm, with each data point representing an average value taken from three measurements of 60–120 s in duration. A moving average was calculated to smooth out anomalous data, and sample errors averaged around 10%, which is comparable to other studies (Myers & Bristow 1989; Davies & Elliot 1996). The outcrop gamma-ray logs are used to calibrate the outcrop descriptions to gamma-ray and V-shale electric log signatures in subsurface reservoirs. Spectral gamma-ray data yielded no additional insights and has not been further documented.

Absolute permeability values derived from outcrop probe permeametry are an order of magnitude greater than those observed in subsurface datasets, this is because the rocks exposed at outcrop are poorly cemented, friable and have undergone significant decompaction during Pleistocene–recent uplift. However, the permeability trends are consistent with trends observed in the subsurface, and highlight lamination-scale heterogeneity. Similar observations have been made in other outcrop probe-permeameter studies (Goggin 1993; Hurst & Goggin 1995). Data were acquired using the Imperial College Portable Probe Permeameter (Wang 1993; Lowden 1993), which measured the steady-state laminar flow of nitrogen gas into the rock. Permeability was derived using standard methods (Goggin *et al.* 1988; Goggin 1993; Sutherland *et al.* 1993). Extensive cleaning of the rock face was required prior to sampling, with each data point representing an average calculated from three samples. The statistical validity of the data was checked using published methods (Goggin 1993) and statistical errors were small (<5%), with no impact on data trends. The unstable nature of the exposures constrained the safe acquisition of probe-permeameter data and also prevented the acquisition of outcrop plugs for comparative permeability measurements using a Hassler cell. It similarly precluded the acquisition of high-quality samples for petrographical study.

Biostratigraphic analysis of the outcrops has revealed distinct palynomorph assemblages that can be linked to specific depositional settings and, hence, have the potential to aid subsurface reservoir description (de Verteuil 2002; Bowman & Johnson 2014). To date, these observations have not been widely applied in the subsurface. In contrast, foraminifera data appear to be of limited use at the reservoir scale (Frampton & Wilson 2002) because sample recoveries are low and assemblages are less ecologically distinct, and are therefore difficult to relate to specific depositional settings recorded in the outcrops.

The terminology proposed by Bowman (2003) is used to describe and locate exposures in the study area. The most southerly exposure is located on the beach at the southern end of Mayaro Bay, approximately 50 m north of the coastguard station. It is termed Mayaro Beach Outcrop 1 (MB1), with further significant outcrops named sequentially from south to north. The most northerly exposure is termed Mayaro Beach Outcrop 9 (MB9). This study mainly focuses on insights from three of these exposures: Mayaro Beach 4 (MB4); Mayaro Beach 5 (MB5); and Mayaro Beach 8 (MB8). Overall, the Mayaro Formation youngs upwards from MB1 to MB9 (Kugler 1961, 1996, 2001): however, there is some ambiguity with regard to the precise stratigraphic relationship between all of the outcrops because of subtle dip variations, patchy exposure, tectonic faults with poorly constrained offsets and the presence of east–west-orientated strike-slip faults in nearby onshore oilfields. Despite this, the stratigraphic profile between MB1 and MB9 is considered robust for two reasons. First, the large Goudron Syncline, within which the exposures occur (Kugler 1996), is not disrupted by large faults. Secondly, the gross upwards coarsening package is comparable in scale and stacking patterns to the succession penetrated by the onshore Mayaro-03 exploration well, located approximately 1 km to the north of the MB8 exposure (Fig. 1) and to reservoir successions penetrated offshore (Leonard 1983; Wood 2000; Sydow *et al.* 2003).

Depositional setting

Previous detailed facies analysis of the Mayaro Formation (Bowman 2003; Urzoa 2008; Bowman & Johnson 2014) has confirmed the dominant role of storm-wave processes in controlling sandstone distribution in the Columbus Basin during the Pliocene, and has also highlighted the presence of distinct units where fluvial and tidal processes significantly influenced sedimentation. The 14 facies associations described in detail by Bowman & Johnson (2014) have been rationalized into seven depositional elements: (1) Pro-Delta; (2) Distal Delta-Front; (3) Delta-Front; (4) Distal Distributary; (5) Axial Delta-Lobe; (6) Medial Delta-Lobe; and (7) Distal Delta-Lobe, which record delta-clinoform and delta-top deposition (Table 1). The 'depositional elements' can be linked and understood in the context of a deltaic depositional setting (Fig. 4), which comprised a lower delta plain bisected by distributary channels that fed fine-grained sandstone to the delta-front. Both fluvial and tidal processes influenced the lower reaches of these distributaries. When accommodation space allowed, a broad subaqueous delta-lobe developed as large storms reworked fine-grained sand from mouthbars offshore and alongshore. It is postulated that a palaeo-Guyana current may have existed and directed sediment alongshore towards the NNE, and created an asymmetric lobate delta with sand bodies preferentially developed towards the NNE. Similar asymmetric patterns of sediment distribution have been widely recognized in modern and ancient wave-dominated deltas (Bhattacharya & Giosan 2003). During periods of reduced delta-top accommodation space, mouthbars were located at, or close to, the shelf-edge and sediment was more efficiently supplied to distal delta-front and pro-delta settings dominated by sediment gravity flows.

The abundant sedimentary evidence for the deltaic origins of the Mayaro Formation suggests that the subsurface interpretation of major Columbus Basin reservoir sand bodies as hyperpycnal flow deposits proposed by Gamero Diaz *et al.* (2011) is incorrect. The assumptions implicit in such a radically different depositional model, which is not supported by the outcrop data, would lead to an oversimplified understanding of reservoir architecture and a limited ability to predict the development of reservoir sandstones at the basin scale. Recognition of a deltaic origin, coupled with an understanding of the high-resolution stratigraphy, enables the context of depositional elements to be properly understood. Since each depositional element records a distinctive range of sandstone–shale ratios and reflects a specific suite of sand- and shale-body geometries. Importantly, these depositional elements are associated with very different production characteristics in the offshore reservoirs (e.g. Hallam & Lumkin 2003; Baillie & James-Romano 2012). The implications of the stratigraphy and architecture of these depositional elements for hydrocarbon reservoirs in the Columbus Basin is described in more detail later.

Stacking patterns and sequence stratigraphy

Sequence stratigraphy provides a framework for evaluating high-order controls on reservoir distribution and heterogeneity (Van Wagoner *et al.* 1990; Emery & Myers 1996; Possamentier & Allen 1999; Ainsworth 2010). In paralic systems, changes in accommodation space control the development of stratigraphically significant erosion surfaces and flooding surfaces (FSs) that define 'systems tracts', and the distribution of facies belts, depositional elements and reservoir architectures. The Mayaro Formation outcrops are at a similar scale to thick stacked reservoir successions in the subsurface, and provide a level of detail about facies, facies juxtapositions and stratigraphic surfaces that subsurface datasets, dominated by well-log and seismic data, cannot.

The Columbus Basin is characterized by very high sediment accumulation rates in the range of 5–10 m ka^{-1} (Wood 2000; Sydow *et al.* 2003), which are amongst the highest described anywhere in the world (Carvajal & Steel 2009). These extraordinary sedimentation rates developed in response to rapid tectonic subsidence within diachronous, growth-fault-bounded depocentres and very high rates of sediment supply (Wood 2000; Sydow *et al.* 2003; Gibson *et al.* 2012). Superimposed on these accommodation space controls were higher frequency autocyclic and allocyclic events (Wood 2000; Bowman 2003; Sydow *et al.* 2003; Urzoa 2008; Bowman & Johnson 2014). Hence, a striking aspect of the Mayaro Formation is the preservation of stratigraphic cycles at multiple scales. These range from flooding-surface-bound upwards-cleaning and -coarsening cycles that are metres thick to the overall approximately 1.75 km-thick regressive–transgressive cyclicity apparent in the Mayaro Formation (Fig. 5).

Wood (2000) presented the first sequence stratigraphic interpretation of the Columbus Basin: integrating a diverse subsurface dataset to recognize approximately 2000–3000 m-thick 'megasequences' associated with the development and evolution of growth-fault-bounded depocentres, which comprised stacked 'parasequences' tens to hundreds of metres thick. Bowman (2003) interpreted the whole Mayaro Formation section as an incomplete record of one of the megasequences described in Wood (2000), and interpreted the MB5 outcrop as the lower, sand-prone, part of a single high-order 'sequence', comprising a falling stage systems tract (FSST), a lowstand systems tract (LST) and a transgressive systems tract (TST). Sydow *et al.* (2003) built on these earlier studies, and integrated subsurface data from producing fields, shallow seismic data acquired over a Holocene LST shelf-edge delta and the Mayaro Formation outcrops (specifically MB5) to develop a sequence stratigraphic model in which the prominent sand-prone intervals, described as parasequences by Wood (2000), are interpreted to be fourth-order sequences. Bowman & Johnson (2014) provided a detailed description of the MB5 succession, interpreting it as a fourth- or fifth-order FSST to LST/TST succession based on the facies and stratigraphic relationships in the exposure and the subsurface context summarized by Sydow *et al.* (2003). In contrast, Urzoa (2008) placed less emphasis on the significance of a prominent erosional surface juxtaposing mouthbar facies onto wave-influenced offshore facies in the MB5 succession than other authors (Bowman 2003; Sydow *et al.* 2003; Bowman & Johnson 2014) and interpreted the MB5 succession as a series of stacked progradational–aggradational parasequences: an interpretation more similar to that of Wood (2000). Urzoa (2008) interpreted the MB5 succession as a shelf-edge HST, which, whilst possible, is considered less consistent with insights gained from similar scale successions in subsurface datasets.

In this study, three sequence stratigraphic units are recognized and discussed: (1) 'sequence sets'; (2) 'sequences'; and (3) 'parasequences'. A sequence set is a stacked succession of high-order sequences (Mitchum & Van Wagoner 1991). The exposed Mayaro Formation is interpreted to form part of a sequence set. Sequences are unconformity, or correlative conformity, bounded units (Van Wagoner *et al.* 1990; Mitchum & Van Wagoner 1991), they can be subdivided into FSST, LST, TST and highstand systems tracts (HST), and are variably exposed in the Mayaro Formation. The MB5 outcrop is interpreted as the FSST to TST of a fourth- or fifth-order sequence (Sydow *et al.* 2003; Bowman & Johnson 2014).

Parasequences are conformable; flooding-surface-bounded units that form the building blocks

Table 1. *Summary of the depositional element recognized in the Mayaro Formation and their constituent facies associations, which have previously been described in detail by Bowman & Johnson (2014)*

Depositional element	Facies associations (Bowman & Johnson 2014)	Sedimentary characteristics	Interpretation	Sand-body dimensions	Reservoir quality	Reservoir potential (assuming dry gas)	Reservoir uncertainties
Pro-Delta	DDFS-1, Silty mudstones DDFS-2, Shallow sandstone scours	• Massive to laminated silty mudstones • Thin graded and rippled sandstone beds, debrites, slumps • Rare thicker fine-grained sandstone-filled scours	• Deposition at the toe of a delta clinoform; common low-density turbidites, debrites, slope-failures against a background of turbiditic and hemi-pelagic mudstones	• Rare isolated sandstone scours are tens of metres wide and up to 2 m deep	• None	None: • Mudstone dominated, few large sand bodies, no significant thin-beds	N/A
Distal Delta-Front	DDFS-1, Silty mudstones DDFS-3, Isolated sandstone chutes DDFS-4, Amalgamated sandstone chutes DDFS-6, Heterolithic channel	• Laminated mudstones with thin graded beds, current, combined flow and wave ripples that are cut by; • metre- to tens of metre-scale chutes filled with variable massive and laminated fine-grained sandstones	• Gravity-driven density-flow processes on delta clinoform • Low-density turbidites mainly occur outside of the chutes • High-density turbidite deposits occur within the chutes	• Isolated sand bodies are <25 m wide and 4 m deep • Amalgamated sand bodies are continuous at outcrop (*c.* 50 m wide and <15 m deep). Seismic data suggest widths of <500 m	• Variable: poor to good. Some thin beds have no reservoir quality. Thicker beds in sandstone chutes are likely to have up to *c.* 15–20% porosity and permeabilities (K) of tens to hundreds of mD	Limited: • Likely laterally discontinuous chute–channel complexes with fine-grained sandstones Very limited: • Thin-bedded reservoir potential in the 'background' sediments	• Lateral continuity of the sandstone-filled chutes • Connectivity between chutes and thin beds
Delta-Front	DDFS-1, Silty mudstones DDFS-5, Silty lobate sandstones PDF-3, SCS	• Lobate erosional sediment bodies comprising tens of centimetre-thick sandstones overlain by stacked thin beds • Current, climbing current, combined flow and wave ripples • Complex erosional geometries • Capped by tubular SCS sandstone beds	• Records the complex interplay of unidirectional currents (river-generated) and oscillatory currents (storm-generated) at the roll-over point separating the delta clinoform from the coeval delta topsets	• Individual sand-prone sediment bodies are *c.* 10–50 m wide and <1.5 m deep • Depositional element is estimated to be 500–2000 m wide based on seismic amplitudes	• Variable: poor to moderate. Some thin beds have no reservoir quality. Some sandier beds likely to have up to *c.* 15–20% porosity and permeabilities (K) of tens to hundreds of mD	Possible: • Thin-bed reservoir potential; complex cross-cutting geometries may help to increase connectivity • Variable reservoir quality likely to be an issue	• Impact of pore-scale variability in clay content • Connectivity of thin beds • Connectivity with overlying tabular SCS sandstones
Distal Distributary	TDF-1, Tidal heterolithics TDF-2, Wavy heterolithics TDF-3, Rippled sandstones	• Mud-draped sigmoidal cross-bedding • Flaser, lenticular and wavy bedding • Current rippled sandstones	• Planar beds record high-energy, fluvial processes – distal mouthbar • Small-scale tide-influenced migrating dunes, sand-flats and mud-flats	• Laterally continuous at outcrop and *c.* 15–20 m thick • Difficult to estimate widths; shallow seismic data suggest that outer distributary	• Good to excellent: >18% porosity and permeabilities (K) of hundreds to thousands of mD	Good: • Clean sandstones, with few mud-drapes • Lack of laterally continuous shales	• Impact on drainage/sweep efficiency of heterolithic beds and mud drapes • Baffling impact of shales

	TDF-4, Cross-bedded sandstone TDF-5, Planar laminated sandstone PDF-2, HCS	• Planar laminated sandstones • Fine-grained, 'clean' friable sandstones • Abundant plant material	• Intercalated HCS and thicker SCS sandstones record large storms and relatively local changes in shoreline shape	mouths could range from 1 to 5 km in width		• Heterolithic beds may reduce drainage/sweep efficiency	• Impact of high-permeability layers • Connectivity between thick-bedded and heterolithic units
Axial Delta-Lobe	PDF-1 Laminated and bioturbated mudstones PDF-3, SCS	• Massive fine-grained sandstone units dominated by SCS with remnant offshore mudstones • Sandstones are fine-grained and quartz rich	• Records reworking of sand from delta-distributaries into a subaqueous delta-lobe	• Laterally continuous at outcrop and tens of metres thick • Subsurface correlations suggest laterally extensive over thousands of metres	• Excellent: >20% porosity permeabilities (K) of hundreds to thousands of mD	Good: • Thick, near-homogenous successions, with very minor mudstone baffles	• Connection to aquifer and susceptibility to aquifer ingress
Medial Delta-Lobe	PDF-1, Laminated and bioturbated mudstones PDF-2, HCS PDF-3, SCS	• Amalgamated SCS sandstones interbedded with thinner HCS sandstone and heterolithic mudstone beds • Sandstones are fine-grained and quartz rich, but may be more clay rich in thinner beds and towards the top of thick beds	• Storm-wave deposition in the middle of a dominated subaqueous delta-lobe • Increased preservation of offshore mudstones and thinner, less-amalgamated, sandstones record more distal deposition	• Individual sand bodies are tens of centimetres to metres thick • Depositional element is laterally continuous at outcrop and tens of metres thick • Subsurface correlations suggest laterally extensive over thousands of metres	• Good–excellent: >18% porosity permeabilities (K) of hundreds to thousands of mD	Moderate – Good: • High-quality laterally extensive sandstones • Relatively predictable arrangement of extensive mudstone beds	• Reservoir quality impact of clays in laminations • K_v/K_h • Connectivity between amalgamated beds and thinner beds
Distal Delta-Lobe	PDF-1, Laminated and bioturbated mudstones PDF-2, HCS	• Thin sandstone and siltstone beds, with HCS and wave ripples, interbedded with heterolithic offshore mudstones	• Records deposition on the distal fringes of a storm-wave-influenced subaqueous delta-lobe	• Individual sand bodies are centimetres to metres thick • Subsurface correlations suggest laterally extensive over thousands of metres	• Variable: poor to good. Some thin beds have no reservoir quality. Others up *c.* 18% porosity and permeabilities (K) of tens to hundreds of mD	Possible: • Thinly bedded reservoir potential, but K_v/K_h is likely to be low	• K_v/K_h • Lateral and vertical connectivity to thicker sandstone beds

These facies associations exert a fundamental control on the distribution of, and connectivity within, reservoir sand bodies in the Columbus Basin.
N/A, not applicable; HCS hummocky-cross stratification; SCS, Swaley-cross stratification.

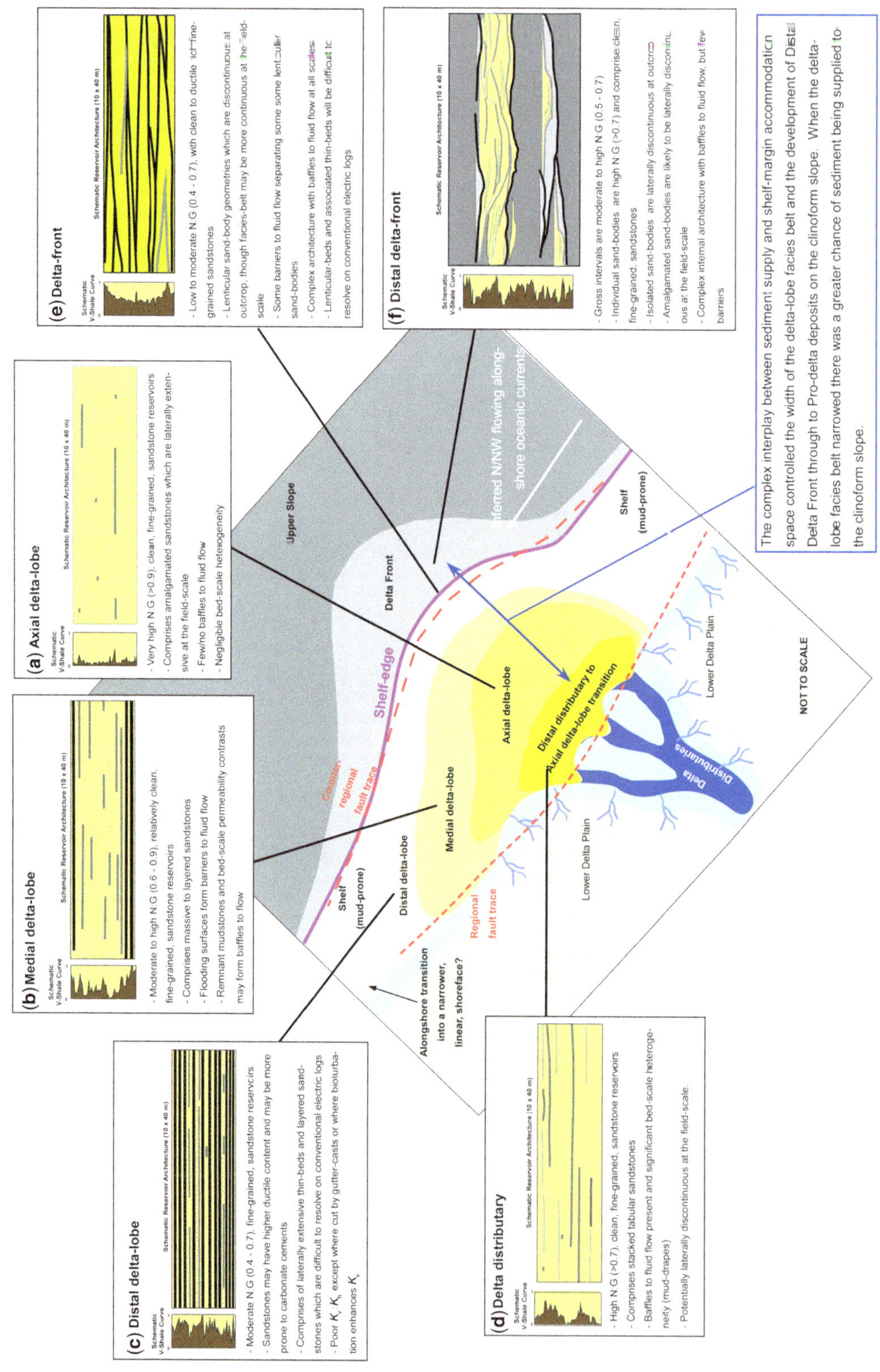
(c) Distal delta-lobe
Schematic V-Shale Curve
Schematic Reservoir Architecture (10 x 40 m)
- Moderate N:G (0.4 - 0.7), fine-grained, sandstone reservoirs
- Sandstones may have higher ductile content and may be more prone to carbonate cements
- Comprises of laterally extensive thin-beds and layered sandstones which are difficult to resolve on conventional electric logs
- Poor K_v:K_h except where cut by gutter-casts or where bioturbation enhances K_v
(b) Medial delta-lobe
Schematic V-Shale Curve
Schematic Reservoir Architecture (10 x 40 m)
- Moderate to high N:G (0.6 - 0.9), relatively clean, fine-grained, sandstone reservoirs
- Comprises massive to layered sandstones
- Flooding surfaces form barriers to fluid flow
- Remnant mudstones and bed-scale permeability contrasts may form baffles to flow
(a) Axial delta-lobe
Schematic V-Shale Curve
Schematic Reservoir Architecture (10 x 40 m)
- Very high N:G (>0.9), clean, fine-grained, sandstone reservoirs
- Comprises amalgamated sandstones which are laterally extensive at the field-scale
- Few/no baffles to fluid flow
- Negligible bed-scale heterogeneity
(e) Delta-front
Schematic V-Shale Curve
Schematic Reservoir Architecture (10 x 40 m)
- Low to moderate N:G (0.4 - 0.7), with clean to ductile fine-grained sandstones
- Lenticular sand-body geometries which are discontinuous at outcrop, though facies-belt may be more continuous at the field-scale
- Some barriers to fluid flow separating some some lent sand-bodies
- Complex architecture with baffles to fluid flow at all scales
- Lenticular-beds and associated thin-beds will be difficult to resolve on conventional electric logs
(f) Distal delta-front
Schematic V-Shale Curve
Schematic Reservoir Architecture (10 x 40 m)
- Gross intervals are moderate to high N:G (0.5 - 0.7)
- Individual sand-bodies are high N:G (>0.7) and comprise clean, fine-grained, sandstones
- Isolated sand-bodies are laterally discontinuous at outcrop.
- Amalgamated sand-bodies are likely to be laterally discontinuous at the field-scale
- Complex internal architecture with baffles to fluid flow, but few barriers
(d) Delta distributary
Schematic V-Shale Curve
Schematic Reservoir Architecture (10 x 40 m)
- High N:G (>0.7), clean, fine-grained, sandstone reservoirs
- Comprises stacked tabular sandstones
- Baffles to fluid flow present and significant bed-scale heterogeneity (mud-drapes)
- Potentially laterally discontinuous at the field-scale.
Upper Slope
Shelf (mud-prone)
Counter-regional fault trace
Shelf-edge
Delta Front
Distal delta-lobe
Alongshore transition into a narrower, linear, shoreface?
Medial delta-lobe
Regional fault trace
Axial delta-lobe
nferred N/NW flowing along-shore oceanic currents
Distal distributary to Axial delta-lobe transition
Lower Delta Plain
Shelf (mud-prone)
Delta Distributaries
Lower Delta Plain
NOT TO SCALE
The complex interplay between sediment supply and shelf-margin accommodation space controlled the width of the delta-lobe facies belt and the development of Distal Delta Front through to Pro-delta deposits on the clinoform slope. When the delta-lobe facies belt narrowed there was a greater chance of sediment being supplied to the clinoform slope.

of sequence stratigraphy and stacking pattern analysis (Van Wagoner *et al.* 1990). They are widely recognizable in the Mayaro Formation, where they range in thickness from approximately 3 to 20 m in thickness and display a range of internal architectures depending on the systems tract in which they occur. The 80–100 kyr time spans attributed to Columbus Basin sequences (Sydow *et al.* 2003) suggest that parasequences record deposition over time periods of thousands of years: hence, they probably record the interplay of autocyclic delta-avulsion processes and growth-fault-controlled changes in accommodation space. They have variable continuity in the subsurface; some can be traced along depositional strike for up to 10 km, whilst others thin laterally over distances of just a few kilometres. In the subsurface, parasequences often define flow units and are typically associated with a limited number of depositional elements. The other, thinner, outcrops in the Mayaro Formation often comprise one or more parasequence, but they are difficult to confidently place into a sequence stratigraphic framework due to the uncertain stratigraphic relationship between each outcrop. The sequence stratigraphic model presented by Sydow *et al.* (2003) suggests that many of the thinner outcrops may record relatively distal or proximal portions of fourth- to fifth-order sequences.

The Mayaro Formation sequence set

The Mayaro Formation exposures (MB1–MB9) form an approximately 1.75 km-thick package that is interpreted as a 'sequence set' that can be divided into three gross intervals: (1) a lower, approximately 400 m-thick, dominantly progradational succession; (2) an approximately 1100 m-thick dominantly aggradational succession; and (3) a thinner, approximately 250 m-thick, aggradational–retrogradational succession (Fig. 5a).

Lower, progradational succession. The most southerly, and the oldest, outcrops of the Mayaro Formation comprise more distal Pro-Delta/Upper Slope facies (Table 1) than are observed in overlying parts of the succession. The basal MB1 exposure is an olistostrome comprising storm-wave-influenced proximal delta-front sandstones olistoliths (*c.* 5 m across) encased within contorted massive and laminated mudstones (Bowman 2003). The poorly exposed MB2 outcrop, which is stratigraphically about 130 m above the MB1 outcrop, comprises thinly bedded turbidites (Bowman 2003). The MB1 to the MB2 succession are interpreted as recording the initial phase of deltaic deposition in the depocentre occupied by the Mayaro Formation. Between the MB2 and MB3 outcrops, transient poor-quality exposures provide evidence (e.g. increased sand content and rare combined flow ripples) of deltaic regression, and a transition from Upper Slope to dominantly Pro-Delta and Distal Delta-Front deposition. It is considered analogous to the deeper parts of Columbus Basin depocentre fills, which are characterized by 'ratty' gamma-ray log signatures forming weakly defined upwards-cleaning and -dirtying packages that pass upwards into more strongly defined, relatively thin (10–30 m) upwards-cleaning regressive successions, like those in the SE Galeota (SEG) area (Fig. 5a). The SEG area includes fields such as Kapok, Immortelle–Amherstia and Cannonball (Fig. 1). These subsurface successions were interpreted as 'slope fan' sediments by Wood (2000): an interpretation consistent with the outcrop observations.

Middle, aggradational succession. The MB3 and MB4 outcrops comprise relatively thin parasequences (10–30 m thick) with evidence of shallow-water storm-wave processes. The overlying MB5 exposure (Fig. 5b, c) is approximately 350 m thick and displays the widest range of facies observed in the Mayaro Formation, which have been interpreted as recording a complete progradational–aggradational shelf-edge delta succession (Sydow *et al.* 2003; Bowman & Johnson 2014). The MB5 succession records evidence of phases of erosion, sediment bypass, delta-lobe avulsion and, ultimately, shelf-edge delta abandonment (Bowman & Johnson 2014). It is overlain by thinner (<30 m thick) stacked dominantly storm-wave-influenced parasequences

Fig. 4. Plan form view of the Pliocene palaeo-Orinoco Delta system: the asymmetry reflects the impact of an inferred NNW-flowing palaeo-Guyana current, mimicking the pattern of present-day oceanic currents (Warne *et al.* 2002). Delta deposition was focused within a shallow subaqueous sand-prone delta-lobe, the location of which was fixed at, or close to, the shelf-edge by basin-scale growth faults. The axial part of this lobe (**a**) was very sand rich, with few heterogeneities; further away from the depositional axis mudstone, bed frequency and continuity increased through the medial delta-lobe (**b**) to the thin-bed-prone distal delta-lobe (**c**), resulting in more complex reservoir architectures. The width of the delta-lobe and, hence, the proximity of the delta distributaries (**d**) to the shelf-edge varied in response to the available shelf-margin accommodation space. During periods of limited shelf-margin accommodation space, distributaries were located at the shelf-edge and supplied sediment directly to a heterolithic delta-front (**e**). During these periods, sand may have reached the distal delta-front (**f**), which was channelized, and provided a conduit through which sand could move onto the pro-delta and, perhaps, the upper slope. N:G, net-to-gross.

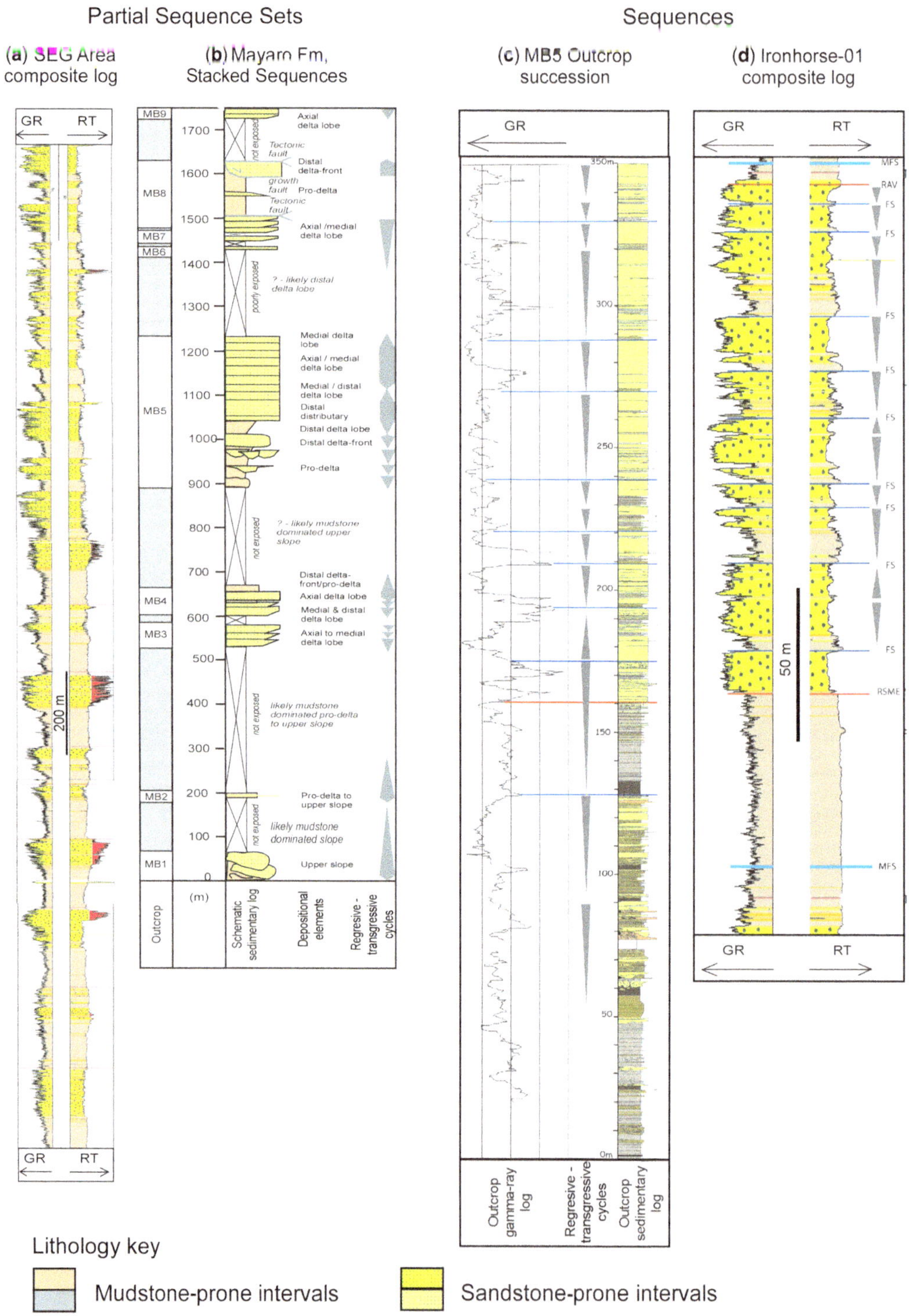

Fig. 5. Sequence sets, sequences and parasequences occur at similar scales in the Mayaro Formation outcrop succession and in subsurface datasets. (**a**) A portion of a composite log from the SE Galeota (SEG) area of the Columbus Basin (adapted from Sydow *et al.* 2003) displayed at the same scale as in (b). (**b**) A summary outcrop

(MB6, MB7 and lower 70 m of the MB8 exposures: Fig. 5b). Overall, the middle part of the Mayaro Formation succession records a switch to a more aggradational stacking pattern that is similar to the stacking patterns documented by Wood (2000) and Sydow *et al.* (2003) (Fig. 5a). These thick aggradational successions, observed at outcrop and in the subsurface, record the 'stalling' of deltas at, or near to, the shelf-edge, in response to accommodation-space creation in the hanging wall of a major depocentre bounded by growth faults (Wood 2000; Sydow *et al.* 2003; Bowman & Johnson 2014).

Upper, retrogradational succession. Within the MB8 exposure, there is a tectonic break in the stratigraphy: a sharp surface with well-developed slickensides is interpreted as a normal fault that juxtaposes Pro-Delta and Distal Delta-Front facies above shallow Proximal to Medial Delta-Lobe facies. Similarly, at the top of the MB8 outcrop, there is another, more poorly exposed, surface that can be interpreted to record another fault. The MB9 outcrop (Fig. 5b) exposes two thin storm-wave-influenced delta-lobe parasequences (Bowman 2003). These are notable for a low-diversity, moderate to high abundance, trace fossil assemblage that is dominated by *Ophiomorpha* and common siderite nodules. These thinner, more bioturbated and cemented parasequence are interpreted to record lower sediment accumulation rates, perhaps in response to reduced accommodation space. They are thought to be analogous to more heterolithic back-stepping successions, like those forming the upper part of the SEG area stratigraphy (Fig. 5a). The tectonic complications and consequent uncertainties in stratigraphic relationships illustrate the care that must be taken to understand the structural framework in the basin before drawing stratigraphic conclusions in the subsurface.

The overall upwards-coarsening and -cleaning succession recorded by the Mayaro Formation is comparable to the megasequences of Wood (2000) and the stacked sequence sets of Sydow *et al.* (2003) (Fig. 5a). It records discrete stages in the evolution and fill of a growth-fault-bounded depocentre. The 'sequence set' stacking pattern influences the gross distribution of potential reservoir and non-reservoir intervals within growth-fault-bounded depocentres. Thinner and more complex reservoir packages, deposited in upper-slope and pro-delta settings, are likely to occur in the deeper parts of a depocentre's fill. Thicker and more sand-prone progradational–aggradational packages, deposited in delta-front and delta-top settings, and which have the potential to form high-quality reservoirs, are more likely to occur in shallower parts of a depocentre's fill. Correctly identifying the lateral transition from high-quality delta-front sandstones deposited inboard of the shelf-edge and mudstone-prone intervals deposited outboard of the shelf-edge is key to reducing exploration risk in the Columbus Basin. Hence, the sequence set stacking patterns observed at outcrop provide high-quality data that can be used to help guide predictive exploration-scale models of sand-body distribution.

Mayaro Formation sequence architecture

At the sequence scale, there is considerable variability in thickness, stratigraphic surfaces and preserved depositional elements. A brief description of, and comparison between, two different partially exposed sequences – the approximately 52 m-thick sequence exposed in the MB4 outcrop and the approximately 350 m-thick sequence exposed in the MB5 outcrop – is made and used to illustrate several important points. First, that the upwards-cleaning and -coarsening successions which occur in the subsurface (e.g. the succession in the SEG area composite log that is tens to hundreds of metres thick: Fig. 5a) are not simply parasequences of variable thickness: instead, they comprise distinct systems tracts that have implications for the distribution of depositional elements and reservoir architecture. Appreciating this enables an improved understanding of reservoir and rock quality distribution, and can influence 'basin-scale' models for reservoir development, hydrocarbon-charge and reservoir connectivity to aquifers, as well as 'field-scale' models of how fluids will flow within a reservoir. Secondly, by illustrating the difficulties in recognizing stratigraphic surfaces and depositional elements, outcrop knowledge can be used to

Fig. 5. (*Continued*) description of the exposed Mayaro Formation from MB1 at the base to MB9 near the top (adapted from Bowman 2003). Both successions comprise stacked deltaic sequences (parts of sequence sets), which are interpreted to record progradational–aggradational and, ultimately, retrogradational stacking patterns as deltas filled in evolving growth-fault depocentres. (**c**) A summary outcrop log of the sequence partially exposed in the MB5 outcrop (adapted from Bowman 2003) and (**d**) shows an interpretation of a comparable scale sequence recorded in a composite well log from the Ironhorse-01 well in the SEG area of the Columbus Basin (adapted from Sydow *et al.* 2003). Flooding-surface-bound parasequences occur at similar scales in both the MB5 and Ironhorse-01 successions. The occurrence of similar scale and hierarchies of stratigraphic units and bounding surfaces in both the outcrops and subsurface data support the interpretation of the Mayaro Formation as a partial sedimentary record of the infill of an early Columbus Basin growth-fault-bounded depocentre. GR, gamma ray; RT, resistivity.

influence data acquisition and surveillance strategies throughout the field life, highlighting the value of high-resolution data such as core, core-calibrated image logs and formation pressure tests.

MB4 sequence stratigraphy

The prograding parasequence set recorded in the MB4 exposure (Fig. 6) is approximately 52 m thick and records a narrower range of storm-wave-influenced delta-lobe deposits than the MB5 exposure (Fig. 6). The uppermost parasequence is notable for two prominent stratigraphic surfaces. The older surface juxtaposes a 9 m-thick unit of sandstone with SCS which occurs on top of 10 m of thinner interbedded, sandstones exhibiting wave ripples and HCS and finely laminated mudstones. Immediately below this surface are anomalously thick gutter casts (Fig. 6). This succession is interpreted to record a suite of 'regressive surfaces of marine erosion' (RSME) and a phase of rapid shoreline progradation in response to decreased accommodation space (Plint & Nummendal 2000). The top of the 9 m-thick amalgamated SCS sandstone is eroded by a younger stratigraphic surface, which at outcrop is observed to erode approximately 1.5 m down into the SCS sandstone. Above this erosive surface is a poorly exposed approximately 18 m-thick, upwards-fining siltstone- and mudstone-prone package, the base of which comprises chaotically bedded 5–30 cm-diameter blocks of Pro-Delta and Distal Delta-Front facies within a contorted silty mudstone matrix (Fig. 6). This younger erosion surface places more distal, remobilized and relatively deep-water facies on top of more proximal shallow-water facies. It is interpreted to record slope failure, and removal of an overlying marine-flood and lower pro-delta succession.

Shelf-margin failure is well imaged in shallow seismic datasets in the basin (Sydow *et al.* 2003; Israel 2005; Moscardelli *et al.* 2012). The succession in the MB4 exposure may have developed in response to: (1) oversteepening of an active pro-delta; (2) a phase of reduced accommodation space, resulting in incision and erosion of underlying shelf-margin sediments; or (3) slope failure associated with a relative sea-level rise. Recognition that erosional features like this exist, and understanding their stratigraphic context and aspects of them like their log-signatures, can help to reduce exploration and appraisal risk when assessing reservoir presence and quality. In parts of the basin lacking high-quality seismic data, they may be difficult to identify because they are associated with a 'ratty' upwards-dirtying gamma-ray log signature that, in isolation, could be mistaken for a thin back-stepping parasequence. However, a unit associated with slope failure is very unlikely to be easy to correlate between wells: furthermore, adjacent wells may have no record of the underlying sand body because of erosion. In contrast, a back-stepping deltaic parasequence will be easier to correlate with confidence in a single field (1–2 km well spacing) and, perhaps, even between fields (>5 km well spacing).

Similarly, the outcrop can be used to illustrate the importance of recognizing stratigraphic surfaces at the field scale. The sharp-based gamma-ray log motif of the RSME and overlying SCS sandstone (Fig. 6) is quite common in the subsurface of the Columbus Basin; how might sand bodies like this behave as hydrocarbon reservoirs? The 9 m-thick SCS sandstone, which has good reservoir quality and little internal heterogeneity, has the potential to form an excellent reservoir. However, in situations where the hydrocarbon column is relatively thin and underlain by a moderate to strong aquifer (a bottom drive), the lack of an underlying medial to distal delta-lobe succession with bed-parallel baffles or barriers means that production wells will lack 'protection' from aquifer influx: a geological scenario that could lead to earlier than desired water-breakthrough. Conversely, because these sand bodies are interpreted as FSST deposits, it can be inferred that they may be relatively isolated and less likely to be connected to a strong aquifer. Neither scenario is necessarily correct; the power of the outcrops is in clearly illustrating the geology of the Columbus Basin, and provoking discussion of the different subsurface scenarios and their possible effects.

MB5 sequence stratigraphy

The MB5 exposure (Fig. 6) is the Mayaro Formation 'type section' and has been previously interpreted to record a complete shelf-edge delta succession (Bowman 2003; Sydow *et al.* 2003; Urzoa 2008; Bowman & Johnson 2014). It is approximately 350 m TST (true stratigraphic thickness), and is comparable in scale and gamma-ray log signature to sequences in the Columbus Basin (Fig. 6) (Wood 2000; Sydow *et al.* 2003). It records the full-range of depositional elements recognized in the Mayaro Formation (Bowman 2003; Bowman & Johnson 2014). A summary of the detailed sequence stratigraphic interpretation of Bowman & Johnson (2014) is provided (Table 2; Fig. 6) to enable comparison with the MB4 outcrop. They recognized eight phases of accommodation-space-controlled deltaic sedimentation and interpreted them to record a falling stage systems tract (FSST), a lowstand systems tract (LST) and a transgressive systems tract (TST), which developed at, or very close to, the shelf-edge in a rapidly subsiding Columbus Basin depocentre. The FSST of Bowman & Johnson (2014) (Table 2; Fig. 6) comprises an upwards-coarsening and -cleaning succession, the lower 125 m of which is dominated by gravity-flow deposits, including small isolated and amalgamated sand-filled gullies (metres thick and tens of metres wide), and records deposition on a delta clinoform. The upper 33 m of the FSST records storm-wave-influenced delta-topset deposition. A sharp erosive surface juxtaposes a gutter-cast-dominated unit associated with the Medial Delta-Lobe and fluvial- and tide-influenced Distal Distributary depositional elements. This facies juxtaposition supports an interpretation of a RSME separating the FSST from a thick aggradational LST (Fig. 6). It is notable that there is no outcrop evidence for subaerial exposure and significant incision. The base of the LST consists of a 17 m-thick fluvial- and tide-influenced facies that forms part of the Distal Distributary depositional element, and is overlain by a 50 m-thick succession dominated by storm-wave-influenced facies that forms parts of the Proximal to Medial Delta-Lobe depositional elements. The majority of the LST comprises a spectacular approximately 125 m-thick aggradational SCS sandstone succession. The TST is around 20 m thick, and comprises a back-stepping parasequence of Medial and Distal Delta-Lobe facies.

The thick MB5 FSST–LST succession and lack of a clear incised 'sequence boundary' record the capture of large volumes of sediment at the shelf-edge. This has been interpreted to reflect high rates of growth-fault-controlled accommodation-space creation (Bowman 2003; Sydow *et al.* 2003; Bowman & Johnson 2014). This suggests that there was only a very limited time period, associated with development of the RSME, during which significant volumes of coarse-grained sediment may have escaped the shelf-edge 'sediment trap' and been supplied to the upper slope. Furthermore, the sandstone chutes that occur in the MB5 FSST are probably not big enough to supply large sediment volumes to the slope, and are similar in scale to relatively small chutes which are observed on seismic data to terminate on the upper slope (Sydow *et al.* 2003; Israel 2005; Moscardelli *et al.* 2012). These outcrop observations suggest that classic models of shelf-edge sediment supply and deposition that predict peak sand supply to the slope during the LST (e.g. Posamentier & Vail 1988; Van Wagoner *et al.* 1988) should not be widely applied to exploration activity in the Columbus Basin and coeval slope. However, seismic data from the deep-water slope to the east of the Columbus Basin demonstrates that there have been periods in the Pleistocene when significant deep-water slope-channel systems linking deltas developed by the palaeo-Orinoco system and the Barbados accretionary prism did develop (Brami *et al.* 2000; Callec *et al.* 2010). An inference that can, perhaps, be drawn from the outcrops is that these deep-water systems seem likely to have developed outboard of thinner, more strongly progradational, shelf-edge sequences associated with more limited shelf-edge accommodation-space creation than those represented by the MB5 outcrop analogue. Perhaps these thin sequences were more like the MB4 outcrop?

The gamma-ray log signature of the MB5 outcrop does not clearly distinguish between the Delta Distributary depositional element and the Medial Delta-Lobe or Distal Delta-Front depositional elements. Nor does the MB5 gamma-ray log signature clearly highlight important stratigraphic surfaces, such as the MB5 RSME or transgressive surfaces of erosion that separate depositional elements with different reservoir architectures. Therefore, care must be taken to consider the range of possible interpretations of depositional elements and stratigraphic architecture when reviewing lower-resolution subsurface data. For example, in the Columbus Basin, the dominant extensional fault trend is orientated roughly NW–SE and strongly influenced the orientation of palaeo-shorelines. Hence, a reservoir interval comprising relatively heterolithic channel and mouthbar sand bodies deposited in a Distal Distributary depositional element along a west–east orientation, perpendicular to the palaeo-shoreline, will require a different approach to the selection of the development well type, orientation and spacing when compared to a reservoir developed in a relatively homogenous Medial Delta-Lobe depositional element orientated parallel to the main palaeo-shoreline and fault trends.

Seismic data are an important subsurface reservoir description tool. High bandwidth seismic data, such as very shallow or site-survey data, will resolve many of the lithological variations highlighted by the lower part of the MB5 exposure, as evidenced by stunning published images (Sydow *et al.* 2003; Israel 2005; Moscardelli *et al.* 2012). In contrast, with lower-frequency seismic data, more typical of Columbus Basin reservoirs buried to depths of approximately 3000–5000 m, it is likely that only the major lithological boundaries at the top and base of the sand-prone upper part of the MB5 succession would be resolved. Nor would the different depositional elements within the sand-prone upper part of the MB5 succession be resolved. At these reservoir depths, the MB4 exposure is likely to be represented by a single wavelet.

In summary, stacked Columbus Basin reservoirs are not simply thickened paralic parasequences or parasequence sets. Both thick Columbus Basin sequences, as exemplified by the MB5 outcrop, and thinner sequences, like the MB4 outcrop, are stratigraphically complex. Within sequences, accommodation space controlled the distribution of systems tracts that are associated with specific depositional

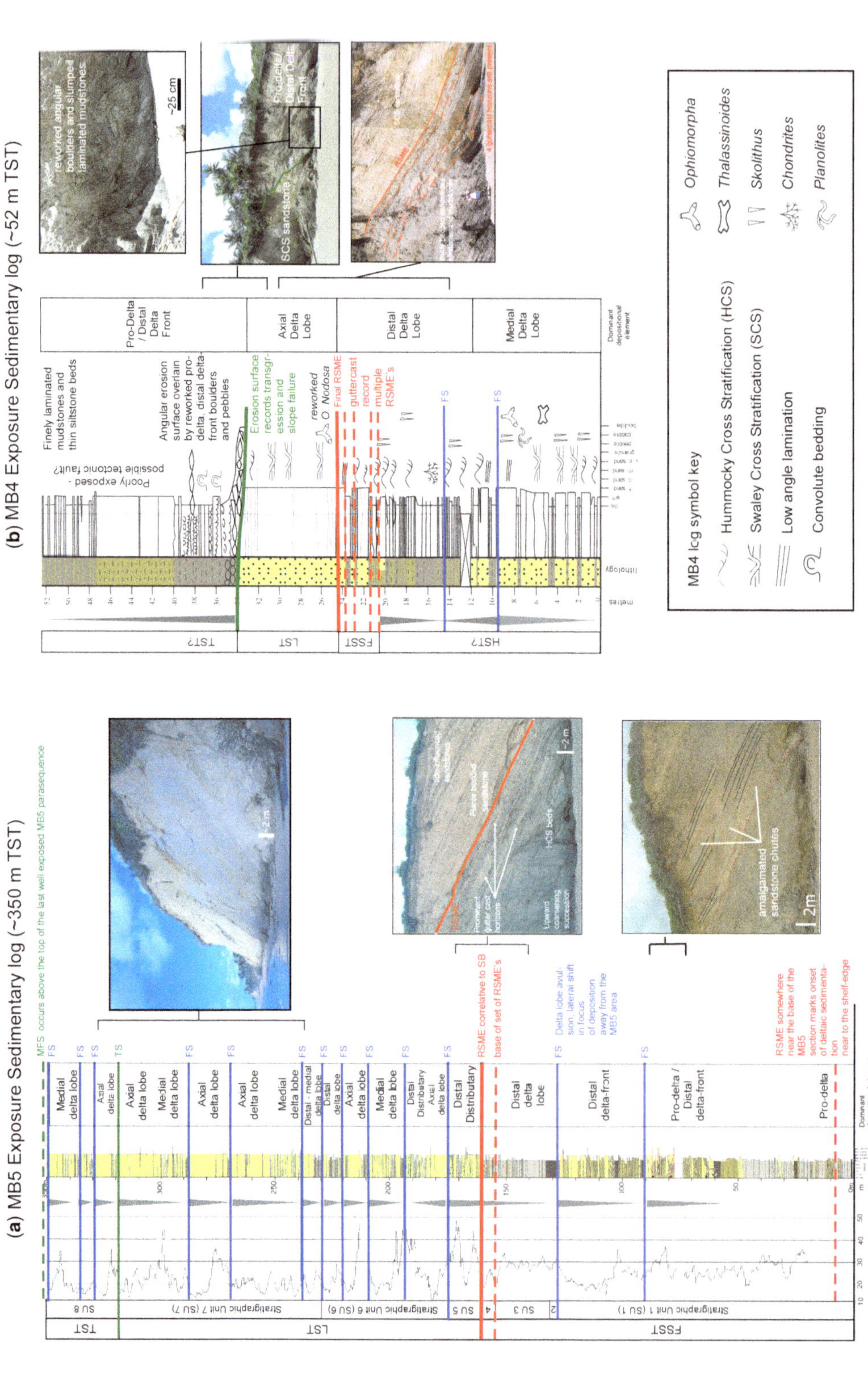
(a) MB5 Exposure Sedimentary log (~350 m TST)
MFS occurs above the top of the last well exposed MB5 parasequence
Medial delta lobe
Axial delta lobe
Medial delta lobe
Axial delta lobe
Distal - medial delta lobe
Distal delta lobe
Distal Distributary
Axial delta lobe
Distal Distributary
RSME correlative to SB
base of set of RSME's
Distal delta lobe
Delta lobe avulsion, lateral shift in focus of deposition away from the MB5 area
Distal delta-front
Pro-delta / Distal delta-front
Pro-delta
RSME somewhere near the base of the MB5 section marks onset of deltaic sedimentation near to the shelf-edge
Dominant depositional element
outcrop gamma ray log (moving average of 10 second total count)
TST
LST
FSST
SU 8
Stratigraphic Unit 7 (SU 7)
Stratigraphic Unit 6 (SU 6)
SU 5
SU 3
Stratigraphic Unit 1 (SU 1)
FS
TS
~2 m
HCS beds
Upward coarsening succession
amalgamated sandstone chutes
2m
(b) MB4 Exposure Sedimentary log (~52 m TST)
reworked angular boulders and slumped laminated mudstones
~25 cm
SCS sandstone
Pro-Delta / Distal Delta Front
Finely laminated mudstones and thin siltstone beds
Poorly exposed - possible tectonic fault?
Angular erosion surface overlain by reworked pro-delta, distal delta-front boulders and pebbles
Erosion surface records transgression and slope failure
reworked O. Nodosa
Final RSME
guttercast record multiple RSME's
Pro-Delta / Distal Delta Front
Axial Delta Lobe
Distal Delta Lobe
Medial Delta Lobe
Dominant depositional element
lithology
metres
TST?
LST
FSST
HST?
MB4 log symbol key
Hummocky Cross Stratification (HCS)
Swaley Cross Stratification (SCS)
Low angle lamination
Convolute bedding
Ophiomorpha
Thalassinoides
Skolithus
Chondrites
Planolites

elements and reservoir architectures. Resolving these subtleties in the subsurface is important in all parts of a field life-cycle, but is likely to be difficult in the absence of high-resolution subsurface datasets such as core or core-calibrated image logs. The outcrops can mitigate this by providing detailed insight into end-member stacking patterns and the distribution of depositional elements that form subsurface reservoirs.

Depositional elements and reservoir architecture

Within parasequences, the primary control on reservoir architecture is depositional process. Storm- and wave-dominated systems tend to be more laterally extensive and layer-cake than fluvial- or tide-influenced systems, which are more laterally discontinuous and typified by jigsaw or labyrinthine architectures (Weber & van Geuns 1990; Reynolds 1999; Ainsworth 2010). In three dimensions, inclined seawards-dipping clinoform surfaces also have the potential to complicate reservoir architecture (e.g. Sech *et al.* 2009). In this section, the depositional elements introduced previously (Table 1; Fig. 4) are described in order from proximal to distal. Each depositional element records a distinctive range of sandstone–shale ratios, and sand- and shale-body geometries, and so each depositional element has specific implications for reservoir architecture and fluid flow in the subsurface.

Facies succession descriptions are integrated with outcrop gamma-ray and probe-permeameter data to develop schematic descriptions of reservoir architecture for six of the seven depositional elements. Since the focus of this study is on those depositional elements with significant reservoir potential, the Pro-Delta depositional element is only briefly reviewed. The schematic descriptions depict heterogeneity at three scales: (1) intra-parasequence scale comparable to a Columbus Basin geocellular model grid cell (*c.* 10 m wide × 2 m deep: Moosai *et al.* 2012); (2) sand-body scale (*c.* 5 m wide × 5 m deep); and (3) bed scale (1 × 1 m thick). Surfaces are described as 'potential barriers' or 'potential baffles', since the extent to which they may influence fluid flow depends upon a reservoir's fluid type and, although the recent history of the basin has been dominated by dry-gas production, there are significant gas-condensate and oil resources in many of the discovered fields, particularly in the NW of the basin (Leonard 1983; Wood 2000; Gibson *et al.* 2004).

Distal Distributary depositional element

This depositional element comprises fine-grained, clean and quartz-rich sandstones, with clay being restricted to discrete mudstone beds or clay drapes. It is only exposed in the MB5 outcrop, where it is bounded by a basal RSME and an upper transgressive surface of erosion (TSE) (*c.* 160–175 m on the sedimentary log in Fig. 6). Six of the facies associations of Bowman & Johnson (2014) are included in this depositional element: TDF-1, Tidal heterolithics; TDF-2, Wavy heterolithics; TDF-3, Rippled sandstones; TDF-4, Cross-bedded sandstone; TDF-5, Planar laminated sandstone; and PDF-2, HCS (Table 1). Two bedding styles occur: (1) flat-lying planar laminated fine-grained sandstones; and (2) a range of sigmoidal cross-bedded and rippled sandstones with mud drapes (Fig. 7). Cross-bed sets are 10–50 cm thick, many metres long, and amalgamated to form composite beds that are approximately 0.3–2.5 m thick and laterally continuous over metres to tens of metres before passing into rippled sandstones, flaser and lenticular beds and wavy heterolithic beds. Palaeocurrents are dominantly orientated towards the NE. Discontinuous, finely laminated or structureless mudstone beds may separate these composite beds. Bioturbation is rare and restricted to small *Skolithos*, *Ophiomorpha* and *Planolites*. Poorly to moderately well-developed *Glossifungites* surfaces occur at various levels. Carbonaceous material, including small branches and fossil leaves, is moderately abundant.

The planar laminated sandstones record erosion and deposition by traction currents in a proximal depositional setting, and are similar to those described in well-exposed shelf-margin mouthbar packages (Plink-Björklund & Steel 2004). They may be genetically related, down depositional dip, to the lobate sand bodies that occur in the Delta-Front depositional element. The mud-draped cross-bedded and rippled sandstone beds, and flaser and lenticular bedding record a range of tide-influenced deposits (Reineck & Wunderlich 1968; Allen 1980),

Fig. 6. A comparison of the different scales of cyclicity in the Mayaro Formation that highlights the important role played by significant stratigraphic surfaces in controlling the distribution of facies associations. (**a**) A summary outcrop sedimentary log and gamma-ray log through the MB5 succession which records eight discrete phases of accommodation-space-controlled shelf-margin deposition in which key stratigraphic surfaces (RSME, TSE and FS) controlled the distribution of facies associations and different reservoir architectures. (**b**) A summary outcrop sedimentary log through the thinner MB4 succession, which highlights the important role of erosional surfaces in juxtaposing facies associations with very different reservoir architectures.

Table 2. *Summary of the stratigraphic units recognized in the MB5 outcrop of the Mayaro Formation (after Bowman & Johnson 2014)*

Systems tract	Stratigraphic unit	Evolution of accommodation space and depositional setting	Sedimentary characteristics	Depositional elements
TST	SU8	• Eustatic sea-level rise creates shelf accommodation space and reduces sediment supply • Back-stepping parasequences are developed as delta retreats landwards away from the shelf-edge	• A 20 m-thick upwards-cleaning and -coarsening succession of interbedded HCS siltstones, and fine-grained sandstones and mudstones	• Medial Delta-Lobe • Distal Delta-Lobe
LST	SU6 and SU7	• Eustatic sea-level stillstand/early rise • Balanced sediment supply and accommodation-space creation • Accumulation of thick aggradational package of proximal delta-front deposits • Steeply rising shelf-edge trajectory	SU7 (125 m thick): • Dominated by SCS sandstones, rare discontinuous mudstones SU6 (50 m thick): • Alternating units of SCS sandstones and mixed wave- and tide-influenced sandstones and mudstones • Dominance of interbedded HCS sandstones and mudstones increases upwards	• Axial Delta-Lobe • Axial Delta-Lobe • Medial Delta-Lobe • Distal Delta-Lobe
	SU5	• Accelerating sea-level fall and increased sediment supply	SU4 (3 m thick): • Anomalously thick HCS beds associated with gutter casts, soft-sediment deformation and slumping	• Distal Distributary (SU5)
FSST	SU4	• Very limited shelf-edge accommodation space • Erosion at the shelf-edge as delta distributaries establish and supply sand to the slope • Falling shelf-edge trajectory	SU5 (17 m thick): • Sharp erosive surface interpreted as a RSME separates the underlying gutter casts from planar-bedded sandstones • Majority of the unit comprises tide-influenced sandstones (cross-bedded and rippled with mud-drapes, evidence of flow reversal and flaser–lenticular beds)	• Medial Delta-Lobe (SU4)
	SU2 and SU3	• Delta avulsion, perhaps in response to subsidence pattern changes • Rapid, short-lived decrease in sediment supply results in local deepening • Renewed sediment supply and renewed delta progradation	SU2 (3 m thick): • Distal deltaic mudstones SU3 (30 m thick): • Upwards-cleaning and -coarsening succession of interbedded HCS siltstones, and fine-grained sandstones and mudstones	• Distal Delta-Lobe • Medial Delta-Lobe
	SU1	• Eustatic sea-level fall reduces shelf accommodation space • High sediment supply and rapid growth-fault-controlled subsidence at/near the shelf-edge • Slightly rising to flat shelf-edge trajectory	• 125 m-thick upwards-coarsening and -cleaning succession consisting of: – interbedded silty mudstones, muddy/silty debrites and thinly bedded very-fine-grained sandstone turbidites overlain by: – isolated and amalgamated chutes filled with turbidite sandstones overlain by; – low-angle scours and slides filled with planar bedded and rippled (current, combined flow and wave) fine-grained sandstones; – SCS sandstones	• Pro-Delta • Distal Delta-Front • Medial Delta-Lobe

Eight stratigraphic units record the evolution of a delta at the shelf-edge and the resulting change in reservoir depositional elements and reservoir architecture.

with thin structureless mudstones recording fluid-mud deposits (MacKay & Dalrymple 2011). The 'clean' nature of the sandstones reflects winnowing by tidal currents. The upwards-coarsening packages record deposition from small migrating bed-forms in a tide-influenced subaqueous setting (Allen 1980; Visser 1980; Tessier & Gigot 1989). The impoverished trace-fossil assemblage records an ecologically stressed environment (Pemberton *et al.* 2002; MacEachern *et al.* 2005). Bowman & Johnson (2014) interpreted this facies association as recording tide-influenced lower distributary channel-fill and fluvial-influenced mouthbar sand bodies, analogous to the lower reaches of tide- and fluvial-influenced distributary channels on the modern Orinoco Delta (Warne *et al.* 2002).

Distal distributary reservoir architecture and subsurface implications. This depositional element is characterized by a more variable gamma-ray log response than might be expected, which superficially resembles the log response observed in the Delta-Front and Medial Delta-Lobe depositional elements. This variability is likely to reflect the impact of mud drapes and thin discontinuous mudstone beds on the gamma-ray log response. The sand–shale ratio is estimated to range from 0.7 to 0.9, well above the 0.28 threshold for poor connectivity in tidal sediments described Jackson *et al.* (2005). At the centimetre to microscopic scale, mudstone heterolithics, flaser and lenticular bedding, and clay drapes within cross-bedded sandstones create permeability heterogeneity at the bed scale (Figs 7 & 8). This is likely to decrease effective permeability (Yoshida *et al.* 2001; Jackson *et al.* 2005). It may also, depending on hydrocarbon phase, well flow rates and heterogeneity orientation, create local capillary pressure gradients, with the potential to trap oil within high-permeability streaks and reduce recovery (Corbett *et al.* 1992).

There are no continuous shales observed in this depositional element at outcrop (Fig. 8); shale layer lengths range from approximately 0.5 to 15 m and will create baffles, but not barriers, to fluid flow. Although the limited extent of the Mayaro Formation outcrops precludes the identification of clinoform surfaces associated with the thin proximal mouthbar deposits in this depositional element, mudstone-prone seawards-dipping clinoform surfaces are described in the literature (e.g. Howell *et al.* 2008) and their distal expression may be the mudstone-prone surfaces bounding lobate heterolithic packages in the Delta-Front depositional element. If present and orientated parallel to depositional strike and at a high-angle to the structural dip, they would have the potential to inhibit aquifer influx, potentially extending the life of the well. Alternatively, if orientated with the structural dip, they could act to cause aquifer waters to over-run dipping mudstones, trapping hydrocarbons in isolated pockets below mudstones and, thus, result in lower recovery.

At outcrop, it is impossible to know the lateral extent of the distributary channels or if they are located within larger incised valleys: they could form jigsaw to labyrinthine architectures (Weber & van Geuns 1990). Shallow seismic data in the Columbus Basin (Israel 2005; Moscardelli *et al.* 2012) reveal a palaeo-geography that is similar to active parts of the present-day Orinoco Delta (Warne *et al.* 2002), which is characterized by 1–3 km-wide distributary channels and NW–SE-orientated palaeo-shorelines, marked tidal creeks that are tens of metres up to 100 m wide, and, offshore, by sand-prone subaqueous delta-lobe deposits. Exposures of Late-Pliocene-age Morne L'Enfer Formation outcrops in SW Trinidad record stacked, tide-influenced, lower delta-plain channels (*c.* 5 m deep × 500 m wide). These data suggest that distributaries within this depositional element were probably 0.5–3 km wide, which is broadly consistent with global analogues (Reynolds 1999). In the subsurface, 'surgical' seismic mapping reveals relatively subtle erosional features in otherwise continuous moderate- to high-amplitude seismic reflectors at these width scales, which when mapped out in three dimensions can be interpreted as channels. Some of these features could represent incised valleys developed on the delta top or close to the shelf edge. Channelized, tide-influenced reservoirs in the Columbus Basin have been previously documented by Bally *et al.* (1999) and Wharton & Muller (2000).

The likely limited lateral extent of distal distributary reservoirs and their orientation, broadly perpendicular to the prevailing fault and palaeo shoreline trend, means that, in comparison to delta-lobe reservoirs, they will tend to have a greater risk of compartmentalization. They are also more likely to be poorly connected to regional aquifers and, hence, may require a greater number of development wells for a given resource density. In delta distributary reservoirs, it may be desirable to drill deviated development wells orientated approximately parallel to channel trends. This approach would maximize the completion length in contact with reservoir sandstones, and help to mitigate stratigraphic compartmentalization risks by linking potentially baffled or barriered sand bodies, leading to an overall increase in draining area per well. Similar oil-bearing compartmentalized reservoirs in the Gulf of Paria, offshore western Trinidad, have been developed using multi-lateral wells (Lougheide *et al.* 2004). In contrast to this, axial and medial delta lobe reservoir segments in the Columbus

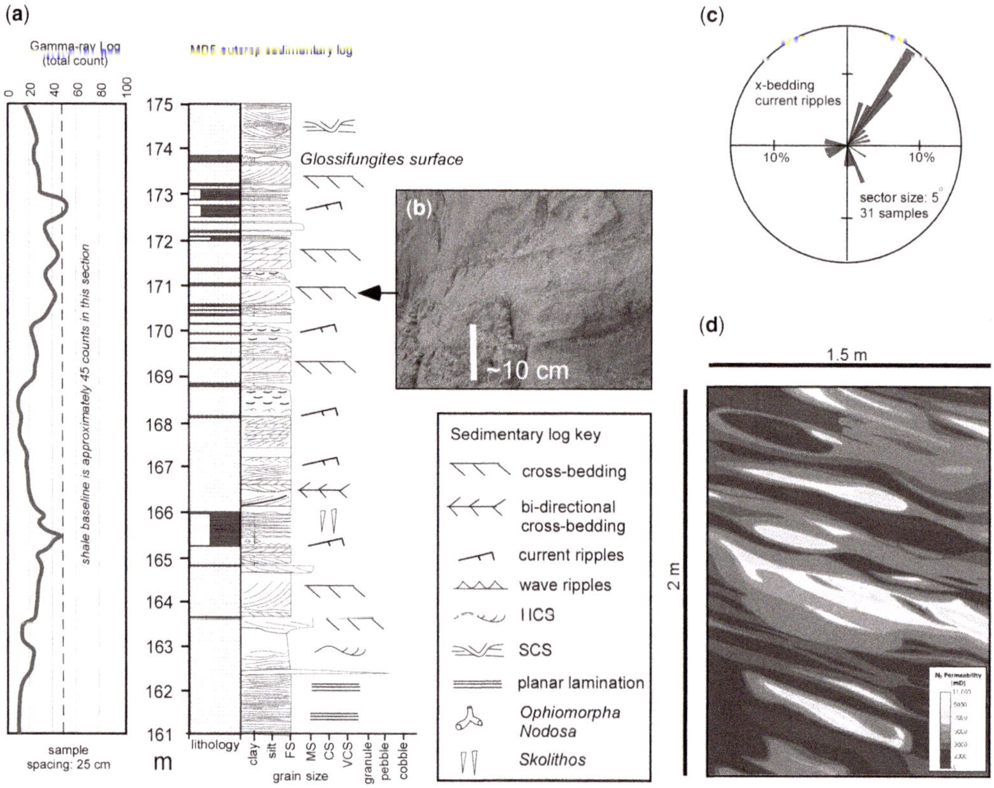

Fig. 7. (**a**) A 14 m-thick sedimentary log and gamma-ray log through the Delta Distributary depositional element (161–175 m; MB5 summary log: Figs 5c & 6) that highlights the strong influence of tidal processes, including mud-draped cross-bedded sandstones such as those shown in (**b**). (**c**) Palaeocurrent measurements collected over the logged interval are dominantly unidirectional towards the NE. (**d**) Permeability map (10 cm data grid) acquired in rippled sandstones and thin wavy heterolithic beds in the Strakers Quarry exposure: facies and sand content to the succession documented between 164.75 and 166 m on the log shown in (a). The permeability map illustrates significant vertical and lateral variations in permeability at short length scales. Note that Strakers Quarry has been built over since these data were acquired in 2001. FS, fine-grained sandstone; MS, medium-grained sandstone; CS, coarse-grained sandstone; VCS, very-coarse-grained sandstone.

Basin, which are less heterogeneous and more laterally extensive, are typically developed through a limited number of horizontal wells located in the crest of structures, and orientated parallel to the major fault trend and palaeo-shoreline trend (e.g. Lumsden *et al.* 2002).

Axial Delta-Lobe depositional element

This depositional element is associated with the Distal Distributary and Medial Delta-Lobe depositional elements, and comprises two facies associations (Table 1): PDF-1, Laminated mudstones and bioturbated mudstones; and PDF-3, SCS (Bowman & Johnson 2014). Amalgamated and laterally continuous beds of friable, very-fine- to fine-grained, quartz-rich SCS sandstones dominate (Fig. 9). Mudstones are absent or rare: where present, they typically comprise centimetre- to decimetre-thick beds of silty mudstone with wave ripples and rare to absent low-diversity trace-fossil assemblages. More common are thin discontinuous beds of mudstone rip-up clasts lining swales. The remnant mudstone beds and rip-up clast horizons occur throughout the thick sand bodies. Bioturbation comprises localized horizons of *Ophiomorpha Nodosa* burrows, with some rip-up clast conglomerates mostly composed of reworked *O. Nodosa* burrows.

The SCS sandstones record high-energy storm-wave deposition, which are commonly interpreted as a record of Upper Shoreface deposition (Leckie & Walker 1982; Walker & Plint 1992; Dumas & Arnott 2006). However, the close association of thick SCS units with remnant mudstones is

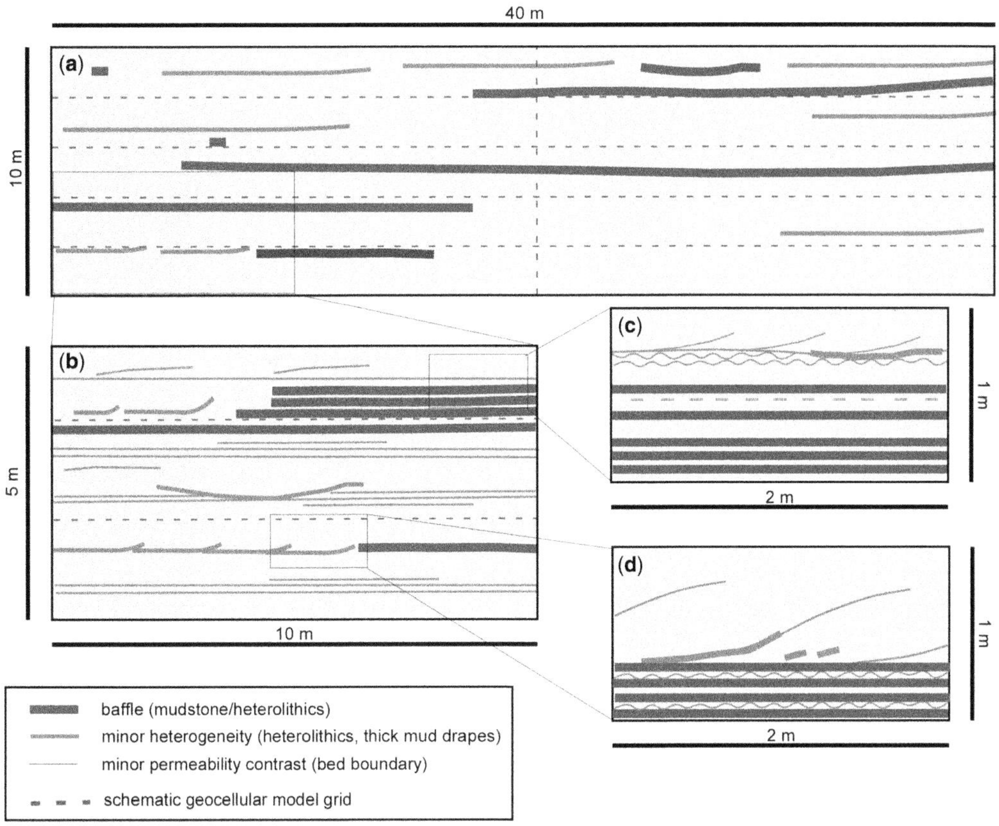

Fig. 8. A schematic representation of reservoir architecture in the Delta Distributary depositional element: there are few baffles at the geocellular grid-cell scale, but, at the bed scale, mud-drapes introduce a level of fine-scale heterogeneity with the potential to impact effective permeability, sweep efficiency and overall recovery.

inconsistent with a shoreface model: instead, this depositional element is interpreted to record storm wave influenced reworking of deltaic sand bodies in the axial part of a subaqueous sand-rich delta-lobe (Bowman & Johnson 2014).

Axial delta-lobe reservoir architecture and subsurface implications. This depositional element is near-homogenous, and represents the simplest architecture and the best reservoir quality encountered in the Columbus Basin. It is characterized by very high sand–shale ratios (>0.95) and 'blocky' low gamma-ray log signatures. Without outcrop knowledge, it is possible that mudstone rip-up clast conglomerates and/or laterally discontinuous shales that are observed in the outcrop could be mistaken for more significant and laterally continuous shales in subsurface well-log data. Similarly, at outcrop, the SCS origin of sandstones in this depositional element is obvious. In subsurface image-log and even core datasets, however, the spectacular erosional swales that are beautifully exhibited in the outcrops may not be obvious and, as a result, units that are tens of metres thick, which in core comprise planar to low angle lamination, may be difficult to interpret with confidence.

At the geocellular model grid-cell scale, there are no barriers to flow: sand bodies are continuous at outcrop and remnant mudstone beds and rip-up clast conglomerate horizons are laterally discontinuous at outcrop, and could only form minor baffles to fluid flow (Fig. 10). Some thick remnant mudstones with inchnofacies indicative of a more distal, offshore depositional setting may represent remnant eroded parasequence-bounding flooding-surface mudstones. Permeability transects through this facies reveal a relatively uniform permeability distribution; small variations in permeability occur in association with subtle changes in grain size at the base of SCS sandstones sets, but these are unlikely to have a significant impact on fluid flow in oil, gas-condensate or dry-gas reservoirs.

Storm-wave processes suggest that these sand bodies are likely to be laterally extensive over

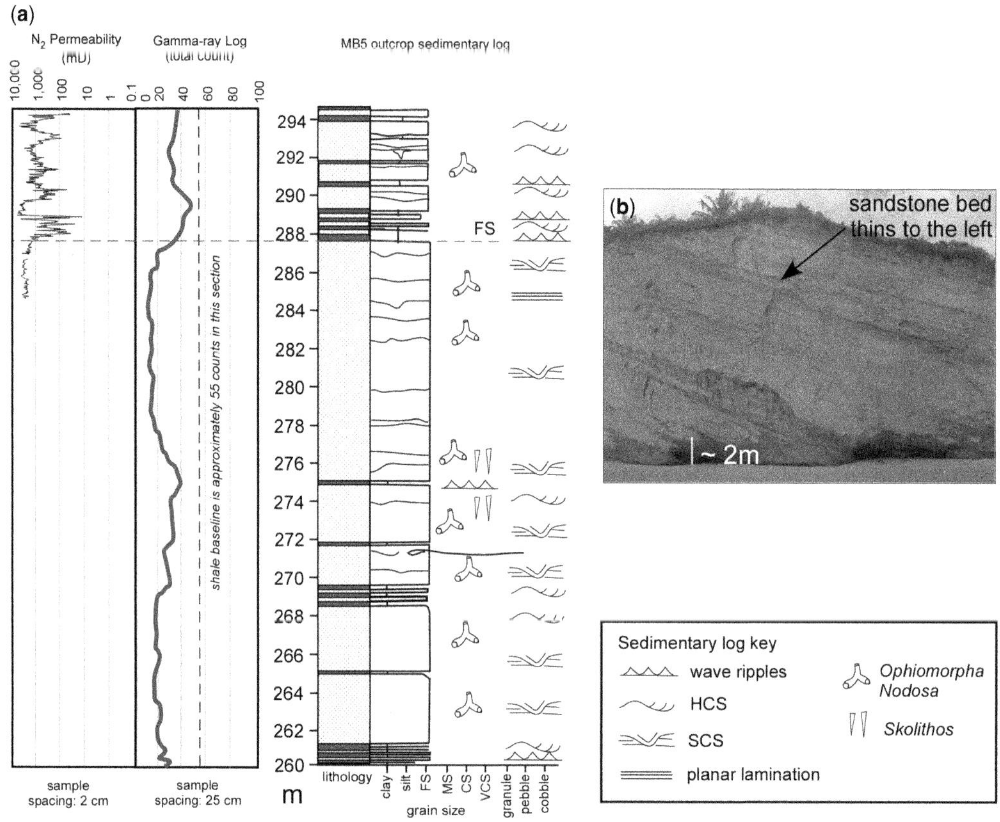

Fig. 9. (**a**) The sedimentary log and electric logs (260–294 m) MB5 summary log (Figs 5c & 6) demonstrate the high sand–shale ratio and low gamma-ray of response characteristic of the Axial Delta-Lobe depositional element. The flooding surface (FS) at approximately 288 m separates Axial Delta-Lobe from Medial Delta-Lobe facies: note the contrast in permeability log character, which reflects the difference between the 'clean' axial delta-lobe sandstones and more clay-rich lamination in the medial delta-lobe and the higher, more serrated, gamma-ray log response. (**b**) Photograph of a sand-prone dominantly axial delta-lobe succession (*c.* 302–336 m; MB5 summary log: Figs 5c & 6): note the very high sand–shale ratio and the scarcity of laterally continuous mudstones. In the upper third of the cliff line there is a prominent set of sandstone beds, comprising large ball and pillow structures that thin towards the south (left-hand side): these are interpreted to record syndepositional faulting. FS, fine-grained sandstone; MS, medium-grained sandstone; CS, coarse-grained sandstone; VCS, very-coarse-grained sandstone.

several kilometres (Reynolds 1999; Howell *et al.* 2008), and subsurface data, such as well logs and 3D seismic imaging, demonstrate lateral continuity along depositional strike for up to about 5 km, after which they gradually pass laterally into more heterolithic Medial Delta-Lobe facies. In addition, pressure communication in axial to medial delta-lobe reservoirs has been documented over distances of up to 10 km (Sydow *et al.* 2003). As a result of their excellent reservoir quality, lack of significant vertical barriers or baffles, lateral extent and stacked nature, field segments comprising this depositional element can be developed with a relatively small number of high-rate (>100 MMSCFD) horizontal wells (Lumsden *et al.* 2002; Hallam & Lumkin 2003; Pucknall *et al.* 2003). This lateral continuity also means that these reservoirs are more likely to extend from the structural crests of the Columbus Basin fields down into the major WSW–ENE-orientated synclines in the basin and are therefore they are more likely to be connected into strong aquifers. This is important because aquifers provide additional reservoir energy that can enhance hydrocarbon production, but can also, if not understood and appropriately managed through appropriate well-placement, completion stand-off and management of well rates, lead to early water-breakthrough ('cresting' and 'coning'), abandonment of wells, reduced recovery and increased cost (Sech 2007).

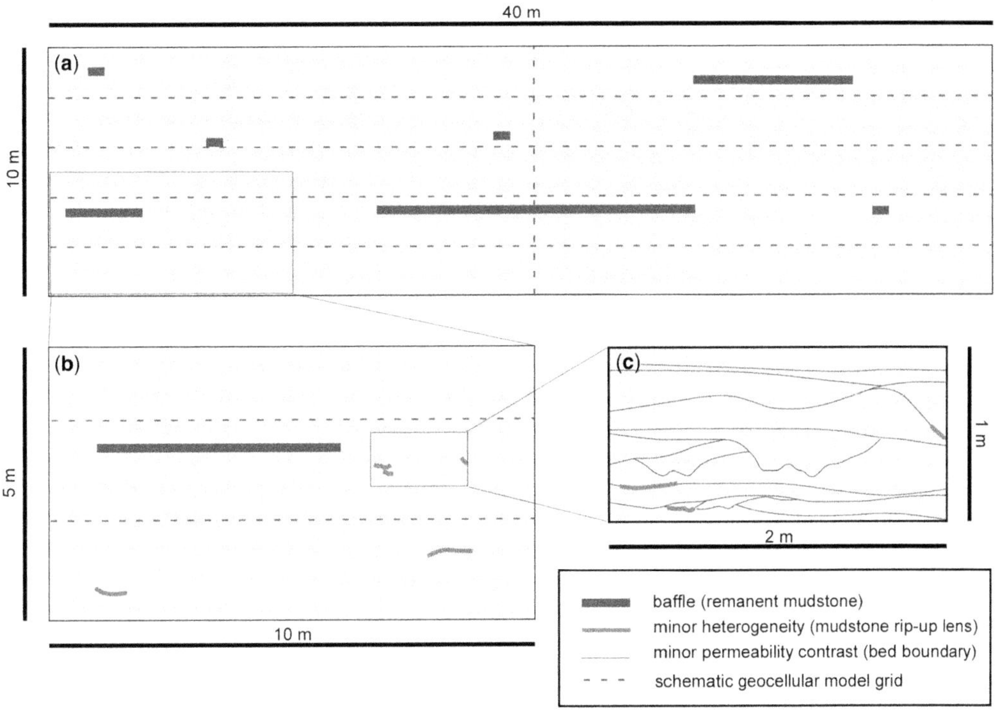

Fig. 10. The schematic representation of the Axial Delta-Lobe depositional element shows the lack of significant barriers and baffles at the geocellular grid-cell scale (**a**), the bed set scale (**b**) and bed scale (**c**).

Medial Delta-Lobe depositional element

Interbedded heterolithic mudstones and friable, very-fine- to fine-grained sandstones, which are laterally continuous at outcrop, make up this depositional element (Fig. 11). It comprises three facies associations (Table 1); PDF-1, Laminated and bioturbated mudstones; PDF-2, HCS; and PDF-3, SCS sandstones (Bowman & Johnson 2014). This depositional element can occur stratigraphically below or, more rarely, above the Axial Delta-Lobe depositional element. Individual sandstone beds are sharp-based and range in thickness from a few to tens of centimetres. They are characterized by hummocky cross-stratification (HCS) and wave ripples, and may either pass up into, or up out of, amalgamated SCS sandstone beds. Laminations in the HCS and SCS sandstone beds are defined by subtle variations in grain size and variable proportions of clay minerals and finely disseminated organic matter. Thicker beds are probably amalgamated and may contain remnant mudstone beds that are centimetres thick, similar to those in the Axial Delta-Lobe depositional element. Gutter casts are often associated with significant sequence stratigraphic surfaces such as RSMEs (e.g. Fig. 6) and may locally increase vertical connectivity between sandstone beds. The heterolithic mudstones separating the sandstones are typically 5–10 cm thick, and exhibit a spectrum of sedimentary structures and bioturbation running from finely laminated silty mudstones with sparse low-diversity bioturbation to less silty beds with thicker wave-rippled laminations and characterized by a moderately abundant, moderately diverse trace-fossil assemblage that may include *Chondrites*, *Anchonichnus*, *Planolites*, *Palaeophycus*, *Rhizocorallium* and *Teichichnus*.

The interbedded HCS sandstone beds record episodic storm-wave-influenced sandstone deposition (Harms *et al.* 1975; Scott 1992) imposed on a lower-energy background of fair-weather wave-influenced deposition (De Raaf *et al.* 1977; Swift *et al.* 1986). Lamination variability in the HCS beds is interpreted to reflect primary depositional processes: 'clean' laminations defined by relatively subtle grain-size variations reflect deposition during the early and peak stages of storms, whilst 'dirtier' laminations marked by more abundant clays and organic material record deposition by either lower-energy storm events or the waning stages of large storms (Scott 1992). The more laminated and bioturbated mudstones record deposition in a less ecologically environment compared to the more silty and

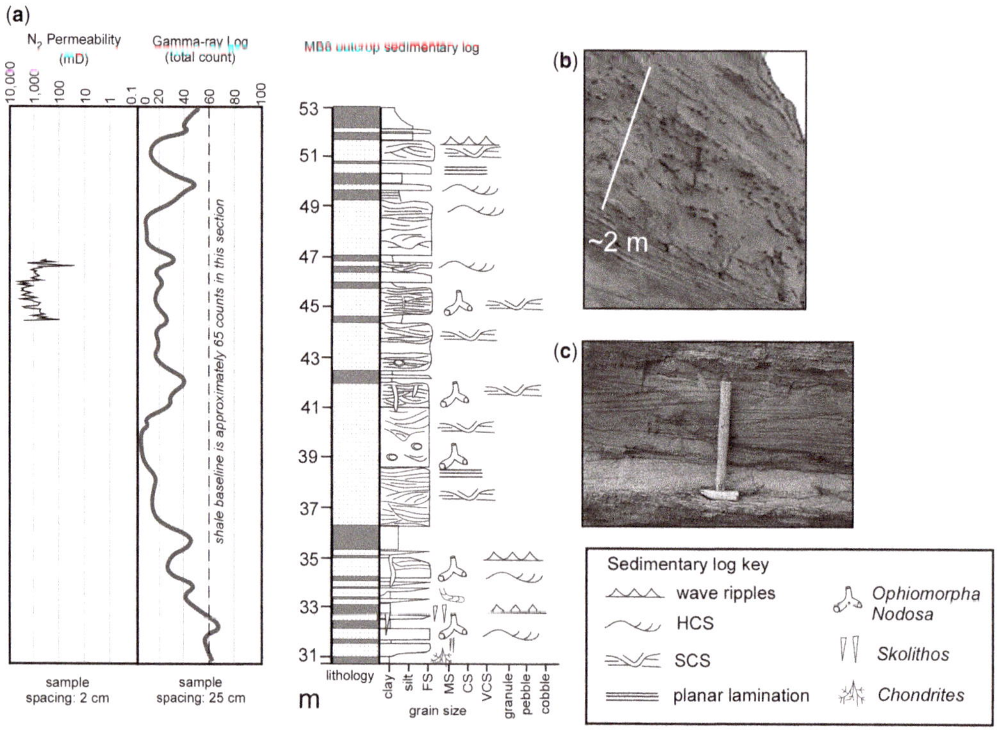

Fig. 11. (**a**) The sedimentary log and electric logs (31–53 m; MB8 summary log: Bowman 2003) demonstrate the sand-prone nature of the Medial Delta-Lobe depositional element that comprises stacked SCS sandstones separated by mudstones which are more abundant and continuous than those in the axial delta-lobe (Fig. 12). Some sandstone beds have more clay-rich laminations than those in the axial lobe: a feature highlighted by the more variable permeability log, and the suppression of the gamma-ray log between 44 and 47 m. (**b**) A photograph of intensely bioturbated SCS sandstones (*c.* 27–29 m; MB8 summary log: Bowman 2003), the hammer is for scale is approximately 30 cm long. (**c**) Detailed view of an HCS sandstone bed located at approximately 32.5 m on the summary log shown in (a). FS, fine-grained sandstone; MS, medium-grained sandstone; CS, coarse-grained sandstone; VCS, very-coarse-grained sandstone.

sparsely bioturbated mudstones: this variability is interpreted to reflect varying proximity to delta distributaries (MacEachern *et al.* 2005). The facies associations and geometries are similar to those described in storm-wave-influenced middle- to lower-shoreface settings (Walker & Plint 1992), but the close association with the axial storm-/wave-influenced depositional element and the occurrence of upwards-coarsening and -fining trends supports deposition on the medial part of the storm-wave-influenced subaqueous delta lobe (Bowman & Johnson 2014).

Medial delta-lobe reservoir architecture and subsurface implications. These rocks are characterized by a lower sand–shale ratio (~0.5–0.7) than axial delta-lobe deposits, and a gamma-ray log response typified by upwards-cleaning or, more rarely, an upwards-dirtying profile. The gamma-ray log struggles to resolve thinner beds (*c.* <1–10 cm). The primary heterogeneity is the interbedded nature of the sandstone and mudstone beds (Fig. 12). Mudstone rip-up clasts and thin remnant mudstones form insignificant intra-sand-body baffles, but, in the absence of outcrop knowledge, they could be mistaken for more significant and extensive mudstone beds in subsurface well-log data. Within sandstones, permeability variation occurs in association with bed-tops, where laminations are more strongly defined by clays and organic material. In these areas, permeability variations approaching an order of magnitude on a logarithmic-scale occur (Fig. 11). Comparison of vertical and horizontal permeability profiles within a composite storm-generated SCS bed indicates only minor permeability anisotropy at the bed scale.

Assessment of mudstone thickness/width relationships in the upper part of the MB5 exposure (Fig. 6: MB5 log; 270–350 m) suggest that most mudstones are discontinuous, and that there is a

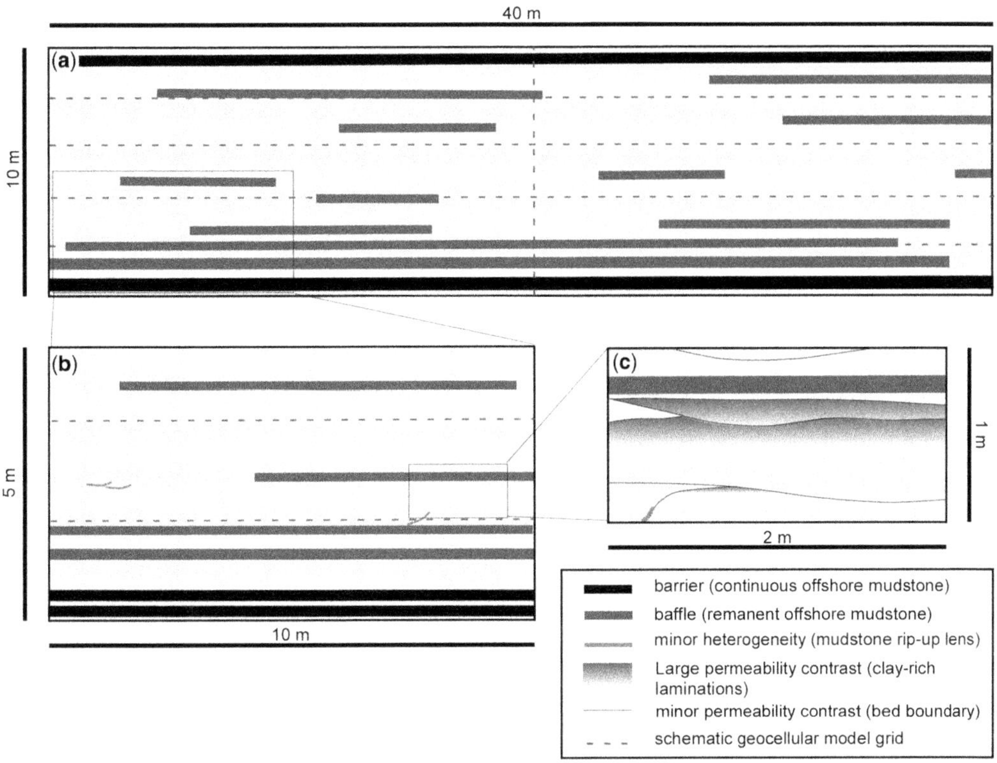

Fig. 12. (**a**) Schematic representation of the Medial Delta-Lobe depositional element that illustrates, at the geocellular grid-cell scale, the presence of continuous mudstones with the potential to form barriers or baffles towards the top and base of parasequences, and the abundance of horizontal baffling mudstones. (**b**) The same barriers and baffles at the bed-set scale. (**c**) The impact of clay-rich laminations that occur towards the top of SCS and HCS sandstone beds, and which create variable permeability contrasts at the bed scale.

weak correlation between shale layer thickness and length (Fig. 13a): 61% of measured mudstones were discontinuous at outcrop; however, 23% of mudstones with a thickness of <30 cm were laterally continuous at outcrop. The method of White & Willis (2000) for estimating a mudstone-length probability density function (PDF), through the analysis of complete mudstones (two terminations), incomplete mudstones (one termination) or laterally continuous mudstones at outcrop, has been applied to 89 mudstones recorded in Axial and Medial Delta-Lobe facies through the interpretation of a photomontage of the upper MB5 exposure (Fig. 6: MB5 summary log; *c.* 210–350 m) (Bowman 2003). A mudstone termination frequency was computed (0.062) and used to calculate a PDF (Fig. 13b) (Bowman 2003) that indicates that a randomly selected point on a randomly selected mudstone within the axial and medial delta-lobe is most likely to be 16 m lateral to a mudstone termination. The scale of the exposures prevented these data being computed for other depositional elements. These data, together with knowledge of sedimentary processes (Harms *et al.* 1975; Scott 1992) and subsurface data (Lumsden *et al.* 2002; Sydow *et al.* 2003), provide numerical support for the geological interpretation that most mudstones within the axial and medial delta-lobe, apart from those associated with flooding surfaces, are likely to be discontinuous and form baffles rather than barriers to fluid flow: a conclusion broadly comparable with larger studies of storm-wave-influenced shoreline deposits from more-extensive outcrops (e.g. Arnott & Good 1997; Eide *et al.* 2014). It is possible that some of these incomplete mudstones could represent strike sections through baffles or barriers associated with dipping clinoform surfaces and, hence, they could have a more significant impact on recovery (Hampson 2000; Jackson *et al.* 2009; Sech *et al.* 2009). The outcrop scale and orientation prevents the recognition of clinoforms in the outcrops, and they have not yet been confidently observed within parasequences in subsurface data. If present, mudstone-prone clinoform surfaces, depending on their

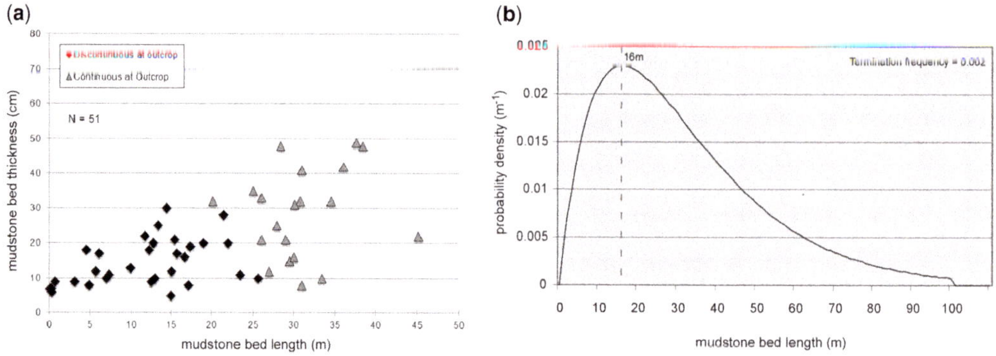

Fig. 13. (**a**) Scatter plot of mudstone bed v. thickness from the medial and axial delta-lobe successions in the MB5 exposure: most mudstones >30 cm thick are continuous at outcrop. (**b**) Graph of a PDF calculated using the methodology of White & Willis (2000) from a termination frequency of 0.062, which indicates that a randomly selected point on a randomly selected axial or medial delta-lobe mudstone is most likely to be 16 m away from a mudstone termination, suggesting that most mudstones in these settings are likely to be laterally discontinuous and form baffles rather than barriers.

orientation relative to structural dip, have the potential to help mitigate the risk of aquifer water influx, or increase the risk of hydrocarbon being bypassed. In the subsurface of the Columbus Basin, mudstone-prone flooding surfaces define flow units and tend to be continuous at the scale of field segments (0.5–3 km), and can sometimes be correlated between fields, along depositional strike, over distances of up to about 10 km. As discussed previously, laterally extensive sand bodies comprising Axial to Medial Delta-Lobe depositional elements can be connected to aquifers. In reservoir segments with aquifers, particularly bottom-drive aquifers, the more mudstone beds preserved in the medial delta-lobe can help to ‘shield’ producing wells from aquifer influx (e.g. Samsundar *et al.* 2007). Medial delta-lobe sand bodies may also be laterally connected to more thinly bedded sandstones (Distal Delta-Lobe depositional element). In the absence of fault compartmentalization, this potentially allows for long-distance drainage of thin-bed prone reservoirs through more proximal, higher net-to-gross and higher reservoir quality, medial delta-lobe reservoirs.

Distal Delta-Lobe depositional element

This depositional element comprises interbedded centimetre-scale sandstone and mudstone beds recording two facies associations (Table 1): PDF-1, Laminated and bioturbated mudstones; and PDF-2, HCS (Bowman & Johnson 2014). The majority of the sandstone and mudstone beds are laterally continuous at outcrop (Fig. 14), and are arranged in upwards-coarsening and -cleaning patterns or, more rarely, fining-upwards or aggradational successions, and bounded by flooding-surface mudstones. This depositional element most often occurs stratigraphically below the Medial Delta-Lobe depositional element. Individual sandstone bed thicknesses range from approximately 2 to 15 cm; they may be massive to subtly graded or contain combined flow and wave-ripple lamination and small-scale HCS. The sandstones are finer grained, silty and, in some case, more clay-rich than those of the axial and medial delta-lobes. Mudstone beds are finely laminated with a varying abundance and diversity of trace fossils, ranging from low-abundance and low-diversity assemblages dominated by *Planolites* and *Chondrites* to more abundant and more diverse *Cruziana*-type assemblages comparable to that developed in the medial delta-lobe. Very rarely, gutter casts are developed linking two–three sandstone beds: these gutter casts are often associated with stratigraphically significant surfaces such as the RSME documented at around 24.5 m on the MB4 summary log (Fig. 6). The facies associations in this depositional element are reminiscent of those interpreted as Lower Shoreface through to Offshore Transition Zone (OTZ) environments (Walker & Plint 1992). A broadly similar setting is envisaged for this depositional element: however, deposition occurred around the fringe of a subaqueous delta-lobe rather than along an elongate shoreline (Bowman & Johnson 2014).

Distal delta-lobe reservoir architecture and subsurface implications. The gamma-ray log signature of this setting is attenuated (Fig. 14) because the beds are too thin to be resolved by standard logging tools (Passey *et al.* 2006). In the subsurface, this results in difficulty in assessing the net-sand thickness and water saturation that results in significant

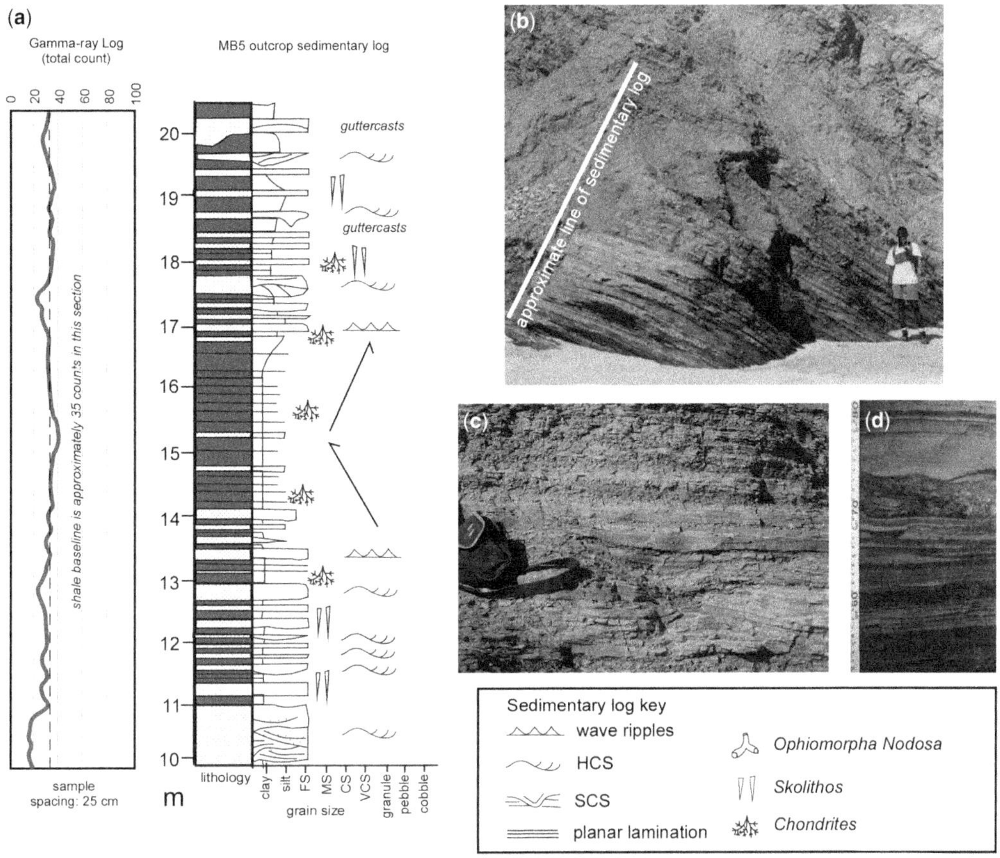

Fig. 14. (**a**) The sedimentary log and gamma-ray log (10–20 m; MB4 summary log: Fig. 6) and the photograph of the logged section (**b**) illustrate the low sand–shale ratio characteristic of the Distal Delta-Lobe depositional element: note how thinly bedded sandstones cannot be resolved by the gamma-ray log. (**c**) Shows the thinly bedded nature of the succession at approximately 13.5 m and (**d**) shows a more detailed and carefully cleaned section through the most mud-prone section at approximately 14.5 m. FS, fine-grained sandstone; MS, medium-grained sandstone; CS, coarse-grained sandstone; VCS, very-coarse-grained sandstone.

uncertainty in the hydrocarbon pore volume of thinly bedded reservoirs (Passey *et al.* 2006; Henderson *et al.* 2012; Taylor *et al.* 2012). The use of advanced logging tools (e.g. image logs and triaxial resistivity logs calibrated to core) can reduce this uncertainty, as can outcrop knowledge. At outcrop, the sand–shale ratio is estimated to range from approximately 0.3 to 0.6 and the reservoir quality of the sandstone prone thin beds is variable. Sandstone beds that are >3 cm thick tend to be quartz rich and are similar to thickly bedded sandstones deposited in more proximal settings. Thinner sandstone beds (<3 cm thick) are more clay rich and, along with finer-grained siltstone beds, which can be clay rich and/or slightly carbonate cemented, are likely to have limited or no reservoir potential. The thin mudstone beds are generally laterally extensive at outcrop (Fig. 15) and will form significant barriers to vertical flow. Hence, there is likely to be significant permeability anisotropy characterized by low vertical-to-horizontal permeability ratios (K_v/K_h) ratios. The sandstone beds, being distal storm-wave deposits, are likely to be less aerially extensive and more patchily distributed at the reservoir scale, a feature recognized in similar distal storm-wave-influenced settings (e.g. Eide *et al.* 2014). Features such as gutter casts and burrows may improve vertical connectivity (Gringras *et al.* 2012), but their overall impact at the reservoir scale is likely to be minimal as neither appears to be extensively developed at outcrop. In this mudstone-prone thinly bedded depositional element, even small faults are likely to be associated with clay smear and, hence, have the potential to form subseismic barriers and baffles to fluid flow over production timescales.

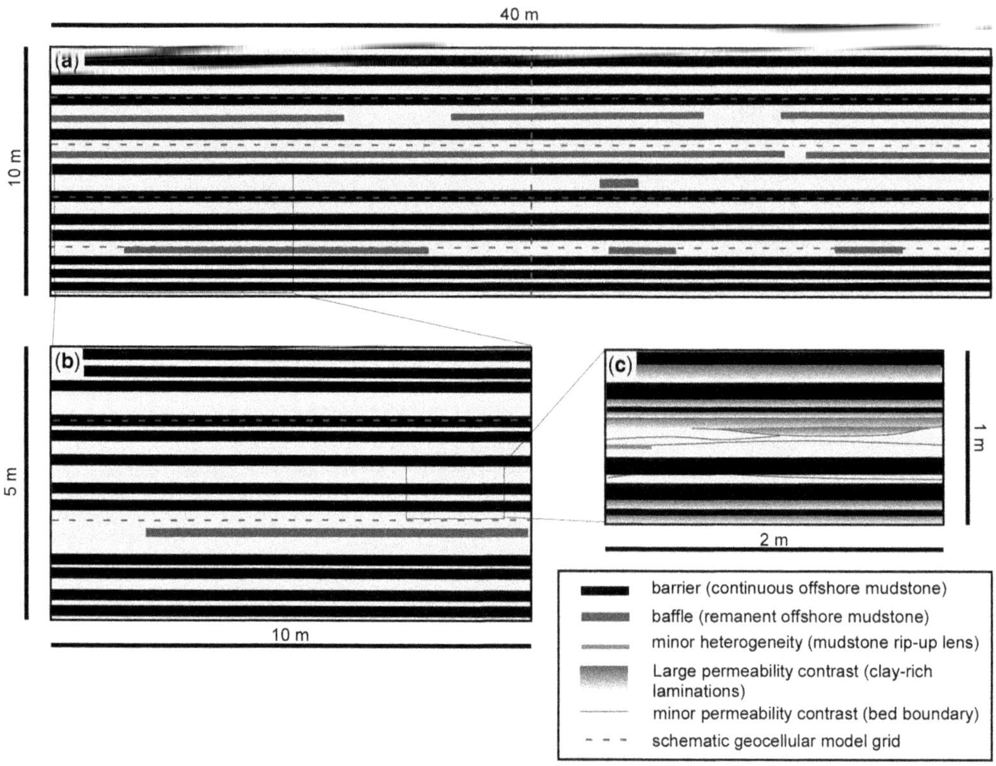

Fig. 15. Schematic representations that highlights the dramatically layered nature of the Distal Delta-Lobe depositional element at the geocellular grid-cell scale (**a**) and the bed-set scale (**b**). Further heterogeneity occurs within individual sandstone beds (**c**) owing to the presence of more silt- or clay-rich laminations, which commonly occur towards the top of storm-beds.

Reservoirs developed in this depositional element will be laterally connected to better-quality reservoir rocks of the medial and axial delta-lobe, but the extensive faulting in the Columbus Basin creates the potential for structural–stratigraphic traps resulting in thin-bed-dominated reservoir segments (Sylvester *et al.* 2005; Baillie & James-Romano 2012; Henderson *et al.* 2012; Taylor *et al.* 2012) that also occur in other deltaic hydrocarbon provinces (Sneider 2003). These reservoirs, with a high proportion of thin beds, are difficult to identify and quantify using standard subsurface data (Passey *et al.* 2006; Baillie & James-Romano 2012; Henderson *et al.* 2012; Taylor *et al.* 2012). Production rates in these thinly bedded Columbus Basin reservoirs are lower than in thickly bedded reservoirs deposited in more proximal settings (10–80 v. 70–250 MMSCFD), and sustained production often requires that wells are repeatedly shut-in and produced at low rates for short periods to enable repressurization of the reservoir (Baillie & James-Romano 2012). This need for repeated reservoir repressurization reflects baffled, tortuous, connectivity to the well bore through thin beds, an inference supported by the recognition of lower depletion rates in thin-bed-prone intervals when they are co-mingled in producing wells with thicker, higher net-to-gross reservoir sandstones (e.g. Taylor *et al.* 2012). The strongly layered reservoir architecture observed at outcrop suggests that future exploitation of thin-bed reservoirs in the Columbus Basin may involve a change from the drilling and completion techniques used to exploit 'thickly-bedded' reservoirs (horizontal, large bore, wells and open-hole gravel-pack completions: e.g. Lumsden *et al.* 2002; Kromah *et al.* 2003; Pucknall *et al.* 2003) towards well designs that enable a higher proportion of the reservoir sandstones to be in contact with the well (e.g. inclined or undulating wellbores or multi-laterals and high-density perforations or stimulated completions).

Delta-Front depositional element

This depositional element occurs stratigraphically above the Distal Delta-Front depositional element and comprises three facies associations (Table 1): DDFS-1, Silty mudstones, DDFS-5, Silty lobate sandstones; and PDF-3, SCS sandstones (Bowman & Johnson 2014). Asymmetric, laterally offset, low-angle scours (tens of metres wide and up to 1 m deep) are infilled by: (1) 5–50 cm-thick beds of very-fine-grained sandstone with low-angle lamination, current ripples and climbing current ripples; and (2) a wide variety of millimetre- to centimetre-scale silty sandstone, siltstone and massive mudstone beds exhibiting current ripples, climbing current ripples and low-angle planar laminations (Fig. 16). Well-developed combined flow and wave ripples occur towards the top of some scour-fill successions. Erosion surfaces, frequently adorned with millimetre- to centimetre-scale mudstone drapes and/or rip-up clasts, are common throughout, as are soft-sediment deformation features. These beds are crudely organized into 2–4 m-thick packages bounded by mudstones, which are locally eroded out at the scale of the outcrop (length scales of tens of metres). The scour-filled sandstone and siltstone units form off-lapping lobate sediment bodies that pass upwards into thicker more laterally continuous sand bodies exhibiting swaley cross-stratification (SCS) (Leckie & Walker 1982; Dumas & Arnott 2006). Within the lobate sediment bodies there is a low-abundance, low-diversity trace-fossil assemblage dominated by small *Skolithos* and *Planolites*. Occasional *Ophiomorpha Nodosa* occur within the SCS sand bodies. Plant material is common and occurs mainly as disseminated carbonaceous flakes. Growth faults and larger-scale slump scars occur (e.g. in the upper MB8 outcrop) (Bowman 2003; Bowman & Johnson 2014).

The low-angle scours are indicative of deposition on an unstable slope. The abundance of sandstone ‘event beds’, massive mudstone beds

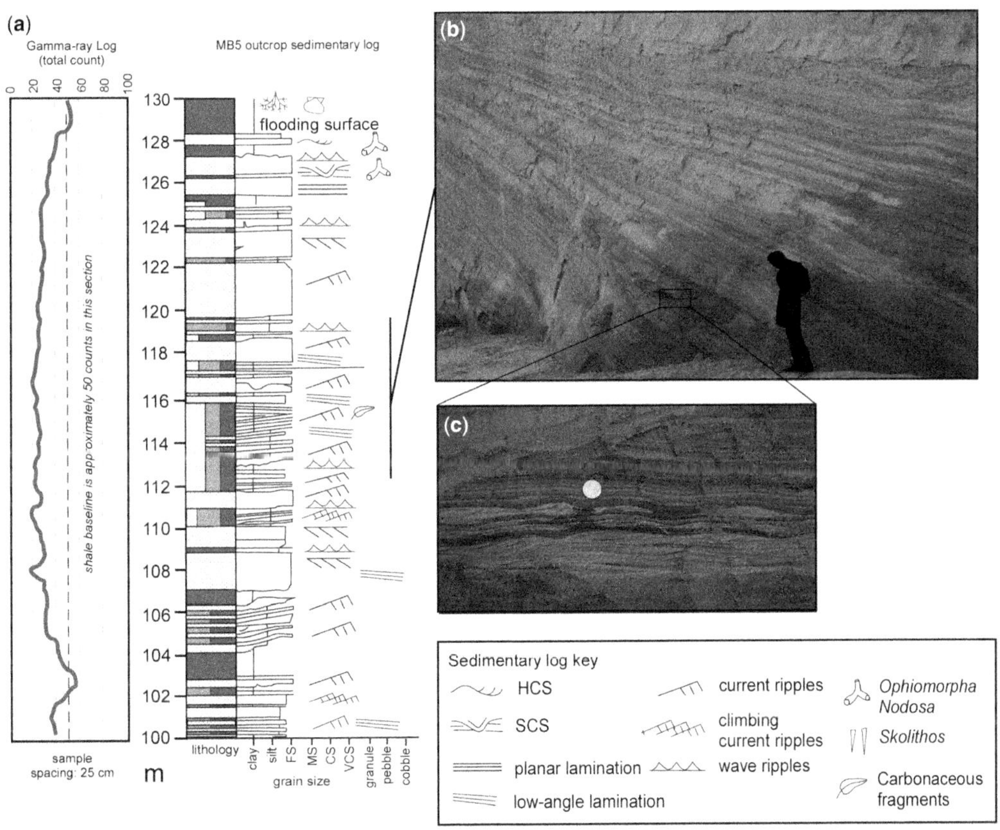

Fig. 16. (**a**) The outcrop sedimentary log (100–128 m; MB5 summary log: Figs 5c & 6) highlights the complex vertical layering that is characteristic of the Delta-Front depositional element: a complexity which the gamma-ray log fails to resolve. (**b**) The photograph highlights the stacked lobate sediment bodies that dominate the Delta-Front depositional element: note the interplay between well-exposed, thicker and laterally discontinuous, sandstones and less well-exposed sections, which are dominated by rippled thin beds such as those shown in (**c**). FS, fine-grained sandstone; MS, medium-grained sandstone; CS, coarse-grained sandstone; VCS, very-coarse-grained sandstone.

interpreted as fluid-mud beds (Rotondo & Bentley 2004), unidirectional flow indicators, sparse, low-diversity bioturbation and abundant plant material is suggestive of deposition close to a significant sediment source. Many of these beds have characteristics consistent with well-described deltaic hyperpycnites (Plink-Björklund & Steel 2004). In contrast, combined flow and wave ripples are indicative of oscillatory flows (Myrow & Southard 1991, 1996; Myrow *et al.* 2002), and SCS is indicative of storm-wave processes (Walker & Plint 1992). Bowman & Johnson (2014) interpreted these sediments as recording the complex interplay of unidirectional and oscillatory currents: probably reflecting the interaction of rivers, waves and large tropical storms on the delta-front. This depositional element may interfinger with more proximal mouthbar deposits updip.

Delta-front reservoir architecture and subsurface implications. The gamma-ray log response associated with this depositional element exhibits an upwards-cleaning to slightly serrated block log signature (Fig. 16) that shares similarities with upwards-cleaning successions associated with storm-wave-influenced delta-lobe successions. Hence, if the stratigraphic context of a succession is poorly understood, then it could be misidentified as a less heterogeneous and complex delta-lobe succession. The reservoir architecture is complex, with laterally discontinuous sandstone beds, erosional scour surfaces, some of which are mudstone lined, and thin beds of variable reservoir quality occurring at scales which mean that they would be very difficult to recognize using standard petrophysical logs and seismic data (Fig. 17). Even with high-resolution well data, such as image logs or core, which enable the recognition of centimetre-scale heterogeneities, it would be very difficult to infer the complex geometric relationships observed at outcrop.

Attempts to acquire probe-permeameter data in the silt-prone and clay-prone thin-bedded facies failed. This is interpreted to reflect lower permeability and poorer rock quality in this facies. An interpretation that is supported by advanced petrophysical analysis (thin section and X-ray tomography) of outcrop samples from the Mayaro Formation by Ramnath (2015), which suggests that silt-prone and clay-prone thin beds have no reservoir potential, but sandstone-prone thin beds may have permeability of the order of tens of mD. Within the lobate sediment bodies, the best porosity and permeability characteristics are likely to occur within the thicker basal sandstone, with the more thinly bedded successions having poorer quality. The basal sandstones in the lobate sand bodies cross-cut one another, increasing bed amalgamation and potentially improving connectivity by eroding mudstone-prone intervals to create tortuous paths for fluid flow. However, in the upper parts of the lobate sediment bodies, the dominance of silty thin beds and cross-cutting, frequently mudstone-lined, erosion surfaces will reduce effective permeability. It is unlikely that there is sufficient connectivity between the lower, sand-prone parts of the lobate sediment bodies and the upper, thin-bed-dominated parts to allow effective drainage of oil or gas-condensate trapped in the thin beds over production timescales. However, if the hydrocarbon phase is dry gas, drainage of these thin beds may be more effective. The larger bounding surfaces visible in the outcrop may be the expression of the distal toes of inclined mouthbar clinoforms. In this case, the 3D distribution of baffles and barriers will be even more complex than is shown in the exposed cliff face. The upper part of this depositional element, comprising cleaner, more tabular storm-wave-influenced sand bodies, is likely to be similar in terms of reservoir properties to the Medial Delta-Lobe depositional element.

This depositional element has the potential to be stratigraphically isolated from overlying, more thickly bedded and better quality sandstones (e.g. as in the MB5 succession), and form hydrocarbon reservoirs. But the relatively poor rock quality (when compared to more proximal storm-wave-influenced depositional elements) and the complexity of the reservoir architecture would create significant development challenges. The fine grain size and poorer rock quality suggest that fluid phase may be an important determiner of developability: dry gas would be more developable than gas-condensate or oil in this complex setting. Assuming a dry-gas hydrocarbon phase, the production behaviour of this depositional element is likely to be reflected in initial moderate rates of gas production as the better rock quality, better connected basal sand bodies are rapidly drained. This would be followed by a rapid rate of production decline, and an extended low-rate production tail as the poorer rock quality and more heterogeneous and baffled thin beds are variably drained. Low-rate production profiles from thinly bedded reservoirs have been documented in Columbus Basin wells by Baillie & James-Romano (2012) and may record production from complex distal delta-front reservoirs. Growth faults and slumps visible at outcrop provide a further level of heterogeneity, which is likely to increase the tortuosity of fluid-flow paths and the potential for complex small-scale structural–stratigraphic compartmentalization. Taylor *et al.* (2012) described discontinuous, thin-bed-dominated intervals in a reservoir in the Dolphin Field in the Columbus Basin that were impacted by listric growth faulting and slumping which had previously been interpreted to record Delta-Front facies (Smart *et al.* 2000).

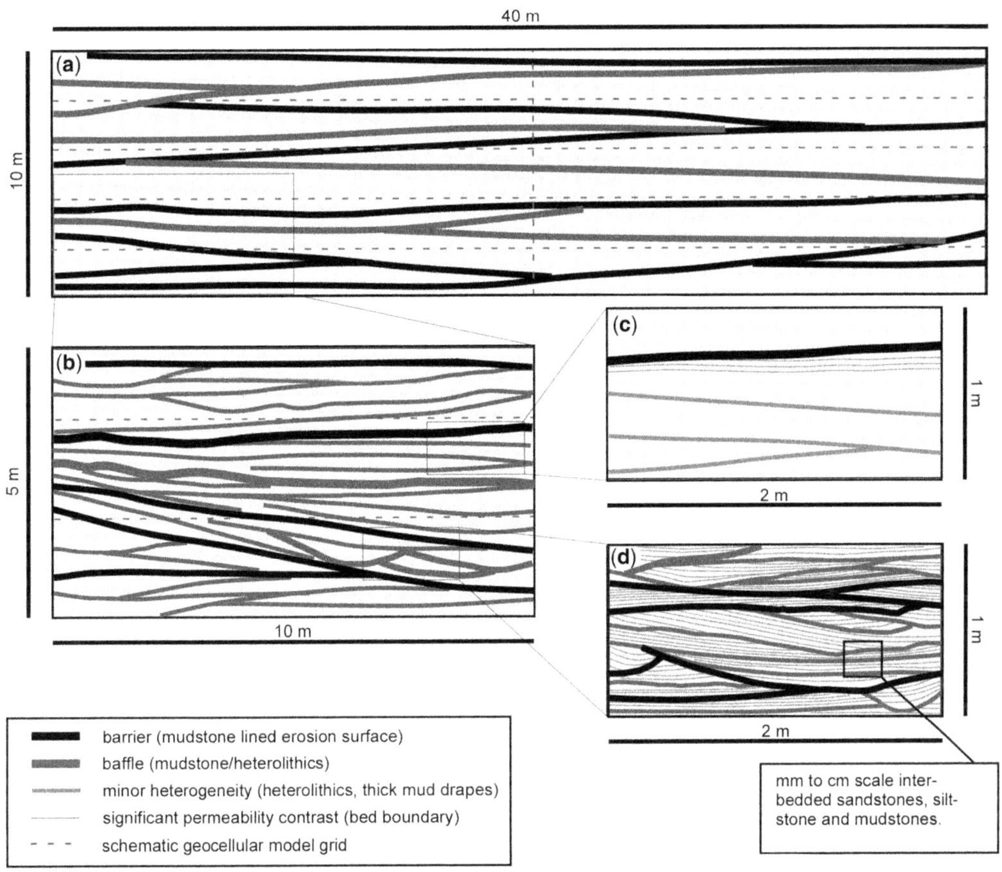

Fig. 17. A schematic representation of reservoir architecture in the Delta-Front depositional element. (**a**) At the scale of a geocellular model grid cell, inclined, frequently mudstone-prone erosion surfaces form barriers and baffles. (**b**) At the bed-set scale, further higher-frequency erosion surfaces occur, creating a range of baffles associated with variable mudstone drapes. (**c**) At the bed scale, within the thicker sandstone beds, there are minor baffles associated with partial mudstone drapes and permeability contrasts controlled by erosion surfaces. (**d**) In the heterolithic intervals, there is a very high degree of heterogeneity where thin beds drape erosion surfaces that are commonly mudstone lined.

Only a single well is reported to have targeted this Dolphins Field example and it is described as producing at 'moderate rates' (Taylor *et al.* 2012). As production in the Columbus Basin matures, these targets may become more appealing to develop. In order to economically access resources trapped in complex reservoirs like these, higher well densities and more innovative well and completion technologies, such as multi-laterals or artificial stimulation technologies, may be required.

Distal Delta-Front depositional element

This depositional element comprises four facies associations (Table 1): DDFS-1, Silty mudstone; DDFS-2, Isolated sandstone chute; DDFS-4, Amalgamated sandstone chutes; and DDFS-6, Heterolithic channel (Bowman & Johnson 2014), which are arranged into weakly developed upwards-coarsening 20–30 m-thick packages bounded by mudstone-prone horizons. Isolated erosional sandstone chutes are up to about 10 m wide and 2 m deep, and may amalgamate to form sand bodies that are up to 25 m wide and 5 m deep. The chute fills comprise very-fine- to fine-grained clean sandstones that exhibit massive to low-angle laminated bedding, and rare small-scale (10–20 cm-thick) cross-bedding and climbing current ripple sets, and are characterized by multiple internal erosion surfaces, some ornamented with angular centimetre-scale rip-up clasts of the background mudstones and muddy siltstones. The chutes occur nested within broader scours filled with laminated silty mudstones and thinly bedded very fine-grained sandstone and

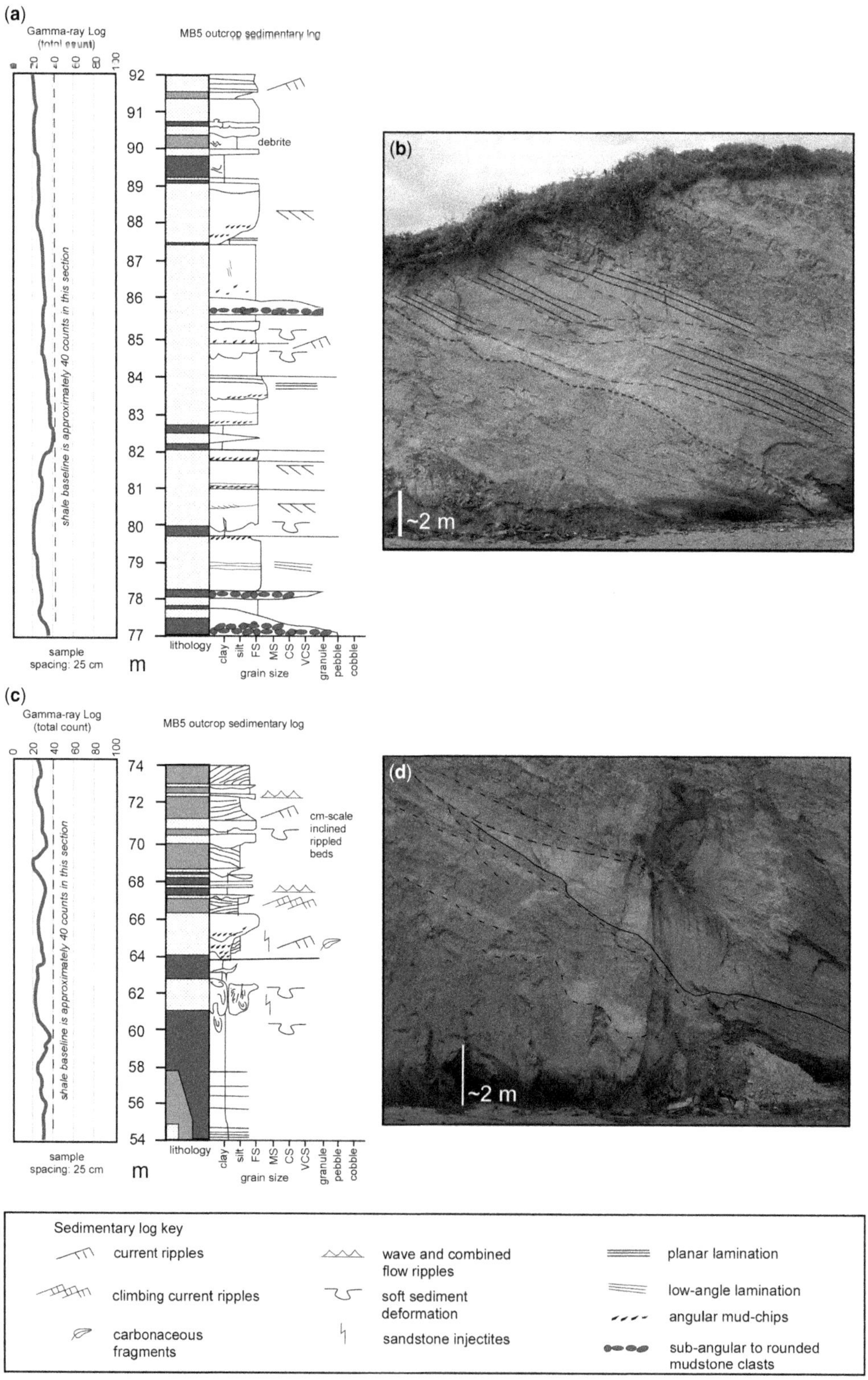
(a)
Gamma-ray Log
(total count)
0 20 40 60 80 100
shale baseline is approximately 40 counts in this section
sample
spacing: 25 cm
MB5 outcrop sedimentary log
92 91 90 89 88 87 86 85 84 83 82 81 80 79 78 77
m
debrite
lithology
clay silt FS MS CS VCS granule pebble cobble
grain size
(b)
~2 m
(c)
Gamma-ray Log
(total count)
0 20 40 60 80 100
shale baseline is approximately 40 counts in this section
sample
spacing: 25 cm
MB5 outcrop sedimentary log
74 72 70 68 66 64 62 60 58 56 54
m
cm-scale
inclined
rippled
beds
lithology
clay silt FS MS CS VCS granule pebble cobble
grain size
(d)
~2 m
Sedimentary log key
current ripples
climbing current ripples
carbonaceous
fragments
wave and combined
flow ripples
soft sediment
deformation
sandstone injectites
planar lamination
low-angle lamination
angular mud-chips
sub-angular to rounded
mudstone clasts

siltstones. The background sediments comprise variable massive to finely laminated mudstones and silty-mudstones (Fig. 18). The background mudstones are variable and comprise: (1) silty mudstones with millimetre- to centimetre-scale laminations; (2) discrete massive, 1–3 cm-thick, clay-rich mudstone beds; and (3) 1–5 cm-thick sharp-based sandstone and siltstone beds. The silty mudstones contain subtly graded beds and 'starved' ripples. Upwards current ripples and climbing current ripples become more common, and, towards the top of the package, combined flow ripples and wave ripples occur. Soft-sediment deformation, such as millimetre- to centimetre-scale sandstone dykes and sills, and small-scale (<10 cm-diameter) slumps and folds, occur throughout the lower part of the succession. Bioturbation is rare. It comprises a low-diversity low-abundance assemblage of *Chondrites*, *Planoliotes* and *Skolithos*. Carbonaceous material, although uncommon, occurs throughout the depositional element.

The background mudstones and siltstones record pro-delta deposition with gravity-driven density-flow processes being dominant. The variability in lamination style and sedimentary structures evident in the upper parts of the successions provide evidence for fluid-mud deposition (Mutti *et al.* 2003; Rotondo & Bentley 2004), pro-delta hyperpycnites (Mutti *et al.* 1996, 2003; Plink-Björklund & Steel 2004), low-density turbidites (Bouma 1962; Mulder & Alexander 2001) and an overprint of storm-wave influence (Myrow & Southard 1996; Bhattacharya *et al.* 2004). The sandstone chutes record either episodic phases of erosion, bypass and back-fill by high-density turbidity currents (Lowe 1982; Mulder & Alexander 2001), or sustained fall-out from steady, waning, low-density turbidites (Kneller 1995; Mulder & Alexander 2001) moving down the clinoform slope of a prograding delta. These gravity-driven density flows were likely to have been initiated by interactions between slope instability, direct river-fed hyperpycnal flows and storm-wave-generated currents (Bowman & Johnson 2014). The trace-fossil assemblage is consistent with rapid deposition in an ecologically stressed pro-delta to distal delta-front environment (Pemberton *et al.* 2002; MacEachern *et al.* 2005). Laterally, away from sediment sources, it is expected that this setting may gradually trend into the more strongly storm-wave influenced thin-bedded successions characteristic of the Distal Delta-Lobe depositional element.

Distal delta-front reservoir architecture and subsurface implications. The isolated and amalgamated sandstone chutes form impressive sand bodies at outcrop within weakly defined upwards-coarsening, mudstone-bounded packages interpreted to record clinoform surfaces (Bowman & Johnson 2014). The sand bodies are characterized by weakly defined, upwards-cleaning to blocky, low gamma-ray log signatures. Mudstone intra-clast horizons cause higher gamma-ray log responses and create a more serrated log signature in parts of the succession (Fig. 18). Hence, in the absence of outcrop knowledge, they could be mistaken for more laterally continuous mudstone beds. Sand–shale ratios in this depositional element are variable. Where this depositional element is exposed in the MB5 exposure (Fig. 6), the lower part of this depositional element, where it is dominated by isolated chutes, has a sand–shale ratio that varies at outcrop from 0.3 to 0.6, depending on whether and where sandstone chutes are intersected. In contrast, where it is dominated by amalgamated chutes, the sand–shale ratio of the depositional element exceeds 0.6.

The large-scale architecture (tens of metres × hundreds of metres) is impossible to assess at outcrop without the integration of subsurface seismic data because the exposures are too narrow to define the edges of the amalgamated chutes. Shallow seismic data from the Columbus Basin shows linear, high-amplitude features, tens to hundreds of metres wide, that are orientated perpendicular to the shoreline and which transit the pro-delta and distal delta-front before terminating at the transition to the upper slope (Sydow *et al.* 2003; Israel 2005; Moscardelli *et al.* 2012). In some cases, these features occur within canyons and are observed to feed larger slope depositional systems (Moscardelli *et al.* 2012). It is likely that these erosional seismic features are the seismic expression of the sand bodies and geometries observed in the Distal Delta-Front depositional element at outcrop.

Fig. 18. A summary of the sedimentary and stratigraphic characteristics of the Distal Delta-Front depositional element. (**a**) A sedimentary and outcrop gamma-ray log through the amalgamated sandstone chutes exposed between 77 and 92 m (MB5 summary log: Figs 5c & 6): note the suppression of the 'clean' gamma-ray log signature by the mud-clast rip-up horizons. (**b**) A photograph of the amalgamated sandstone chute: note the presence of multiple erosion surfaces. (**c**) A sedimentary and outcrop gamma-ray log through the isolated sandstone chutes exposed between 54 and 74 m (MB5 summary log: Figs 5c & 6): note the slightly serrated log-signature. (**d**) A photograph of the logged isolated sandstone chute summarized in (c): note the sand-prone 'gull-wing' chute that is nestled with a broader scour surface filled with thinly bedded sandstones and mudstones. FS, fine-grained sandstone; MS, medium-grained sandstone; CS, coarse-grained sandstone; VCS, very-coarse-grained sandstone.

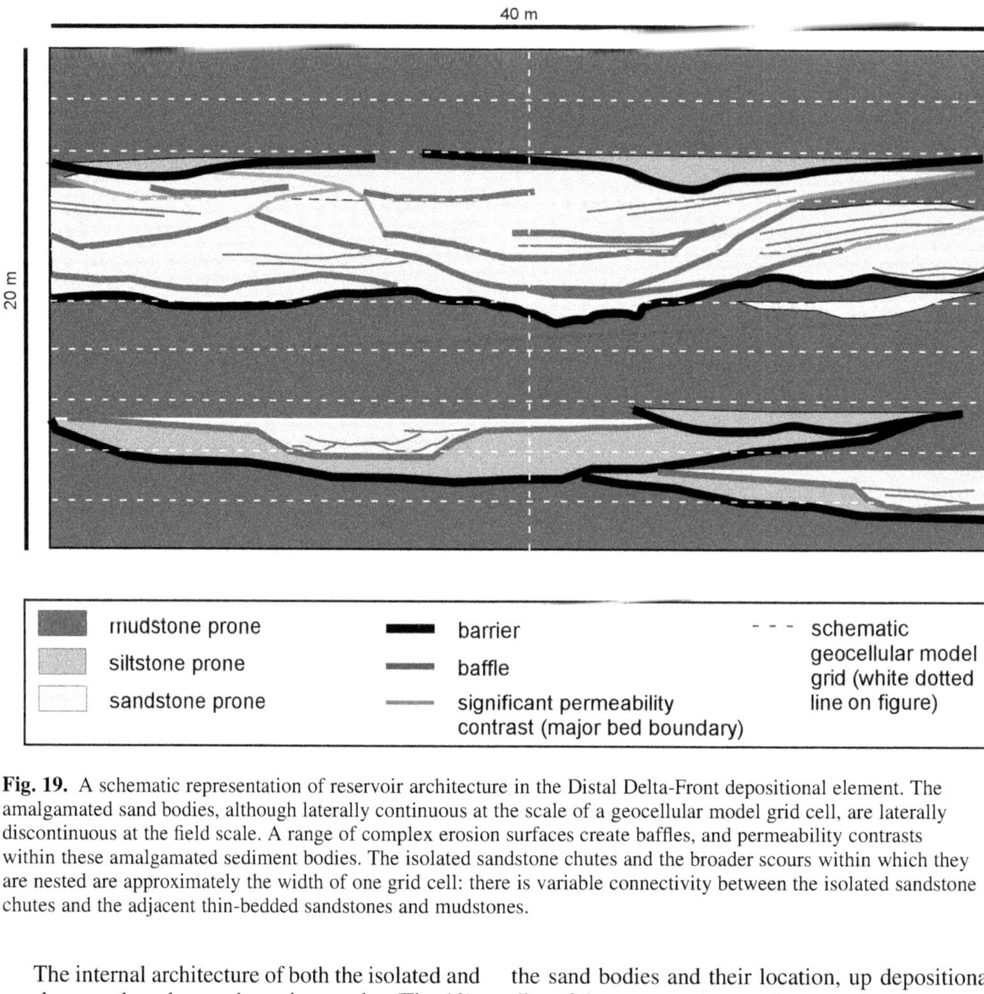

Fig. 19. A schematic representation of reservoir architecture in the Distal Delta-Front depositional element. The amalgamated sand bodies, although laterally continuous at the scale of a geocellular model grid cell, are laterally discontinuous at the field scale. A range of complex erosion surfaces create baffles, and permeability contrasts within these amalgamated sediment bodies. The isolated sandstone chutes and the broader scours within which they are nested are approximately the width of one grid cell: there is variable connectivity between the isolated sandstone chutes and the adjacent thin-bedded sandstones and mudstones.

The internal architecture of both the isolated and amalgamated sandstone chutes is complex (Fig. 19). Common erosion surfaces, which may be wholly or partially ornamented with mudstone or siltstone intra-clast rip-ups, have the potential to form baffles or, less likely, barriers to fluid flow. Lamination-scale variations in permeability within sand bodies are assumed to be minor owing to the clean nature of the sandstones and the relative scarcity of well-defined sedimentary structures. The fine-grained background sediments do contain thinly bedded (<2 cm thick) very-fine-grained sandstones and siltstones. They are laterally discontinuous, rarely intersect one another and are thought likely to have poorer reservoir quality. As a result, they are not expected to have significant thin-bedded pay potential. Small-scale sandstone injection features and thin sandstone beds may allow pressure and fluid communication between sandstone beds over geological timescales, but are less likely to enable fluid communication over production timescales. The relatively stratigraphically isolated nature of the sand bodies and their location, up depositional dip, of deep-water slope sediments means that they could be predisposed to being overpressured, and could, therefore, represent a drilling hazard. The difficulties in recognizing this depositional element using relatively low resolution and sparse subsurface data, and the complexity of its reservoir architecture are significant exploration and development challenges.

Pro-Delta depositional element

This depositional element occurs above and below the main sequence-scale reservoir intervals, and forms subregional seals to the main reservoirs. Since it has no reservoir potential and is not associated with significant intra-reservoir heterogeneity, its description is brief. It consists of two facies associations: DDFS-1, Silty mudstone; and DDFS-2, Shallow sandstone scours (Bowman & Johnson 2014). The DDFS-1 silty mudstones can be arranged into weakly developed upwards-cleaning or -dirtying cycles, particularly in intervals where they contain thin-bedded turbidites with current rippled or planar bedding. Muddy/silty debrites occur, often in close association with slumped units. Shallow sandstone scours (tens of metres wide and <50 cm thick) also occur towards the top of successions, as this depositional element transitions upwards into the Distal Delta-Front depositional element. The facies in this depositional element are interpreted to record distal pro-delta to upper slope deposition.

Discussion

The stratigraphic surfaces and depostional architectures revealed in the Mayaro Formation outcrops can be directly applied to improve the understanding of how different reservoirs or different reservoir segments in the Columbus Basin can be explored for, developed and produced. The most proximal depositional element comprises fluvial- and tide-influenced distributary channel and mouthbar deposits. Seawards of the shoreline, this depositional element interfingers with storm-wave-influenced deposits of the axial delta-lobe. In the absence of high-quality core and/or seismic data, it may be difficult to distinguish deposits associated with delta distributaries from adjacent axial delta-lobe deposits. Developing reservoirs comprising the Distal Distributary depositional element may require higher well counts, as the orientation of linear distributary sand bodies oblique to, or perpendicular to, delta-lobe sand bodies and fault trends increases the potential for reservoir compartmentalization.

Reservoirs comprising the Axial Delta-Lobe depositional element are characterized by very thick aggradational high net-to-gross successions. Heterogeneities are limited; shale-layer terminations and shale-layer thickness–width relationships suggest that, apart from preserved flooding surfaces, most mudstones are likely to be discontinuous at reservoir scales (2–5 km). Medial delta-lobe deposits are more heterogeneous than those of the axial delta-lobe: this reflects increased deposition and preservation of mudstones away from the axis of delta-deposition, but, again, most non-flooding-surface mudstones are not laterally extensive and will probably form baffles rather than barriers to fluid flow. In the subsurface, axial and medial delta-lobe reservoirs form laterally extensive high net-to-gross intervals that are commonly connected to regional aquifers and host much of the current production in the Columbus Basin. They are often developed by a limited number of horizontal or high-angle wells that are characterized by sustained high- (80–100 MMSCFD) to very-high-rate (>200 MMSCFD) production rates (Hallam & Lumkin 2003; Lumsden *et al.* 2002; Pucknall *et al.* 2003). In medial delta-lobe reservoirs, local mudstones can form local baffles that help to 'protect' wells from early water breakthrough. Whilst in axial delta-lobe reservoirs, which lack vertical baffles/barriers, this is not the case. As a result, careful well placement (e.g. ensuring sufficient stand-off from hydrocarbon–water contacts) and reservoir management is required in order to benefit from additional reservoir energy, whilst mitigating the risks of early aquifer influx into production wells.

Distal delta-lobe deposits are much more heterogeneous than those of the axial delta-lobe; centimetre-scale layering of sandstones, siltstones and mudstones results in a strongly layered reservoir architecture with poor vertical connectivity. Reservoir quality is variable; the thickest sandstone beds (3–15 cm) are comparable to sandstones in thick-bedded depositional elements. Thinner sandstone beds, siltstone beds and clay-rich sandstones are much poorer quality, and are unlikely to form developable reservoirs offshore. The interfingering of the distal and medial delta-lobe sediments may be of critical importance to the developability of these thin-bed reservoirs, as thicker medial delta-lobe sand bodies could potentially provide improved lateral and vertical connectivity into, and between, more distal thin beds: allowing them to be more effectively drained and developed via medial delta-lobe sand bodies.

Delta-front reservoirs are dominated by lenses of thinly bedded sandstones, siltstones and clay-rich mudstones. The stacking of these lenses is complicated, leading to variably amalgamated, internally complex sand bodies, which may be linked updip to delta mouthbars. Low-angle scours, slump scars and growth faults may occur, and result in increased reservoir complexity, increasing the potential for the development of stratigraphic baffles and barriers at inter-well scales. Reservoirs comprising this depositional element will be challenged by a high-proportion of shaley sandstones and thin beds, very variable reservoir quality, relatively low effective permeability and low K_v/K_h. Distal Delta-Front and Pro-Delta depositional elements probably have a limited reservoir potential, although thicker sandstone chutes developed in the Distal Delta-Front depositional element may be connected to overlying or adjacent delta-front sand bodies.

Distal Delta-Lobe and Delta-Front depositional elements represent one of the most significant future development opportunities and challenges in the Columbus Basin. This is because recent interpretations of dynamic production and pressure data in several large mature fields in the basin, such as the Pliocene-age Cassia Field in the SW of the basin (Baillie & James-Romano 2012) and the

Pleistocene-age Dolphin Field in the east of the basin (Taylor *et al.* 2012), have indicated that some reservoir segments are characterized by more reservoir energy than was predicted by original descriptions based on static data. This phenomena has been interpreted to reflect additional hydrocarbon resources trapped in thinly bedded and shaley sandstones, which are difficult to resolve using standard subsurface tools (Baillie & James-Romano 2012; Henderson *et al.* 2012; Taylor *et al.* 2012). The outcrops, through their unique high-resolution record of Distal Delta-Lobe and Delta-Front depositional elements, provide important geological insights that can help to overcome this development challenge and realize the opportunity. The outcrops demonstrate that diagnosis of distal delta-lobe and delta-front reservoir architectures will require the acquisition of high-resolution geological data, such as triaxial resistivity logs, high-resolution image logs and, crucially, core data. Pilot wells and associated side-track wells could provide useful subsurface insights into the lateral continuity and spatial distribution of distal delta-lobe and delta-front sandstones. Studies such as Baillie & James-Romano (2012) and Taylor *et al.* (2012) show that static geological data should be supported by the acquisition of pre-production pressure data, ideally followed by the acquisition of repeat surveillance data (e.g. well tests and production logging tools (PLTs)). The highly layered nature of these reservoirs means that, in the future, a shift away from large well-bore high-angle/horizontal wells, which are commonly used to develop axial and medial delta-lobe reservoirs in the Columbus Basin, may be required. This is because high-angle/horizontal wells can only intersect a limited number of thin beds. In contrast, vertical wells can intersect many thin beds but limit the reservoir area exposed to the well bore. Therefore, deviated wells, which combine the ability to intersect many thin beds whilst maximizing the reservoir area exposed to the well bore, may prove an effective development well option for these reservoirs. In order to effectively drain these heterogeneous reservoirs, wells may need to be spaced more closely together and, given the offshore setting, multi-lateral wells may prove an effective way of increasing the number of off-take points in a reservoir. The highly layered nature of these reservoirs means that they are unlikely to be at high risk of water-influx and so there may be scope to complete wells closer to fluid contacts in order to maximize well rates and ultimate recovery. Stimulated completions may help to enhance vertical connectivity and result in improved well rates. As the basin matures successful development of these more complex depositional elements may be aided by denser subsurface data and the availabilty of existing facilities.

Conclusions

The Mayaro Formation is the key outcrop analogue for the prolific Columbus Basin hydrocarbon province. The studied outcrops have provided crucial insights into the basin's stratigraphy and depositional setting, helping to support a substantial growth in hydrocarbon production, particularly gas and gas-condensate production, in the last 15 years. As the development of hydrocarbon resources in the Columbus Basin matures and more geologically complex reservoirs, deposited in distal delta-lobe and delta-front settings, form a larger proportion of the remaining resource base, the importance of the outcrops as an aid to understanding uncertainties in reservoir distribution and architecture will grow. The most important insights from the Mayaro Formation outcrops can be summarized as:

- The recognition that the scales of cyclicity, parasequence and sequence stacking patterns, and stratigraphic surfaces that occur at outcrop, also occur in the subsurface reservoirs offshore. Therefore, the detailed stratigraphic insights provide by the outcrops can be more confidently applied to the subsurface.
- An appreciation that successions that are tens to hundreds of metres thick in the subsurface are not simply thickened parasequences. They are likely to be stratigraphically complex sequences comprising a range of different depositional elements, the distribution and development of which reflects the evolution of shelf and shelf-edge accommodation space at the time of deposition.
- The identification of six depositional elements with reservoir potential in the Mayaro Formation. Exposures of these six depostional elements reveals the range and stratigraphic distribution of reservoir facies likely to be encountered in the subsurface, enabling the dominant sedimentary processes to be recognized and the implications of sedimentary processes for reservoir architecture to be better understood.
- A clearer understanding of the uncertainty associated with the use of key subsurface tools such as gamma-ray and V-shale log signatures to recognize important reservoir depositional elements.
- Recognition that the integration of outcrop and subsurface data can help to drive the development of alternate subsurface scenarios, which can help to mitigate exploration and development risks associated with inherent uncertainties in identifying depositional elements using limited, often relatively low-resolution, subsurface data.

More broadly, through its complete record of pro-delta to delta-top shelf-edge facies, stratigraphy and reservoir architecture, the Mayaro Formation also provides insights into hydrocarbon reservoirs developed in shelf-margin settings elsewhere in the world, such as the Gulf of Mexico (Mayall *et al.* 1992), the Niger Delta (Damuth 1994; Saugy & Eyer 2003) and the Mahakam Delta (Roberts & Sydow 2003). The unusual thickness of the stacked deltaic successions, which reflects the uncommonly high sediment supply/high-accommodation setting of the Columbus Basin (Carvajal & Steel 2009), means that the exposures are likely to be of particular value in evaluating reservoirs developed in similar high-accommodation settings, such as the Baram Delta (Back *et al.* 2001, 2005; Hiscott 2001; Hodgetts *et al.* 2001; Saller & Blake 2003; Cullen 2010).

BP is thanked for permission to publish this paper; the opinions and views expressed here are those of the author and not necessarily those of BP. Much of the content derives from A. Bowman's 2003 PhD thesis, which was supervised by H.D. Johnson at Imperial College London, and funded by NERC Award GT4/99/1490, BP Venezuela and BP Trinidad & Tobago The author is grateful for discussions with BP colleagues, particularly D. Bodnar, R. Dixon, J. Finneran, J. Sydow, D. Tasker and H. Vincent. The original manuscript was improved following reviews by N. Cross, R. James and H. Smyth.

References

AINSWORTH, R.B. 2005. Sequence stratigraphic-based analysis of reservoir connectivity: influence of depositional architecture – a case study from a marginal marine depositional setting. *Petroleum Geoscience*, **11**, 257–276, http://doi.org/10.1144/1354-079304-638

AINSWORTH, R.B. 2010. Prediction of stratigraphic compartmentalization in marginal marine reservoirs. *In*: JOLLEY, S.J., FISHER, Q.J., AINSWORTH, R.B., VROLUK, P.J. & DELISLE, S. (eds) *Reservoir Compartmentalization*. Geolological Society, London, Special Publications, **347**, 199–218, http://doi.org/10.1144/SP347.12

ALGAR, S.T. 1993. *Structural, stratigraphic and thermochronological evolution of Trinidad*. PhD thesis, Dartmouth College, Hanover, NH.

ALLEN, J.R.L. 1980. Sand waves; a model of origin and internal structure. *Sedimentary Geology*, **26**, 281–328.

ARNOTT, M.J. & GOOD, T.R. 1997. Shales in shallow marine sandstones: quantitative outcrop data and its applications in reservoir simulation studies. *In*: *Shallow Marine and Non-marine Reservoirs: Sequence Stratigraphy, Reservoir Architecture and Production Characteristics*. Poster presented at the GCSSEPM Foundation 18th Annual Bob F. Perkins Research Conference, 7–11 December 1997, Houston, Texas, USA, 7–11.

BABB, S. & MANN, P. 1999. Structural and sedimentary development of a Neogene transpressional plate boundary between the Caribbean and South American plates in Trinidad and the Gulf of Paria. *In*: MANN, P. (ed.) *Caribbean Basins: Sedimentary Basins of the World, Volume 4*. Elsevier Science, Amsterdam, 495–557.

BACK, S., MORLEY, C.K., SIMMONS, M.D. & LAMBIASE, J.J. 2001. Depositional environment and sequence stratigraphy of Miocene deltaic cycles exposed along the Jerudong Anticline, Brunei Darussalam. *Journal of Sedimentary Research*, **71**, 913–921.

BACK, S., TIOE, H.J., THANG, T.X. & MORLEY, C.K. 2005. Stratigraphic development of synkinematic deposits in a large growth-fault system, onshore Brunei Darussalam. *Journal of the Geological Society, London*, **162**, 243–258, http://doi.org/10.1144/0016-764903-006

BAILLIE, K.A. & JAMES-ROMANO, J. 2012. Identifying and quantifying thin bedded pay (TBP) part B: the use of dynamic data to evaluate productivity potential in gas reservoirs. *Paper SPE 133535-MS presented at the SPE Trinidad and Tobago Energy Resources Conference*, 27–30 June 2012, Port of Spain, Trinidad.

BALLY, Y.K., STAINES, M., FINNERAN, J., ALI-NANDALAL, J. & GAMERO, H. 1999. Integrated 3-dimensional reservoir modelling to develop a thin oil column with horizontal wells in the Mahogany Field, offshore Trinidad. *In*: *Extended Abstracts Volume 1999 AAPG International Conference and Exhibition*, 12–15 September, Birmingham, UK. American Association of Petroleum Geologists, Tulsa, OK, 33–36.

BHATTACHARYA, H.N., BHATTACHARYA, B., CHATRABORTY, I. & CHARABORTY, A. 2004. Sole marks in storm event beds in the Permo-Carboniferous Talchir Formation, Raniganj Basin, India. *Sedimentary Geology*, **166**, 209–222.

BHATTACHARYA, J.P. & GIOSAN, I. 2003. Wave-influenced deltas: geomorphological implications for facies reconstruction. *Sedimentology*, **50**, 187–210.

BOUMA, A.H. 1962. *Sedimentology of Some Flysch Deposits*. Elsevier, Amsterdam.

BOWMAN, A.P. 2003. *Sequence stratigraphy and reservoir characterisation in the Columbus Basin, Trinidad, West Indies*. PhD thesis, Imperial College London.

BOWMAN, A.P. & JOHNSON, H.D. 2014. Storm-dominated shelf-edge delta successions in a high accommodation setting: The palaeo-Orinoco Delta (Mayaro Formation), Columbus Basin, South-East Trinidad. *Sedimentology*, **61**, 792–835.

BRAMI, T.R., PIRMEZ, C., ARCHIE, C., HEERALAL, S. & HOLMAN, K.L. 2000. Late Pleistocene deep-water stratigraphy and depositional processes. *In*: WEIMER, P., SLATT, R.M. ET AL. (eds) *Deep Water Reservoirs of the World. GCSSEPM 20th Bob F. Perkins Research Conference*. Gulf Coast Section of the Society of Economic Paleontologists and Mineralogists, Houston, TX, 104–115.

CALLEC, Y., DEVILLE, D. ET AL. 2010. The Orinoco turbidite system: tectonic controls on sea-floor morphology and sedimentation. *American Association of Petroleum Geologists Bulletin*, **94**, 869–887.

CARR-BROWN, D. 1998. Revision of the late Neogene stratigraphy of southeast Trinidad (abstract with figures). *In*: ALI, W., PAUL, A. & YOUNG, O.V. (eds) *Transactions of the Third Geological Conference of the Geological Society of Trinidad and Tobago and the 14th Caribbean Geological Conference, Port of Spain, Volume 1*. Geological Society of Trinidad and Tobago, La Romaine, 757.

CARVAJAL, C. & STEEL, R.J. 2009. Shelf-edge architecture and bypass of sand to deep water: influence of shelf-edge processes, sea level, and sediment supply. *Journal of Sedimentary Research*, **79**, 652–672.

CORBETT, P.W.M., RINGROSE, P.S., JENSEN, J.L. & SORBIE, K.S. 1992. Laminated clastic reservoirs: the interplay of capillary pressure and sedimentary architecture. *In*: *SPE Annual Technical Conference and Exhibition*, 4–7 October 1992, Washington, DC. Society of Petroleum Engineers, Richardson, TX, 365–376.

CULLEN, A.B. 2010. Transverse segmentation of the Baram-Balabac Basin, NW Borneo: refining the model of Borneo's tectonic evolution. *Petroleum Geoscience*, **16**, 3–29, http://doi.org/10.1144/1354-079309-828

DAMUTH, J.E. 1994. Neogene gravity tectonics and depositional processes on the deep Niger Delta continental margin. *Marine and Petroleum Geology*, **11**, 320–346.

DAVIES, S.J. & ELLIOT, T. 1996. Spectral gamma ray characterisation of high-resolution sequence stratigraphy: examples from Upper Carboniferous fluvio-deltaic systems, County Clare, Ireland. *In*: HOWELL, J.A. & AITKEN, J.F. (eds) *High Resolution Sequence Stratigraphy: Innovations and Applications*. Geological Society, London, Special Publications, **104**, 25–35.

DE RAAF, J.F.M., BOERSMA, J.R. & VAN GELDER, A. 1977. Wave generated structures and sequences from a shallow marine succession, Lower Carboniferous, County Cork, Ireland. *Sedimentology*, **4**, 1–52.

DE VERTEUIL, L. 2002. *Appendix I: Palynostratigraphy of the Type Section of the Mayaro Formation*. Latinum Geological Consultants Ltd, Port of Spain, Trinidad. *In*: BOWMAN, A.P. 2003. *Sequence stratigraphy and reservoir characterisation in the Columbus Basin, Trinidad, West Indies*. PhD thesis, Imperial College London.

DIAZ DE GAMERO, M.L. 1996. The changing course of the Orinoco River during the Neogene: a review. *Palaeogeography, Palaeoclimatology, Palaeoecology*, **123**, 385–402.

DI CROCE, J., BALLY, A.W. & VAIL, P. 1999. The sequence stratigraphy of the East Venezuela Basin. *In*: MANN, P. (ed.) *Caribbean Basins: Sedimentary Basins of the World, Volume 4*. Elsevier Science, Amsterdam, **4**, 419–476.

DIXON, R.J. 2005. Field trip to examine outcrops of the Mayaro Formation, Mayaro Beach, SE Trinidad. *In*: *Guide to Selected Geological Localities on Trinidad and Tobago*. Geological Society of Trinidad and Tobago, La Romaine, http://archives.datapages.com/data/gstt/gstt_field15.htm

DUMAS, S. & ARNOTT, R.W.C. 2006. Origin of hummocky and swaleyy cross-stratification – the controlling influence of unidirectional current strength and aggradation. *Geology*, **34**, 1073–1076.

EIDE, H.C., HOWELL, J. & BUCKLEY, S. 2014. Distribution of discontinuous mudstone beds within wave-dominated shallow-marine deposits: Star Point Sandstone and Blackhawk Formation, Eastern Utah. *American Association of Petroleum Geologists Bulletin*, **98**, 1401–1429.

EMERY, D. & MYERS, K.J. 1996. *Sequence Stratigraphy*. Blackwell Science, Oxford.

FARRELLY, J.J. 1987. *Depositional setting and evolution of the Pliocene–basal Pleistocene section of southeast Trinidad*. Master's thesis, University of Texas at Austin.

FRAMPTON, J. & WILSON, B. 2002. *Appendix II: Foraminiferal Analyses of Subsurface Samples from the Mayaro Bay, Southeast Trinidad*. Biostratigraphic Associates (Trinidad) Ltd., Port of Spain. *In*: BOWMAN, A.P. 2003. *Sequence stratigraphy and reservoir characterisation in the Columbus Basin, Trinidad, West Indies*. PhD thesis, Imperial College London.

GAMERO DIAZ, H., CONTRERAS, C., LEWIS, N., WELSH, R. & ZAVALA, C. 2011. Evidence of shelfal hyperpycnal depositon of Pliocene Sandstones in the Oilbird Field, southeast coast, Trinidad: impact on reservoir distribution. *In*: SLATT, R.M. & ZAVALA, C. (eds) *Sediment Transfer from Shelf to Deep Water – Revisiting the Delivery System*. American Association of Petroleum Geologists, Studies in Geology, **61**, 193–214.

GIBSON, R.G. 1994. Fault-zone seals in a siliciclastic strata of the Columbus Basin Offshore Trinidad. *American Association of Petroleum Geologists Bulletin*, **78**, 1372–1385.

GIBSON, R.G., DZOU, L.I.P. & GREELY, D.F. 2004. Shelf Petroleum system of the Columbus basin, offshore Trinidad, West Indies. I. Source rock, thermal history, and controls on product distribution. *Marine and Petroleum Geology*, **21**, 97–108.

GIBSON, R.G., MEISLING, K.M. & SYDOW, J.C. 2012. Columbus basin offshore Trinidad: a detached pull-apart basin in a Transpressional Foreland Setting. *In*: ROBERTS, D.G. & BALLY, A.W. (eds) *Regional Geology and Tectonics: Phanerozoic Passive Margins, Cratonic Basins and Global Tectonic Maps*, 1st edn. Elsevier, Amsterdam, 636–656.

GOGGIN, D.J. 1993. Probe permeametry: is it worth the effort? *Marine and Petroleum Geology*, **10**, 209–308.

GOGGIN, D.J., THRASHER, R.L. & LAKE, L.W. 1988. A theoretical and experimental analysis of mini permeameter response including gas slippage and high velocity flow rates. *In Situ*, **12**, 79–116.

GoRTT MEEA 2013. *Petroleum Geology of Trinidad and Tobago*. Online publication by the Ministry of Energy and Energy Affairs, Government of the Republic of Trinidad and Tobago, http://www.energy.gov.tt/wp-content/uploads/2014/02/petroleum_geo.pdf

GRINGRAS, M.K., BANIAK, G. ET AL. 2012. Porosity and permeability in bioturbated sediments. *In*: KNAUST, K. & BOTTOMLEY, R. (eds) *Trace Fossils as Indicators of Sedimentary Environments*. Developments in Sedimentology, **64**. Elsevier, Amsterdam, 837–868.

HALLAM, R.J. & LUMKIN, M. 2003. Performance of Trinidad gas reservoirs (Cassia, Immortelle, Flamboyant, Mahogany, Amherstia and Teak). *Paper SPE 81010 presented at the SPE Latin American and Caribbean Petroleum Engineering Conference*, 27–30 April 2003, Port-of-Spain, Trinidad and Tobago.

HAMPSON, G.J. 2000. Discontinuity surfaces, clinoforms and facies architecture in wave-dominated shoreface-shelf parasequences. *Journal of Sedimentary Research*, **70**, 325–340.

HARMS, J.C., SOUTHARD, J.B., SPEARING, D.R. & WALKER, R.G. 1975. Depositional environments as interpreted from primary sedimentary structures and stratification sequences. *SEPM Short Course*, **9**, 242.

HENDERSON, K., ROSE, H.J. & WINTER, R. 2012. Identifying and quantifying thin-bedded pay (Part A): log characteristics and reservoir quality. *Paper SPE 133534-MS presented at the Trinidad and Tobago Energy Resources Conference*, 27–30 June 2012, Port of Spain, Trinidad.

HISCOTT, R.N. 2001. Depositional sequences controlled by high rates of sediment supply, sea-level variations, and growth faulting: the Quaternary Baram Delta of northwestern Borneo. *Marine Geology*, **175**, 67–102.

HODGETTS, D., IMBER, J. ET AL. 2001. Sequence stratigraphic responses to shoreline perpendicular growth faulting in shallow marine reservoirs of the Champion Field, offshore Brunei Darussalam, South China Sea. *American Association of Petroleum Geologists Bulletin*, **85**, 433–458.

HOORN, C., GUERRO, J., SARMIENTO, G.A. & LORENTE, M.A. 1995. Andean tectonics as a cause for changing drainage patterns in Miocene northern South America. *Geology*, **23**, 237–240.

HOWELL, J.A., SKORSTAD, A., MACDONALD, A., FORDHAM, A., FLINT, S., FJELLVOLL, B. & MANZOCCHI, T. 2008. Sedimentological parameterization of shallow-marine reservoirs. *Petroleum Geoscience*, **14**, 17–34, http://doi.org/10.1144/1354-079307-787

HURST, A. & GOGGIN, D.J. 1995. Probe permeametry: an overview and bibliography. *American Association of Petroleum Geologists Bulletin*, **79**, 463–473.

ISRAEL, R. 2005. Depositional model of a paleo-Orinoco shelf margin delta system, Columbus Basin, Trinidad. Extended Abstract. *Paper presented at the AAPG International Conference*, 16–19 June 2005, Calgary, Alberta.

JACKSON, M.D. & MUGGERIDGE, A.H. 2000. Effects of discontinuous shales on reservoir performance during horizontal waterflooding. *SPE Journal*, **5**, 446–455.

JACKSON, M.D., YOSHIDA, S., MUGGERIDGE, A.H. & JOHNSON, H.D. 2005. Three dimensional reservoir characterisation and flow simulation of heterolithic tidal sandstones. *American Association of Petroleum Geologists Bulletin*, **89**, 507–528.

JACKSON, M.D., HAMPSON, G.J. & SECH, R.P. 2009. Three dimensional modelling of a shoreface-shelf parasequence reservoir analog: Part 2. Geologic controls on fluid flow and hydrocarbon recovery. *American Association of Petroleum Geologists Bulletin*, **93**, 1183–1208.

JONES, R.W. 1998. Palaeoenvironmental interpretation of the late Miocene and Pliocene of Trinidad based on micropalaeontological data. *In*: ALI, W., PAUL, A. & YOUNG, O.V. (eds) *Transactions of the Third Geological Conference of the Geological Society of Trinidad and Tobago and the 14th Caribbean Geological Conference, Port of Spain, Volume 1*. Geological Society of Trinidad and Tobago, La Romaine, 86–101.

KNELLER, B. 1995. Beyond the turbidite paradigm: physical models for deposition of turbidites and their implications for reservoir prediction. *In*: HARTLEY, A.J. & PROSSER, D.J. (eds) *Characterisation of Deep Marine Clastic Systems*. Geological Society, London, Special Publications, **94**, 31–49, http://doi.org/10.1144/GSL.SP.1995.094.01.04

KROMAH, M.J., LUMSDEN, P.J., HENNINGTON, E.R. & BRAYSHAW, A.C. 2003. Trinidad's first 500 MMscfd well: fact or fiction? *Paper SPE 81045. SPE Latin American and Petroleum Engineering Conference*, 27–30 April, Port of Spain, Trinidad and Tobago.

KUGLER, H.G. 1961. Part 2: 32 geological map of Trinidad 1:100 000; geological sections through Trinidad 1:100 000. *In*: BOLLI, H.M. & KNAPPERTSBUSCH, M. (eds) *Treatise on the Geology of Trinidad*. Museum of Natural History, Basel.

KUGLER, H.G. 1996. Part 3: 32 detailed geological maps and sections with explanatory notes. *In*: BOLLI, H.M. & KNAPPERTSBUSCH, M. (eds) *Treatise on the Geology of Trinidad*. Museum of Natural History, Basel.

KUGLER, H.G. 2001. Part 4: the palaeocene to Holocene Formations of Trinidad. *In*: BOLLI, H.M. & KNAPPERTSBUSCH, M. (eds) *Treatise on the Geology of Trinidad*. Museum of Natural History, Basel.

LECKIE, D.A. & WALKER, R.G. 1982. Storm and tide dominated shorelines in Late Cretaceous Moosebar–Lower Gates Interval – outcrop equivalents of deep basin gas traps in western Canada. *American Association of Petroleum Geologists Bulletin*, **66**, 138–157.

LEONARD, R. 1983. Geology and hydrocarbon accumulations, Columbus Basin, offshore Trinidad. *American Association of Petroleum Geologists Bulletin*, **67**, 1081–1093.

LOUGHEIDE, D., LUTCHMAN, K. & ELRED, A. 2004. Trinidad's first multilateral well successfully integrates horizontal openhole gravel packs. *Paper OTC 16244 presented at the Offshore Technology Conference*, 3–6 May 2004, Houston, TX.

LOWDEN, B.D. 1993. *A methodology for the quantification of outcrop permeability heterogeneities through probe permeametry*. PhD thesis, Department of Earth Science and Engineering, Imperial College, University of London.

LOWE, D.R. 1982. Sediment gravity flows II: depositional models with special reference to the deposits of high-density turbidity currents. *Journal of Sedimentary Petrology*, **52**, 279–297.

LUMSDEN, P.J., BALGOBIN, C.J. ET AL. 2002. The Kapok Field: a step change for Trinidad gas developments. *In*: *Proceedings of the SPE Gas Technology Symposium (SPE75670)*. Calgary, Canada, April 2002, 1–10.

MACEACHERN, J.A., BANN, K.L., BHATTACHARYA, J.P. & HOWELL, C.D. 2005. Ichnology of deltas: organism responses to the dynamic interplay of rivers, waves, storms and tides. *In*: GIOSAN, L. & BHATTACHARYA, J.P. (eds) *River Deltas – Concepts, Models, and Examples*. SEPM (Society of Economic Paleontologists and Mineralogists), Special Publications, **83**, 45–85.

MACKAY, D.A. & DALRYMPLE, R.W. 2011. Dynamic mud deposition in a tidal environment: the record of fluid-mud deposition in the Cretaceous Bluesky Formation, Alberta, Canada. *Journal of Sedimentary Research*, **81**, 901–920.

MAYALL, M.J., YIELDING, C.A., OLDROYD, J.D., PULHAM, A.J. & SAKURAI, S. 1992. Facies in a shelf-edge delta: an example from the susbsurface of the Gulf of Mexico, Middle Pliocene, Mississippi Canyon, Block 109(1). *AAPG Bulletin*, **76**, 435–448.

MITCHUM, R.M., JR & VAN WAGONER, J.C. 1991. High-frequency sequences and their Stacking Patterns: sequence stratigraphic evidence of high-frequency eustatic cycles. *Sedimentary Geology*, **70**, 131–160.

Moosai, R.S., Alleyne, B., Thomas, S., Smith, D., Cavaleri, N.J. & Anthony Quinn, A.F. 2012. Race for liquids v. gas: how to redesign a depletion strategy to maximize fluids recovery in a mature thin-oil rim reservoir. *Paper SPE 157949 presented at the SPETT 2012 Energy Conference and Exhibition*, 11–13 June 2012, Port-of-Spain, Trinidad.

Moscardelli, L., Wood, L. & Dunlap, D. 2012. Shelf-edge deltas along structurally complex margins: a case study from eastern offshore Trinidad. *American Association of Petroleum Geologists Bulletin*, **96**, 1483–1522.

Mulder, T. & Alexander, J. 2001. The physical character of subaqeous sedimentary flows and their deposits. *Sedimentology*, **48**, 269–299.

Muller, D.S., Wharton, S.R. & Hoffman, K.S. 1999. Recognizing, targeting and developing bypassed oil in a mature, complexly faulted, multi-pay oil field, Teak Field, Trinidad and Tobago. *Paper OTC 10742 presented at the Offshore Technology Conference*, 3–6 May 1999, Houston, TX.

Mutti, E., Davoli, G., Tinterri, R. & Zavala, C. 1996. The importance of fluvio-deltaic systems dominated by catastrophic flooding in tectonically active basins. *Memorie di Scienze Geologiche*, **48**, 233–291.

Mutti, E., Tinterri, R., Benevelli, G., de Biase, D. & Ernst, M. 2003. Deltaic, mixed and turbidite sedimentation of ancient foreland basins. *Marine and Petroleum Geology*, **20**, 733–755.

Myers, K.J. & Bristow, C.S. 1989. Detailed sedimentology and gamma ray characteristics of a Namurian Deltaic succession, part II: Gamma-ray logging. *In*: Whately, M. & Pickering, K.T. (eds) *Deltas: Sites and Traps for Fossil Fuels*. Geological Society, London, Special Publications, **41**, 81–88, http://doi.org/10.1144/GSL.SP.1989.041.01.07

Myrow, P.M. & Southard, J.B. 1991. Combined-flow model for vertical stratification sequences in shallow marine storm-deposited beds. *Journal of Sedimentary Research*, **66**, 875–887.

Myrow, P.M. & Southard, J.B. 1996. Tempestite deposition. *Journal of Sedimentary Research*, **66**, 875–887.

Myrow, P.M., Woodward, F. & Goodge, J.W. 2002. Wave-modified turbidites; combined flow shoreline and shelf deposits, Cambrian, Antartica. *Journal of Sedimentary Research*, **72**, 641–656.

Passey, Q.R., Dahlberg, K.E., Sullivan, K.B., Yin, H., Brackett, R.A., Xiao, Y.H. & Guzman-Garcia, A.G. 2006. *Hydrocarbon Pore-Thickness in Thinly Bedded Clastic Reservoirs*. American Association of Petroleum Geologists, Archie Series, **1**.

Pemberton, G.S., Spila, M., Pulham, A.J., Saunders, T., MacEachern, J.A., Robbins, D. & Sinclair, I.K. 2002. *Ichnology and Sedimentology of Shallow Marine to Marginal Marine Systems: Ben Nevis and Avalon Reservoirs, Jeanne d'Arc Basin*. Geological Association of Canada, Short Course Notes, **15**.

Pindell, J.L. & Kennan, L. 2009. Dextral shear, terrane accretion and basin formation in the Northern Andes: best explained by interactions with a Pacific-derived Caribbean Plate. *In*: James, K., Lorente, M.A. & Pindell, J.L. (eds) *Origin and Evolution of the Caribbean Plate*. Geological Society, London, Special Publications, **328**, 487–531, http://doi.org/SP328.20

Plint, A.G. & Nummendal, D. 2000. The falling stage sytems tract: recognition and importance in sequence stratigraphic analysis. *In*: Hunt, D. & Gawthorpe, R.L. (eds) *Sedimentary Responses to Forced Regressions*. Geological Society, London, Special Publications, **172**, 1–17, http://doi.org/10.1144/GSL.SP.2000.172.01.01

Plink-Björklund, P. & Steel, R.J. 2004. Initiation of turbidity currents: outcrop evidence for Eocene hyperpycnal flow turbidites. *Sedimentary Geology*, **165**, 29–52.

Pocknall, D.T., Wood, L.J. & Green, A.F. 1999. Depositional facies in the Pliocene-Pleistocene section, offshore eastern Trinidad. *In*: Ali, W., Paul, A. & Young, O.V. (eds) *Transactions of the Third Geological Conference of the Geological Society of Trinidad and Tobago and the 14th Caribbean Geological Conference Volume 1*. Geological Society of Trinidad and Tobago, La Romaine, 713–732.

Pocknall, D.T., Wood, L.J., Geen, A.F., Harry, B.E. & Hedlund, R.W. 2001. Integrated paleontological studies of Pliocene to Pleistocene deposits of the Orinoco Delta, eastern Venezuela and Trinidad. *In*: Goodman, D.K. & Clarke, R.T. (eds) *Proceedings of the IX International Palynological Congress, Houston, Texas, U.S.A.*, 1996. American Association of Stratigraphic Palynologists Foundation, 319–326.

Possamentier, H.W. & Allen, G.P. 1999. *Siliciclastic Sequence Stratigraphy – Concepts and Applications*. SEPM (Society of Economic Paleontologists and Mineralogists), Concepts in Sedimentology and Palaeontology, **7**.

Posamentier, H.W. & Vail, P.R. 1988. Eustatic controls on clastic deposition I – sequence and systems tracts models. *In*: Wilgus, C.K., Hastings, B.S., Kendall, C.G.St.C., Posamentier, H.W., Ross, C.A. & Van Wagoner, J.C. (eds) *Sea-level Changes; An Integrated Approach*. Spec. Publ. SEPM **42**, 109–124.

Pucknall, J., Holder, G. & Seesahai, A. 2003. Amherstia Field, Trinidad: breaking barriers in the New Millenium. *Paper SPE 81012 presented at the SPE Latin American and Caribbean Petroleum Engineering Conference*, 27–30 April 2003, Port-of-Spain, Trinidad and Tobago.

Ramnath, M.M. 2015. *Thin-bedded reservoirs in the Plio-Pleistocene of the Columbus Basin, offshore Trinidad: challenges of reservoir architecture, quantification and characteristics*. PhD thesis, University of Manchester, UK.

Rampersad, T., Cegna, A., Steiner, C. & Kommedal, J. 2014. A new wave of illumination –ISS OBC acquisition in the Columbus Basin. *Paper SPE 169981 presented at the SPE Energy Resources Conference*, 9–11 June 2014, Port of Spain, Trinidad and Tobago.

Reineck, H.E. & Wunderlich, F. 1968. Classification and origin of flaser and lenticular bedding. *Sedimentology*, **11**, 99–104.

Reynolds, A.D. 1999. Dimensions of paralic sandstone bodies. *American Association of Petroleum Geologists Bulletin*, **83**, 211–229.

Roberts, H.H. & Sydow, J.C. 2003. Late Quaternary stratigraphy and sedimentology of the offshore Mahakam Delta, East Kalimatan (Indonesia). *In*: Sid, F.H., Nummendal, D., Imbert, P. & Posamentier, H.W.

(eds) *Tropical Deltas of Southeast Asia – Sedimentology, Stratigraphy and Petroleum Geology*. SEPM Spec. Publ. **76**, 125–145.

ROHR, G.M. 1991. Exploration potential of Trinidad and Tobago. *Journal of Petroleum Geology*, **14**, 343–354.

ROTONDO, K.A. & BENTLEY, S.J. 2004. Marine dispersal of fluvial sediments as fluid muds: old concept, new significance. *In*: SCOTT, E.D. & BOUMA, A.H. (eds) *Depositional Processes and Reservoir Characteristics of Siltstones, Mudstones and Shales*. SEPM (Society of Economic Paleontologists and Mineralogists), Miscellaneous Publications, **2**, 24–49.

SALLER, A. & BLAKE, G. 2003. Sequence stratigraphy and sydepositional tectonics of Upper Miocene and Pliocene deltaic sediments, offshore Brunei Darussalam. *In*: SIDI, F.H., NUMMEDAL, D., IMBERT, P., DARMAN, H. & POSAMENTIER, H.W. (eds) *Tropical Deltas of Southeast Asia – Sedimentology, Stratigraphy, and Petroleum Geology*. SEPM (Society of Economic Paleontologists and Mineralogists), Special Publications, **76**, 219–234.

SAMSUNDAR, K. & CHUNG, R. 2006. From reservoir to well: using technology for world class results in Trinidad and Tobago. *Paper SPE 99408 presented at the SPE Intelligent Energy Conference and Exhibition*, 11–13 April 2006, Amsterdam, The Netherlands.

SAMSUNDAR, K., MOOSAI, R. & CHUNG, R. 2007. Surveillance planning: the key to managing a mature gas reservoir. *Paper SPE 107279 presented at the Latin American & Caribbean Petroleum Engineering Conference*, 15–18 April 2007, Buenos Aires, Argentina.

SAUGY, L. & EYER, J.A. 2003. Fifty years of exploration in the Niger Delta (West Africa). *In*: HALBOUTY, M.T. (ed.) *Giant Oil and Gas Fields of the Decade 1990–1999*. American Association of Petroleum Geologists, Memoirs, **78**, 211–226.

SCOTT, E.S. 1992. The palaeoenvironments and dynamics of the Rannoch–Etive nearshore and coastal succession, Brent Group, northern North Sea. *In*: MORTON, A.C., HASZLEDINE, R.S., GILES, M.R. & BROWN, S.R. (eds) *Geology of the Brent Group*. Geological Society, London, Special Publications, **61**, 129–147, http://doi.org/10.1144/GSL.SP.1992.061.01.08

SECH, R. 2007. *Quantifying the impact of geological heterogeneity on gas recovery and water cresting, with application to the Columbus Basin Fields, offshore Trinidad*. PhD Thesis, Imperial College, London.

SECH, R.P., JACKSON, M.D. & HAMPSON, G.J. 2009. Three dimensional modelling of a shoreface-shelf parasequence reservoir analog: Part 1. Surface based modelling to capture high-resolution facies architecture. *American Association of Petroleum Geologists Bulletin*, **93**, 1155–1181.

SMART, E., STARCHER, M., SADLER, S. & WACH, G. 2000. Dolphin Field: sedimentology and reservoir architecture of a supply-dominated, accommodation-controlled system, ECMA, Trinidad. *Paper SPE RC05 presented at the GSTT SPE Conference*, 10–13 July 2000, Port of Spain, Trinidad.

SNEIDER, R.M. 2003. Worldwide examples of low resistivity pay. *Houston Geological Society Bulletin*, **45**, 47–59.

SUTHERLAND, W.J., HALVORSEN, C., HURST, A., MCPHEE, C.A., ROBERTSON, G., WHATTLER, P.R. & WORTHINGTON, P.F. 1993. Recommended practice for probe permeametry. *Marine and Petroleum Geology*, **10**, 309–317.

SWIFT, D.J.P., HANS, G. & VINCENT, V.E. 1986. Fluid processes and sea-floor response on a modern storm dominated shelf: Middle Atlantic shelf of North America. Part 1: the storm current regime. *In*: KNIGHT, R.J. & MCCLEAN, J.R. (eds) *Shelf Sands and Sandstones*. Mem. Can. Pet. Geol, **11**, 99–119.

SYDOW, J.C., FINNERAN, J. & BOWMAN, A.P. 2003. Stacked shelf edge delta reservoirs of the Columbus Basin, Trinidad, West Indies. *In*: ROBERTS, H.H., ANDERSON, J.B. & FILLON, R.H. (eds) *Shelf Margin Deltas and Linked Downslope Depositional Systems: Global Significance and Future Exploration Potential. GCSSEPM Foundation 23rd Annual Bob F Perkins Research Conference*. Gulf Coast Section of the Society of Economic Paleontologists and Mineralogists, Houston, TX, 441–465.

SYLVESTER, I.F., COOK, R., SWIFT, R., PRITCHARD, T. & MCKEEVER, J. 2005. Integrated reservoir modelling enhances the understanding of reservoir performance of the Dolphin gas field, Trinidad and Tobago. *Paper SPE 94343 presented at the SPE Europe/EAGE Annual Conference*, 13–16 June 2005, Madrid, Spain.

TAYLOR, K.S., KALITA, N., DOWLATH, J.F., JEGEDE, O.M. & AMEERALI, R.A. 2012. A greater Dolphin Area case study part 1: defining geological uncertainty. *Paper SPE 158545 presented at the SPETT 2012 Energy Conference and Exhibition*, 11–13 June 2012, Port-of-Spain, Trinidad.

TESSIER, B. & GIGOT, P. 1989. A vertical record of different tidal cyclicities: an example from the Miocene Marine Molasse of Digne (Haute Provence, France). *Sedimentology*, **36**, 767–776.

URZOA, C.A. 2008. *Process and architectures of deltas in shelf-break and ramp platforms: examples from the Eocene of West Spitsbergen (Norway), the Pliocene palaeo-Orinoco Delta (SE Trinidad), and the Cretaceous Western Interior Seaway (S Wyoming & NE Utah)*. PhD thesis, The University of Texas at Austin.

VAN WAGONER, J.C., MITCHUM, R.M., CAMPION, K.M. & RAHMANIAN, V.D. 1990. *Silicilastic Sequence Stratigraphy in Well Logs, Cores and Outcrops: Concepts for High-resolution Correlation of Time and Facies*. AAPG Methods in Exploration Series No. **7**. AAPG, Tulsa, Oklahoma, USA, 1–55.

VAN WAGONER, J.C., POSAMENTIER, H.W., MITCHUM, R.W., VAIL, P.R., SARG, J.F., LOUTIT, T.S. & HARDENBOL, J. 1988. An overview of the fundamentals of sequence stratigraphy and key definitions. *In*: WILGUS, C.K., HASTINGS, B.S., KENDALL, C.G.St.C., POSAMENTIER, H.W., ROSS, C.A. & VAN WAGONER, J.C. (eds) *Sea-level Changes: An Integrated Approach*. Spec. Publ. SEPM **42**, 109–124.

VISSER, M.J. 1980. Neap-spring cycles reflected in Holocene sub-tidal largescale bedform deposits: a preliminary note. *Geology*, **8**, 543–546.

WALKER, R.G. & PLINT, A.G. 1992. Wave and storm dominated shallow marine systems. *In*: WALKER, R.G. & JAMES, N.P. (eds) *Facies Models: Responses to Sea Level Change*. Geological Association of Canada, Ottawa, 219–238.

WANG, J.T.M. 1993. *Permeability description in the presence of small-scale heterogeneity*. PhD thesis, Department of Earth Science and Engineering, Imperial College, University of London.

WARNE, A.G., MEADE, R.H. *ET AL.* 2002. Regional controls on geomorphology, hydrology, and ecosystem integrity in the Orinoco Delta, Venezuela. *Geomorphology*, **44**, 273–307.

WEBER, K.J. & VAN GEUNS, L.C. 1990. Framework for constructing clastic reservoir simulation models. *Journal of Petroleum Technology*, **42**, 1248–1253 & 1296–1297.

WHARTON, S.R. & MULLER, D. 2000. Exploitation of 'Incised Valley Fill' reservoir sequences in the mature Teak Field, Columbus Basin, Trinidad and Tobago. *Paper SPE RC02 presented at the GSTT 2000 SPE Conference*, 10–13 July 2000, Port of Spain, Trinidad.

WHITE, C.D. & WILLIS, B.J. 2000. A method to estimate length distributions from outcrop data. *Mathematical Geology*, **32**, 389–419.

WINTER, R. 2006. *Reservoir characterisation of the Cruse Formation, Southern Trinidad*. Master's thesis, The University of Texas at Austin.

WOOD, L.J. 2000. Chronostratigraphy and tectonostratigraphy of the Columbus Basin, eastern offshore Trinidad. *American Association of Petroleum Geologists Bulletin*, **84**, 1905–1928.

YOSHIDA, S., JACKSON, M.D., JOHNSON, H.D., MUGGERIDGE, A.H. & MARTINUS, A.W. 2001. Outcrop studies of tidal sandstones for reservoir characterisation (Lower Cretaceous Vectis Formation, Isle of Wight, southern England). *In*: MARTINSEN, O.J. & DREYER, T. (eds) *Sedimentary Environments Offshore Norway — Palaeozoic to Recent, Proceedings of the Norwegian Petroleum Society Conference*. Norwegian Petroleum Society, Special Publications, **10**, 233–257.

ZIPPI, P.A. 2007. Appendix 2: Micropalaeontology of outcrop samples from the Pliocene of East Trinidad. *In*: URZOA, C.A. 2008. *Process and architectures of deltas in shelf-break and ramp platforms: examples from the Eocene of West Spitsbergen (Norway), the Pliocene palaeo-Orinoco Delta (SE Trinidad), and the Cretaceous Western Interior Seaway (S Wyoming & NE Utah)*. PhD thesis, The University of Texas at Austin.

Making outcrops relevant for an unconventional source rock play: an example from the Eagle Ford Group of Texas

ARTHUR D. DONOVAN

Texas A&M College of Geosciences, 797 Lamar St., College Station, Texas 77840, USA (e-mail: art.donovan@tamu.edu)

Abstract: The utility of outcrops to geoscientists working the subsurface is typically limited by provincial nomenclature, outcrop-based lithostratigraphic subdivisions, which commonly differ in name and numbers among researchers, and often a basic indifference to the subsurface stratigraphy. In the subsurface of South Texas, the Eagle Ford Group is commonly divided into an organic-rich Lower Eagle Formation and a carbonate-rich Upper Eagle Ford Formation. In contrast, coeval strata that crop out nearby in West Texas are traditionally referred to as the Boquillas Formation. Adding further complexity to the outcrop stratigraphy is that previous workers divided the Boquillas into two to five informal lithostratigraphic units whose names and boundaries differ among researchers. By convolving information from a handheld gamma ray spectrometer, outcrop geochemical data and sequence stratigraphy, the Lower and Upper Eagle formations, as defined in the subsurface of South Texas, can now be defined in the coeval Eagle Ford Group outcrops of West Texas. Integrating basic geochemical and petrophysical data to the updated stratigraphy on these outcrops was transformative. By doing this, these West Texas exposures became portals to examine, as well as explain and predict, the distribution and thickness variations of specific chronostratigraphic units that are critical to the economic success of exploiting the Eagle Group across the unconventional source rock play fairway in South Texas.

The Upper Cretaceous Eagle Ford Group is a prolific source rock in the Gulf of Mexico coastal plain and an unconventional source rock reservoir in the subsurface of South Texas (Fig. 1). In Lozier Canyon, located in Terrell County (West Texas), the Eagle Ford Group crops out in numerous spectacular cut-bank exposures, each thousands of feet long and hundreds of feet high (Fig. 2). These Eagle Ford exposures provide a unique opportunity to study an unconventional source rock reservoir in outcrop.

Even with the best of exposures, certain challenges exist when trying to tie established subsurface stratigraphy with traditional outcrop stratigraphy. These challenges include (1) different and/or regionally changing provincial outcrop nomenclature, (2) regional facies changes, (3) the fundamentally different way stratigraphic units are commonly defined in the subsurface (by surfaces, cycles, and log patterns) v. outcrops (by simple lithology) and (4) variations in the lithostratigraphic boundaries and/or terminology used by different researchers. Fortunately, with the advent of handheld radiation detection units, geoscientists working on outcrops now have the ability to generate total gamma ray profiles of outcrops, as well as collecting spectral gamma ray data providing basic chemo-stratigraphic information. This information, coupled with more traditional geochemical and biostratigraphic data from laboratory analysis of outcrop samples, along with utilizing basic sequence stratigraphic concepts, facilitates identifying the subsurface stratigraphic units in outcrop.

In the subsurface of South Texas, the Eagle Ford Group is commonly divided into an organic-rich Lower Eagle Formation and a carbonate-rich Upper Eagle Ford Formation (Fig. 3). In contrast, coeval strata that crop out nearby in West Texas are traditionally referred to as the Boquillas Formation. Adding further complexity to the outcrop stratigraphy is that previous workers divided the Boquillas into two to five informal lithostratigraphic subunits whose names and boundaries differ among researchers (Fig. 4). By convolving information from a handheld gamma ray spectrometer, outcrop geochemical data, biostratigraphic information and sequence stratigraphic concepts, the Lower and Upper Eagle formations, as defined in the subsurface of South Texas, can now be defined in the outcrops of West Texas. Integrating basic geochemical and petrophysical data to the updated stratigraphy on these outcrops was transformative. By doing this, these West Texas exposures became portals to examine, as well as explain and predict, the distribution and thickness variations, of specific chronostratigraphic units that are critical to the economic success to exploiting the Eagle Group across the unconventional source rock play fairway in South Texas.

From: Bowman, M., Smyth, H. R., Good, T. R., Passey, S. R., Hirst, J. P. P. & Jordan, C. J. (eds) 2016. *The Value of Outcrop Studies in Reducing Subsurface Uncertainty and Risk in Hydrocarbon Exploration and Production*. Geological Society, London, Special Publications, **436**, 193–206.
First published online September 1, 2016, http://doi.org/10.1144/SP436.12

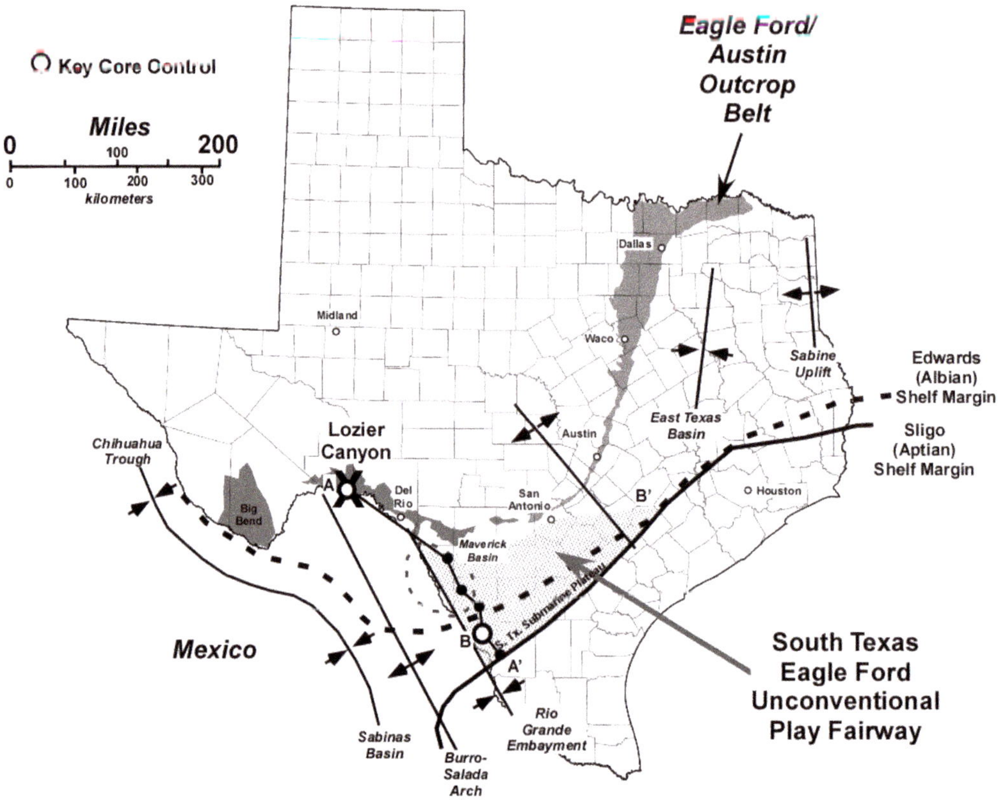

Fig. 1. Major structural and physiographic features of Texas. Please note location of outcrop belt (green), cross-section, key cores, Lozier Canyon, Type Well in South Texas, and Eagle Ford play fairway (grey).

Geological setting

The major structural features affecting the Coastal Plain succession of Texas consist from east to west of the Sabine Uplift, East Texas Basin, San Marcos Arch and Rio Grande Embayment (Fig. 1). Other key physiographic features affecting coastal plain deposition across Texas include the shelf margins of the Lower Cretaceous (Aptian) Sligo Group and younger (Albian) Edwards Group (Fig. 1). Within the lower portions of the Upper Cretaceous succession in the South Texas Coastal Plain, the vertical succession of units from the base up consists of the Del Rio Formation, the Buda Formation, the Eagle Ford Group and the Austin Chalk (Fig. 5). Within this succession the Eagle Ford is a major source rock, as well as unconventional source rock play, south of the San Marcos Arch in South Texas. Along the eastern flank of the East Texas Basin, Woodbine strata are the primary reservoirs in East Texas Field sourced by hydrocarbons migrated out of the overlying Eagle Ford from deeper portions of the basin (Halbouty 1991). Classic Eagle Ford exposures are present around Dallas, Waco and Austin along the western flank of the East Texas Basin, and in the Lozier Canyon region in West Texas (Fig. 1). In terms of the Eagle Ford, a regional palaeogeographic map is offered as Figure 6. This map illustrates a change from more proximal organic-rich argillaceous offshore mudstones in East Texas to more distal organic-rich offshore carbonate mudstones in south and west Texas.

Methods

The focus of our efforts has been to demonstrate how the outcrops of the Eagle Ford Group in West Texas could be most effectively used to better understand the distribution and thickness patterns of the Eagle Ford Group unconventional reservoirs in the subsurface of South Texas (Fig. 1).

A key aspect of the research was to collect spectral gamma ray (SGR) profiles of the measured sections using a handheld gamma ray spectrometer. SGR information was sampled at the stratigraphic vertical spacing of 1 foot. The SGR data was used to generate a gamma ray profile that could be

Fig. 2. (**a**) Uninterpreted image of the cut-bank exposure of Upper Cretaceous strata on the Scott Ranch in Lozier Canyon. Note location of BP/SLB Lozier Canyon #1 research borehole from rig position on outcrop. (**b**) Interpretation of the strata exposed along the south flank of the Scott Ranch outcrop.

directly tied to Eagle Ford wells in the subsurface. The data also provided key chemo-stratigraphic information on the vertical variability in uranium (U), thorium (Th) and potassium (K) within the succession. U, Th and K content are commonly used respectively as generalized proxies for organic matter, bentonite and clay content.

A handheld XRF (X-ray fluorescence) tool, calibrated with XRD (X-ray diffraction) data, was used to provide compositional information. This calibration offers low-cost, easy-to-collect XRF data to provide elemental information. The XRD data were then used to semi-quantitatively measure mineral content; this was critical for calibrating XRF-based mineral models.

Samples for total organic carbon and biostratigraphy (foraminifera and calcareous nanno-fossils) were also collected and analysed for the Lozier Canyon outcrops in West Texas and the cored well in South Texas. Carbon isotope ($\delta^{13}C$) analysis, for comparison with the geochemical signature at the Cenomanian–Turonian global strata-type section and point located near Pueblo, Colorado, was also conducted. This stage boundary is unique in having a distinct positive $\delta^{13}C$ excursion, interpreted as Ocean Anoxic Event 2 (OAE2), which is associated with this boundary. The maximum peak of the positive $\delta^{13}C$ excursion associated with this event is taken as the geochemical proxy for the base of the Turonian (Ogg & Hinnov 2012).

Another key aspect of our research was a borehole located a few hundred feet behind the measured outcrop section in Lozier Canyon (Fig. 1). This well was cored and a full suite of modern geophysical logs collected with the help of our research partner, Schlumberger. Detailed description of this work is beyond the scope of this paper. It included a platform express, CMR log, elemental capture spectroscopy log, sonic scanner and formation micro-image log, together with other advanced logging data.

The following terminology is used throughout this paper. In terms of the geophysical logs, the term 'bell-shaped' log pattern refers to intervals of upwardly decreasing gamma log values, while

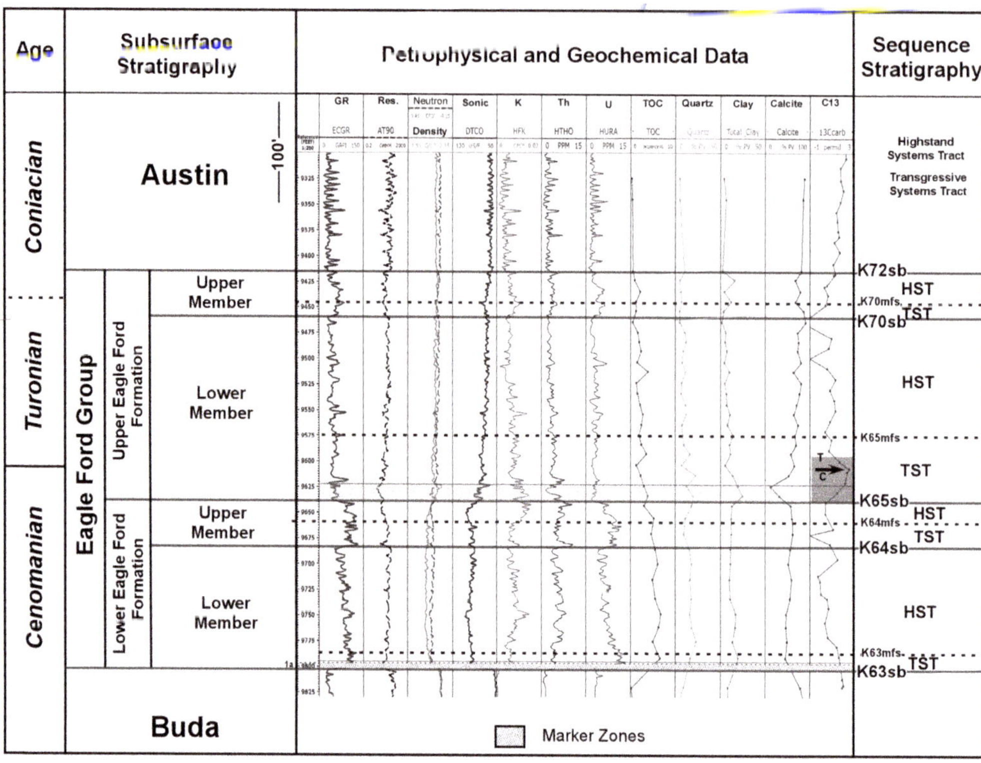

Fig. 3. Eagle Ford type well from Webb County in South Texas. Please note petrophysical, geochemical and biostratigraphic characteristics of formations, members and sequence stratigraphic surfaces proposed in this study. The base of the Upper Eagle Ford is marked by a distinct sharp GR decrease driven a drop in U and Th content at this boundary. A distinct positive $\delta^{13}C$ isotope excursion also occurs at the base of the Upper Eagle Ford. The peak inflection point of this excursion is the interpreted proxy for Cenomanian/Turonian stage boundary. Geochemical and petro-physical data presented courtesy of Core Laboratory and Swift Energy. Well location on Figure 1.

'funnel-shaped' log pattern refers to intervals of upwardly increasing gamma log values. For defining and mapping regional surfaces, this paper utilizes an alphanumeric scheme that starts with an abbreviation for each geological period (K for Cretaceous) and then goes from 0 to 99 from the base up. The modifiers sb (sequence boundary), ts (transgressive surface) and mfs (maximum flooding surface) are also added to provide clarity to the surface type. In terms of the Eagle Ford Group, it is bounded by the K63sb at its base and the K72sb at its top. Throughout this paper the following abbreviations are also used: U (uranium), K (potassium), Th (thorium), TOC (total organic carbon), GR (gamma ray), SGR (spectral gamma ray), LST (lowstand systems tract), TST (transgressive systems tract) and HST (highstand systems tract). This paper follows the bed classification scheme of Campbell (1967) of lamina, laminaset, bed and bedset; here very thin beds are those less than 3 cm thick, thin beds are 3–10 cm thick, medium beds are 10–30 cm thick, thick beds 30–100 cm thick and very thick beds are those greater than 100 cm thick.

In terms of the chrono-stratigraphic diagram depicted in Figure 5, this chart, in general, follows the times, fossil zonations and stage boundaries outlined in the Cretaceous section (Ogg & Hinnov 2012) of the 2012 Geological Time Scale (Gradstein *et al.* 2012). However, on the 2012 time scale a proposed (non-approved) placement for the Turonian/Coniacian stage boundary was used. In Texas, the new proposed criteria, the lowest occurrence of the inoceramid *Cremnoceramus deformas erectus*, occurs well into the Austin Group. Because this proposed boundary placement offers no clear lithological or petro-physical proxies, it was rejected for our work. In contrast, the traditional macro-faunal boundary for the base of the Turonian, the first occurrence of the ammonite *Prionocyclus germani* or inoceramid *M. incertus*, coincides with the base of the Austin Group in the outcrops of West (Cobban *et al.* 2008) and Central (Kennedy & Cobban

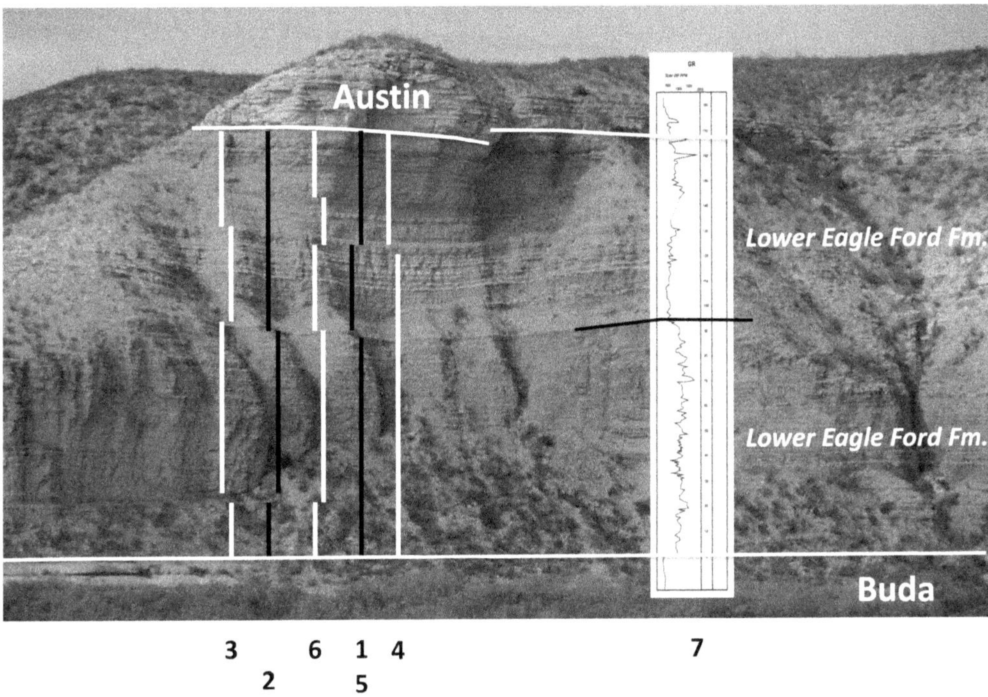

Fig. 4. Summary of the various lithostratigraphic subdivisions proposed for the Eagle Ford/Boquillas strata in west Texas, all from the base up: (1/5) Lower, Middle and Upper (Hazzard 1959; Ruppel *et al.* 2012); (2) Lower, Middle and Upper (Lock & Peschier 2006; Lock *et al.* 2010); (3) Pinch and Swell, Flagstone, Ledgy, and Laminated (Freeman 1961, 1968); (4) Rock Pens and Langtry (Pessagno 1969); (6) A–E (Donovan & Staerker 2010); and (7) Lower Eagle Ford Formation and Upper Eagle Ford Formation (Donovan *et al.* 2012).

1990) Texas. Unfortunately, this macro-faunal criterion is less useful in the subsurface and as such the micro-flora CC12/13 boundary, which occurs at or near the interpreted mfs of the upper member of the Upper Eagle Ford, was used as a close mid-Upper Turonian proxy. This biostratigraphic proxy can also be readily defined in outcrop and subsurface samples by industry biostratigraphers, and has an associated well log signature that most geoscientists can readily define and map as well (Fig. 3).

Eagle Ford stratigraphy

South Texas

Between the cities of Austin in Central Texas and Del Rio in South Texas outcrops of the Austin and underlying Eagle Ford Group are rare and most of our knowledge comes from the subsurface. In the subsurface of South Texas, the classic work is Grabowski (1995), who was the first to introduce the concept of an organic-rich, Cenomanian-aged, Lower Eagle Ford, and a carbonate-rich, generally-Turonian-aged, Upper Eagle Ford (Fig. 3). As illustrated on Figure 3, the base of the Upper Eagle Ford Formation is marked by an abrupt GR decrease, driven by a distinct drop in U and Th content. TOC content also drops at the base of the Upper Eagle Ford, while sonic and density logs record velocity and density increases. Another characteristic of the base of the Upper Eagle Ford is the presence of a distinct positive $\delta^{13}C$ excursion, whose peak inflection point is taken as the geochemical proxy for the Cenomanian/Turonian boundary.

West Texas

The Devils River, which enters the Rio Grande just northwest of Del Rio, marks the classic geographic and geological boundary between South and West Texas (Fig. 1). In West Texas, the strata situated between the Buda Formation and Austin Group have typically been referred to as Boquillas Formation (Udden 1907; Freeman 1961, 1968; Lock & Peschier 2006; Ruppel *et al.* 2012). Based on utility, as well as nomenclature precedence, Donovan *et al.* (2012) proposed use of the term Eagle Ford Group for these strata, and it is this nomenclature that we will follow in this paper.

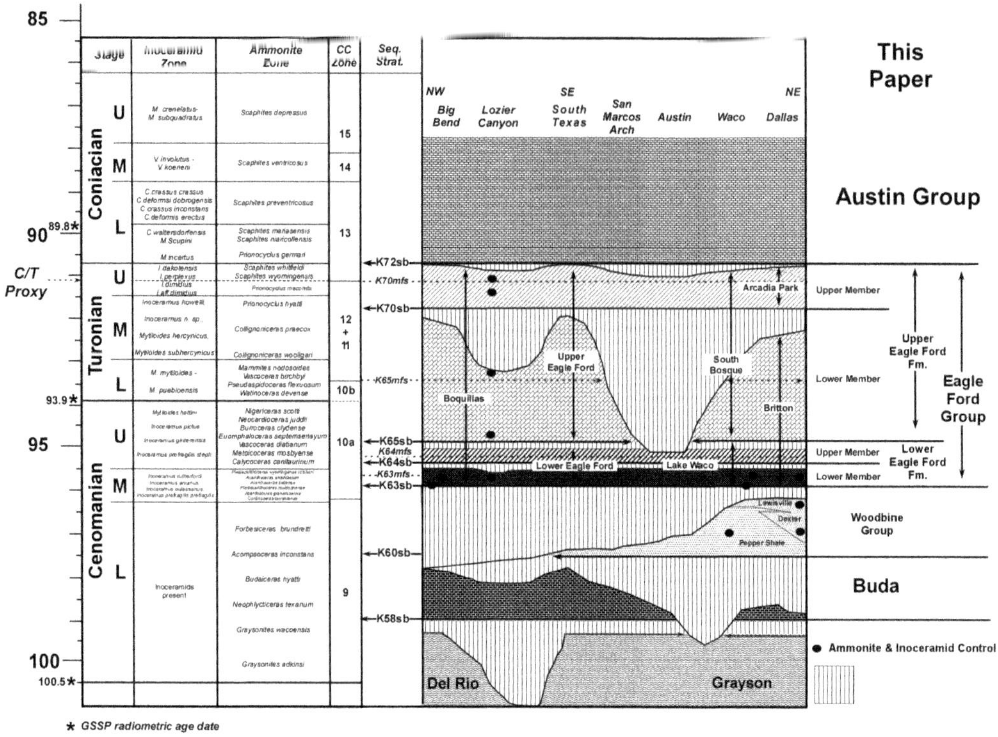

Fig. 5. Chronostratigraphy of the Cenomanian through Coniacian stratigraphic succession from West Texas to the East Texas Basin based on published molluscan and nannofossil data (after Donovan *et al.* 2015).

In the Lozier Canyon region in Terrell County (Fig. 4), Hazzard (1959) outlined a three-fold division for the Eagle Ford: (1) a lower shaly member; (2) middle interbedded calcareous mudstones and limestone member; and (3) an upper shaly member. Freeman (1961, 1968) outlined a four-fold division for these strata which from the base he named: (1) pinch and swell; (2) flagstone; (3) ledgy; and (4) laminated. Freeman's upper two members were equivalent to Hazzard's middle and upper members, while his lower two units represented a simple subdivision of Hazarrd's lower shaly member (Fig. 4). Based mainly on biostratigraphic criteria, Pessagno (1969) proposed a two-member subdivision (Rock Pens & Langtry) for the Eagle Ford Group in West Texas. Pessagno's lower Rock Pens Member contained Freeman's three lowermost units, while his upper Langtry Member equated to Freeman's uppermost (laminated) unit (Fig. 4). Pessagno's Rock Pens and Langtry terms were also used in subsequent USGS work in this area (Smith 1981). More recently the terms lower, middle and upper Eagle Ford (Boquillas) were used by Lock & Peschier (2006) and Lock *et al.* (2010) as well as Ruppel *et al.* (2012). However, what was defined as lower, middle and upper by each was different (Fig. 4). In order to more easily convey superposition of the various lithostratigraphic subunits defined by previous researchers, Donovan & Staerker (2010) introduced a simple A–E designation from the base up (Fig. 4). Their lower three units (A–C) were identical to Freeman's lower three, while their upper two units were simply a division of Freeman's uppermost (laminated) member into a lower mudstone-prone (unit D), and an upper very thinly interbedded (unit E).

With the collection of spectral gamma ray and geochemical data for a type section in Lozier Canyon, Donovan *et al.* (2012) were finally able to port the traditional subsurface stratigraphy of South Texas into the outcrops of West Texas, and defined a Lower Eagle Ford Formation and Upper Eagle Ford Formation (Figs 4 & 7). As illustrated on Figure 7, the same distinct GR drop, driven by decreases in U and Th content, marks the base of the Upper Eagle Ford in outcrop. A drop off in TOC content, along with the presence of the distinct positive $\delta^{13}C$ excursion, whose peak inflection point is taken as the geochemical proxy for the Cenomanian/Turonian boundary, also marks the base of the Upper Eagle Ford in outcrop.

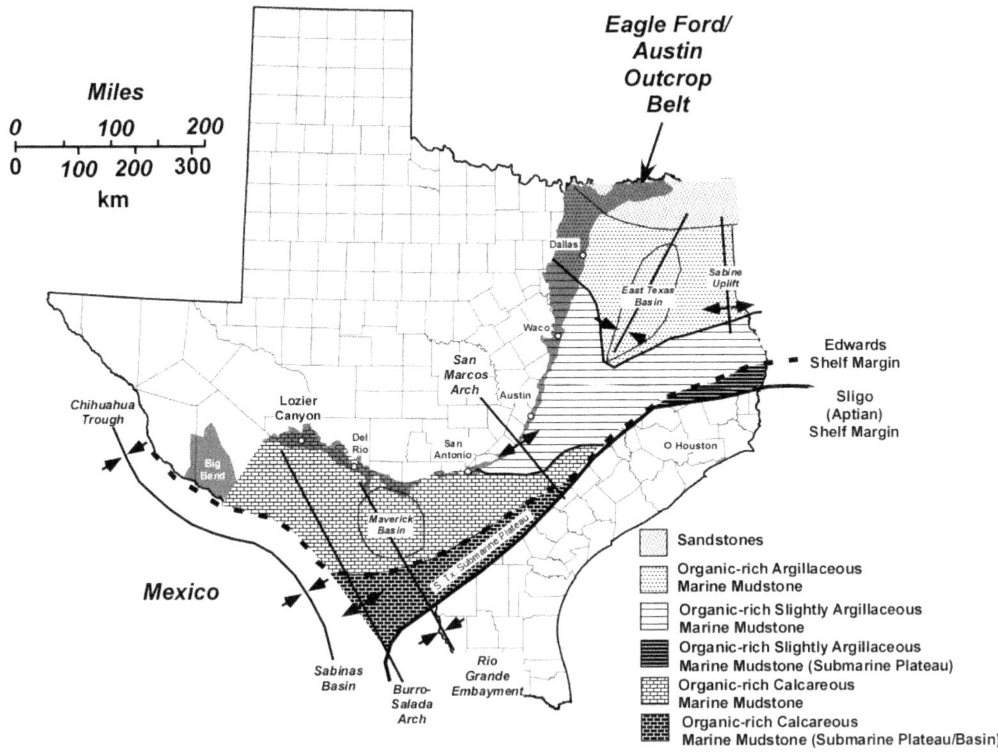

Fig. 6. Regional palaeogeographic reconstruction of Late Cenomanian of Texas. Overall change from more organic-rich argillaceous offshore mudstones in the East Texas to organic-rich carbonates offshore mudstones in south and west Texas.

Eagle Ford sequence stratigraphic units and surfaces

Introduction

Donovan *et al.* (2012) also identified four genetically related depositional sequences (allostratigraphic members), two within the Lower Eagle Ford Formation and two within the Upper Eagle Ford Formation in the outcrops of West Texas (Fig. 5). From the base up, Donovan *et al.* (2013) named these units the K63 (Lozier Canyon) member, the K64 (Antonio Creek) member, the K65 (Scott Ranch) member and the K70 (Langtry) member. Each of these depositional sequences has distinct petrophysical and geochemical characteristics, as well as chronostratigraphic properties, which make them particularly useful for regional correlations. In order to more easily convey superposition and stratigraphic relationships, as well as to gain regional terminology portability, Donovan *et al.* (2015) introduced the terms Lower and Upper Members of the Lower Eagle Ford Formation and Lower and Upper Members of the Upper Eagle Ford Formation for these four depositional sequences (allo-members). This updated nomenclature is used in this paper.

Lower Member of the Lower Eagle Ford Formation

The Lower (Lozier Canyon) Member of the Lower Eagle Ford Formation in West Texas is an organic-rich (typically >4%), bentonite-poor, mudstone-dominated depositional sequence (Figs 7, 8 & 9a). It is characterized by a bell-shaped GR profile in its lower portions and funnel-shaped GR profile in its upper portions (Figs 7 & 8). Geophysical logs from the research borehole illustrate that this member (depositional sequence) is characterized by an overall high resistivity. A thin (2.5 feet) clay-rich low-resistivity marker zone denotes its base (Figs 7 & 8).

The K63sb at the base of this depositional sequence is marked by a change from (Buda) wackstones with low GR values, along with low U and TOC content (below) to a mudstone-dominated unit (above) with elevated U and TOC content and higher GR values. The K64sb at the top is marked by a change from high-resistivity TOC-rich

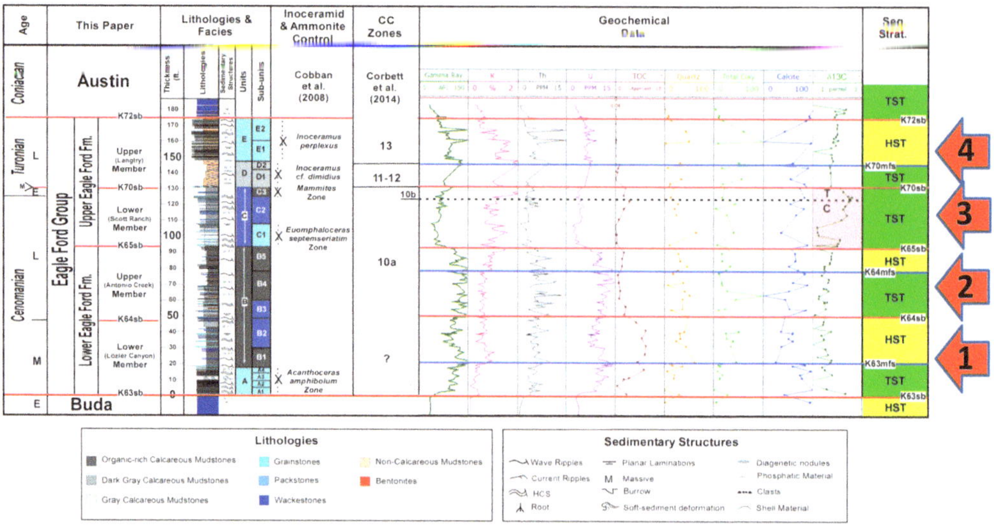

Fig. 7. Summary of the lithostratigraphic, biostratigraphic and geochemical data collected at the first measured section in Lozier Canyon at the Scott Ranch. Please note the distinct GR decrease at the base of the Upper Eagle Ford Formation driven by drop in U and Th content. Similar to the subsurface of South Texas, please note the positive $\delta^{13}C$ isotope excursion that occurs at the base of the Upper Eagle Ford. The peak inflection point of this excursion is the interpreted proxy for Cenomanian/Turonian stage boundary. Please note the four depositional sequences and their geochemical characteristics defined within the Eagle Ford Group of West Texas. The basal Lower Member of the Lower Eagle Ford Formation (1) contains the most organic-rich mudstones. This chronostratigraphic unit is primary unconventional source rock reservoir interval in the subsurface of South Texas.

mudstones with moderate U content (below) to U- and Th- (bentonite) rich mudstones with more moderate TOC content, higher GR and decreasing resistivity values (Figs 7 & 8). An interpreted mfs, labelled the K63mfs, is placed at the maximum GR inflection, which also marks the change from a bell-shaped GR profile (below) to an overall funnel-shaped GR profile (above). This interpreted surface also corresponds to the most mudstone-prone portion of this sequence in both outcrop and core (Figs 7 & 8). Transgressive deposits are interpreted below the mfs and highstand deposits above (Figs 7 & 8).

Work by Corbett *et al.* (2014) on calcareous nanno-flora for this interval were indeterminate. However, Cobban *et al.* (2008) reported ammonites of the late middle Cenomanian *Acanthoceras amphibolum* Zone from unit A at Lozier Canyon (Figs 5 & 7). Thus based on the present biostratigraphic control, the Lower (Lozier Canyon) Member of the Lower Eagle Ford Formation is interpreted to be late middle Cenomanian in age.

Upper Member of the Lower Eagle Ford Formation

The Upper (Antonio Creek) Member of the Lower Eagle Ford Formation in West Texas is a uranium-, thorium- and bentonite-rich, high-GR mudstone-dominated depositional sequence with moderate (1–2%) TOC content (Figs 7, 8 & 9b). The distinctive high GR signature of this sequence is clearly driven by its high uranium and bentonite (Th) content (Figs 7 & 8). The high abundance of bentonite beds is the most distinctive aspect of this depositional sequence in outcrop and core, and their unique abundance provides a distinct chronostratigraphic aspect to this unit. Geophysical logs from a research borehole illustrate that this member is also characterized by an overall resistivity decrease, at least across the basal portions of this sequence, as well as more subtle changes on the neutron, density and sonic logs (Fig. 8).

The K64sb at the base of this depositional sequence (allomember) is marked by a change from TOC-rich, high-resistivity mudstones with an overall funnel-shaped GR profile (below), to U- and Th- (bentonite) rich mudstones with high GR values and decreasing resistivity values from the base up (Figs 7 & 8). The K65sb at its top (Figs 7 & 8) is marked by a change from U- and Th- (bentonite) rich, high-GR mudstones below to U- and TOC-poor, low-GR thickly interbedded bioturbated packstones and mudstones (above). An interpreted mfs, labelled the K64mfs (Figs 7 & 8), is placed at the maximum GR inflection which also marks the

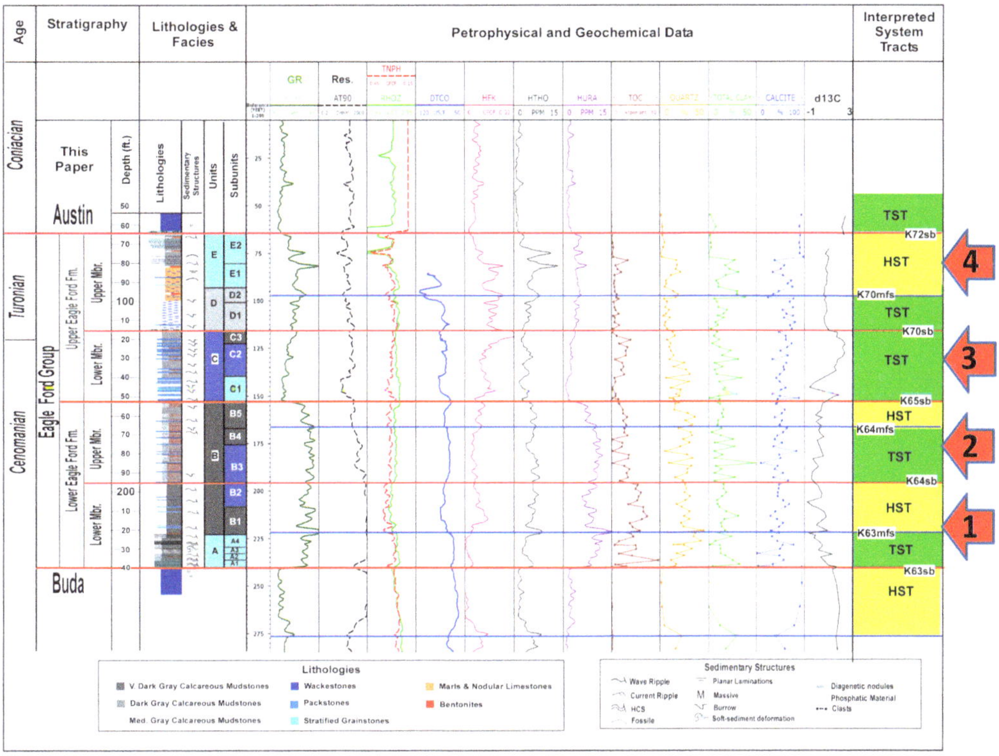

Fig. 8. Summary of the lithologic, petro-physical and geochemical data, as well as the sequence stratigraphic units and surfaces identified in the BP/SLB Lozier Canyon #1 test well. Well location on Figure 1. Please note the high resistivity log signature of the organic-rich Lower Member of the Lower Eagle Ford Formation, as well as the U increase, TOC decrease and resistivity drop-off at the K64sb which marks the top of this unit. Also note the GR decrease driven by U and Th content, as well as the onset of the positive $\delta^{13}C$ isotope excursion, at the base of the Upper Eagle Ford.

change from a bell-shaped GR profile (below) to an overall funnel-shaped (decreasing) GR profile (above). Transgressive deposits are interpreted below the mfs and highstand deposits above (Figs 7 & 8). Superposition, as well as work by Corbett *et al.* (2014) on calcareous micro-flora, suggests a late Cenomanian age for this sequence (Fig. 5). Presently, however, there is no molluscan age control for this sequence (allo-member) from Lozier Canyon.

Lower Member of the Upper Eagle Ford Formation

The Lower (Scott Ranch) Member of the Upper Eagle Ford Formation in West Texas is a uranium-poor (low-GR) depositional sequence made up of thickly inter-bedded bioturbated white packstones and grey mudstones (Figs 7, 8 & 9c). Besides its overall blocky GR profile, one of the most diagnostic aspects of this member is the presence of a positive $\delta^{13}C$ isotope excursion, the peak of which is the interpreted proxy for the Cenomanian/Turonian stage boundary, in the basal (± 30 feet) of this depositional sequence in outcrop and core (Figs 7 & 8). A distinctive thin (± 10 feet) clay-rich, low-TOC, low-resistivity zone also marks the base of this sequence on geophysical logs (Fig. 8). The K70sb at its top is marked by a change from thickly interbedded packstones and carbonate mudstones below to bioturbated mudstones above. A slight increase in U and Th content, as well as a change to more bell-shaped log GR and resistivity patterns, occurs above this surface. At Lozier Canyon, the positive $\delta^{13}C$ isotope excursion is restricted to this member, and the excursion also ends at the K70sb (Fig. 7).

In terms of age control in the Lozier Canyon outcrops, Cobban *et al.* (2008) report middle late Cenomanian *Euomphaloceras septemseriatum* fauna in unit C1 at the base of this member and Early Turonian *Mammites* fauna from unit C3 at the top of this

Fig. 9. Lozier Canyon region outcrop images of the (**a**) organic-rich Lower Member of the Lower Eagle Ford Formation; (**b**) bentonite-rich Upper Member of the Lower Eagle Ford Formation; (**c**) interbedded and U-poor Lower Member of the Upper Eagle Ford Formation; and (**d**) mudstone-prone and bentonite-bearing Upper Member of the Upper Eagle Ford Formation. B, Bentonite Bed.

member. Corbett and others also identified latest Cenomanian calcareous micro-flora from subunits C1 and C2 and earliest Turonian micro-flora from subunit C3 (Fig. 7). This is consistent with the associated positive $\delta^{13}C$ isotope excursion profile in this member representing the OAE2, and the maximum positive excursion of this profile being the geochemical proxy for the base of the Turonian (Figs 7 & 8). Thus in Lozier Canyon the Upper Member of the Lower Eagle Ford is interpreted as Latest Cenomanian to Earliest Turonian in age.

Upper Member of the Upper Eagle Ford Formation

The Upper (Langtry) Member of the Upper Eagle Ford Formation is a bentonite-bearing depositional sequence that in West Texas consists of bioturbated mudstones in its lower half and thinly bedded mudstones and grainstones in its upper half (Figs 7, 8 & 9d). Not surprisingly, its GR profile is bell shaped in its lower half and funnel shaped in its upper half (Figs 7 & 8). The interpreted K70 mfs occurs within a clay-rich low-resistivity and low-velocity marker zone within this depositional sequence. The micro-flora CC12/13 boundary also occurs at or near the K70mfs. Cobban *et al.* (2008) report *Inoceramus cf. dimidius* in the lower half of this depositional sequence and *Inoceramus perplexus* fauna in the upper half. This is consistent with the micro-flora findings of Corbett *et al.* (2014), making the interpreted transgressive deposits of this depositional sequence early late Turonian in age and the highstand latest Turonian in age.

In terms of bounding surfaces, Donovan *et al.* (2015) reported rip-up clast lags of underlying strata along the basal K70sb, as well as overlying K72sb. The K70sb at the base is also marked by a change from thickly interbedded packstones and carbonate mudstones below to bioturbated mudstones above. A slight increase in U and Th content, as well as a change to more bell-shaped log GR and resistivity patterns, occurs above this surface (Figs 7 & 8). The K72sb at the top, which also marks the contact with the overlying Austin Group, is marked by a distinct drop in U and Th content and change to an overall blocky GR and resistivity pattern above the K72sb (Figs 7 & 8).

West Texas to South Texas sequence stratigraphy

A type well for the Eagle Ford Group in South Texas (Webb County) is presented as Figure 3. Based the petrophysical and geochemical data provided for this well, the same basic allo-members (depositional sequences) defined in the outcrops in West Texas can also be readily defined in this well (Fig. 3). A quick look at the sonic and TOC columns on this well plot highlights the TOC-rich Lower Eagle Ford Formation and the carbonate-rich Upper Eagle Ford Formation. Distinct low-resistivity marker zones denote the base of both formations as well (Fig. 3). High-GR (U-enriched) zones at the top of both formations define their upper members. Furthermore, the CC12/13 microflora boundary occurs at or near the interpreted mfs within the Upper Member of the Upper Eagle Ford Formation. In the Lower Eagle Ford Formation, organic-rich, high-resistivity carbonate mudstones define the Lower Member of the Lower Eagle Ford Formation, while in the Upper Eagle Ford Formation, an overall blocky (low) GR interval, related to low U content, marks the Lower Member of the Upper Eagle Ford (Fig. 3). The basal portion of the Lower Member of the Upper Eagle Ford is also marked by the presence of the distinct positive $\delta^{13}C$ isotope excursion, the peak of which is the interpreted proxy for the base of the Turonian (Fig. 3).

Figure 10 is a two-well cross-section, datumed on the base of the Upper Eagle Ford, tying the research borehole in West Texas to the type well in South Texas (Fig. 1). The sequence boundaries (red lines) and maximum flooding surfaces (blue lines) of all four sequences (allo-members) are correlated between these two wells (Fig. 10). In each depositional sequence (allo-member), strata between the basal sequence boundary and mfs represent the transgressive deposits, while the strata between the mfs and overlying sequence boundary represent highstand deposits. In both of these wells, the thicknesses of the Upper Member of the Lower Eagle Ford Formation (light grey), as well as the Upper Member of the Upper Eagle Ford Formation (light purple), are similar (Fig. 10). In contrast, the Lower Member of the Lower Eagle Ford Formation (dark grey), as well as the Lower Member of the Upper Eagle Ford Formation (light blue), are both thicker in the well in South Texas than in the borehole in West Texas (Fig. 10). For both sequences, it is the thickness of the preserved highstands that varies between the two areas.

Figure 11 is a more complete well-log cross-section from West Texas to South Texas. It reveals how the Lower Member of the Lower Eagle Ford Formation, the primary source rock reservoir in the subsurface, thins: (1) onto the Central Platform to the west; (2) across the Edwards Margin; and (3) to the Sligo Margin to the east. The thinning of this depositional sequence towards the Central

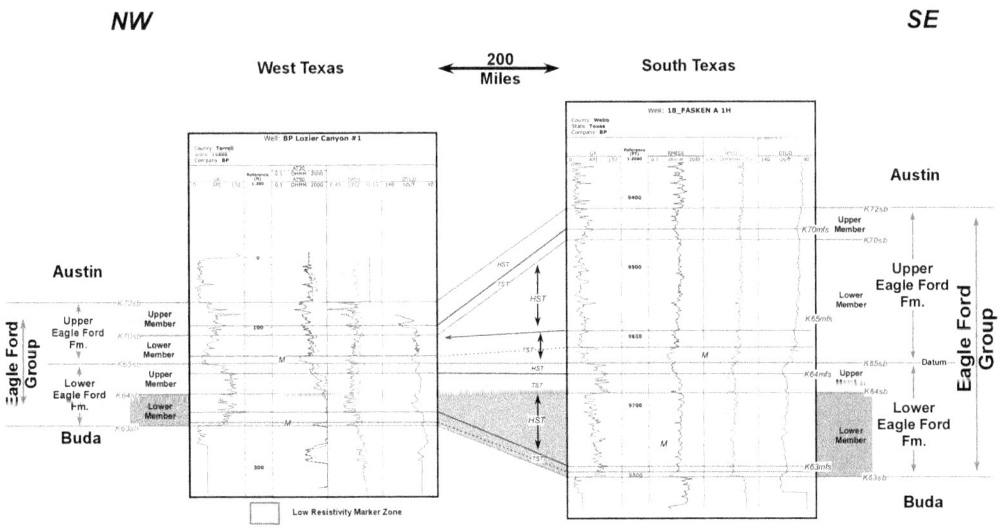

Fig. 10. Correlation of the key surfaces and depositional sequences between the test well in West Texas and the type well in South Texas. The well locations are noted on Figure 1. Thickness variations occur mainly within the interpreted highstand systems tracks of the K63 and K65 depositional sequences, essentially the lower members of the Lower and Upper Eagle Ford.

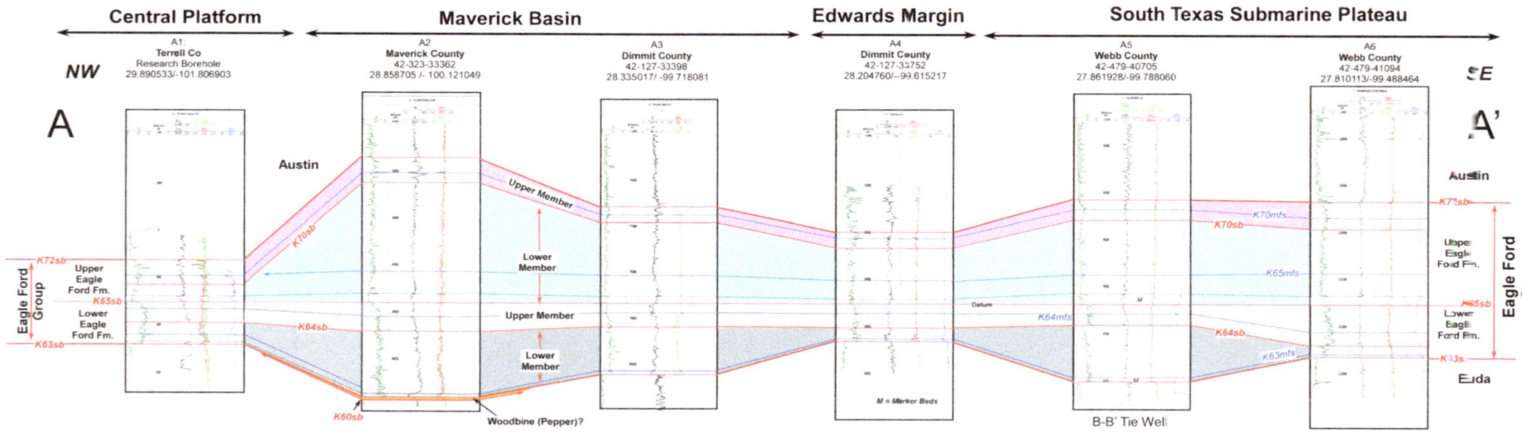

Fig. 11. Well–log cross-section illustrating the thickness variations of the four proposed Eagle Ford Members from West Texas to South Texas. Please note the thickness variations of the Lower Member of the Lower Eagle Ford towards the Central Platform region to the west, across the Edwards Margin, and towards the Sligo Margin to the east. This unit is the primary source rock reservoir zone in the subsurface of South Texas.

Platform, as well as across the Edwards Margin, can be attributed to primary inherited accommodation variations. However, the thinning to the east most likely represents the easterly depositional limit of this sequence, and an off-lapping relationship with the overlying K64 sequence, the Upper Member of the Lower Eagle Ford Formation. In contrast, thickness variations within the Upper Eagle Ford are probably related to tectonically enhanced truncation by the K70sb at the base of the Upper Member of the Upper Eagle Ford Formation.

Conclusions

With the aid of spectral gamma ray information, as well as geochemical and biostratigraphic data, collected on the outcrops in Lozier Canyon in West Texas, the Lower Eagle Ford Formation and Upper Eagle Ford Formation as traditionally defined in the subsurface of South Texas can now be identified in the outcrops of West Texas. Integrating basic geochemical and petrophysical data to the updated stratigraphy on these outcrops was transformative. By doing this, these West Texas exposures became portals to examine, as well as explain and predict, the distribution and thickness variations, of specific chronostratigraphic units that are critical to the economic success to exploiting the Eagle Group across the unconventional source rock play fairway in South Texas.

In these West Texas outcrops, four surface-bounded allostratigraphic members (depositional sequences) were defined within the outcrops of Eagle Ford Group of West Texas. These depositional sequences were then correlated back into the subsurface of South Texas, to get an improved understanding of the distribution and thickness variations of each chronostratigraphic unit. For nomenclature simplicity the four depositional sequences (allostratigraphic units) were simply named the Lower and Upper Members of the Lower Eagle Ford Formation, and the Lower and Upper Members of the Upper Eagle Formation.

The Lower Member of the Lower Eagle Ford Formation is an organic-rich (3–5%TOC), high-resistivity sequence with a clay rich, low resistivity marker zone at its base. This interval is the primary unconventional source rock reservoir in the subsurface of South Texas. The thickness and distribution of this key sequence appears to be controlled by inherited accommodation patterns across the Central Platform and Edwards Margin, but thinning towards the Sligo Margin to the east, probably marks the eastward depositional limit of this sequence. The Upper Member of the Lower Eagle Ford Formation is a uranium- and thorium-rich high-GR mudstone-dominated sequence with more moderate (1–3%) TOC values. On geophysical logs it is marked by an overall upward decreasing resistivity values, as well as higher density and velocity values. While the bentonite abundance within this sequence makes it a distinct chronostratigraphic unit, this characteristic probably reduces the effectiveness of this sequence as an unconventional source-rock reservoir in the subsurface of South Texas.

The Lower Member of the Upper Eagle Ford Formation is low- to moderate-TOC, uranium-poor sequence with a more blocky GR log profile, which is characterized by the presence of a distinct positive $\delta^{13}C$ excursion in the basal 30 feet of the unit. A clay-rich low resistivity zone marks its base. The lower overall TOC values in this sequence diminishes its value as a primary exploitation target. The Upper Member of the Upper Eagle Ford Formation is a TOC-poor and a thorium- and bentonite-rich sequence containing a clay-rich high-GR, low-velocity and -resistivity zone located at or near the interpreted mfs of this sequence. The K70sb at the base of the Upper Member of the Upper Eagle Ford appears to be a tectonically enhanced sequence boundary that can affect the distribution and thickness variations of underlying Eagle Ford strata, in particular the Lower Member of the Upper Eagle Ford Formation.

The author would like to thank the BP Exploration (Global Unconventionals) for permission to publish this work, as well as their support of this research. The author is indebted to the Crane, Goebel, Foster and Scott families for access to the spectacular Eagle Ford outcrops on their properties in West Texas. The cooperation of Swift Energy and Core Laboratory for their use of core data and laboratory results in support of the graduate student research is also most appreciated. Schlumberger was the strategic partner for the research borehole at Lozier Canyon and their support for this aspect of the research is also acknowledged. The data summarized in this paper is also the result of the research efforts of many of my colleagues from BP (Scott Staerker, Rand Gardner, Aris Pramudito, Matt Corbett) and TAMU (Mike Pope, Matthew Wehner) whose hard work made this topical summary paper possible. The author is also indebted to the manuscript reviewers whose comments and suggestions made vast improvements to the content and flow of the final product.

References

Campbell, C.V. 1967. Lamina, laminaset, and bedset. *Sedimentology*, **8**, 7–26.

Cobban, W.A., Hook, S.C. & McKinney, K.C. 2008. Upper Cretaceous Molluscan record along a transect from Virden, New Mexico, to Del Rio, Texas. *New Mexico Geology*, **30**, 75–92.

Corbett, M.J., Watkins, D.K. & Pospichal, J.J. 2014. A quantitative analysis of calcareous nannofossils bioevents of the Late Cretaceous (Late Cenomanian–Coniacian) Western Interior Seaway and their reliability in established zonation schemes. *Marine Micropaleontology*, **109**, 30–45.

Donovan, A.D. & Staerker, S. 2010. Sequence stratigraphy of the Eagle Ford (Boquillas) Formation in the subsurface of South Texas and the outcrops of West Texas. *Gulf Coast Association of Geologic Societies Transactions*, **60**, 861–899.

Donovan, A.D., Staerker, T.S. et al. 2012. The Eagle Ford outcrops of West Texas: a field laboratory for understanding heterogeneities within unconventional mudstone reservoirs. *GCAGS Journal*, **1**, 162–185.

Donovan, A.D., Staerker, T.S. et al. 2013. A 3-D outcrop perspective of an unconventional carbonate mudstone Reservoir. *In*: *Technical Program Unconventional Resources Technology Conference*, 12–14 August 2013, Denver, CO, URTeC 1580954.

Donovan, A.D., Gardner, R.D. et al. 2015. Chronostratigraphic relationships of the woodbine and Eagle Ford Groups across Texas. *GCAGS Journal*, **4**, 67–87.

Freeman, V.L. 1961. Contact of the Boquillas flags and Austin Chalk in Val Verde and Terrell Counties, Texas. *American Association of Petroleum Geologists Bulletin*, **45**, 105–107.

Freeman, V.L. 1968. *Geology of the Comstock–Indian Wells Area Val Verde, Terrell, and Brewster Counties, Texas*. USGS Professional Papers, **594-K**.

Grabowski, G.J. 1995. Organic-rich chalks and Calcareous mudstones of the Upper Cretaceous Austin Chalk and Eagle Ford Formation, South-Central Texas, USA. *In*: Katz, B.J. (ed.) *Petroleum Source Rocks*. Springer, Heldelberg, 209–234.

Gradstein, F.M., Ogg, J.G., Schmitz, M. & Ogg, G. 2012. *The Geologic Time Scale*. Elsevier.

Halbouty, M.T. 1991. *East Texas Field-U.S.A.* East Texas Basin, Texas, AAPG Stratigraphic Traps II, 189–206.

Hazzard, R.T. 1959. Measured sections. *In*: Cannon, R.L., Hazzard, R.T., Young, A. & Young, K.P. (eds) *Geology of the Val Verde Basin: West Texas Geological Society Guidebook*, 33–37.

Kennedy, W.J. & Cobban, W.A. 1990. Cenomanian ammonite fauna from the Woodbine Formation and lower part of the Eagle Ford Group, Texas. *Palaeontology*, **33**, 75–154.

Lock, B.E. & Peschier, L. 2006. Boquillas (Eagle Ford) upper slope sediments, West Texas: outcrop analogs for potential shale reservoirs. *Gulf Coast Association of Geological Society Transactions*, **56**, 1491–1508.

Lock, B.E., Peschier, L. & Whitcomb, N. 2010. The Eagle Ford (Boquillas Formation) of Val Verde County, Texas – a window on the South Texas play. *Gulf Coast Association of Geological Societies Transactions*, **60**, 419–434.

Ogg, J.G. & Hinnov, L.A. 2012. Cretaceous. *In*: Gradstein, F.M., Ogg, J.G., Schmitz, M. & Ogg, G. (eds) *The Geologic Time Scale*. Elsevier, Oxford, 793–853.

Pessagno, E.A. 1969. *Upper Cretaceous Stratigraphy of the Western Gulf Coast Area of Mexico, Texas, and Arkansas*. Geological Society of America, Boulder, CO, Memoirs, **111**.

Ruppel, S.C., Loucks, R.G. & Frebourg, G. 2012. Guide to the field exposures of the Eagle Ford-equivalent Boquillas Formation and Related Upper Cretaceous units in Southwest Texas. *BEG Mudrock Consortium Field Trip*, 8–12 March 2012, Austin, Texas, p10.

Smith, C.C. 1981. *Calcareous nannoplankton and stratigraphy of Late Turonian, Coniacian, and Early Santonian age of the Eagle Ford and Austin groups of Texas*. US Geological, Survey Professional Papers, **1075**.

Udden, J.A. 1907. A sketch of the geology of the Chisos country, Brewster County, Texas. *University of Texas Bulletin*, **93**, 29–33.

Field-based structural studies as analogues to sub-surface reservoirs

DAVID J. SANDERSON[1,2]

[1]*Faculty of Engineering and the Environment, University of Southampton, Southampton SO17 1BJ*

[2]*Reservoir Development, BP, Chertsey Road, Sunbury-on-Thames TW16 7BP*

(e-mail: d.j.sanderson@soton.ac.uk)

Abstract: This paper examines a number of field analogues to structures in the sub-surface, based mainly on examples from SW Britain. Whereas sedimentological analogues are largely interpreted in terms of a clearly defined sequence of depositional events related to stratigraphy, structures can develop throughout the history of the reservoir and their timing is poorly constrained stratigraphically. This leads to fundamental differences in how the two types of analogues are used. The simplest and most direct use of structural analogues comes from comparison of field structures with those sampled in the sub-surface, where they can guide conceptual development of reservoir models. Many examples of this approach exist from the development of stratabound fracture systems to the formation of linked fault systems and fracture networks. Field analogues also play an important role in the study of sub-seismic structures, providing a link to evidence from core. One new area is the application of topology to the study of fault and fracture networks. Not only does this provide insight into fundamental relationships between structures, such as their relative age and connectivity, but it also provides new methods of analysis of great relevance to the spatial variation and scaling of structures.

Structural geology is largely concerned with the geometry, topology, kinematics and mechanics of post-depositional features at a wide range of scales. Structural features include folds and faults that affect the shape and architecture of the reservoir, but also fabrics and discontinuities that affect rock properties. Thus, structural studies impact significantly at all stages in exploration and production. Regional tectonics defines different play types and, together with the deformation history, defines source rock maturation and the subsequent charge history. The development and interaction of individual structures determine the size and shape of traps, and their sealing potential. Smaller-scale structures control many reservoir properties and, hence, the recovery and production history of reservoirs.

Field studies have always played an important role in the interpretation of the sub-surface. Our observations in the laboratory, field and sub-surface are made at different scales and are usually scale limited. Detailed analysis of exposures and regional mapping provide important links between these scales, and generally provide greater opportunity to study the spatial and temporal arrangement between features.

Thus structural geology plays an important role in both the development of conceptual models and the testing of sub-surface techniques. This paper will review examples from both conventional and unconventional reservoirs based mainly around the development of faults and fractures in SW Britain.

Differences between structural and sedimentological analogues

In sedimentology the main focus is on depositional, biological and diagenetic processes. There are well established principles upon which to build sedimentological models, ranging from simple application of Steno's 'Law of superposition' to Walther's Law relating facies sequences to adjacency of depositional environments and processes. Today, we interpret such sequences in terms of sequence stratigraphy, which provides a clear link between the surface and sub-surface records of these depositional processes and their control by sea-level changes, sediment supply and development of accommodation space.

Structures are produced by a set of boundary conditions (stresses and displacements) that operate on rocks with a range of material properties. Both the boundary conditions and material properties change with location, depth and time. Structures can form 'anywhere' and at 'any time' (Fig. 1), with cross-cutting relationships providing information

From: BOWMAN, M., SMYTH, H. R., GOOD, T. R., PASSEY, S. R., HIRST, J. P. P. & JORDAN, C. J. (eds) 2016. *The Value of Outcrop Studies in Reducing Subsurface Uncertainty and Risk in Hydrocarbon Exploration and Production*. Geological Society, London, Special Publications, **436**, 207–217.
First published online December 17, 2015, http://doi.org/10.1144/SP436.5

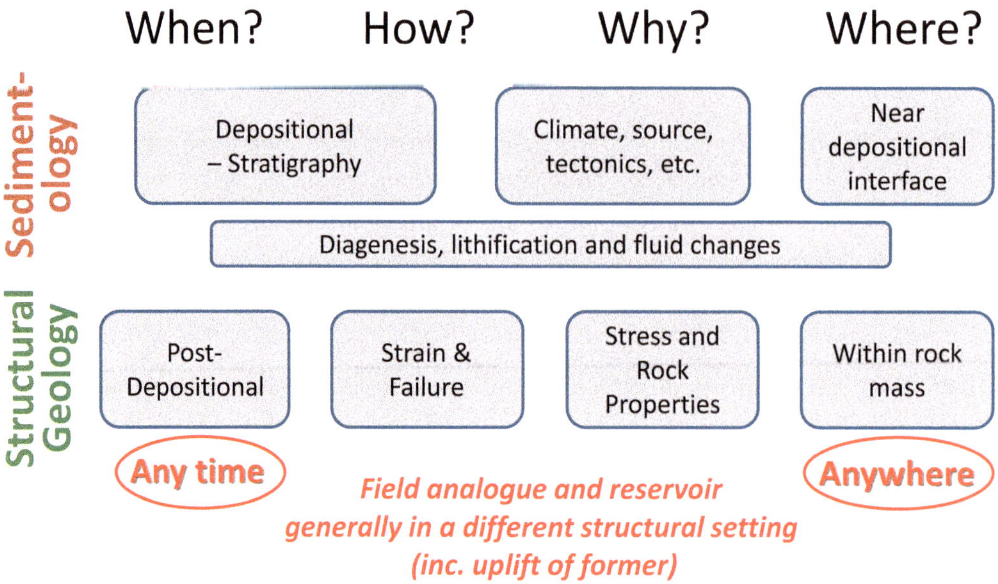

Fig. 1. Comparison of rationale for use of field analogues in structural geology and sedimentology.

about their relative age. They are rarely linked to surface depositional processes and events as in sedimentology and, hence, are not easily related to stratigraphy, except at unconformities. This requires a very different approach to the study and interpretation of analogues, since the history and conditions of the development of structures may be very different at a surface exposure from those at a sub-surface location. Thus, structural analogues are largely used to establish structural 'styles', to build conceptual models and to test and calibrate sub-surface techniques.

Geometry, topology, kinematics and mechanics

To understand the strengths and limitations of structural analogues, we need to consider how structural geologists work (Fig. 2). Observation and recording of exposures provide much data on the *geometry* of structures. This, together with mapping of more extensive exposures and regions, yields valuable information on the relationships between structural features – the *topology*. Based on such observations, and a knowledge of the primary depositional fabrics and textures, structural geologists then interpret the geometry, kinematics and history of deformation. Variation in the geometry and kinematics, such as displacement variation along and across faults, provides information on variation in strain, both in relation to larger structures and regionally. Knowledge of the stress state (based on estimation of depth, pore-pressure, etc.) and the likely rheological behaviour (including estimates of material properties) allows discussion of the *mechanics*, usually in the light of geomechanical data and models. A key test of any geomechanical model is the geometry and strain distribution it produces. Thus, the progression from geometry → topology → kinematics → mechanics → geometry is recursive, with the need to test and modify ideas at all stages. Integration of field analogues with sub-surface observations involves checking against data provided by observation and measurement.

In the following sections I will illustrate these ideas using some examples from field studies and relate these to the investigation of both conventional and unconventional reservoirs.

Use of analogues

Demonstration and reassurance

The simplest and most direct use of structural analogues comes from direct comparison of field structures with those sampled in the sub-surface. In this case we use field examples to develop models, mainly of the geometry and kinematics of structures. Figure 3 shows a simple example that compares deformation bands (e.g. Fossen *et al.* 2007) sampled from sandstones in the field with observations from core. At a simple level, the field analogues are used to 'illustrate' equivalent structures

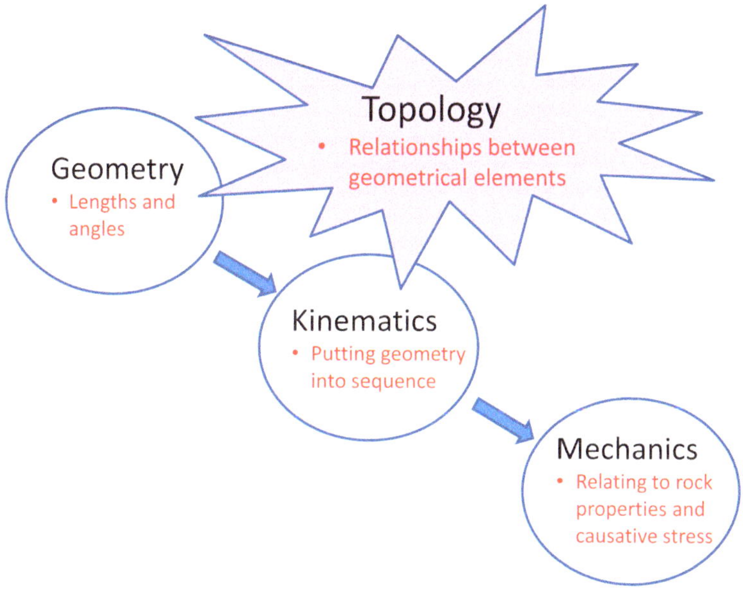

Fig. 2. Generalized workflow in structural geology.

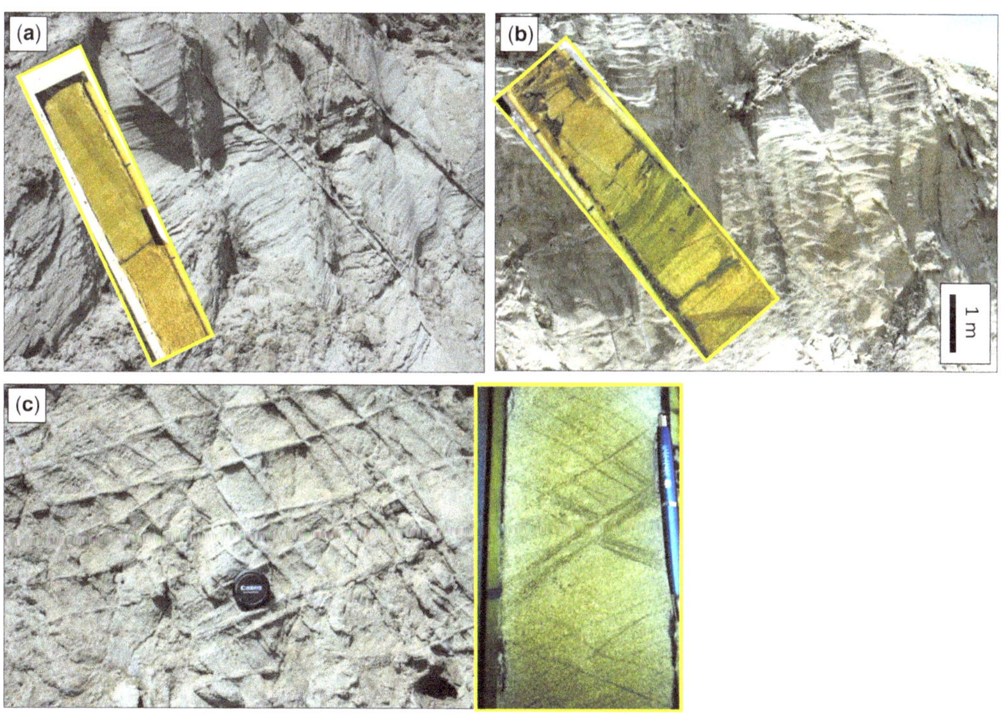

Fig. 3. Increasing intensity of deformation bands exposed in Yasamal Valley, west of Baku, Azerbaijan, compared with those in slabbed core from similar aged sandstones from off-shore ACG field, Azerbaijan.

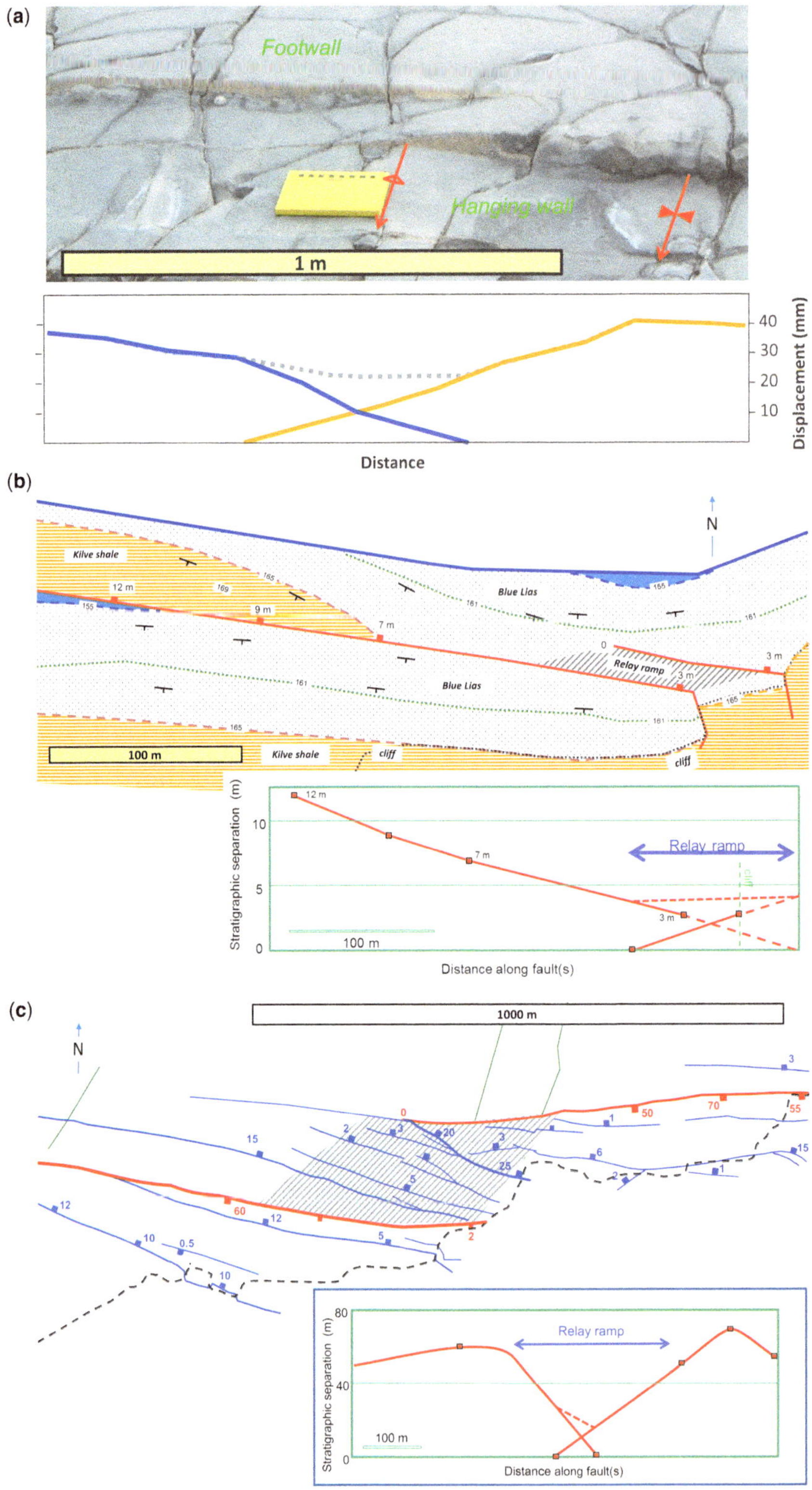
(a)
Footwall
Hanging wall
1 m
40
30
20
10
Displacement (mm)
Distance
(b)
N
Kilve shale
12 m
9 m
7 m
Blue Lias
Relay ramp
3 m
3 m
0
Blue Lias
100 m
Kilve shale
cliff
cliff
Stratigraphic separation (m)
12 m
7 m
3 m
Relay ramp
100 m
Distance along fault(s)
(c)
1000 m
N
0
50
70
55
60
2
Relay ramp
Stratigraphic separation (m)
80
40
0
100 m
Distance along fault(s)

in the sub-surface, often providing reassurance as a geoscientist seeks to convince others of a particular aspect of a reservoir model.

Whilst these simple uses of field observations are both widespread and useful, the field geologist needs to remember that the boundary conditions, rock properties and structural history of the field exposures may be very different to that in the sub-surface. Thus, one must seek clear evidence for the existence of an analogue feature; in the case of the examples in Figure 3, we see similar deformation bands in core, in similar intensities and orientations, and in the same stratigraphic units and sedimentary facies. Even then there may be uncertainties in ascribing features, such as the connectivity of the deformation bands and their impact on fluid flow that needs to be assessed carefully.

Developing structural models

Most geometrical and kinematic models stem from field observations, especially where there is extensive exposure of both individual structures and of the relationships between different structures. The literature is full of many fine examples of this use, but I will illustrate this with reference to the role that the exposures of the Lower Jurassic of the North Somerset coast have played in our concept of relay ramps in extensional settings. The *c.* 18 km-long coast section between Hinkley Point and Blue Anchor Bay is probably one of the most visited in the world, being the location of many academic and industry training courses, and the subject of many publications over the last 25 years.

Individual normal faults are well exposed and provide clear illustrations of the geometry in both cliff sections and wave-cut platforms, allowing a level of 3D interpretation normally lacking in field analogues. The whole outcrop provides an excellent analogue to aid interpretation of 3D seismic, with the cliff sections being equivalent to vertical profiles or dip lines and the wave-cut platform being equivalent to a time slice.

The high-resolution stratigraphy provides an essential template for the interpretation and quantification of the structures (Whittaker & Green 1983). Fault separations can be measured to centimetre-scale precision, and the consistency of the lithostratigraphy provides better than metre-scale accuracy, even for faults with over 50 m separation. This allows study of small (sub-seismic) structures in relation to large (seismic) structures (Glen *et al.* 2005; Putz-Perrier & Sanderson 2008) and an evaluation of their role during basin extension.

One feature of the extensional faulting in north Somerset is the segmented nature of the faults and the development of relay ramps (Peacock & Sanderson 1991, 1994). A *relay ramp* is an area of re-orientated bedding that links two normal fault segments that overstep in map-view and have the same dip direction (Larsen 1988). A relay ramp connects the footwall with the hanging-wall of a fault zone (Fig. 4a), the continuity of bedding giving rise to the ramp-like geometry. This is very obvious for exposed small relays (Fig. 4a), but similar structures can be mapped at different scales (Fig. 4b, c).

In the field, one can map the strata on both sides of the fault trace and produce a separation diagram (Fig. 4, insets). This is shown for two relay ramps on the wave-cut platform to the east of Kilve Pill (Fig. 4b, c). The key to such mapping is to use a log of the local stratigraphic sequence to determine the stratigraphic separation of beds at various points along the faults. Where dips of bedding are low, stratigraphic separation $\approx$ throw, or one may use the orientation of the bedding and fault plane to calculate any separation (e.g. throw), and the slip vector to estimate displacement.

This process is essentially that of displacement mapping in 3D seismic volumes, where separation (usually throw) is mapped onto the fault plane. In practice it is useful to plot the footwall and hanging-wall cut-offs against distance along the strike of the faults – the Allan diagram. The wave-cut platform at Kilve provides an excellent analogue to a time slice and hence useful training in these techniques.

Relay ramps are important in evaluating trapping by fault sealing as they may facilitate flow across the fault zone (spill points). They also transfer displacement between the overstepping segments of the fault (Peacock & Sanderson 1991, 1994), and may represent the sites of localized minima in the displacement (Fig. 4 - insets).

Variation in displacement along fault segments and transfer across relay ramps require deformation of the surrounding wall-rocks. For example at Kilve, hanging-wall synclines develop at higher displacement parts of the faults, with less common hanging-wall anticlines being associated with lower displacements, as at relay ramps. Given the importance of faulting as a deformation mechanism during the extension, it is not surprising that some of the wall-rock deformation is accommodated by faulting.

Fig. 4. Field examples of relay ramps in normal fault at Kilve, North Somerset, with insets showing plots of stratigraphic separation against distance, with broken line showing cumulative displacement on the two faults: (**a**) small scale faults (separation <50 mm); (**b**) faults with 0–15 m separation; and (**c**) fault with 0–55 m separation. Values on faults are stratigraphic separation; small numbers on beds identify limestones (after Whittaker & Green 1983).

One manifestation of this is seen in Figure 4c, where the large relay ramp between two overlapping south-dipping faults is the site of many smaller north-dipping faults. These are more intensely developed in the ramp, where they appear to link the two bounding faults and have a slightly more clockwise orientation. Note that these faults mainly have displacements of 10 m or less and would not be resolved on commercial seismic. Thus, this subtle geometry might easily be unresolved in sub-surface studies, indeed the whole relay ramp may easily be 'missed' in an equivalent sub-surface interpretation. If the smaller faults link across the ramp, they may act as a baffle to flow and, hence, prevent or reduce spill from the footwall to hanging-wall. Such a situation has been studied by Rotevatn *et al.* (2007, 2008), who suggest that a network of cataclastic deformation bands causes compartmentalization of a ramp in Arches National Park, Utah. They estimate that deformation band clusters could reduce along-ramp effective permeability by more than three orders of magnitude.

Testing hypotheses

As well as the lessons learnt from mapping in the field, we can also use field observations to construct hypotheses that can subsequently be tested in both field and sub-surface investigation, as in the following simple example.

Many field studies have shown that fracture intensity (number of fractures per metre) decreases with increasing bed thickness (e.g. McQuillan 1973). Often fracture spacing (1/intensity) is approximately equal to bed thickness, a situation that is often referred to as fracture saturation. This can be explained by the modification of stress by fractures in a layer (Bai & Pollard 2000). There are many exceptions to this relationship, but if established it serves as a useful 'rule of thumb' in constructing models of sub-surface fracture distribution (e.g. Wennberg *et al.* 2007).

Figure 5 shows an example of this based on observations from Upper Carboniferous clastic sequences to the west of Westward Ho!, North Devon. Figure 5a–c shows that the spacing of a set of quartz veins in adjacent turbidites increases with increasing bed thickness. These analogues can be used to establish a hypothesis that the layer-bound fractures develop with a spacing ≈ layer thickness (Fig. 5d). This hypothesis may be tested from sub-surface observation where the thinner layers support the hypothesis and the thicker layer does not contradict the hypothesis (Fig. 5e).

Direct investigation of sub-seismic structures

A common issue in the evaluation and production of hydrocarbons is the need to characterize the reservoir at length-scales below that imaged seismically. Well logs and core may provide important data, but such data are both scarce and generally biased, usually with little data from faults and complex parts of the reservoir. Field analogues play an important role in bridging the gap between seismic and core data.

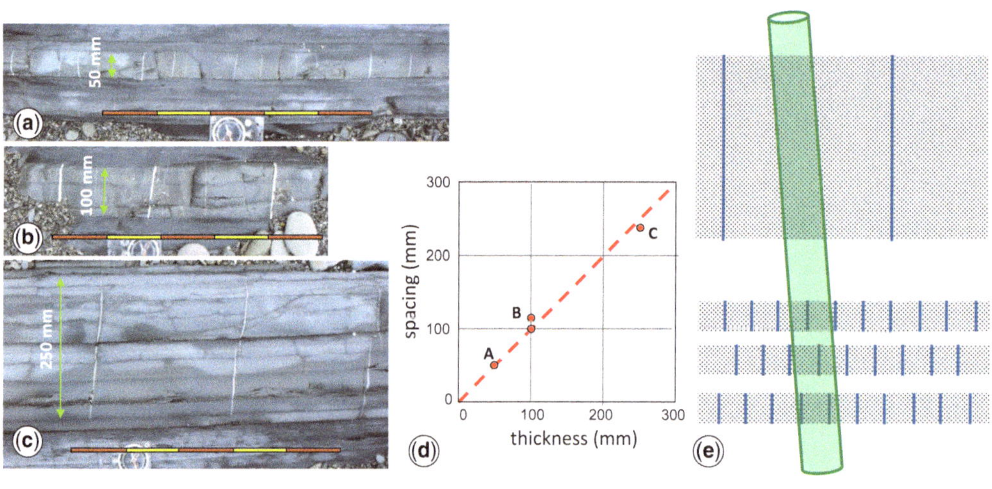

Fig. 5. Examples of spacing of veins developed in Upper Carboniferous turbidites of the Bideford Group, Westward Ho!, North Devon; at bed thicknesses of (**a**) 50 mm, (**b**) 100 mm and (**c**) 250 mm; scale bar 1 m. (**d**) Spacing is approximately equal to bed thickness. (**e**) How a well might be expected to encounter extensional fractures in thinner layers, where the diameter of the well is less than the fracture spacing, but not necessarily in a thicker layer where fracture spacing exceeds well diameter.

A good example of this is the study of fault rocks and damage zones (sub-seismic features) related to faults and their wall-rocks that can be imaged seismically. Figure 6a shows a normal fault, again in the Liassic of North Somerset. The main fault has a stratigraphic separation of 8 m, corresponding to dip-slip of 10.5 m, and would thus be just at or below the resolution of a commercial 3D seismic survey. The down-throw is to the north and the surrounding beds dip at *c.* 15° to the south. The fault is interpreted as one of a set of domino-faults with the fault planer and beds back-rotated during extension.

The fault is associated with a number of sub-seismic features. A narrow core (0–20 cm) is formed by a combination of clay smearing and brecciation of limestone beds and contains several slip surfaces with crystal-fibre slickensides whose lineation indicated down-dip displacement. There is a *c.* 1 m-wide zone of normal drag, in which the limestone beds become increasingly fractured. Surrounding this is a 2–3 m-wide damage zone consisting of a number of synthetic and antithetic faults that root into the main fault. These are usually associated with centimetre-wide pull-aparts as they cut steeply through the limestone beds (Fig. 6b).

These sub-seismic fabrics can be shown to have a profound effect on the flow of fluids associated with the faulting that can only be appreciated from the detailed observation possible in the field. The clay smearing would contribute to cross-fault sealing, but the mineralized slip-surfaces indicate fluid flow parallel to the fault plane has occurred. The pull-aparts in the damage zone develop parallel to the strike of the fault, and together with the fracturing in the limestones would facilitate along-strike flow. This, together with the displacement variation along such faults, would suggest good communication along the faults, with the potential to breach the main slip surfaces at spill points and relay ramps.

In the absence of wells penetrating faults, such observations may go undetected in the sub-surface, but the analogues provide a valuable reminder of the potential importance of sub-seismic structures in building reservoir models. This may provide incentive for the collection of high-spatial-resolution geometric, kinematic and geomechanical data from the sub-surface and/or provide models for how to distribute properties within the reservoir.

Developing techniques

Many techniques used in the sub-surface have their origins in field procedures. An illustration of this is the use of topology in characterizing fracture networks, again with an example from the Liassic of North Somerset.

On a bedding surface (Fig. 7), a fracture network can be considered to form a system of branches and nodes (Sanderson & Nixon 2015). When extensively exposed, the branches and nodes can be used to define and measure the length and orientation of the fractures, but they also define the topological relationships between the elements of the network. The proportions of isolated (I), abutting (Y) and crossing (X) nodes provide a basis for describing the topology (Manzocchi 2002; Mäkel 2007; Sanderson & Nixon 2015). In this case the field exposures indicate a well-connected network of four sets of fractures, dominated by Y-nodes as later fractures abut earlier ones.

Using the mapped network, we can simulate sampling by a horizontal well (Fig. 7) with a hole diameter of 20 cm (i.e. *c.* 8.5 inch hole). Obviously the orientation of the fractures and the fracture intensity can be measured and corrections applied for the bias introduced by the orientation of the 'well'. The trace lengths of the branches and their connectivity are less obviously sampled in the well. There is, however, some topological information contained in the 'well' sample in the form of occasional sampling of the nodes. Although this may be scale limited and subject to orientation and size bias, the nodes and branches provide additional information about the development of the fracture network, allowing us to deduce the following (for more discussion see Sanderson & Nixon 2015):

(1) From the abutting relationships (Y-nodes) we can deduce the relative age of the fractures, in a similar manner to that obtained from field exposures, although separation of the first two sets (1 and 2 in Fig. 7) is not possible from the 'well data'.

(2) It is easy to measure the fracture intensity – the fracture length per unit area. This is about 3 m^{-1}, equivalent to an average spacing of 0.33 m.

(3) Most of the fractures encountered extend beyond the 'well', so how do we count the number of fractures? Following Mauldon *et al.* (2001), an unbiased estimate of number of branches or traces encountered in the 'well' can be obtained by counting the number of terminations in an area – each fracture will have two terminations. Let N_I, N_Y and N_X represent the number of I-, Y- and X-nodes, respectively, then the number of traces (N_L) and branches (N_B) is given by (Sanderson & Nixon 2015):

$$N_L = \frac{(N_I + N_Y)}{2}$$

$$N_B = \frac{(N_I + 3N_Y + 4N_X)}{2}$$

(a)

Fig. 6. (**a**) Photograph of north-dipping normal fault at Kilve, North Somerset (UK Grid: ST14484446). The fault has a stratigraphic separation 8 m, with a dip-slip displacement of 10.5 m. Small numbers on beds identify limestones (after Whittaker & Green 1983); notebook (circled) is 200 mm in length. (**b**) Details of the fault core and the structures developed within the damage zone. Protractor (circled) is 100 mm in length.

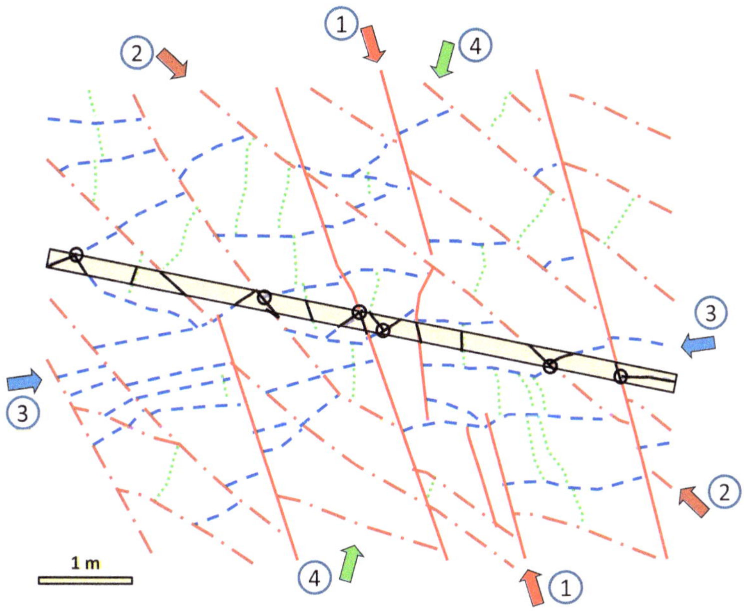

Fig. 7. Fracture network mapped on top surface of a 0.4 m-thick limestone bed, between Kilve and East Quantoxhead (UK Grid: ST13764431). Dashed lines identify fractures belonging to four sets; arrows indicate inferred direction of maximum horizontal stress; numbers in circles indicate relative order of development based on consistent abutting relationships. The long rectangle represents a hypothetical horizontal well with the location of fracture traces and nodes used to assess the geometry and topology of the network.

In Figure 7 there are six *Y*-nodes in the 7 m-long well section, giving estimates of trace and branch frequency of *c.* 2.2 and *c.* 6.5 m^{-2}, respectively.

(4) This then allows us to calculate the average (or characteristic) line and branch length by dividing the total fracture length by these numbers. Note that the branch lengths seen in the 'well' are very heavily censored by the sampling and, hence, cannot be measured directly, but the use of the unbiased frequency estimate allows an estimation of the characteristic trace and branch lengths of *c.* 1.4 and *c.* 0.46 m that agrees well with estimates from the 'field' exposures of the same network.

(5) The nodes can be used to classify the topology. In this case the dominance of *Y*-nodes suggests that we are dealing with a well connected fracture network.

Discussion

The examples discussed above illustrate just some of the ways that field analogues can be used to investigate reservoir structure. There are several observations that can be drawn from these examples.

The detail provided by field analogues clearly contrasts with the low spatial resolution of most seismic data. This mainly affects the precision of measurement, with the accuracy of field surveys being dependent on the quality of the exposure and care in mapping and that of seismic surveys being dependent on acquisition and processing. One could argue that seismic surveys could be as accurate as some field mapping, but never attain the same precision. Thus a major contribution of field analogues is to allow sub-seismic features to be incorporated into reservoir models. Seismic allows imaging of 3D volumes, whereas most field observation is on irregular, 2D surfaces, although small-scale observation can be 3D.

Comparison of field analogues with well data yields somewhat different conclusions. Arguably the precision and accuracy in measuring features in the field and in core or image logs are broadly similar, with both also allowing sampling of rock for more detailed study (optical microscopy, SEM, etc.; e.g. Fig. 3). The main limitation of well data is its restricted spatial scale – we essentially have a very narrow cylinder of data in one orientation through a large 3D volume of rock, whereas field exposures provide more extensive areas through the rock mass, although this is not truly 3D nor always continuous.

Thus field, seismic and well data lead to clear and obvious differences in the length scales sampled and hence in the measurements produced. With appropriate correction, the orientations of structures and the angular relationships between structural elements can be compared, but comparison of the size and spacing of features needs to be very carefully analysed (Figs 4 & 5).

An important aspect of structural geology is the analysis of the spatial relationships between different structural elements, although this is often not emphasized in the way we discard important spatial information (from the map) in seeking to present orientation data using stereograms, etc. Many relationships are topological rather than geometric, meaning that they are invariant to continuous transformations. We have seen this with the importance of abutting fractures in determining their relative ages (Fig. 7). At first sight topological analysis in the sub-surface might seem impractical (or even impossible), but some topological information is contained in such well data. The limited information of node types in core and image logs can allow the relative dating of fractures, assessment of their size and spacing even if these cannot be measured directly, and even assessment of the connectivity of a network of fractures. Many fault systems mapped from seismic account for the strains involved in basin formation, but appear to form poorly connected networks. This is almost certainly due to the poor resolution of smaller faults and low displacement parts of the fault network. A lot more work needs to be done in understanding the geometry, topology and scaling within networks, although recent studies begin to address such issues (Nixon *et al.* 2011, 2012).

A major limitation to the use of field analogues in the interpretation of the sub-surface concerns the differing structural history, boundary conditions and physical properties within the crust. We must recognize that structures seen in the field may be represented very differently or not developed at all in the sub-surface. One very common element of all surface exposures is their pervasive fracturing owing to uplift. Thus the open fractures (joints) seen in North Somerset (Fig. 7) almost entirely postdate the last phase of faulting (Peacock 2001), which from regional considerations was probably in the late Eocene or Oligocene (Gale *et al.* 1999). All the fractures associated with faulting are mineralized, mainly by calcite from circulating basin fluids. The joints cross-cut these, locally reactivating them in places, and have no mineralization. Thus it is very unlikely that the well-connected fracture system in Figure 7 would be present in any reservoir at depth.

Since the Liassic in southern Britain is a potential shale-gas/oil play, what would the joints in the field exposures be analogues of? Clearly not the natural fracture network although veins associated with early faults would almost certainly be present in the sub-surface. If the joints formed over the last 35 myr, owing to reduction in effective stress due to uplift, then they could provide valuable analogues to the nature of the fracture system developed owing to induced hydraulic fracturing!

Conclusions

The use of field analogues in the interpretation of sub-surface structure differs fundamentally from sedimentological analogues of depositional processes and systems. The simplest and most direct use of structural analogues comes from direct comparison of field structures with those sampled in the sub-surface. These can guide development of reservoir models, where they can both provide reassurance and help bracket uncertainty in a model.

Examination of the geometry, topology and kinematics of well-exposed structures, and integration of geomechanical properties, has allowed structural geologists to develop models of a wide range of structures. These are invaluable in guiding the collection and interpretation of sub-surface data and the testing of hypotheses. Many examples of this approach exist from the development of stratabound fracture systems to the formation of linked fault systems and fracture networks. A common issue in the evaluation and production of hydrocarbons is the need to characterize the reservoir at length scales below that imaged seismically. Field analogues play an important role in bridging the gap between seismic and core data.

New models and techniques are often developed from field investigations. One example is the application of topology to the study of fault and fracture networks. Not only does this provide insight into fundamental relationships between structures, such as their relative age and connectivity, it also provides new methods of analysis of great relevance to the spatial variation and scaling of structures.

Since structures can develop in a wide range of settings and at various stages in the history of a basin, field analogues can be used to address many different issues. In many ways the value of a structural analogue is not to help characterize the form and shape of a reservoir, but more to provide insight into process. For example, late, uplift-related fractures may be telling us more about the response of the reservoir to development and production through changes in effective stress, than to the presence of natural fractures in the reservoir.

References

BAI, T. & POLLARD, D. D. 2000. Closely spaced fractures in layered rocks: initiation mechanism and propagation kinematics. *Journal of Structural Geology*, **22**, 1409–1425.

FOSSEN, H., SCHULTZ, R. A., SHIPTON, Z. K. & MAIR, K. 2007. Deformation bands in sandstone: a review. *Journal of the Geological Society*, **164**, 755–769.

GALE, A. S., JEFFREY, P. A., HUGGET, J. M. & CONNOLLY, P. 1999. Eocene inversion history of the Sandown Pericline, Isle of Wight, southern England. *Journal of the Geological Society*, **156**, 327–339.

GLEN, R. A., HANCOCK, P. L. & WHITTAKER, A. 2005. Basin inversion by distributed deformation: the southern margin of the Bristol Channel Basin, England. *Journal of Structural Geology*, **27**, 2113–2134.

LARSEN, P.-H. 1988. Relay structures in a Lower Permian basement-involved extension system, East Greenland. *Journal of Structural Geology*, **10**, 3–8.

MÄKEL, G. H. 2007. The modelling of fractured reservoirs: constraints and potential for fracture network geometry and hydraulics analysis. *In*: JOLLEY, S. J., BARR, D., WALSH, J. J. & KNIPE, R. J. (eds) *Structurally Complex Reservoirs*. Geological Society, London, Special Publications, **292**, 375–403.

MANZOCCHI, T. 2002. The connectivity of two dimensional networks of spatially correlated fractures. *Water Resources Research*, **38**, 1162, http://doi.org/10.1029/2000 WR000180

MAULDON, M., DUNNE, W. M. & ROHRBAUGH, M. B. 2001. Circular scanlines and circular windows: new tools for characterizing the geometry of fracture traces. *Journal of Structural Geology*, **23**, 247–258.

MCQUILLAN, H. 1973. Small-scale fracture density in the Asmari Formation of southwest Iran and its relation to bed thickness and structural setting. *American Association of Petroleum Geologists Bulletin*, **57**, 2367–2385.

NIXON, C. W., SANDERSON, D. J. & BULL, J. M. 2011. Deformation within a strike-slip fault network at Westward Ho!, Devon U.K.: domino vs conjugate faulting. *Journal of Structural Geology*, **33**, 833–843.

NIXON, C. W., SANDERSON, D. J. & BULL, J. M. 2012. Analysis of a strike-slip fault network using high resolution multibeam bathymetry, offshore NW Devon U.K. *Tectonophysics*, **541–543**, 69–80.

PEACOCK, D. C. P. 2001. The temporal relationship between joints and faults. *Journal of Structural Geology*, **23**, 329–341.

PEACOCK, D. C. P. & SANDERSON, D. J. 1991. Displacements, segment linkage and relay ramps in normal fault zones. *Journal of Structural Geology*, **13**, 721–733.

PEACOCK, D. C. P. & SANDERSON, D. J. 1994. Geometry and development of relay ramps in normal fault systems. *American Association of Petroleum Geologists Bulletin*, **78**, 147–165.

PUTZ-PERRIER, M. W. & SANDERSON, D. J. 2008. Spatial distribution of brittle strain in layered sequences. *Journal of Structural Geology*, **30**, 50–64.

ROTEVATN, A., FOSSEN, H., HESTHAMMER, J., AAS, T. E. & HOWELL, J. A. 2007. Are relay ramps conduits for fluid flow? Structural analysis of a relay ramp in Arches National Park, Utah. *In*: LONERGAN, L., JOLLY, R. J. H., SANDERSON, D. J. & RAWNSLEY, K. (eds) *Fractured Reservoirs*. Geological Society, London, Special Publications, **270**, 55–71.

ROTEVATN, A., BUCKLEY, S. J., HOWELL, J. A. & FOSSEN, H. 2008. Overlapping faults and their effect on fluid flow in different reservoir types: a LIDAR-based outcrop modeling and flow simulation study. *American Association of Petroleum Geologists Bulletin*, **93**, 407–427.

SANDERSON, D. J. & NIXON, C. W. 2015. The use of topology in fracture network characterization. *Journal of Structural Geology*, **72**, 55–66.

WENNBERG, O. P., AZIZZADEH, M. *ET AL*. 2007. The Khaviz Anticline: an outcrop analogue to giant fractured Asmari Formation reservoirs in SW Iran. *In*: LONERGAN, L., JOLLY, R. J. H., SANDERSON, D. J. & RAWNSLEY, K. (eds) *Fractured Reservoirs*. Geological Society, London, Special Publications, **270**, 23–42.

WHITTAKER, A. & GREEN, G. W. 1983. *Geology of the country around Weston-super-Mare*. Memoir of the Geological Survey of GB, Sheet 279 and parts of 263 and 295.

Developing digital fieldwork technologies at the British Geological Survey

COLM J. JORDAN* & BRUCE NAPIER

British Geological Survey, Environmental Science Centre, Nicker Hill, Keyworth, Nottingham NG12 5GG, UK

**Corresponding author (e-mail: cjj@bgs.ac.uk)*

Abstract: Geological Surveys are faced with budget constraints and calls for efficiency gains; the effective application of digital techniques is often seen as a route to meeting these demands while increasing the value of outcrop studies and reducing the inherent subsurface uncertainty. The British Geological Survey may be the oldest national Survey in the world (established in 1835); however, developing and implementing new, innovative and efficient technologies for fieldwork is a high priority. Efficient tools for capturing, integrating, manipulating and disseminating outcrop data and information are imperative to enable geoscientists to increase their understanding of geological processes and therefore to reduce subsurface uncertainty and risk. Systems for capturing structured digital field data and for visualizing and interacting with large datasets are increasingly being utilized by geoscientists in the UK and internationally. Augmented reality and unmanned aerial vehicles are amongst the developing technologies being explored for future operational implementation. This paper describes the digital field mapping (BGS·SIGMAmobile) and visualization (GeoVisionary) systems and refers to a case study outlining their contribution to reducing uncertainty and risk in hydrocarbon exploration.

Geological Surveys including the British Geological Survey (BGS) have primarily utilized conventional analogue mapping techniques until relatively recently. Analogue field techniques had changed little since the original days of fieldwork by pioneers such as William Smith, who created the geological map of Britain in 1815. Smith documented his field records in a notebook and marked symbols and linework on paper fieldslips in much the same way as mapping geologists did until very recently. The continuing prevalence of paper techniques was highlighted as recently as 2007 when the fourth edition of *Basic Geological Mapping* (Barnes & Lisle 2007) listed the field equipment that a geologist should possess, and described in detail the type of field notebook that is recommended, that is, 'it should have good quality "rain-proof" paper, a strong hard cover and good binding' whilst 'at least three pencils are needed'. There is no mention of digital field mapping techniques or the prospect of their introduction in the future, although current educators are trying to address this (e.g. England *et al.* 2010; Pavlis *et al.* 2010). BGS has endeavoured to develop and apply technological solutions to geoscience data and visualization challenges; systems such as BGS·SIGMAmobile and GeoVisionary are providing solutions to geoscientists for data acquisition, field mapping and data visualization.

Developing a digital field system

Digital field mapping has been an aim of Geological Surveys for many years (Brodaric 1997; Farrant *et al.* 2001; De Donatis & Bruciatelli 2006). The concept is that the process of acquiring digital data in the field would have several benefits including increased efficiency for capturing, manipulating, integrating, understanding and disseminating field and outcrop data. Specific gains should include: (a) ensuring obligatory data collection – guaranteeing that vital data (e.g. location information such as map grid reference) are collected either manually or automatically from a Global Positioning System (GPS); (b) standardized recording – making sure that the same sets of data are collected by each member of a field team at each outcrop; (c) using standardized nomenclature, for example, making use of drop-down menus where appropriate so that field teams are constrained to standard dictionaries such as the 'Munsell Color' chart for recording colour; (d) enabling inter-operability with other data, for example, structure data collected in the field can be used on-the-fly to create structure contours; and (e) the system should allow access to 'prior information' that results in 'smarter mapping', for example, aerial/satellite imagery, existing geological fieldslips and geophysics data (Jordan *et al.* 2005).

From: Bowman, M., Smyth, H. R., Good, T. R., Passey, S. R., Hirst, J. P. P. & Jordan, C. J. (eds) 2016. *The Value of Outcrop Studies in Reducing Subsurface Uncertainty and Risk in Hydrocarbon Exploration and Production*. Geological Society, London, Special Publications, **436**, 219–229.
First published online December 16, 2015, http://doi.org/10.1144/SP436.6

Furthermore, rather than simply replicating the tasks that can be completed with the analogue pen/paper/map routine, the digital system must offer additional functionality. Examples of this include the ability to enter a dip/strike measurement and to automatically create structure contours using an underlying Digital Terrain Model (DTM), or to provide the facility to compile all of the field data collected (text, measurements, photos etc.) into an MS Word report. Crucially, the field system must provide an efficiency gain. Time may not necessarily be saved at the outcrop; however, transferring validated digital field data directly to corporate databases allows more time for manipulation and modelling rather than transposing/digitizing, which takes resources at the office and can also potentially lead to errors.

In 2001, BGS published the user requirements for digital field data collection (Farrant *et al.* 2001). The specifications document outlined in broad terms what the system should collect for each mapping terrain. The main objective was to develop an integrated system that would not only collect point structural geological data, but would also collect the full range of data required by a Geological Survey, including Quaternary geology (Lee & Booth 2006), landform descriptions, landslide pro formas and photographs. The system should be constrained by drop-down menus or tick boxes where appropriate to save time, but must also provide the functionality to draw sketches and write 'free text' so that the process of field mapping is not constrained, some data are not collected or mapping becomes a 'box ticking' exercise rather than a meaningful scientific endeavour. Ultimately, it was vital that the system was developed with significant input from field geoscientists, who are the end users.

BGS first explored the concept of digital field data collection in 1989, with the conclusion that the mobile computing hardware at that time was not suitable. The development was therefore postponed. External reviews of BGS such as Walton & Lee (2001) recommended revisiting the prospect of digital field mapping and in the same year an international workshop on digital field data capture was hosted by BGS in the knowledge that other organizations such as the Geological Survey of Canada had also begun to explore digital field mapping (Brodaric 1997). North American and European Geological Surveys attended and presented the status of mapping systems, if they existed, in their countries. Similarly, software and hardware suppliers were invited to demonstrate their products. At that time, the available systems were capable of limited point data capture using personal digital assistants (PDAs) and, as this was not sufficient for BGS geoscientists, the organization set about designing and developing a bespoke system.

In 2001 BGS started testing Husky *fex21* hardware with the PocketGIS® software primarily for a granite and landscape mapping project in the Cairngorm Mountains of Scotland (Thomas *et al.* 2004). Challenges with the hardware (primarily related to the battery life and screen quality) resulted in a move to PDAs operating Windows CE™ in the same year. A customized version of ESRI ArcPad™ served as the front end, whilst a bespoke BGS eMbedded Visual Basic (eVB) application containing hard-coded data structure links in a compact database format was used to collect and hold additional relational data that would otherwise have been stored in the geoscientist's notebook. Hierarchical input forms were constructed to collect various levels of data; index level data were added for each field site and an 'Open Notebook' button gained access to more detailed forms for various mapping modules. At the time, similar systems were in development in the USA (Pavlis & Little 2001). The small screen size of the PDA (*c.* 6 × 8 cm) was sufficient to display a small map area along with the user's position, which was derived from a GPS grid reference that was served via a Bluetooth device. The field staff were equipped with a 'digital toolbox' containing the PDA, a Bluetooth GPS, a digital camera and various accessories.

Whilst this was a significant advance from a paper field system, the screen size was a major limiting factor. While it was arguably sufficient for point sample collection, feedback from the majority of field geologists stated that it was not suitable for geoscientists working with maps. The screen was too small to visualize enough of the mapface to gain spatial context, and furthermore, annotating the visible area of the maps with lines, polygons and text proved problematic because scrolling beyond the current view was required to delineate even the smallest of landscape features. Furthermore, each release of new PDA hardware (BGS was primarily using the Hewlett Packard iPAQ platform) brought Open CE updates, which often required time-consuming modifications to the eVB code. Clegg *et al.* (2006) came to the same conclusion when comparing tablet PC and PDA hardware, stating that the tablet devices were more suitable for a wide range of geological data collection tasks when using their 'MAP IT' software. Nevertheless, some organizations adopted the small screen size devices and still like them for their portability (Pavlis pers. comm. 2015).

Fortuitously, by 2002, the first rugged tablet PCs entered the market, and while the early incarnations were too expensive to equip field teams and often too heavy to carry for long periods

(>2.5 kg), they provided a solution to both the screen size and the operating system issues. BGS began experimenting with, and developing software for, tablet PCs in the expectation that the hardware would become more widely available, more affordable and more fit-for-purpose. By 2004 the software was migrated to a ruggedized tablet PC system operating on Microsoft XP for Tablet Edition. Training courses for staff in this new 'BGS·SIGMAmobile' (System for Integrated Geoscience Mapping) began in 2005 (Fig. 1). At this time, a similar digital field system (GeoMap) was also being developed in the USA on tablet PCs based on the Strata Software's PenMap (Birmhall *et al.* 2002).

BGS·SIGMAmobile is a heavily customized version of ESRI ArcMap as the front end with relational data held in a bespoke MSAccess2007 database. Additional functionality is provided by linking modified versions of InkWriter (software that enables handwriting recognition). Software development of BGS·SIGMAmobile system was done using a variety of languages including VBA and .NET. With each new release of ESRI software, BGS has also updated the field system and it currently operates with ArcMap10.1.

BGS·SIGMAmobile is an integrated field system that enables a broad array of geoscientific data to be recorded using tick boxes, sketches, drop-down lists, tagged free text and photographs where appropriate (Fig. 2). Spatial location and navigation are managed by built-in GPS whilst the stylus enables points, lines, polygons and comments to be added to the digital map face. As with the preceding PDA system, additional relational information is added using customized forms and a selection of interfaces. The system is modular, with tabs for various themes or domains of geological data such as structural readings, landslide information and auger/section recording. Furthermore, there are additional tools including the ability to draw sketches, annotate photographs, produce structure contours and navigate using bearings. All of the data collected in BGS·SIGMAmobile are tagged with a Unique User IDentifier, enabling them to be queried and tracked through corporate repositories.

It has been proposed that the preference in geological mapping is to interpret observations during the mapping process (Jones *et al.* 2004; McCaffrey *et al.* 2005); however, BGS·SIGMAmobile primarily records observations, separate from interpretations, in order to improve traceability in the derived outputs and to separate, where possible, observed data with interpreted information. It is expected that this increases confidence in the data and therefore reduces the uncertainty of the derived maps and models.

A choice of tools for adding text, lines and polygons to the map face is provided, ranging from a basic tool that replicates the pencil and paper routine through to tools that enable topologies and attributed lines to be created in the field. Advanced handwriting software is used extensively to deliver legible field notes (even on the map face); however, cursive text can still be used for rough notes where appropriate. Drop-down menus and tick boxes are used where possible and efficient to ensure that entries conform to accepted standards and that the agreed nomenclature is used. Areas for free-form text are also provided to allow flexibility. Novel systems have been developed and employed to ensure that the data recorded is unambiguous; for example, rather than asking a geologist to tick a box to note if they are using the right-hand rule when recording a structural measurement, a compass is provided that is ticked to identify the dip direction. This reduces confusion regarding how the data were recorded.

The tablet PC hardware was a challenge for the early releases of the system, for example, weight, screen visibility in bright light and shorter-than-advertised battery life were all problematic. However technology has advanced and platforms now exist that are generally suitable for use in the field by geoscientists. Following trials with a SunscreenPC in 2001, the favoured hardware was the Itronix GoBook and, subsequently, the GoBook DuoTouch followed by the Xplore iX104 series, the GETAC V100 and currently the Panasonic ToughPad. Pavlis *et al.* (2010) record a similar history of equipment trials. Non-rugged systems are also available, such as the Microsoft SurfacePro. In general the non-rugged systems are lighter and cheaper, but their screens are not as readable in variable light conditions, and they are more prone to every-day wear-and-tear and fatal damage from water/dust ingress and from drops and knocks. There are a range of ruggedness ratings, for example, IP54 or IP67, so there is a choice to make between the level of hardware resilience required and the cost the organization is willing to pay. There is an option to use a non-rugged system and to use a weather-proof case.

While BGS·SIGMAmobile was originally designed purely as a data collection tool, a significant part of its power and functionality comes from the ability to bring a wide range of data (e.g. satellite imagery, aerial photography, geophysical data, historic field slips and topographic maps) to the field. This has also been recognized elsewhere, for example, Carver *et al.* (1995), when a Geographic Information System (GIS) was used in the field. This improved access to data and information in the field significantly increases the knowledge-based decisions of the geologists, leading to reduced uncertainty in the data collected and the information derived from it.

Fig. 1. (**a**) Digital field mapping training course and (**b**) digital field mapping in the United Arab Emirates.

Aside from bedrock and Quaternary geologists, the ability to record features such as landslides, dimension stone information and mine information, has broadened the user base significantly (Jordan 2010*a*; Jordan & Pennington 2011; Evans *et al.* 2013).

In 2009 BGS·SIGMAmobile was released at no cost to both academic and commercial users, downloadable from the BGS website. The premise for free distribution was to promote its use and to encourage the growth of a developer community. The only stipulation prior to download is that new developments must be supplied to BGS for inclusion in subsequent free releases. This free release has led to the use of the system for teaching in university departments (e.g. England *et al.* 2010) and also by other Geological Surveys (e.g. Jordan 2010*b*; Henderson & Guilio 2011). Over 2000 licences have been downloaded worldwide. The system also gained recognition when it won the 2007 ESRI Central Government GIS Excellence Innovation Award.

The BGS·SIGMAmobile system was the sole component in the first two releases; however, the version released in 2013 integrated the BGS·SIGMAdesktop functionality that had previously

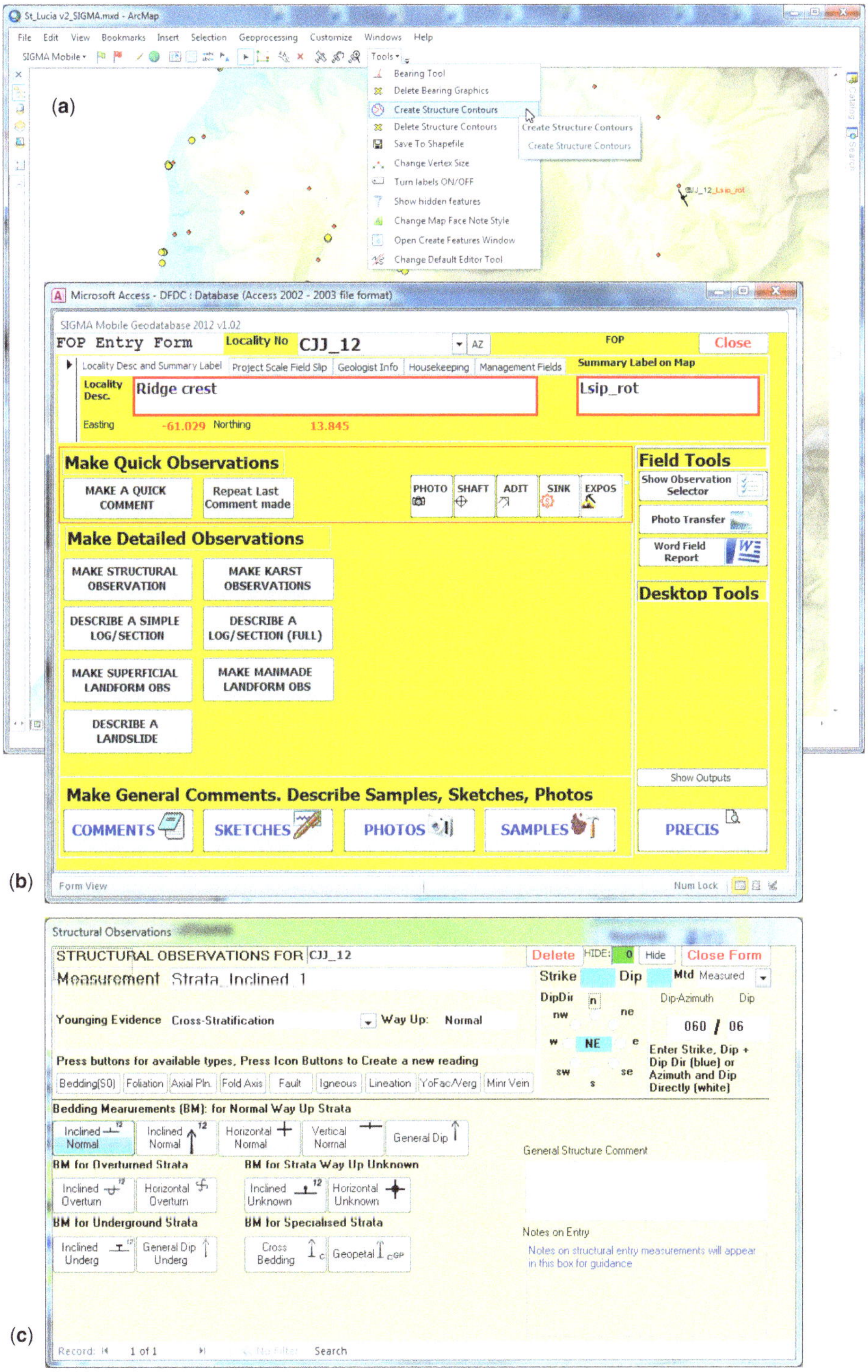

Fig. 2. Sample Graphical User Interface to BGS·SIGMAmobile. (**a**) ArcMap front end; (**b**) top-level forms for structured data capture; and (**c**) sub-form for collecting bedrock structural data. NEXTMap Britain elevation data from Intermap Technologies.

not been released outside BGS. This provides tools for routine transformation of field data into corporate standard geological models and derivative map outputs. Development of the system is still ongoing as a result of both user feedback and the changing face of technology. Investigations into the development of a BGS·SIGMA smartphone app are currently taking place alongside system developments such as a new and more streamlined data entry system.

UAV field data collection

A growing area of interest internationally is the capability of collecting geoscience field data using unmanned aerial vehicles (UAVs), also called unmanned aerial systems or remotely piloted aerial systems. It is argued here that UAVs can be considered as field data acquisition systems because the equipment can be taken to a field site in a standard vehicle (or in a rucksack, depending on the system) and generally operated by a geologist, with suitable safety and regulatory training. In this respect it is no different from routine geoscience instruments such as terrestrial laser scanning, and in fact the latest UAV systems incorporate laser scanning technology.

UAVs, in the form of parafoils, have been used in BGS since 1986, and in the last 10 years these have been added to with kites, fixed wing and rotary systems (Hobbs *et al.* 2010). The use of ground control and differential GPS ensures that calibrated and validated outputs can be delivered, including orthorectified aerial photography, point clouds, triangular irregular network models and gridded elevation models. Desktop software using structure for motion technology has put stereo aerial photography processing and DEM extraction in the hands of the masses. Elevation/motion and volume changes can be calculated, for example, for landslides, and BGS is now routinely using UAV technology to acquire multi-temporal data over landforms such as landslides (Fig. 3). RGB cameras are now the ‘elder statesmen’ of sensor systems on UAVs while multi-spectral and thermal cameras are becoming ubiquitous. Miniaturized sensors such as laser scanners, gas monitors and geophysical equipment are breaking into the commercial market and their systematic use will significantly expand the data collection opportunities available in the field at this scale.

Developing a virtual field reconnaissance system

The BGS Virtual Field Reconnaissance (VFR) project was developed to allow geologists to immerse themselves in a virtual landscape, providing the ability to ‘bring the field into the office’. Teams gather in an immersive virtual environment and discuss complex field outcrops, followed by fieldwork focused on addressing specific issues that have arisen in the office.

The initial challenge set for the VFR project was to build on existing project-based virtual field trip and geoscience visualization applications that had been created by BGS during and after the Digital Geoscience Spatial Model programme from 2000 to 2004 (Riddick *et al.* 2005), and to create systematic efficiency gains in fieldwork. Primarily, this would be achieved through use of newly acquired national high-resolution datasets, such as the Nextmap Great Britain 5 m Digital Surface Model and 5 m Digital Terrain Model, and aerial photography, along with the wealth of digital geological data held by the BGS. A Virtalis Activewall single channel active stereo visualization system, known in BGS as the immersive 3 Dimensional Visualization Facility (i3DVF) is used by teams while the system can also be used by individuals on their PCs (or a suitably equipped laptop) to ensure that virtual fieldwork is available to all BGS geoscientists.

In a review of existing software with 3D visualization capability such as ESRI, Google and NASA, or 3D geological modelling software like GOCAD and GSI3D, BGS staff decided that no single software package could meet the user requirement for a BGS VFR system. Essential elements of the user requirement that could not all be addressed by any one of those software packages included handling the volume of data and graphics output on PC workstations, ease of use, use in the i3DVF, interaction with the virtual landscape and integration with BGS·SIGMA. BGS’s solution was to work with Virtalis Ltd, who adapted their engineering model visualization software to work with geoscience datasets and allow the user to interact with them to the VFR specification. The initial pilot version was judged a success so the project was continued, and since 2007 BGS and Virtalis Ltd have worked together to create commercially available software for visualization and interpretation of large geospatial datasets from multiple sources. That software, GeoVisionary, which was first released in July 2008, allows users to visualize terabytes of surface and subsurface data in high-powered immersive 3D visualization systems, as well as on desktop PCs and laptop workstations.

GeoVisionary provides tools for digitizing points, poly lines and polygons that allow the user to map geological features limited only by the resolution of the terrain model and imagery. Lighting angles can be changed to help identify features from the virtual terrain. Structural measurements from oriented planes can be drawn in three

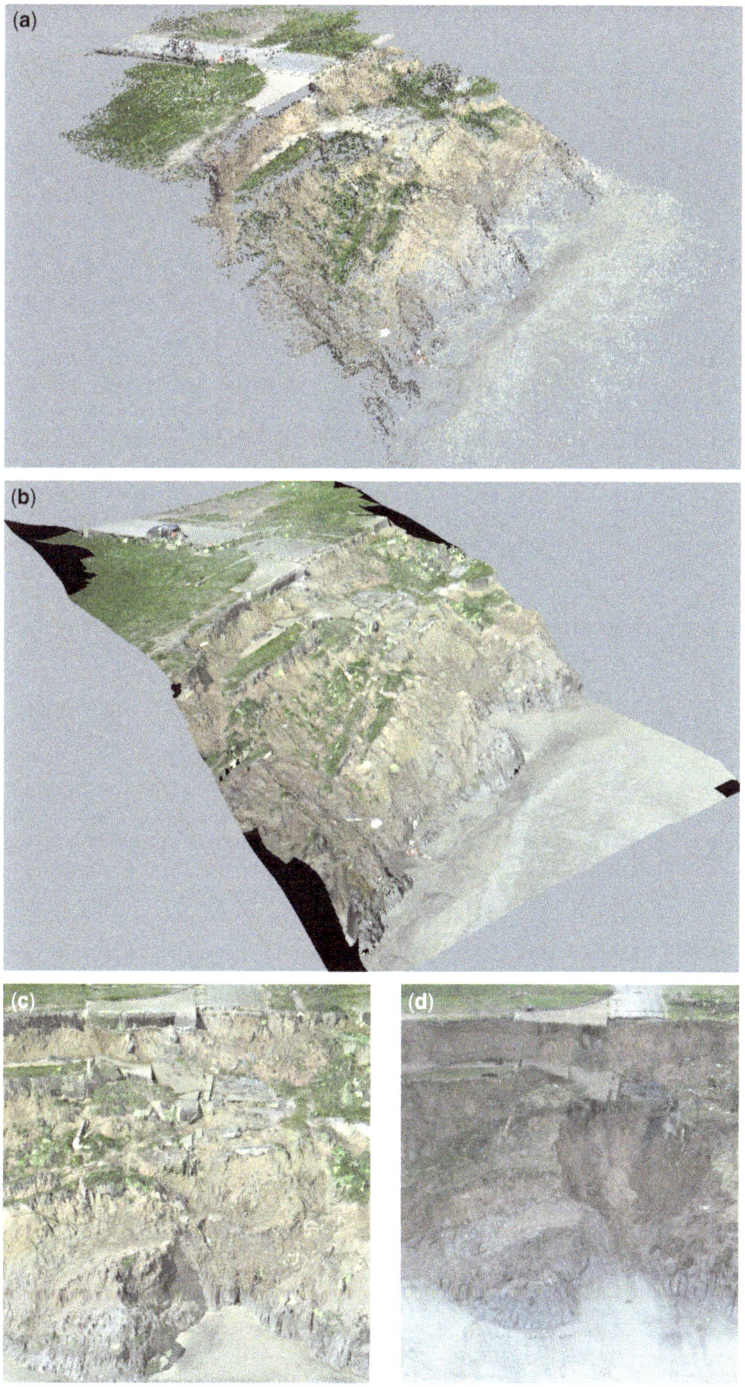

Fig. 3. Outputs derived from BGS UAV photography. (**a**) Point cloud; (**b**) textured DEM derived; (**c**) 3D model of coastal landslide in 2013; and (**d**) 3D model of same coastal landslide in 2014.

dimensions, calculated from three points picked from the terrain model. The user can compare the existing geological interpretation of an area with that gained from the virtual environment and decide whether or not they agree with that interpretation and therefore make better decisions on where to target field work. All of the data collected in GeoVisionary can be saved as ESRI 3D shapefiles for use in GIS and 3D modelling software.

BGS created custom software for ESRI ArcMap, the Arc2GV Toolbar, which links GeoVisionary on the i3DVF PC and BGS·SIGMA on a tablet PC. Location data, sent wirelessly from GeoVisionary, is used by BGS·SIGMA to match the 2D GIS view with the virtual landscape. The data collected in the virtual field environment is immediately transferred to the BGS·SIGMA device with the Arc2GV Toolbar and can be taken directly to the field. On return from fieldwork, the Arc2GV link is restored and the newly collected data from the field can be visualized and interpreted in the i3DVF. Using virtual reality, fieldwork and GIS together in this way, has been shown to bring better-quality results from time spent in the field, increase the accuracy of interpretation and help build better team understanding, communication and confidence (Ford *et al.* 2013; Calazans *et al.* 2014). A degree of computer literacy is required by the users of the system, and it is strongly advised to back up one's work in case of equipment failure or loss.

Virtalis have developed a streaming data engine, fully utilizing the latest graphics card technology from nVidia that has helped to overcome one of the biggest technological problems in geoscience visualization: how to smoothly visualize huge data volumes of multiple resolution data in a convincing virtual reality environment. It goes a long way towards answering many of the problems with multi-scale geoscience model visualization identified by Jones *et al.* (2009). In 2012 visualization of LiDAR point cloud data and volumetric (voxel) models was added to the GeoVisionary functionality list. The volume and density of terrestrial LiDAR have rapidly increased in recent years. In response, the point cloud capability was increased in 2014, enabling visualization and interpretation (measurement, digitization, structural measurement) of billions of points, simultaneously with all of the other data in a single GeoVisionary project (Fig. 4).

Case studies

Published case studies describe the use of the BGS digital systems for field mapping in terrains such as the UK (Evans *et al.* 2013; Leslie *et al.* 2014) Ghana, Madagascar and the United Arab Emirates (Jordan 2010*a*) and Norway (Henderson & Guilio 2011), and petroleum exploration in Tajikistan (Jordan *et al.* 2009). These studies have highlighted how the systems have been applied to various BGS mapping projects and also by geologists from other organizations. Experience from the case studies has demonstrated that time at the outcrop is often limited (owing primarily to cost) and it is desirable to derive the most value from fieldwork, for example, by (a) having as much appropriate ancillary data to hand as possible to promote more informed decision-making in the field, (b) ensuring that geologists collect the full suite of mandatory data at each outcrop, (c) standardizing the nomenclature that is used and (d) providing on-the-fly functionality such as deriving structure contours. All of these factors contribute significantly to reducing uncertainty in the decisions made at the outcrop. The BGS studies also highlight that a well-documented workflow is a prerequisite in order to ensure (a) adequate preparation of data prior to fieldwork, because it is often not possible to have data sent to the field area if it has been forgotten, (b) staff fully trained in the use of the systems and (c) protocols for data transfer, manipulation and long-term management/storage.

It has also been demonstrated that the large amount of data now available to geologists using digital techniques in the field, along with the capacity to collect new structured digital data, makes the 'field mapping process much more efficient and increases the reliability and repeatability of collected data' (Pavlis *et al.* 2010).

The case studies above emphasize the impact that new technologies have made to geological mapping, and how they have contributed to reducing uncertainty. The Tajikistan case study (Jordan *et al.* 2009) specifically relates to outcrop studies for petroleum exploration undertaken by Tethys Petroleum and BGS when a Production Sharing Contract was signed and a short timescale was available to start exploration drilling in the 40 000 km^2 Bokhtar area. GeoVisionary was used to compile a 3D model using existing conventional oil company data in the area, consisting of mainly elderly Soviet era geological maps, well logs and very sparse dubious seismic, all on paper. Tethys had 18 months for the initial phase of geological studies, seismic acquisition and reprocessing, and field rehabilitation trials, with exploration drilling in the second 18 month phase and a first relinquishment after 7 years. Remote sensing data were acquired and analysed in order to study the large remote area efficiently. They were used to plan field geology and seismic acquisition in order to complete the first phase of the exploration programme on time. A model was built in GeoVisionary from Landsat images, SRTM and DTM data, and loaded

Fig. 4. A terrestrial LiDAR scan of a cliff, consisting of more than 300 million points. The red and green and cubes show where a point has been selected for digitization as a poly-line. At the top of the image, distance, incline and bearing are shown as readings from the GeoVisionary Terrain Measuring Tool.

into the 3D visualization facility for stereo viewing by the team of BGS and Tethys geologists. The geological and structural model was further improved using higher spatial and spectral resolution ASTER satellite imagery. Cross-sections were prepared and seismic and well logs incorporated where appropriate.

A reconnaissance field trip assessed the quality of this remote work and identified areas of specific interest for the remainder of the exploration work programme. As part of the reconnaissance, seismic lines were planned and the routing was checked in the field using BGS·SIGMAmobile. Outcrops encountered in the field were recorded in the digital system, and information from them was fed back into the 3D model. The combination of the pre-field 3D visualization and the digital field data allowed Tethys to conduct their exploration on schedule and to plan the seismic campaign with confidence (Jordan *et al.* 2009).

It is fitting to re-evaluate the Tajikistan case study in light of new technologies and to consider what might be done differently now. Firstly, UAV technology was not widely available in 2008 and therefore they were not incorporated into the project. The technology is readily available today to collect a suite of site-specific high-resolution data including stereo aerial photography, thermal, hyperspectral, geophysical and LiDAR data. The field tablet PCs have reduced in weight and increased in processing and graphics power, so not only can they be used to collected data, but those data can be visualized at the outcrop using augmented reality such as iGeology 3D (http://www.bgs.ac.uk/igeology/3d.html). 3D visualization systems, such as GeoVisionary, are now able to incorporate a wider range of datasets such as 3D point clouds and multi-scalar DTMs, enabling the geologist to add more detail to the mapping and further reduce the outcrop uncertainty. It is debatable whether the digital field mapping systems have encountered a step change in technology since 2008; however, the user interfaces are more streamlined, the systems are more stable (hardware and software), and the protocols to prepare and manage the data are more complete. The range of visualization and digital mapping systems now on the market is testament to their increased integration into routine mapping.

Discussion and conclusions

The strategy for most Geological Surveys (including BGS) has been to develop and implement digital systems that increase our understanding of the subsurface and to move from printing paper geological maps to delivering focused outputs such as 3D and 4D models. Validated digital field data capture provides a streamlined route for populating corporate databases, from which an array of outputs can be delivered, including paper/digital maps, 3D models and smartphone applications.

The culture of geoscience field mapping has changed in Geological Surveys, and the introduction of digital field systems has had a large input to this (Jordan *et al.* 2008). It is generally accepted that field mapping in Geological Surveys encompasses, and benefits greatly from, digital techniques (Leslie *et al.* 2014; McCaffrey *et al.* 2005), and students are also benefiting from structured digital techniques (England *et al.* 2010; Pavlis *et al.* 2010). Systems developed in BGS are helping to integrate and collect complex digital data in the field and subsequently to transfer those data immediately and efficiently to corporate databases, thereby making them instantaneously available for downstream uses such as 3D modelling (Henderson & Guilio 2011).

Visualization systems are also bringing field sites into the office and ensuring best use of geologists' time through virtual field reconnaissance prior to and post fieldwork. Time spent waiting for computers to load/transfer spatial data for geographic areas of interest has been reduced; GeoVisionary has advanced geoscience visualization technology by placing large volumes of data in the hands of the user in near real time. Resources are freed to focus on visualizing and interpreting huge volumes of raster and vector data in a single environment rather than transferring them from archives or servers to visualize in separate software systems in i3DVF or on desktop PC workstations.

Digital data collection and visualization have also been put into the hands of the public, for example, through smartphone applications such as *my*Volcano and *i*Geology (http://www.bgs.ac.uk/igeology/3d.html). Furthermore, the *i*Geology 3D application is an augmented reality system that projects 3D geology onto the smart phone screen, overlaid onto the landscape via the camera on the device. This new level of interaction with the data is currently used to promote geoscience to the public, but can equally be used by professionals as an additional knowledge tool to decrease uncertainty at the outcrop.

The breadth of digital tools being made available to field geoscientists by a wide range of providers internationally is significant; for example, 10 years ago there were few integrated digital geological capture systems that had the functionality of point data recording and polygon mapping tools with an underlying relational database. Some of the credit for the arrival of digital mapping systems goes to the timely delivery to market of the rugged tablet PC; however, the availability of the hardware is more than balanced by the foresight of those who developed software in the expectation that these types of hardware would become available.

Modern Geological Surveys also routinely utilize systems such as UAVs, although the differing levels of sensor use are still stark, for example, the contrast between a basic digital camera and a laser scanner. These systems are now delivering truly valuable data and their use is predicted to proliferate, although care must be taken to ensure that the systems are operated safely and that the results are calibrated and validated. Looking to the future, there is scope to further streamline the input systems; voice recognition is still under-used and the day will come when geoscientists will be able to verbally describe the outcrop and a digital system will tag the words and automatically populate the database, symbolize a map and deliver the data back to base, where it can be used instantaneously.

Sincere thanks are due to the dedicated team of GIS and database developers, for building the systems described here, and to Virtalis Ltd for co-developing the GeoVisionary system. The constructive feedback from field geoscientists and the support of management throughout development and implementation are gratefully acknowledged. We acknowledge E. Haslam and P. Hobbs for UAV photo acquisition and L. Jones for TLS data acquisition. The authors thank two reviewers for their constructive comments. This paper is published with the permission of the Executive Director, British Geological Survey (NERC).

References

Barnes, J. W. & Lisle, R. J. 2007. *Basic Geological Mapping*, 4th edn. John Wiley and Sons, Chichester.

Birmhall, G. H., Vanegas, A. & Lerch, D. 2002. GeoMapper program for paperless field mapping with seamless map production in ESRI ArcMap and GeoLogger for drill-hole data capture: applications in geology, astronomy, environmental remediation, and raised-relief models. *In*: Soller, D. R. (ed.) *Digital Mapping Techniques 2002*. US Geological Survey Open-file Report **02-370**, http://pubs.usgs.gov/of/2002/of02-370

Brodaric, B. 1997. Field data capture and manipulation using GSC FIELDLOG v3.0. *In*: Soller, D. R. (ed.) *Digital Mapping Techniques: Methods for Geologic Map Capture, Management and Publication*. US Geological Survey, Open File report, 77–81.

Calazans, P. P., Büchi, A., Napier, B., Reinhardt, P. M., Orlandi, P., Dias, L. S. O. 2014. Sistema Integrado de Mapeamento Geológico (SIGMA) com aplicação de realidade virtual e captura de dados digitais em campo. *In*: *SIMEXMIN, 2014, Ouro Preto – MG*. SIMEXMIN, ADIMB, Brazil, 2014.

Carver, S., Heywood, S. C. & Sear, D. 1995. Evaluating field-based GIS for environmental modelling, characterization and decision support. *International*

Journal of Geographical Information Systems, **9**, 475–486.

CLEGG, P., BRUCIATELLI, L., DOMINGOS, F., JONES, R. R., DE DONATIS, M. & WILSON, R. W. 2006. Digital geological mapping with tablet PC and PDA: a comparison. *Computers & Geosciences*, **32**, 1682–1698.

DE DONATIS, M. & BRUCIATELLI, L. 2006. MAP IT: the GIS software for field mapping with a tablet PC. *Computers & Geosciences*, **32**, 673–680.

ENGLAND, R., SMITH, S., TATE, N. & JORDAN, C. J. 2010. *Digital Geological Mapping for Earth Science Students*. EGU 2010–2225. EGU General Assembly, Vienna, Austria, **12**.

EVANS, H., PENNINGTON, C., JORDAN, C. J. & FOSTER, C. 2013. Mapping a nation's landslides: a novel multistage methodology for landslide mapping and inventory within the UK. *In*: MARGOTTINI, C., CANUTI, P. & SASSA, K. (eds) *Landslides Science and Practice*, Springer, Berlin, Special Publications, **1**, 21–27, http://doi.org/10.1007/978-3-642-31325-7_2

FARRANT, A. R., JORDAN, C. J. & THOMAS, C. W. 2001. *A User Requirement for Land Survey Digital field Data Entry*. British Geological Survey Internal Report, **IR/01/137**.

FORD, Jon, DEARDEN, Rachel, NAPIER, Bruce & KESSLER, Holger. 2013. 3-D visualisation at the British Geological Survey. [Lecture.] In: *GSA, Denver Annual Meeting, Denver, USA, October 2013*. Geological Society of America, http://nora.nerc.ac.uk/503929/

HENDERSON, I. H. C. & GUILIO, V. 2011. Implementering av digital berggrunnskartlegging ved Norges Geologiske Undersøkelse – NGU. *Mineral Produksjon*, **1**, 83–101.

HOBBS, P. R. N., HASLAM, E. & JORDAN, C. J. 2010. Applications for UAVs at the British Geological Survey. *In*: *Proceedings of RSPSoc Annual Conference*, Cork.

JONES, R. R., MCCAFFREY, K. J. W., WILSON, R. W., & HOLDSWORTH, R. E. 2004. Digital field acquisition: towards increased quantification of uncertainly during geological mapping. *In*: CURTIS, A. & WOOD, R. (eds) *Geological Prior Information*. Geological Society, London, Special Publications, **239**, 43–56.

JONES, R. R., MCCAFFREY, K. J. W. *ET AL*. 2009. Integration of regional to outcrop digital data: 3D visualisation of multi-scale geological models. *Computers & Geosciences*, **35**, 4–18.

JORDAN, C. J. 2010*a*. BGS·SIGMAmobile; the BGS Digital Field Mapping System in Action. *In*: SOLLER, D. R. (ed.) *Digital Mapping Techniques 2009, May 10–13 Morgantown West Virginia*. US Geological Survey Open-file Report, **2010-1335**, http://pubs.usgs.gov/of/2010/1335

JORDAN, C. J. 2010*b*. A new era for digital geological mapping with the 2nd release of the BGS digital mapping system. *Geoconnexion: International*, November, 9, 44–35.

JORDAN, C. J. & PENNINGTON, C. 2011. Digital field systems for landslides research; an insight from the latest British Geological Survey (BGS) product development. *Geoconnexion: UK*, **9**, January/February, 8–9.

JORDAN, C. J., BEE, E. J., SMITH, N. A. S., LAWLEY, R. S., FORD, J., HOWARD, A. S. & LAXTON, J. L. 2005. The development of digital field data collection systems to fulfil the British Geological Survey mapping requirements. *In*: *Proceedings of the International Association of Mathematical Geology 2005: GIS and Spatial Analysis*, Toronto, **2**, 886–891, http://nora.nerc.ac.uk/8434

JORDAN, C. J., HOWARD, A. S. & BEE, E. 2008. The British Geological Survey digital field data capture system: better than pen and paper? *In*: *Proceedings of the 33rd International Geological Congress*, Oslo, http://ww.cprm.gov.br/33IGC/1306002.html

JORDAN, C. J., BATESON, L., BOW, J., NAPIER, B. & JOHNSON SABINE, R. 2009. GeoVisionary™ software for 3D visualisation and petroleum exploration in southern Tajikistan. *In*: *RSPSoc 2009 Annual Conference and Proceedings*, Leicester, UK, http://nora.nerc.ac.uk/8854

LEE, J. R. & BOOTH, S. J. 2006. *Quaternary field mapping: Lowland Britain*. British Geological Survey Internal report, **IR/06/099**, http://nora.nerc.ac.uk/7401/1/IR06099.pdf

LESLIE, G., SMITH, N. & JORDAN, C. J. 2014. Digital field mapping with the British Geological Survey. *Geophysical Research Abstracts*, **16**, EGU 2014–4172.

MCCAFFREY, K. J. W., JONES, R. R. *ET AL*. 2005. Unlocking the spatial dimension: digital technologies and the future of geoscience fieldwork. *Journal of the Geological Society, London*, **162**, 927–938.

PAVLIS, T. L. & LITTLE, J. 2001. Using handheld personal computers as field data collection tools: some lessons learned in the school of hard knocks in the Wingate Wash project and related projects using Fieldlog/Field-worker software exported to ArcInfo. *In*: SOLLER, D. R. (ed.) *Digital Mapping Techniques '01 Workshop Proceedings*. US Geological Survey Open-File Report, **01-233**, 115–122, http://pubs.usgs.gov/of/2001/of01-223/

PAVLIS, T. L., LANGFORD, R., HURTADO, J. & SERPA, L. 2010. Computer-based data acquisition and visualization systems in field geology: results from 12 years of experimentation and future potential. *Geosphere, 2010*, **6**, 275–294. http://doi.org/10.1130/GES00503.1

RIDDICK, A., LAXTON, J. *ET AL*. 2005. *Digital geoscience spatial model project final report*. Keyworth, British Geological Survey (BGS Occasional Publication 9).

THOMAS, C. W., GILLESPIE, M. R., JORDAN, C. J. & HALL, A. M. 2004. *Geological structure and landscape of the Cairngorm Mountains*. Scottish Natural Heritage Commissioned Report **064** (ROAME No. F00AC103).

WALTON, G. & LEE, M. K. 2001. *Geology for our diverse economy: Report of the Programme Development Group for Onshore Geological Surveys*. Keyworth, British Geological Survey (NERC).

Linking outcrop analogue with flow simulation to reduce uncertainty in sub-surface carbon capture and storage: an example from the Sherwood Sandstone Group of the Wessex Basin, UK

ANDREW J. NEWELL[1]* & SEYED M. SHARIATIPOUR[2]

[1]*British Geological Survey, Maclean Building, Wallingford, Oxfordshire OX10 8BB, UK*

[2]*Coventry University, Priory Street, Coventry CV1 5FB, UK*

**Corresponding author (e-mail: ajn@bgs.ac.uk)*

Abstract: Modelling the behaviour of carbon dioxide (CO_2) injected into sub-surface reservoirs as part of carbon capture and storage (CCS) strategies is often performed using models that incorporate very limited geological detail, particularly at the subseismic (metre to decametre) scale. Those modelling studies that incorporate varying degrees of geological realism show the inherent risks and uncertainties that can result from neglecting heterogeneity and reservoir–caprock topography along the migration path of an injected CO_2 plume. A key problem is that detailed geological data are often not available for the relatively deep saline aquifers that are an important target for CCS. Deep saline aquifers fall between the relatively data-rich environments of shallow freshwater aquifers and hydrocarbon reservoirs and it is in these settings that outcrop analogues may play an important part in reducing the risks and uncertainties associated with CCS. This study uses an example from the Sherwood Sandstone Group (Otter Sandstone Formation) of the Wessex Basin to show how an outcrop study can impart a much greater understanding of heterogeneity in critical reservoir–caprock zones. Here the transition from the Sherwood Sandstone Group (fluvial sandstone reservoir) to the Mercia Mudstone Group (playa lacustrine mudstone seal) is not simple, but includes a major change in fluvial style that introduces considerable heterogeneity at the top of the reservoir. The study shows how laser-scanned outcrop can be used to rapidly construct static geological models that are taken through to flow simulations. In combination, the use of appropriate outcrop analogues and flow modelling can reduce the risks and uncertainties associated with CCS.

Capturing carbon dioxide (CO_2) from anthropogenic sources and containing it in sub-surface geological formations is one option for the reduction of greenhouse gas emissions into the atmosphere (IPCC 2005; Boot-Handford *et al.* 2014). The world's first commercial-scale carbon capture and storage (CCS) facility is now operational at SaskPower's Boundary Dam coal-fired power plant in Estevan, Saskatchewan, Canada (Watson 2014). A variety of sub-surface storage scenarios are being considered around the world, including operational or depleted oil and gas reservoirs, deep saline aquifers and unmineable coal seams (Bachu *et al.* 2007). Secure storage of CO_2 is achieved through a combination of physical and chemical trapping mechanisms that are effective over different timeframes and scales (Holloway 2009). Aside from one operational plant, CCS is still at an early stage of development with only a few large-scale field trials worldwide (Hosa *et al.* 2011). Moreover, none of these practical trials involves volumes, rates and time scales that approach the anticipated requirements for CCS at an industrial scale (Nilsen *et al.* 2012). Model-based forecasts are therefore critically important in understanding the risks and uncertainties associated with injecting CO_2 into the sub-surface (Mackay 2013).

Modelling the behaviour of CO_2 injected into sub-surface reservoirs is often performed using models that incorporate very limited geological detail, particularly at the subseismic (metre to decametre) scale (Nilsen *et al.* 2012). However, those modelling studies that incorporate varying degrees of geological realism show the inherent risks and uncertainties that can result from neglecting heterogeneity and reservoir–caprock topography along the migration path of an injected CO_2 plume (Lindeberg & Bergmo 2003; Nilsen *et al.* 2012; Goater *et al.* 2013; Shariatipour *et al.* 2013). A key problem is that detailed geological data are often simply not available for the relatively deep saline aquifers that are an important target for the industrial-scale sequestration of CO_2 (Holloway 2009). Deep saline aquifers fall between the relatively data-rich environments of shallow freshwater aquifers and hydrocarbon reservoirs and it is in these settings that outcrop analogues (Hodgetts 2013; Howell *et al.* 2014) may play an important part in

From: Bowman, M., Smyth, H. R., Good, T. R., Passey, S. R., Hirst, J. P. P. & Jordan, C. J. (eds) 2016. *The Value of Outcrop Studies in Reducing Subsurface Uncertainty and Risk in Hydrocarbon Exploration and Production*. Geological Society, London, Special Publications, **436**, 231–246.
First published online January 13, 2016, http://doi.org/10.1144/SP436.2

reducing the risks and uncertainties associated with CCS.

This study uses an example from the Sherwood Sandstone Group (Otter Sandstone Formation) of the Wessex Basin to show how an outcrop study can improve our understanding of how rock heterogeneity controls the migration and trapping of buoyant CO_2 plumes injected into dipping saline aquifers. Here the transition from the Sherwood Sandstone Group (fluvial sandstone reservoir) to the Mercia Mudstone Group (playa lacustrine mudstone seal) is not simple, but includes a major change in fluvial style that introduces considerable heterogeneity at the top of the reservoir. The study shows how a laser-scanned outcrop can be used to rapidly construct static geological models that are taken through to flow simulations. In combination, the use of carefully selected outcrop analogues and flow modelling may reduce the risks and uncertainties associated with CCS in frontier dipping saline aquifers.

Mechanisms of sub-surface CO_2 storage

Dipping saline aquifers v. structural traps

Conventional structural traps, which generally comprise a porous reservoir rock folded or faulted into a structural high and covered by an impermeable seal, are probably the most obvious way to trap buoyant CO_2 injected into a sub-surface geological formation (Fig. 1) (Williams *et al.* 2013). Conventional structural traps that are operational or former hydrocarbon reservoirs are particularly attractive because of an abundance of pre-existing geological data, the potential for economic benefits from enhanced oil or gas recovery and their proven sealing capacity, subject to injected CO_2 not exceeding the interfacial threshold pressure for caprock breakthrough or reaching a reservoir spill point (Holloway 2009; Watson 2014). Dipping saline aquifers with no large-scale structural closure (Fig. 1) are a less obvious target for the sub-surface storage of CO_2 although, simply because of their great abundance in sedimentary basins throughout the world, they are potentially more important than conventional structural traps (Holloway 2009; Ogawa *et al.* 2011). In dipping saline aquifers, it is envisaged that long-term storage is largely achieved through the dissolution of injected CO_2 into the pore fluid and the development of an irreducible 'residual' CO_2 saturation in pore spaces as the plume migrates through the aquifer (Holloway 2009). Counter-intuitively, although dipping aquifers are open structures, they are arguably a more secure form of storage than closed structural traps, because in the latter the risk will always remain that free CO_2 could escape from the reservoir if the seal is breached (Mackay 2013). However, a problem

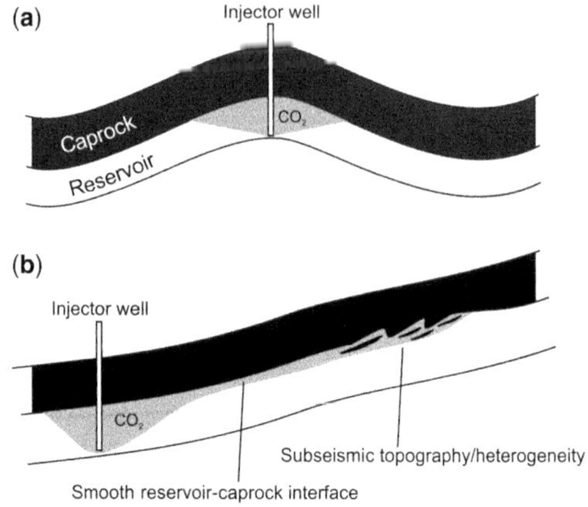

Fig. 1. Schematic diagram showing two potential CO_2 storage mechanisms: (**a**) injection into an aquifer or hydrocarbon reservoir that is sealed by a caprock and has a large-scale structural closure; and (**b**) injection into a dipping open aquifer where much of the long-term storage results from dissolution of the CO_2 into the pore fluids or by residual trapping. Small-scale structural and stratigraphic traps ('bumps') along the migration path of the CO_2 plume may have a significant impact on storage as the buoyant plume migrates upwards along the interface (Goater *et al.* 2013).

common to most dipping saline aquifers is that hard geological data from boreholes, seismic and other sources are sparse, particularly in the critical zone between the reservoir and the caprock. The morphology of this interface is one of the key areas of uncertainty in CCS modelling studies, particularly regarding the amount of structural trapping that might be anticipated and the general long-term migration of the plume (Nilsen *et al.* 2012; Shariatipour *et al.* 2013). Geological features controlling the morphology of the reservoir–seal interface are commonly only included in reservoir models if they are resolvable in seismic surveys (Nilsen *et al.* 2012). This is likely to exclude much of the small-scale (metre to decametre) topography and heterogeneity created by sedimentary depositional and erosional processes.

The importance of the reservoir–caprock interface

CO_2 is mobile and once injected it will typically migrate over large distances before becoming fully trapped. Movement is primarily driven by its greater buoyancy relative to the pore waters and, once injected, generally as a supercritical fluid, it will tend to accumulate and migrate up-dip as thin layers beneath the caprock (Fig. 1). Migration will cease once all of the CO_2 has been structurally trapped, dissolved, left as an irreducible fraction in the pore space or chemically reacted with the host rock (Bachu *et al.* 2007). Until this immobile state has been achieved, the morphology of the reservoir–caprock interface can have an important influence on the migration of the plume. Modelling studies have shown that minor structural and stratigraphic traps ('bumps') and sedimentary heterogeneity along the migration path of the CO_2 can have a significant effect on the amount of physical trapping and the general form and migration distance of the plume (Bachu *et al.* 2007; Pickup *et al.* 2011; Goater *et al.* 2013; Shariatipour *et al.* 2013).

The importance of the Sherwood Sandstone Group to UK CCS

In southern Britain, Triassic rocks are predominantly terrestrial deposits that accumulated within a series of linked rift basins which connected southern sediment sources in the Variscan fold-belt with major sediment sinks in the East Irish Sea and Southern North Sea basins to the north (McKie & Williams 2009; Morton *et al.* 2013; Fig. 2). Excellent storage prospects for CO_2 within the Triassic basin-fill are created by an upward transition in the Triassic stratigraphy from the reservoir sands of the Sherwood Sandstone Group (and Bunter Formation) at the base, into sealing mudstones and evaporites of the Mercia Mudstone (and Haisborough) groups (Fig. 3). The Triassic Sherwood Sandstone Group, and broadly correlative units such as the Bunter Formation in the Southern North Sea (Williams *et al.* 2013), are some of the key formations being considered for CO_2 storage in the UK (Armitage *et al.* 2013). The switch from reservoir sandstone to sealing mudstone is complex, both in terms of its temporal distribution and in the sedimentological architecture of the transition interval. Biostratigraphy and magnetostratigraphy show that the Sherwood Sandstone Group–Mercia Mudstone Group boundary is strongly diachronous, becoming younger southwards as coarse-grained fluvial systems retreated towards their sources in the Variscan mountain belt (Hounslow & McIntosh 2003).

Throughout much of the sub-surface, particularly in areas of saline aquifers, data on the Sherwood Sandstone Group are sparse and here outcrop analogues may have a considerable significance in terms of gaining greater understanding of sub-seismic heterogeneity and reservoir–caprock morphology. In this paper we examine some large-scale outcrop of the Triassic Sherwood Sandstone Group (Otter Sandstone Formation) in the western part of the Wessex Basin in south Devon, England (Fig. 2). The paper aims to show how outcrop can improve our understanding of sedimentary heterogeneity at reservoir–caprock interfaces beyond what might be achieved by using sparse sub-surface datasets alone. It also aims to demonstrate how modern field tools such as terrestrial laser scanners make it a relatively straightforward task to extend field

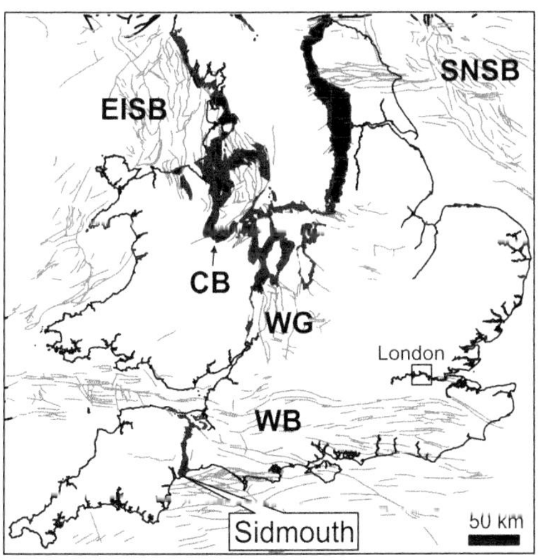

Fig. 2. Map showing the distribution of the major Triassic basins in southern onshore Britain and adjacent offshore areas: WB (Wessex Basin), WG (Worcester Graben), CB (Cheshire Basin), EISB (East Irish Sea Basin) and SNSB (Southern North Sea Basin). Mesozoic faults are shown as grey lines (Pharaoh *et al.* 1996) and the outcrop of the Sherwood Sandstone Group as black shading. The location of the study area is at Sidmouth in SW England.

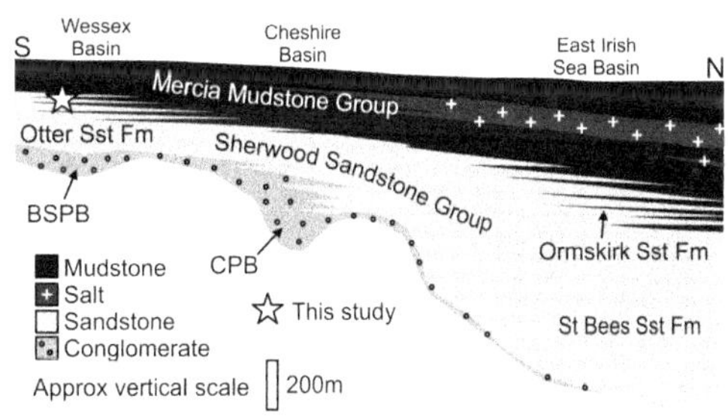

Fig. 3. Schematic Triassic stratigraphy and lithofacies of southern Britain (after McKie & Williams 2009). In general terms gravelly and sandy fluvial systems of the Sherwood Sandstone Group flowed to the north while the diachronous transgression of Mercia Mudstone Group playa lacustrine environments was towards the south. The stratigraphic position of this study at the contact between the Otter Sandstone Formation and the Mercia Mudstone Group is shown by a star. BSPB, Budleigh Salterton Pebble Beds; CPB, Chester Pebble Beds.

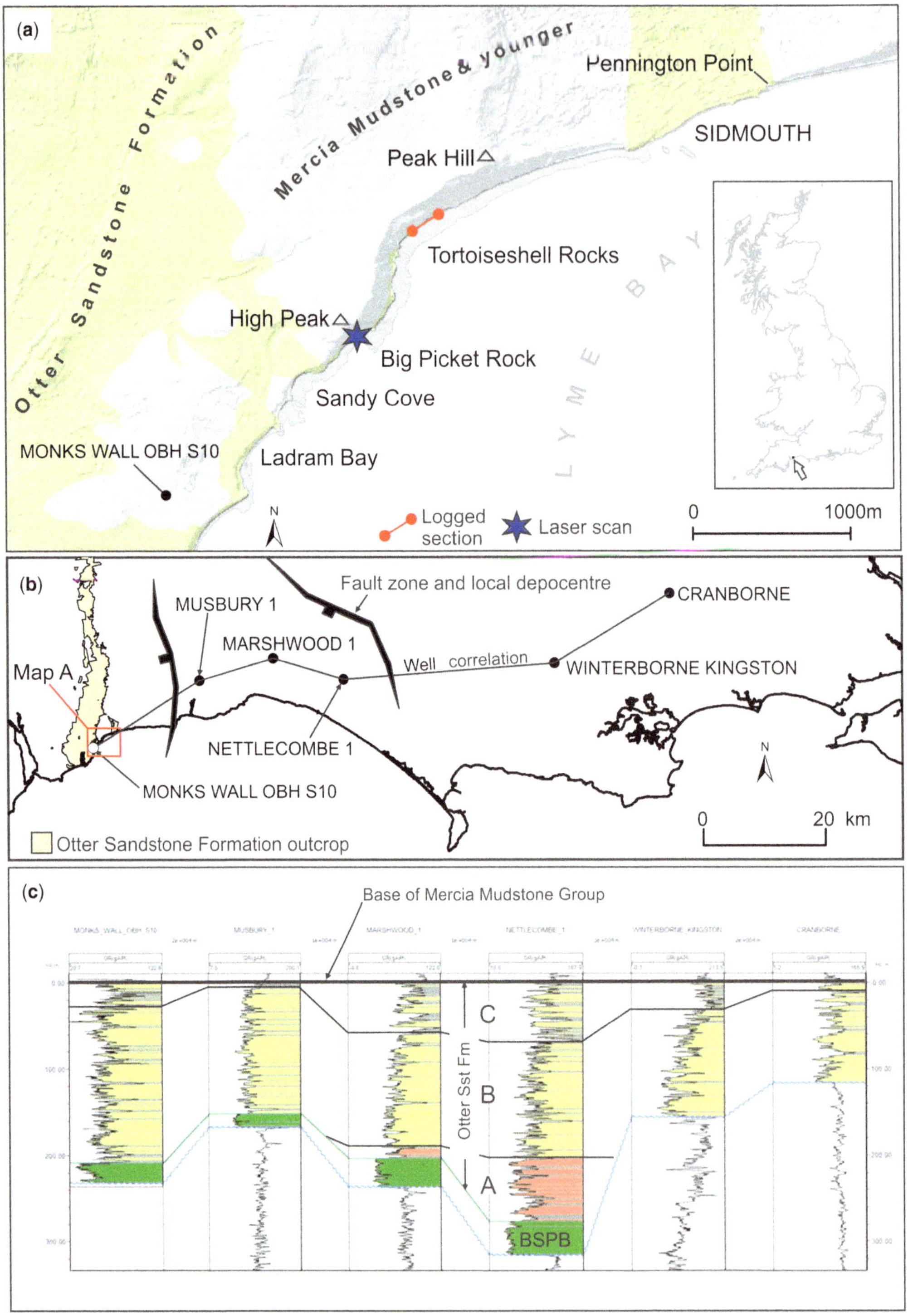

Fig. 4. (**a**) Map showing the location of the contact between the Sherwood Sandstone (Otter Sandstone Formation) and Mercia Mudstone groups in south Devon. Good exposures occur below High Peak to the west of Sidmouth and at Pennington Point just to the east. (**b**) Map showing the outcrop of the Otter Sandstone Formation. The

studies into dynamic flow simulations that may increase the value of the outcrop work still further.

The Sherwood Sandstone Group of the Wessex Basin

The Triassic Sherwood Sandstone Group crops out along the western margin of the Wessex Basin, an area of extensional faulting in southern England that was episodically active between the Permian and Cretaceous periods (Newell 2000) (Fig. 4). The broadly north–south orientated outcrop of the Sherwood Sandstone Group intersects the NW–SE trending coastline between Budleigh Salterton and Sidmouth in south Devon where virtually the entire succession can be seen in spectacular sea-cliff exposures (Fig. 5). The Sherwood Sandstone Group comprises the conglomeratic Budleigh Salterton Pebbles Beds (around 30 m thick) overlain by the Otter Sandstone Formation, which is around 200 m thick. Approximately 8 m of aeolian sandstone occurs at the base of the Otter Sandstone Formation (Wright *et al.* 1991), but the bulk of the formation is composed of vertically and laterally amalgamated fluvial sheet sandstones (Newell 2006). Mudstone mostly occurs as occasional isolated channel plugs, apart from the uppermost 15 m of the Otter Sandstone Formation, where it increases greatly in abundance in a subdivision called the Pennington Point Member, immediately beneath the Sidmouth Mudstone Formation, the lowermost unit of the Mercia Mudstone Group (Gallois 2005). From its western outcrop, the Otter Sandstone Formation dips below the thick Jurassic and Cretaceous fill of the Wessex Basin and it forms an important hydrocarbon reservoir at Wytch Farm (Bowman *et al.* 1993). Boreholes provide evidence for some fault-control on the facies and thickness patterns of the Otter Sandstone Formation, particularly towards the base of the formation where thick intervals of anhydritic mudstone are locally present (Fig. 4; Holloway *et al.* 1989).

The primary focus of this study is on the uppermost part of the Otter Sandstone Formation (Pennington Point Member) because in a carbon capture and storage scenario this region would form the reservoir–caprock interface under which buoyant CO_2 would accumulate and spread. In south Devon this part of the stratigraphy is superbly exposed under High Peak to the west of Sidmouth where it occurs at an elevation of around 30 m above sea-level (Fig. 5). By a combination of a northeasterly structural dip and several down-to-the-east faults, the Pennington Point Member reaches sea-level at Tortoiseshell Rocks on Sidmouth beach (Fig. 4). Faulting repeats the unit at Pennington Point, immediately east of Sidmouth (Gallois 2005).

The Pennington Point Member at the reservoir–caprock interface

The Pennington Point Member is around 15–20 m thick and has an erosive contact with the underlying amalgamated sheet sandstones (Chiselbury Bay Member) of the Otter Sandstone Formation (Fig. 6). Localized scours on this basal erosion surface are variably infilled with sandstone, mudstone or a combination of both (Fig. 6). The upper contact of the Pennington Point Member with the massive reddish-brown mudstones of the Mercia Mudstone Group is gradational, often with a progressive decrease in sandstone bed thickness (Fig. 7). Sandstone and mudstone occurs in approximately equal proportions in the Pennington Point Member, although there is lateral variation owing to the thickening and thinning of the beds. The sandstones are weakly cemented, often micaceous, very fine- to fine-grained and moderately sorted. A number of distinct lithofacies can be recognized within the Pennington Point Member, which are summarized below.

Amalgamated channel-bar and channel-plug complexes

These sandstone-dominated units range up to 3 m thick and have a sheet-like geometry. The base of each sheet is an erosion surface, often marked by an abundance of mudstone rip-up clasts. The internal structure of sheets is complex and most are built from laterally and vertically amalgamated sandstone bodies up to 1 m thick which are separated by erosion surfaces and thin mudstone drapes (Fig. 6). Individual sandstone bodies tend to have flat bases and concave-down tops. Where internal structure is discernible, the sandstone bodies are predominantly trough cross-bedded, although tabular cross-bedding and ripple cross-lamination can

Fig. 4. (*Continued*) formation subcrops throughout much of the Wessex Basin to the east. Well data indicate a number of well-defined fault-controlled depocentres. (**c**) Well correlation panel from outcrop into the sub-surface. Section is flattened on the top of the Otter Sandstone Formation. The conglomeratic Budleigh Salterton Pebble Beds (BSPB) are shown by green shading. Overlying this formation, the Otter Sandstone Formation can be subdivided into a number of units, an anhydrite-rich unit at the base (a) overlain by a fluvial sheet sandstone (b) becoming mud-prone towards the top (c).

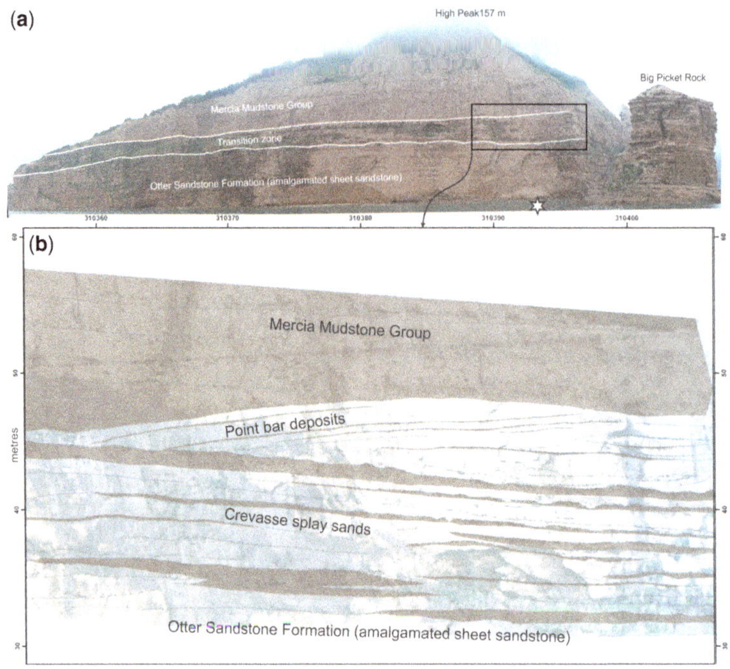

Fig. 5. (**a**) Large cliff exposure of the Otter Sandstone Formation and Mercia Mudstone Group contact below High Peak near Sidmouth, Devon (see Fig. 4 for location). Between the amalgamated sheet sandstones of the Otter Sandstone Formation and the mudstones of the Mercia Mudstone Group is a 15–20 m-thick transition zone of interbedded sandstone and mudstone. The structure of this interval, which here was impossible to access, was investigated using a laser scanner (scan site indicated by star on Figs 5a & 4a). The scanning produced a 3D digital outcrop model as well as orthorectified base maps (**b**) which could be used to accurately map sedimentary structure without the distortion of conventional photographs.

also occur. Traced laterally, individual sandstone bodies commonly wedge out into metre-thick mudstone plugs with flat tops and concave-up bases.

These complex sheet-like deposits probably represent an amalgamation of fluvial channel-bar and channel-plug deposits (Bridge 2003). An abundance of internal erosion surfaces indicates multiple episodes of channel migration and avulsion within a broad channel belt. Mudstone plugs represent the fill of former channels.

Isolated point-bar deposits

Isolated point-bar deposits are best observed in the cliffs at High Peak near Big Picket Rock (Fig. 5) where they occur immediately below the Mercia Mudstone Group. This part of the cliff cannot be accessed directly but a terrestrial laser scanner with an integrated digital camera (Hodgetts 2013) was used to create a virtual digital outcrop model and an undistorted base-map (Fig. 5). The point bar deposits are up to 6 m thick and comprise a lower bar core composed of trough cross-bedded sandstone which passes upwards and laterally into a series of inclined sandstone and mudstone bed couplets that downlap onto a relatively horizontal basal erosion surface.

These deposits represent the relatively complete preservation of a fluvial point bar. Trough cross-bedded sandstone represents lower point bar deposits deposited close to the channel floor by sinuous-crested dunes while the overlying and offlapping inclined sandstone–mudstone couplets represent lateral accretion of the bar-top (Bridge 2003). These remarkably preserved point bar deposits probably owe their preservation to the relatively passive drowning associated with the transgression of the overlying Mercia Mudstone Group. The undulating contact between the Otter Sandstone Formation and Mercia Mudstone Group is effectively a drowning unconformity.

Splay sandstones

These are represented by intervals of relatively thin (<1 m) tabular or lenticular sandstone beds interbedded with mudstone. Sandstone beds generally have erosive bases, often with distinct gutters

Fig. 6. Photograph showing the Pennington Point Member (PPM) at the western end of Sidmouth Beach. The base of the member is an erosion surface which cuts downward from left to right into the Chiselbury Member (CM) of the Otter Sandstone Formation (OSF). A sandstone body with flat top and concave-up base infills the scour; above this the sandstones tend to have flat bases and concave-down tops. The top of the Penningon Point Member is a rapid transition into the Sidmouth Mudstone Formation (SMF) of the Mercia Mudstone Group (MMG). White arrows highlight the base and top of the Pennington Point Member.

along their base, and a discontinuous lenticular geometry (Fig. 5). Sandstone beds are commonly massive, but can show small-scale trough cross-bedding or ripple cross-lamination. Mudstone beds are massive, dark reddish brown and lenticular with occasional discontinuous sand laminae and sand-filled desiccation cracks, burrows and water escape structures (Fig. 8).

These thin sandstone beds probably represent small-scale overland flood events that were unconfined or only weakly confined within shallow channels. Interbedded mudstones were probably deposited from suspension following each flood event. Given the evidence for the co-existence of large river channels, flood events could have been initiated from lateral overbank flood events, which would have built up levees adjacent to the main river channel, or from frontal splays downstream from a distributary channel (McKie & Williams 2009).

Floodbasin muds

Floodbasin mudstones are dominated by reddish brown mudstones in laterally extensive beds, which are typically up to 1 m thick. The mudstones generally appear massive, but include sandstone as thin lenses and laminae, within fluid escape structures and as the fill to subvertical cylindrical burrows up to 10 mm in diameter (Fig. 8). These muds were deposited from suspension on floodplains and in temporary lakes. The lack of palaeosols suggests relatively continuous sedimentation.

Taken together, the sedimentary lithofacies points towards a meandering fluvial environment for the Pennington Point Member of the Otter Sandstone Formation with sandy channel belt deposits passing laterally (and probably also in a downstream direction) into thinly bedded splay sandstones and muddy flood-basin deposits. Relative to the amalgamated braided fluvial sheet sandstones of the underlying Otter Sandstone Formation (Chiselbury Bay Member), this interval of muddy meandering fluvial deposits is heterogeneous and introduces considerable complexity into the reservoir–caprock transition zone. In particular, a profusion of inclined or concave-down sandstone beds terminating up-dip against mudstone drapes or more laterally extensive floodplain deposits (Fig. 5) have the potential to form stratigraphic traps for buoyant CO_2. Numerous mudstone baffles and complex connectivity between sandstone bodies created by basal scour, sand-filled water escape structures and burrows (Fig. 8) are likely to create a tortuous flow path.

Building a static geological model

Base maps from laser scanning

Laser scanning geological exposures using ground-based instruments is now a routine technique in field geology where direct access is difficult or dangerous

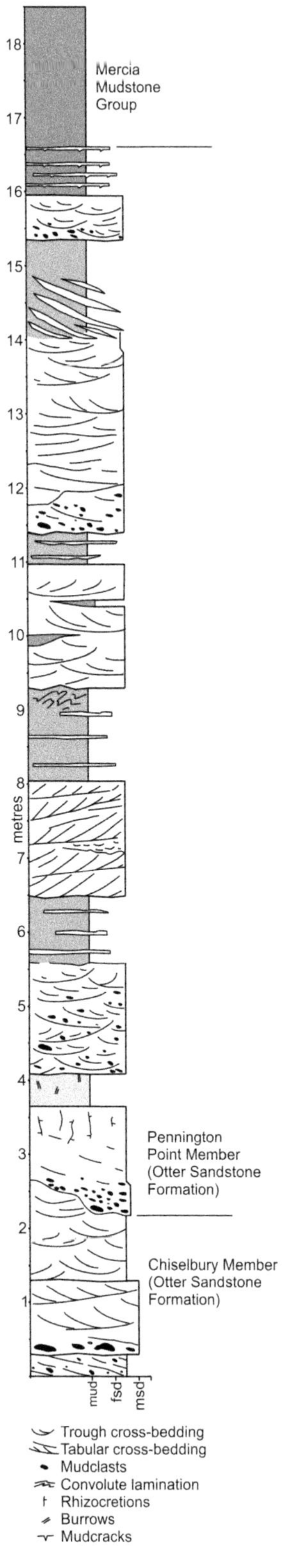

or where an accurate base-map is required for the mapping and measurement of geological structures (Hodgetts 2013; Howell *et al.* 2014). Laser scanning generates a dense cloud of points, each with a Cartesian coordinate, which together replicate the topography of the geological exposure with millimetre accuracy. Points can be migrated into real-world coordinates by establishing the location of the laser scanner using differential Global Positioning System. Moreover, by mounting a digital camera on the laser scanner each point can be painted with Red, Green, Blue (RGB) intensity to produce a photo-realistic digital outcrop model (Hodgetts 2013). Laser-scanned cliff sections were the starting point for the construction of static and dynamic flow models in this study (Fig. 9).

Three laser scans of the Otter Sandstone Formation–Mercia Mudstone Group transition were obtained at Big Picket Rock, Tortoiseshell Rocks and Pennington Point (Fig. 4). The process of converting laser scans into static geological models is illustrated using data from Peak Hill (Fig. 5). Here laser scanning was the only option to obtain scaled information on the shape and dimensions of sedimentary deposits which were otherwise inaccessible. This section was also important because, unlike the more accessible exposures, it offered a cross-section of the deposits that was cut at a high angle to the regional (north to NE) palaeoflow direction. Laser scanning using Riegl LPM2K equipment was undertaken at low tide to maximize the distance between the laser scanner and the target cliff. Tides are clearly an important consideration when undertaking laser scanning on intertidal marine platforms as the process is relatively slow. The dense point clouds generated by the scanner was migrated into real world (British National Grid) coordinates and each point was attributed with an RGB value using digital photographs from the scanner-mounted camera. The point clouds were edited to remove unwanted points before meshing into a continuous triangulated irregular network. RGB values were transferred as a property onto the mesh and interpolated to improve the rendering of virtual cliff face. The digital outcrop model was used as a base map for mapping sedimentological features such as sandstone and mudstone beds (Fig. 5).

Extrapolating mapped structures into 3D

It is very usual that the full 3D geometry of a sedimentary deposit can be determined, even in areas

Fig. 7. Logged section of the transition zone (Pennington Point Member) between the Otter Sandstone Formation and the Mercia Mudstone Group showing the approximately equal proportions of sandstone and mudstone.

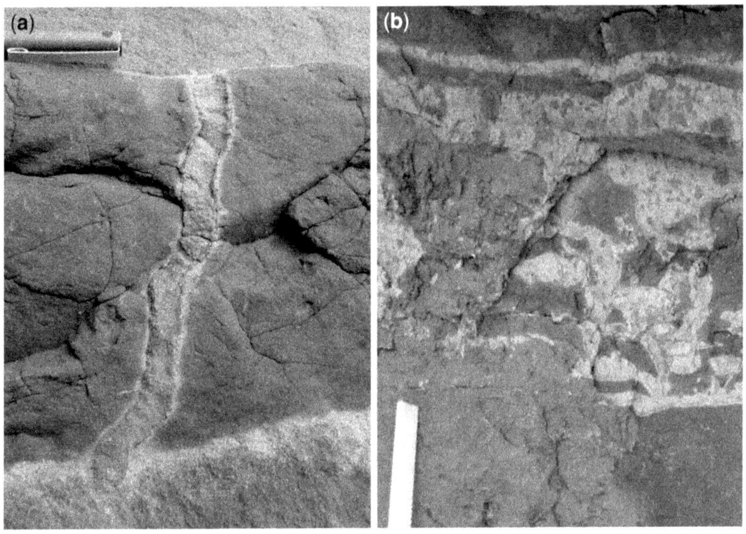

Fig. 8. (**a**) Sand-filled vertical burrow cross-cutting a mudstone bed and providing flow connectivity between two sandstone beds (pen lid is 30 mm long). (**b**) Mudstone bed with thin interbedded sandstones and sand-filled fluid escape structures (ruler scale is 20 mm across).

of exceptional exposure where erosion provides multiple cross-sections in different orientations. An understanding of the full 3D geometry of a sedimentary deposit generally requires careful sequential excavation, although geophysical techniques such as ground-penetrating radar provide a useful and non-destructive alternative (Kostic & Aigner 2007). In the absence of hard physical data, qualitative or quantitative sedimentary process models can assist greatly in the extension of mapped 2D sections and 1D boreholes into an unknown rock volume (Willis & Tang 2010). The approach taken in this study was simply to extrude the mapped sedimentary facies boundaries (Fig. 5) perpendicular to the cliff section (Fig. 9a). This approach was not only rapid and technically simple, but considered reasonable in this case given (a) the relatively small volume of the cube, (b) supporting observations on the geometry and lengths of bedding units in flow parallel and flow transverse section and (c) a desire to maintain the integrity of mapped observations rather than introduce assumptions on bed length, geometry and the overall 3D form of the deposit. This approach is clearly not applicable to all situations, particularly where the geological model is much larger than the typical dimensions of beds or discrete sedimentary building blocks. Here the modelled volume had dimensions of 50 × 35 × 30 m and was discretized into 100 × 70 × 150 regular cells (Fig. 9a). The grid was subdivided into three main regions, a lower region representing the amalgamated sheet sandstones (Chiselbury Bay Member) of the Otter Sandstone Formation, a middle region representing the heterogeneous mudstone and sandstone of the Pennington Point Member and an upper region representing the Mercia Mudstone Group. The position of horizontal and inclined mudstone baffles was incorporated within the Pennington Point Member together with the undulating topography created by the point-bar deposits beneath the drowning unconformity at the base of the Mercia Mudstone Group (Fig. 5). The structural tilt of the formations was 6° towards the SE.

Porosity and permeability values

Porosity and permeability values for the sandstone were based on core plugs taken from outcrop and neighbouring borehole core of the Otter Sandstone Formation (Allen *et al.* 1997; Newell 2006; Fig. 10). Porosity values range from approximately 5 to 35% and permeability from 0.001 to 10 000 mD. In the model, the relatively coarser grained braided fluvial sheet sandstones of the Chiselbury Bay Member of the Otter Sandstone Formation were given a fixed porosity value of 30% while the finer-grained sandstones of the Pennington Point Member were assigned a range of porosity values using sequential Gaussian simulation based on a normal distribution with mean of 20% and a standard deviation of 2%. The correlation between porosity and horizontal permeability was derived from core plug data (Fig. 10). Vertical permeability was set at 80% of the horizontal

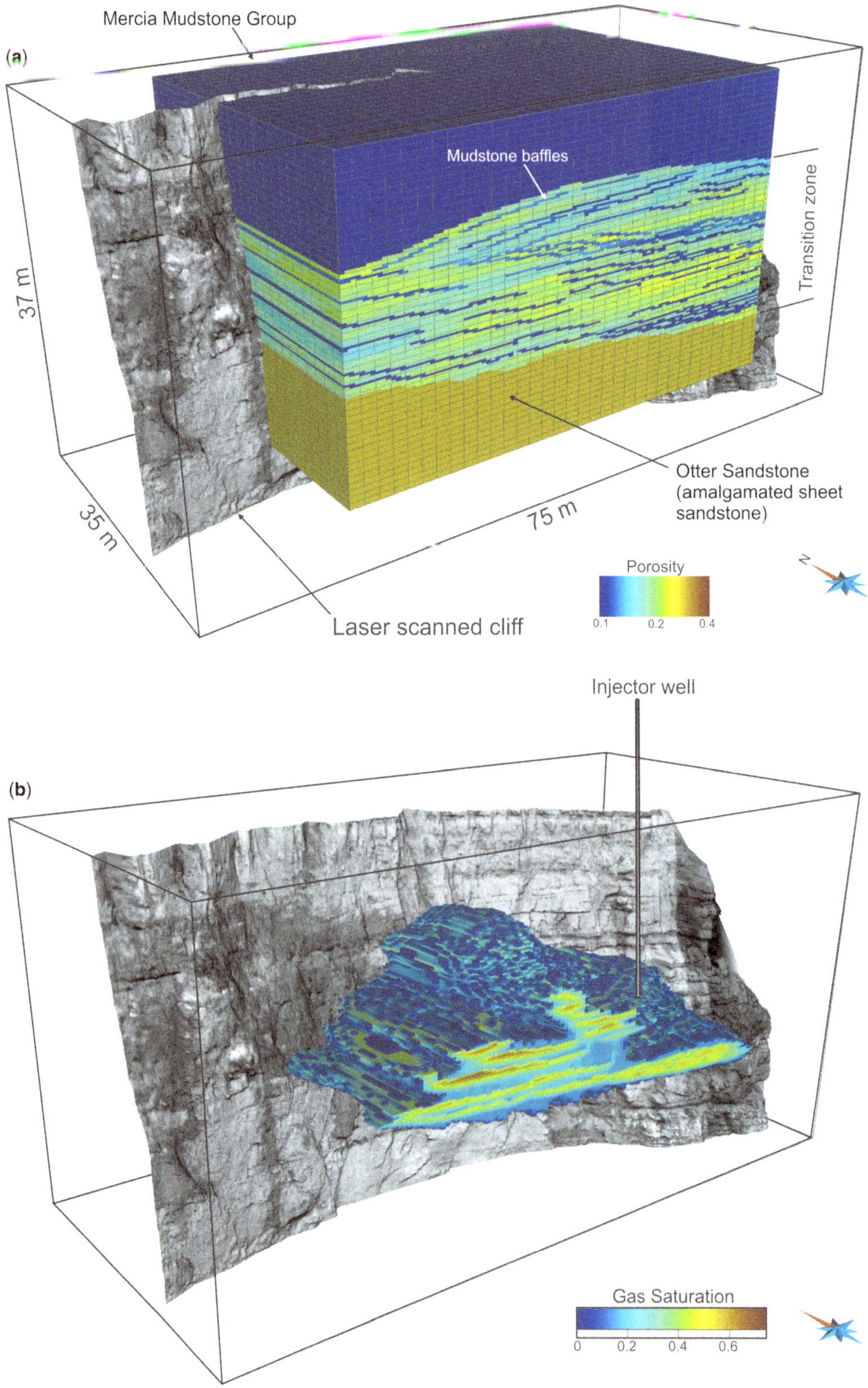
Mercia Mudstone Group
(a)
Mudstone baffles
Transition zone
37 m
35 m
75 m
Otter Sandstone
(amalgamated sheet
sandstone)
Laser scanned cliff
Porosity
0.1
0.2
0.4
N
Injector well
(b)
Gas Saturation
0
0.2
0.4
0.6

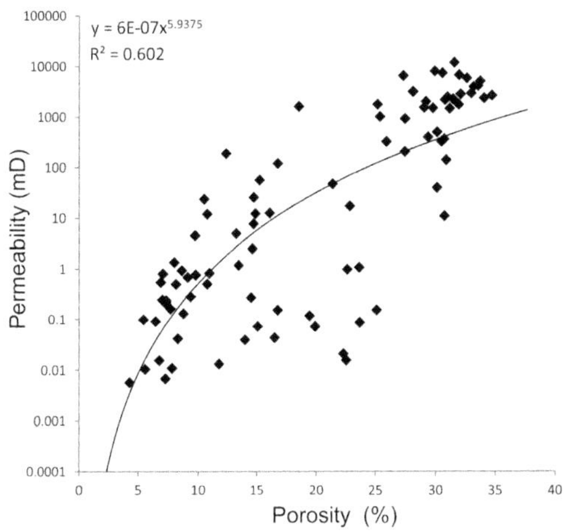

Fig. 10. Cross-plot showing the porosity (%) v. the log of gas permeability (mD) for plug samples taken from outcrop of the Otter Sandstone Formation and neighbouring shallow boreholes. In the geological model the permeability parameters are mapped from porosity through the empirical correlation between porosity and permeability.

permeability based on the typical variation between horizontal and vertical core plug measurements (Allen *et al.* 1997). Mudstone was assigned constant values of 10% porosity, a typical value for the Mercia Mudstone Group (Armitage *et al.* 2013) and a constant permeability of 0.02 mD. This permeability is considerably higher than direct measurements from core of the Mercia Mudstone Group, where values can be as low as 10 nD (Armitage *et al.* 2015). A value higher than those measured from core plugs was selected to account for the presence of sand-filled burrows (Fig. 8a), sand-filled fluid injection structures (Fig. 8b) and sandy laminae within the lowermost part of the Mercia Mudstone Group in south Devon. However, it would clearly be of interest to undertake additional modelling runs where the permeability contrast between the sandstones and mudstones was progressively increased to reflect the measured Mercia Mudstone Group values of Armitage *et al.* (2015). Disparities in the permeability of the Mercia Mudstone Group determined from direct measurement of core- and field-derived hydraulic conductivity values, which are generally much higher (Armitage *et al.* 2015), show the desirability of using a quantitative approach (e.g. Pickup & Hern 2002) to estimate the effective permeability of representative volumes of reservoir and caprock. It is also appreciated that the generalized outcrop and shallow aquifer porosity and permeability values selected for this model will not necessarily be representative of the Otter Sandstone Formation at greater depths where calcite, dolomite and anhydrite may occlude some or all of the porosity within the sandstones (Holloway *et al.* 1989). However, it might be anticipated that, where porosity is not entirely occluded by pore-filling cement, the contrast in the porosity and permeability of the sandstone and mudstone beds, which is the primary aspect being explored in this modelling study, will still be valid.

Dynamic model setup

The grid for the dynamic flow model had dimensions of 50 × 35 × 30 m and was discretized into 100 × 70 × 150 cells. One injector was placed on the down-dip (eastern) side of the model and CO_2 was injected through perforations at the bottom of the aquifer (bottom 25 layers) (Fig. 9b). The model was taken down to 1 km depth and the pore volume of the outer column of cells on the left- and right-hand side of the model were multiplied by a factor of 1000 to take account of this.

The well was controlled by the surface rate with a maximum pressure limit of 190 bars. CO_2 was injected during 6 years (around 3 Mt of CO_2 in total), and after that the model was run for 100 years to study the CO_2 plume migration. Initial pressure and initial temperature were set to 100 bar and 45°C at 1000 m depth. Initially, it was assumed that the reservoir contained 100% brine, with mole fraction of 0.967 and 0.033 for water (H_2O) and sodium chloride (NaCl), respectively The simulations were performed using ECLIPSE 300 compositional simulator with the CO2STORE option (Schlumberger 2013). Re-Studio and Petrel were used for pre and post-processing respectively.

Dynamic modelling results

In most geological storage cases being considered, CO_2 is injected in a supercritical (SC) state, which is above 31.1°C and 73.9 bar. Supercritical CO_2 is a fluid with a liquid-like density and a gas-like

Fig. 9. (**a**) A simple porosity model of the outcrop created by the extrusion of the sandstone and mudstone units perpendicular to the digital outcrop model. While unrealistic, this method has the advantage of speed, and more importantly it maintains the spatial relationship of known mapped structures throughout the cube. (**b**) Morphology of the CO_2 plume after 6 years of injection from an injector well located on the down dip (eastern) part of the static geological model (Fig. 8). Colour ramp shows percentage CO_2 saturation at 6 years from the start of injection.

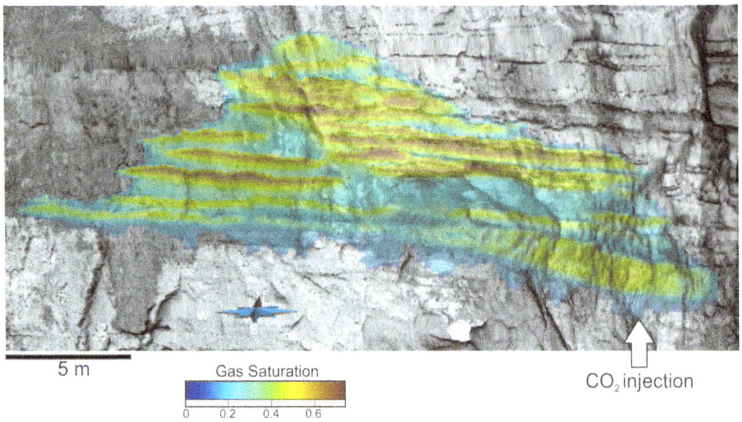

Fig. 11. Results of the Eclipse dynamic flow simulation shown on the vertical plane of laser-scanned digital outcrop model. Colour ramp shows percentage CO_2 saturation at 6 years from the start of injection. Red colours indicate a high gas saturation caused by pooling of the buoyant CO_2 below mudstone baffles.

viscosity, which generally occurs at a depth of storage of more than 800 m. In this study, CO_{2SC} is also injected into the saline aquifer model and will stay supercritical because the minimum pressure and temperature in the model are above the least pressure and temperature for the supercritical CO_2 conditions.

When CO_2 is injected into saline aquifers, it is retained through four basic trapping mechanisms: structural/stratigraphic, dissolution, residual trapping and migration trapping (Holloway 2009). Structural/stratigraphic trapping is the most significant trapping mechanism in the short term (IPCC 2005) before the CO_2 dissolves in the pore fluids and reacts with mineral components of the rock. Our short-duration model runs were primarily concerned with short-term structural/stratigraphic trapping.

The density of free phase CO_2 is much lower than that of the formation brine. Therefore, when free CO_2 is injected into the model, it migrates upwards under buoyancy until it reaches low-permeability layers and then spreads out laterally (Fig. 11). This lateral migration depends on the extension of the low-permeability layers and also the capillary entry pressure. Because of the discontinuity of these mudstone baffles in the modelled volume (Fig. 9a) and also the relatively high permeability (0.02 mD) assigned to them, a proportion of the injected CO_2 approaches, but does not reach, the main reservoir–caprock interface at the base of the Mercia Mudstone Group (Fig. 11). The simulation results clearly show that the existence of multiple mudstone baffles in the upper part of the Otter Sandstone Formation aquifer enhances the security of CO_2 storage by preventing vertical migration of CO_2 towards the main caprock. Subsequently, more CO_2 migrates laterally and more CO_2 dissolves in fresh brine. As CO_2-saturated brine is denser than pure brine, it therefore sinks into the main storage formation represented by the more homogeneous braided sheet sandstones of the Otter Sandstone Formation (Fig. 11), and this can also initiate the convective mixing process (Lindeberg & Bergmo 2003), which could accelerate CO_2 dissolution in brine. The large-scale point-bar deposits that are seen immediately beneath the Mercia Mudstone Group are likely to form particularly effective stratigraphic traps for CO_2 because of their inclined to concave-down geometry (Fig. 5). The modelling shows that they can cause a reversal in the migration direction of the CO_2 plume where the stratal dip of the inclined heterolithic stratification is counter to that of the regional structural dip (Fig. 11).

In order to investigate the effect of grid size effect, the base case model is modified by coarsening the model in *X*, *Y* and *Z* by factors of 2, 2 and 5, respectively. Coarsening the model reduces the simulation run time significantly owing to the decrease in the number of cells from 1 050 000 cells to 52 500 cells. However, the accuracy of flow simulation of CO_2 storage drops owing to two factors. Firstly, it diminishes the effect of small-scale heterogeneities in the model. Secondly, it increases the numerical dispersion errors (Shariatipour 2013) and thus increases the amount of CO_2 dissolution in the model and shrinks plume migration (Fig. 12a, b).

Farhat *et al.* (2015) investigated the effect of capillary pressure on plume migration and CO_2 dissolution using the same model. In this study,

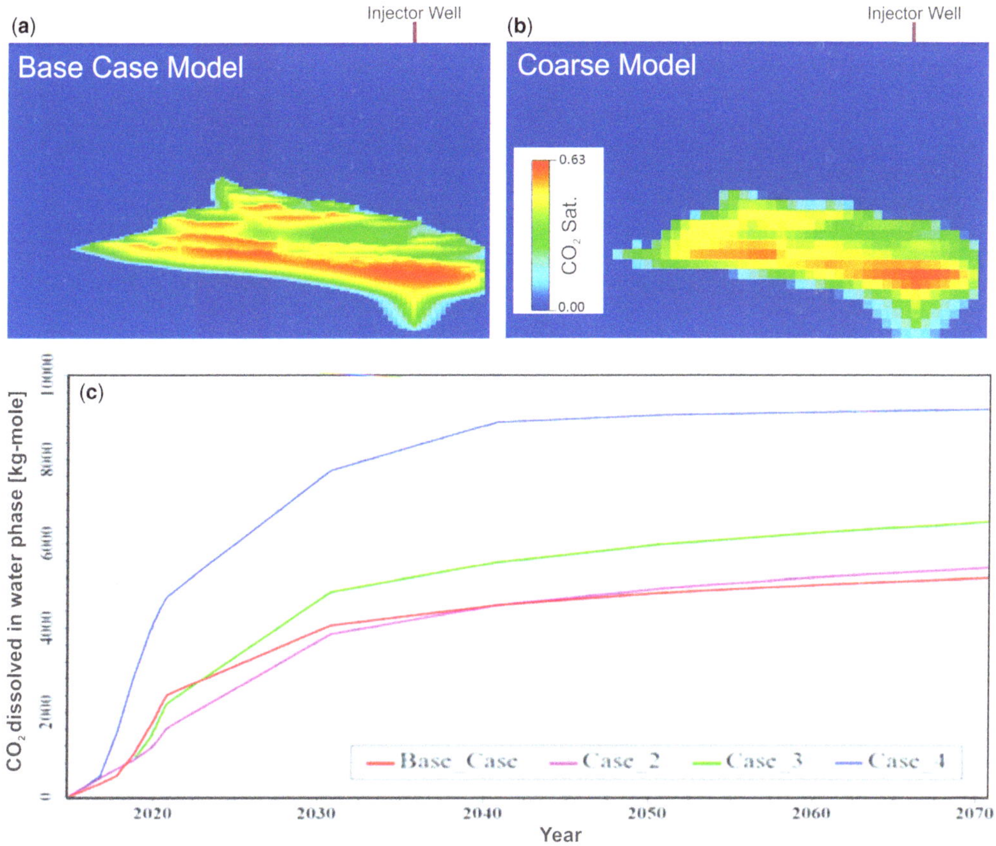

Fig. 12. Plume migration at the end of the injection period (6 years) showing the effect of grid size on plume migration: (**a**) shows the base case model and (**b**) shows the coarse model. (**c**) Plot showing CO_2 dissolved in the water phase against time. Base case, one type of capillary pressure (Pc) was used; case 2, two types of Pc curves were used; case 3, three types of Pc curves were used; case 4, four types of Pc curves were used.

the permeability range used for the Sherwood Sandstone and Mercia Mudstone groups was modified and different capillary pressure curves were also used. The results showed that using different capillary pressure curves for different regions (permeability variation) has an effect on the plume migration and CO_2 dissolution (Fig. 12c). That means using more accurate data or small-scale characterization of the storage formation, especially just beneath the caprock, is very important for CO_2 storage.

Discussion

Outcrop work provides information at a resolution that lies between borehole core and seismic data (Howell *et al.* 2014) and may be particularly important for CCS prospects in saline aquifers where existing sub-surface geological data are sparse and new sub-surface data are extremely expensive to acquire.

Great care must clearly be taken to match each sub-surface situation with an appropriate outcrop analogue. Even within the same region or basin there may be great variability at a particular lithostratigraphic level. This is readily illustrated by the diachronous and sedimentologically variable contact between the Sherwood Sandstone and Mercia Mudstone groups (McKie & Williams 2009). In the Wessex Basin (as discussed here) the contact between the Sherwood Sandstone Group and the Mercia Mudstone Group is marked by a shift from amalgamated 'braided' channel sandstones towards muddy meandering channel and floodplain deposits. However, in the East Irish Sea Basin the Mercia Mudstone Group transgression was preceded by the development of aeolian sand sheet and sabkha deposits of the Omskirk Sandstone Formation

(Herries & Cowan 1997). Further south in the Cheshire Basin, trace fossils provide evidence for the development of intertidal sand and mud flats in the Tarporley Siltstone Formation at the base of the Mercia Mudstone Group (Pollard 1981). Elsewhere back-stepping fluvial systems at the top of the Sherwood Sandstone Group appear to have simply dispersed their water and sandy sediment load as thin diffuse sheets across low-gradient mudflats (McKie & Williams 2009). A minimum of sub-surface data is clearly required to match the reservoir with the potential outcrop analogue. The correct analogue may not come from the same time-frame or sedimentary basin as the reservoir under consideration.

Despite the operational status of the world's first commercial-scale CCS facility in Canada (Watson 2014), sub-surface carbon storage is still at an early stage of development with relatively few large-scale field trials: modelling studies are still at the forefront of understanding the risks and uncertainty associated with injecting CO_2 into the subsurface (Mackay 2013). Most modelling studies operate on coarse (decametre) grids and over large (kilometre) areas, but the importance of a multiscalar approach has been highlighted by some workers (Nilsen *et al.* 2012). Modern fieldwork methods such as terrestrial laser scanning now make it a relatively straightforward process to build digital outcrop models in a form suitable for undertaking flow simulation, although challenges still remain with extrapolating hard data from outcrops into a 3D volume (Willis & Tang 2010). However, even with such limitations, flow models based on outcrop are useful for understanding how small-scale geological heterogeneity influences the movement of CO_2. This study clearly shows how the addition of localized stratigraphic traps and closures below the main reservoir–caprock interface causes the immobilization of high-saturation CO_2. This has been shown to be an important factor in increasing storage efficiency because it prevents the CO_2 plume rising and spreading along the top of the reservoir, thus bypassing much of the available pore volume (Lindeberg & Bergmo 2003; Ambrose *et al.* 2008). Moreover, mudstone baffles improve the lateral sweep of CO_2 after injection, leading to greater reservoir contact and lower migration velocities (Goater *et al.* 2013). This would be expected to lead to more residual trapping and dissolution (Juanes *et al.* 2006; Lengler *et al.* 2010).

Although methods such as laser scanning greatly increase the accuracy of facies mapping from large inaccessible cliff sections, there are still many uncertainties in the input parameters (e.g. mudstone shape and dimension, porosity and permeability) for both the static geological modelling and the dynamic flow modelling. The results presented in this study illustrate a single realization of the static geological model and a limited number of flow simulations. It would clearly be desirable to extend this work towards multiple model realizations so that the sensitivity of the results to porosity, permeability, mudstone length and other variables can be fully understood before they are applied to real-world situations.

Possibly of equal importance to increased technical understanding is that outcrop studies, when carried through to flow simulations, serve as a means of promoting dialogue between geologists and flow modellers. They provide a very tangible sense of how much geological heterogeneity might be contained within a single cell of a simulation grid.

Conclusions

- Facies mapping of laser-scanned cliff faces in SW England shows that the contact between the Sherwood Sandstone (Otter Sandstone Formation) and Mercia Mudstone Group (Sidmouth Mudstone Formation) is not a simple planar surface but a complex transition through heterogeneous point-bar sandstones and sheet-like splay deposits.
- With some understanding of the porosity and permeability of the constituent facies it is possible to convert facies maps derived from laser-scanned cliff faces into attributed digital outcrop models for the purposes of flow simulation.
- Flow simulation shows that the arrangement of inclined point-bar sandstones, horizontal splay sandstones and mudstone baffles in the upper part of the Sherwood Sandstone Group produces complex flow paths with much greater potential for dispersed physical trapping of buoyant fluids such as injected CO_2. Flow simulations based on the assumption of a simple, inclined planar interface between reservoir and caprock are liable to be incorrect.
- Combining outcrop analogue with flow simulation provides a cost-effective means of providing addition insight into what might control the morphology and magnitude of injected CO_2 plumes in the sub-surface. Care must be taken that the outcrop is appropriately matched to the subsurface geology.

S. M. Shariatipour thanks Schlumberger for the use of Eclipse 300 and Petrel and Amarile for the use of the Re-Studio. Marcus Dobbs and David Morgan (BGS) are thanked for their hard work and expertise with the terrestrial laser scanner. Antoni Milodowski, Richard Worden and Simon Passey are thanked for their constructive comments and editorial work on the manuscript. A. J. Newell publishes with permission of the Executive Director of the British Geological Survey.

References

ALLEN, D. J., BREWERTON, L. J., COLEBY, L. M., GIBBS, B. R., MACDONALD, A. M. & WILLIAMS, A. T. 1997. *The physical properties of major aquifers in England and Wales*. British Geological Survey, Keyworth, WD/397/034.

AMBROSE, W. A., LAKSHMINARASIMHAN, S., HOLTZ, M. H., NÚÑEZ-LÓPEZ, V., HOVORKA, S. D. & DUNCAN, I. 2008. Geologic factors controlling CO_2 storage capacity and permanence: case studies based on experience with heterogeneity in oil and gas reservoirs applied to CO_2 storage. *Environmental Geology*, **54**, 1619–1633.

ARMITAGE, P. J., WORDEN, R. H., FAULKNER, D. R., APLIN, A. C., BUTCHER, A. R. & ESPIE, A. A. 2013. Mercia Mudstone Formation caprock to carbon capture and storage sites: petrology and petrophysical characteristics. *Journal of the Geological Society*, **170**, 119–132.

ARMITAGE, P. J., WORDEN, R. H., FAULKNER, D. R., BUTCHER, A. R. & ESPIE, A. A. 2015. Permeability of the Mercia Mudstone: suitability as caprock to carbon capture and storage sites. *Geofluids*, http://doi.org/10.1111/gfl.12134

BACHU, S., BONIJOLY, D., BRADSHAW, J., BURRUSS, R., HOLLOWAY, S., CHRISTENSEN, N. P. & MATHIASSEN, O. M. 2007. CO_2 storage capacity estimation: methodology and gaps. *International Journal of Greenhouse Gas Control*, **1**, 430–443.

BOOT-HANDFORD, M. E., ABANADES, J. C. *ET AL*. 2014. Carbon capture and storage update. *Energy and Environmental Science*, **7**, 130–189.

BOWMAN, M. B. J., MCCLURE, N. M. & WILKINSON, D. W. 1993. *Wytch Farm oilfield: deterministic reservoir description of the Triassic Sherwood Sandstone*. Geological Society, London, Petroleum Geology Conference Series, **4**, 1513–1517.

BRIDGE, J. S. 2003. *Rivers and Floodplains: Forms, Processes, and Sedimentary Record*. Blackwell Science, Oxford.

FARHAT, M. M., SHARIATIPOUR, S. M. & ANDREW, J. N. 2015. *The Effect of Capillary Pressure on the CO_2 Storage at the Interface between Caprock and Storage Formation*. 3rd International Smposium on Energy Challenges & Mechanics, Aberdeen, Scotland.

GALLOIS, R. W. 2005. The type section of the junction of the Otter Sandstone Formation and the Mercia Mudstone Group (mid Triassic) at Pennington Point, Sidmouth. *Geoscience in South-west England – Proceedings of the Ussher Society*, **11**, 51–58.

GOATER, A. L., BIJELJIC, B. & BLUNT, M. J. 2013. Dipping open aquifers – the effect of top-surface topography and heterogeneity on CO_2 storage efficiency. *International Journal of Greenhouse Gas Control*, **17**, 318–331.

HERRIES, R. D. & COWAN, G. 1997. *Challenging the 'sheetflood' myth: the role of water-table-controlled sabkha deposits in redefining the depositional model for the Ormskirk Sandstone Formation (Lower Triassic), East Irish Sea Basin*. Geological Society, London, Special Publications, **124**, 253–276.

HODGETTS, D. 2013. Laser scanning and digital outcrop geology in the petroleum industry: a review. *Marine and Petroleum Geology*, **46**, 335–354.

HOLLOWAY, S. 2009. Storage capacity and containment issues for carbon dioxide capture and geological storage on the UK continental shelf. *Proceedings of the Institution of Mechanical Engineers, Part A: Journal of Power and Energy*, **223**, 239–248.

HOLLOWAY, S., MILODOWSKI, A. E., STRONG, G. E. & WARRINGTON, G. 1989. The Sherwood Sandstone Group (Triassic) of the Wessex Basin, southern England. *Proceedings of the Geologists' Association*, **100**, 383–394.

HOSA, A., ESENTIA, M., STEWART, J. & HASZELDINE, S. 2011. Injection of CO_2 into saline formations: benchmarking worldwide projects. *Chemical Engineering Research and Design*, **89**, 1855–1864.

HOUNSLOW, M. W. & MCINTOSH, G. 2003. Magnetostratigraphy of the Sherwood Sandstone Group (Lower and Middle Triassic), south Devon, UK: detailed correlation of the marine and non-marine Anisian. *Palaeogeography, Palaeoclimatology, Palaeoecology*, **193**, 325–348.

HOWELL, J. A., MARTINIUS, A. W. & GOOD, T. R. 2014. *The application of outcrop analogues in geological modelling: a review, present status and future outlook*. Geological Society, London, Special Publications, **387**, 1–25.

IPCC 2005. *Intergovernmental panel on climate change*. Special Report, Carbon Dioxide Capture and Storage, Summary for Policymakers Montreal, Canada.

JUANES, R., SPITERI, E. J., ORR, F. M. & BLUNT, M. J. 2006. Impact of relative permeability hysteresis on geological CO_2 storage. *Water Resources Research*, **42**, W12418.

KOSTIC, B. & AIGNER, T. 2007. Sedimentary architecture and 3D ground-penetrating radar analysis of gravelly meandering river deposits (Neckar Valley, SW Germany). *Sedimentology*, **54**, 789–808.

LENGLER, U., DE LUCIA, M. & KÜHN, M. 2010. The impact of heterogeneity on the distribution of CO_2: numerical simulation of CO_2 storage at Ketzin. *International Journal of Greenhouse Gas Control*, **4**, 1016–1025.

LINDEBERG, E. & BERGMO, P. 2003. The long-term fate of CO_2 injected into an aquifer. *Greenhouse Gas Control Technologies*, **1**, 489–494.

MACKAY, E. J. 2013. Modelling the injectivity, migration and trapping of CO_2 in carbon capture and storage (CCS). *In*: GLUYAS, J. & MATHIAS, S. (eds) *Geological Storage of Carbon Dioxide (CO_2)*. Woodhead, Cambridge, 45–70.

MCKIE, T. & WILLIAMS, B. 2009. Triassic palaeogeography and fluvial dispersal across the northwest European Basins. *Geological Journal*, **44**, 711–741.

MORTON, A., HOUNSLOW, M. W. & FREI, D. 2013. Heavy-mineral, mineral–chemical and zircon-age constraints on the provenance of Triassic sandstones from the Devon coast, southern Britain. *Geologos*, **19**, 67–85.

NEWELL, A. J. 2000. Fault activity and sedimentation in a marine rift basin (Upper Jurassic, Wessex Basin, UK). *Journal of the Geological Society*, **157**, 83–92.

NEWELL, A. J. 2006. *Calcrete as a source of heterogeneity in Triassic fluvial sandstone aquifers (Otter Sandstone Formation, SW England)*. Geological Society, London, Special Publications, **263**, 119–127.

NILSEN, H. M., SYVERSVEEN, A. R., LIE, K.-A., TVERANGER, J. & NORDBOTTEN, J. M. 2012. Impact of top-surface morphology on CO_2 storage capacity. *International Journal of Greenhouse Gas Control*, **11**, 221–235.

OGAWA, T., NAKANISHI, S., SHIDAHARA, T., OKUMURA, T. & HAYASHI, E. 2011. Saline-aquifer CO_2 sequestration in Japan – methodology of storage capacity assessment. *International Journal of Greenhouse Gas Control*, **5**, 318–326.

PHARAOH, T. C., MORRIS, J., LONG, C. B. & RYAN, P. D. 1996. *The Tectonic Map of Britain, Ireland and adjacent areas.* Sheet 1. 1:1 500 000 scale. British Geological Survey and Geological Survey of Ireland.

PICKUP, G. E. & HERN, C. Y. 2002. The development of appropriate upscaling procedures. *Transport in Porous Media*, **46**, 119–138.

PICKUP, G. E., JIN, M. ET AL. 2011. Geological storage of CO_2: site appraisal and modelling. *Energy Procedia*, **4**, 4762–4769.

POLLARD, J. E. 1981. A comparison between the Triassic trace fossils of Cheshire and South Germany. *Palaeontology*, **24**, 555–588.

SCHLUMBERGER 2013. ECLIPSE Reservoir Simulation Software, Technical Description.

SHARIATIPOUR, S. M. 2013. *Modelling of CO_2 Behaviour at the Interface between the Storage Formation and the Caprock.* Doctoral dissertation, Heriot-Watt University.

SHARIATIPOUR, S. M., PICKUP, G. E. & MACKAY, E. J. 2013. The effect of aquifer/caprock interface on geological storage of CO_2. *Energy Procedia*, **63**, 5544–5555.

WATSON, R. 2014. Building the future. *Greenhouse Gases: Science and Technology*, **4**, 259–261.

WILLIAMS, J. D. O., JIN, M., BENTHAM, M., PICKUP, G. E., HANNIS, S. D. & MACKAY, E. J. 2013. Modelling carbon dioxide storage within closed structures in the UK Bunter Sandstone Formation. *International Journal of Greenhouse Gas Control*, **18**, 38–50.

WILLIS, B. J. & TANG, H. 2010. Three-dimensional connectivity of point-bar deposits. *Journal of Sedimentary Research*, **80**, 440–454.

WRIGHT, V. P., MARRIOTT, S. B. & VANSTONE, S. D. 1991. A ‘reg’ palaeosol from the Lower Triassic of south Devon: stratigraphic and palaeoclimatic implications. *Geological Magazine*, **128**, 517–523.

Do technical studies reduce subsurface risk in hydrocarbon exploration: and if not, how do they add value?

FRANK J. PEEL[1]* & JONATHAN WHITE[2]

[1]*National Oceanography Centre, University of Southampton Waterfront Campus, European Way, Southampton SO14 3ZH, UK*

[2]*Perenco UK Limited, 1 Franklins Row, London SW3 4SW, UK*

**Corresponding author (e-mail: Frank.peel@noc.ac.uk)*

Abstract: The title 'The Value of Outcrop Studies in Reducing Subsurface Uncertainty and Risk' might suggest that we expect new information to improve prospect risk, but this is not correct. Gaining new information generally does change our estimate of prospect risk: the change may be up or down, and the average of all possibilities is zero change. You cannot acquire data for the purpose of increasing the probability of success. We should expect that: (a) the expected value of the risk of a single prospect, post-data, is equal to the prior (pre-data) value; and (b) risk should become worse for the majority of prospects. New information adds value, not from changing pre-drill risk, but from decisions made as a consequence. The main value added is from identifying prospects not to drill, thereby saving the cost of likely dry holes, and by choosing the ones to test first, accelerating revenue from a successful outcome. Further value is added if the new information leads to the identification of new prospects or new plays, or suggests follow-on potential elsewhere in the region. Value may also be added if the new information is negative for a whole region, enabling us to focus our attention elsewhere.

Readers with experience in the petroleum industry may have heard statements along the lines of 'we will acquire 3D data and this will reduce our prospect risk', or 'your task is to work up these three prospects, improve their risk, and promote them to drillable status'. Implicit to such statements is a belief that it is possible to design a work programme that will change our estimated chance of prospect success for the better. This article explores the possible root of this belief in the Labour Theory of Value, and shows that it is not valid when applied to prospect risks.

Why we believe that doing work adds value: and why this does not apply to prospect risk

The Labour Theory of Value (Smith 1776) is founded on the observation that effort commonly results in increase of value. This concept influences philosophy, economics and, to some extent, hydrocarbon exploration. Bertrand Russell (1946) traced its origin to the statement by Thomas Aquinas (1265) that 'value can, does and should increase in relation to the amount of labor which has been expended in the improvement of commodities'. This was key to the economic philosophy of Adam Smith and others; Karl Marx (1898) went further, stating 'the value of a commodity is determined by the quantity of labour bestowed upon its production'.

The value we assign to an undrilled prospect portfolio is dependent on many factors, which include the expected number of discoveries and the risked volume of hydrocarbons we believe they contain. Thus the perceived value of a portfolio depends critically on the chance of success we assign to the prospects within it. Applying the Labour Theory of Value to the business of oil and gas exploration gives rise to a belief that is common, seductive and wrong; namely that by acquiring information (whether in the form of seismic data acquisition, analogue field studies or whatever) we expect to increase the value of a prospect, or a portfolio of prospects, by improving the risk (geological chance of success). Indeed, the title of the 2014 Geological Society meeting 'Reducing Subsurface Uncertainty and Risk through Field-based Studies' (Geological Society 2014) can be read as an expression of this expectation – if we carry out studies (in this case field-based studies) we expect to reduce subsurface risk.

This belief can have a severe impact on companies and on individuals. We have seen cases where performance contracts require a team to improve the chance of success of key prospects by remapping

From: Bowman, M., Smyth, H. R., Good, T. R., Passey, S. R., Hirst, J. P. P. & Jordan, C. J. (eds) 2016. *The Value of Outcrop Studies in Reducing Subsurface Uncertainty and Risk in Hydrocarbon Exploration and Production*. Geological Society, London, Special Publications, **436**, 247–259.
First published online December 16, 2015, http://doi.org/10.1144/SP436.4

and further geological studies; companies have committed to 3D seismic programmes with the expectation that the data will lead to improved chance of success of existing prospects.

The Labour Theory of Value appeals to the human desire that our efforts are worthwhile. The contrary view is that 'all the work one cares to add will not turn a mud pie into an apple tart; it remains a mud pie, value zero' (Heinlein 1959). Our problem in exploration is that we do not know whether the undrilled prospect is a figurative apple tart or a mud pie; all we have is an estimate of the relative likelihood of the good outcome v. the bad outcome, based on our current state of knowledge and the information available to us at the time. This article explores the implications that acquiring new information has on subsurface (prospect) risk.

The meaning of 'expectation'

How we use words in common language can be very different from how we use the same words in formal science, and this can cause confusion when we use these words to discuss exploration risk. In common language, we 'expect' an outcome that commonly occurs, such as the birth of a child (Murkoff & Mazel 1984). It has a different meaning in probability and statistics, where the expectation is the predicted average outcome of an experiment (or process). The expected value is the sum of ((each of the possible outcomes) × (the probability of the outcome occurring)). The expected value may not occur frequently, or may indeed never occur.

The expected number of children in a family in the UK is 1.7 (UK Office of National statistics, 2012) but even a statistician would be surprised to have 1.7 children. The expected value (in the formal sense) of a fair, six-sided, dice is 3.5, being the average of all outcomes $(1 + 2 + 3 + 4 + 5 + 6)/6$. to actually land a score of 3.5 on one toss would be most unexpected (in the informal sense).

In petroleum exploration, we commonly use the word in its formal sense, for example EV (Expected Value) is the probability-weighted average value of all identified possible outcomes, and we will use the words 'expectation' and 'expected value' in that sense throughout this article.

The Law of Total Probability and its impact on prospect risk

The Law of Total Probability is a fundamental part of modern probability theory and it is of vital significance to this discussion. Salkind (2010) noted that the law is implicit to Bayes's Theorem (Bayes 1764). The Law is widely stated in modern texts (e.g. Bean 2001; Cyganowski *et al.* 2002; Salkind 2010) without an original attribution, but it can traced back to the Theory of Inverse Probability (Laplace 1774). When applied to the business of oil and gas exploration, it creates some non-intuitive results, which run counter to the Labour Theory of Value. Let us consider the case of an oil prospect, based on 2D seismic data, for which we have a prior chance of success, and over which we are acquiring 3D seismic data. This is illustrated graphically in Figure 1a.

The Law of Total Probability can be expressed as

$$P(B) = \sum_{i=1}^{n} P(Ai) * P(B|Ai)$$

where $P(B)$ is our prior probability of an outcome B (such as an exploration well success); this is the initial estimate made prior to gaining new information. A represents some event (such as acquiring and

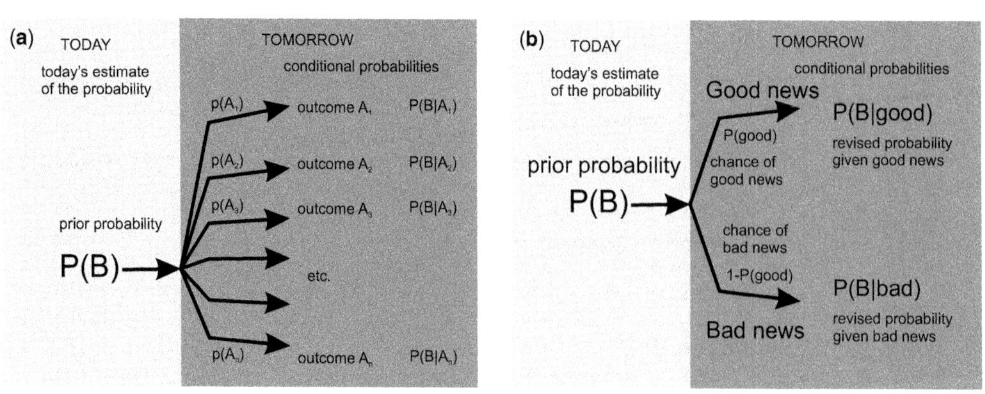

Fig. 1. (**a**) general case of the Law of Total Probability, with many possible outcomes considered; (**b**) a special case where the new information is binary – only two possible outcomes.

interpreting new 3D seismic data) that will provide new information that may cause us to revise that prior estimate. That event A has n possible outcomes, such as:

A_1 = 3D confirms closure exists, and shows amplitudes that conform to closure;
A_2 = 3D confirms closure but has no amplitudes;
A_3 = 3D confirms closure does not exist, etc.

$P(Ai)$ represents the likelihood of each of those outcomes happening and $P(B|Ai)$ is the revised probability of outcome B (the new estimate of chance of success, made using the new information).

The Law of Total Probability states that the mean outcome – the expected value of the chance of success after we have the new information – is equivalent to the prior probability $P(B)$, estimated prior to obtaining that information. In other words, in our example, the expected value of the prospect risk after we have acquired 3D data should be exactly the same as the prior; some outcomes changed the risk for the better, some for the worse, but the mean outcome is no change. This can be expressed by the adage 'you can't put oil in the ground by shooting seismic data'.

This law can be generalized to describe a situation in which the new information does not deliver n discrete possible outcomes, but instead delivers a continuum of possible results, represented by a continuous distribution; again the mean of the outcomes, weighted according the probability of each outcome, is equivalent to the prior chance.

A simple case (Fig. 1b) in which the information can only deliver two possible outcomes – good news or bad news – gives a simpler equation:

$$P(B) = P(\text{good}) * P(B|\text{good}) + (1 - P(\text{good})) * (P(B|\text{bad}))$$

In effect, this law means that the expected outcome of new information is no change. We should not expect that acquiring new information should improve our chance of success: it might do so, or might not. We emphasize that we are using the word 'expected' in its formal sense, as described above; in informal language, we will be surprised at any outcome where value does not change, but the average of all possible outcomes is no change.

Because this law can appear quite counter-intuitive, and because it is so important to our case, we will illustrate this by two simple examples of binary outcomes, in which we provide the numerical values and demonstrate the validity of the law in these cases.

In the simple game of dice shown in Figure 2, we win if the dice are both showing one spot. Today, the dice have been cast, but they are both covered, and we do not know which way they have fallen. Today, we estimate the probability of winning as 1 in 36; this is called the prior probability. If winning has a value of \$100, and there is no cost to losing – our expected value of the game is \$100/36 = \$2.78.

Tomorrow, we will obtain better information: one of the cups will be removed. Our estimate of the chance of success will change, depending on what we see. This new chance of success is called the conditional probability (Popper 1959; Hájek 2003). It may be better, or it may be worse, than today's estimate; if the exposed dice has one spot, we are still in the game, and only the second dice has to have the winning number; the new probability will be 1/6. From today's perspective, we describe this as the conditional probability if the exposed dice shows one spot. However, if the exposed dice has more than one spot, there is no way we can win, and the new probability will be zero. In this simple game, we know exactly what the chance of change in either direction is; there are five ways it can be bad news (anything but one spot showing) and one way it can be good news. The conditional probability will improve to 1/6 in 1/6 of the outcomes, and it will get worse, decreasing to zero, in 5/6 of the outcomes. The expected value of the conditional probability is the probability-weighted average of these, $(5/6 \times \text{zero}) + 1/6 \times 1/6) = 1/36$. The expected value of the game, from today's standpoint, remains \$100/36 = \$2.78.

Note that this value is unchanged from today's estimate of the chance of success: the Total Probability Rule, or Average Rule, which derives from Law of Total Probability, effectively states that the probability weighted average of possible outcomes (in this case the conditional probabilities) is equal to the prior expectation (in this case the prior probability).

We emphasize that, in each possible outcome in this game of dice, gaining new data does change the probability of success, but you cannot know before hand whether the probability of success will go up or go down. Thus you cannot acquire data for the purpose of increasing the probability of success. In the case of the simple game of dice shown here, the application of Law of Total Probability is intuitively obvious: in more complex scenarios, typical of petroleum exploration, the same rule applies, but it is less intuitively obvious.

Exactly the same principle applies in the game of petroleum exploration, as shown in Figure 3. We can consider a prospect for which today's state of knowledge is limited to a sparse 2D grid. Contouring the depth values seen on the 2D lines indicates the possibility of a closure, and we estimate that the chance of this trap geometry existing is 0.4. Our expected value of the game is 0.4 of a discovery. Tomorrow, we will receive a 3D dataset that we think will either confirm or deny the existence

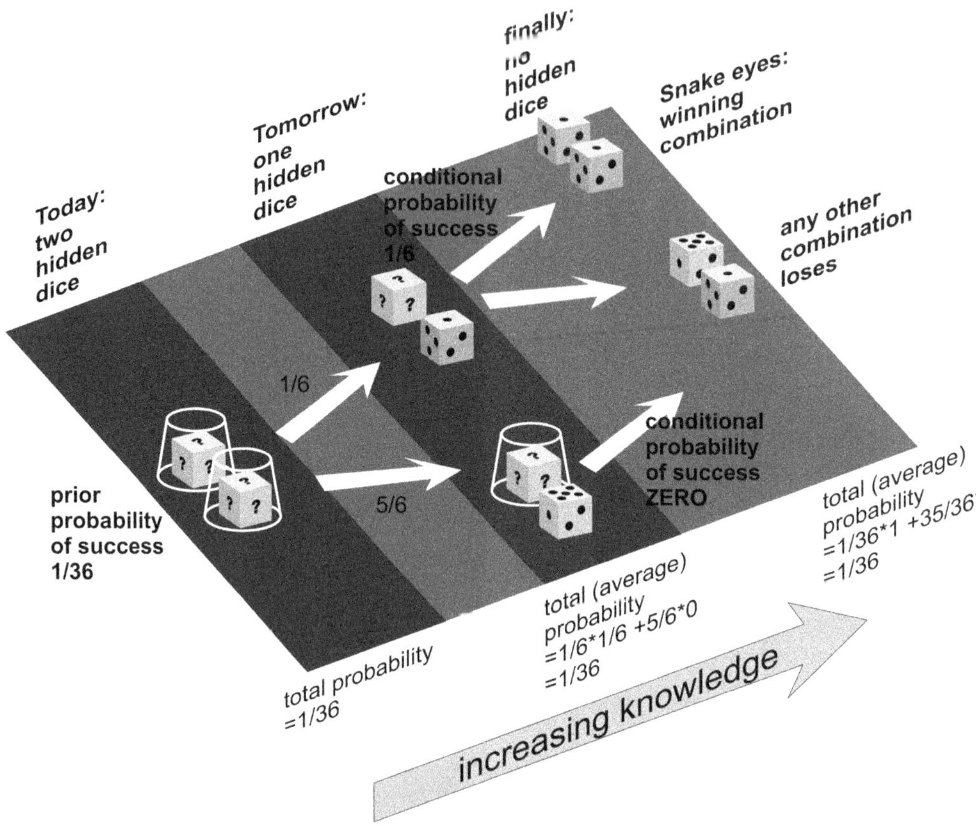

Fig. 2. The Law of Total Probability illustrated by a game of dice. The total probability of winning, considering all the possible outcomes, is 1/36 at all stages in the game. Some outcomes give good results, some give bad, but the expected odds (the average of all outcomes) is unchanged.

of the closure. We should expect the average probability after receiving the new information to be the same as the prior probability; from today's perspective, the expected value of the game is still 0.4 of a discovery. In other words, our new estimate of the chance of success of the prospect may become better or worse, but the average of all possible outcomes is unchanged.

In these two games, the result of the new information was binary (there were only two, discrete outcomes). The same principle applies when the new information can give many different outcomes, or the possible outcomes form a continuous distribution: the expected value of the conditional probability is the same as today's prior probability.

What is prospect risk?

Prospect risk estimation is a vital part of the management of an exploration portfolio (e.g. Allais 1956; Newendorp 1975; Megill 1977; Rose 1987). Prospect risk (also known as *geological risk*, or the *chance of geological success* of a prospect) describes our current opinion of the likelihood that the success case geological model is correct, and that the parameter ranges associated with that success case scenario are appropriate. This estimate is based on our current understanding and experience and the information currently available to us. It is, therefore, not an attribute of the geology of the prospect; instead it is attribute of the observer. The estimate will change with time, dependent on the information available about the prospect, the level of understanding about the geological model, and to some extent on the world view of the observer.

'Prospect risk' commonly refers to the chance of success, but it has also been used to mean the chance of failure, generating confusion as to whether 'high risk' means likely failure or likely success. To avoid this confusion, it is advisable to avoid the term 'risk' other than for informal purposes, and refer specifically to the chance of geological success. There is no consistent terminology for this concept.

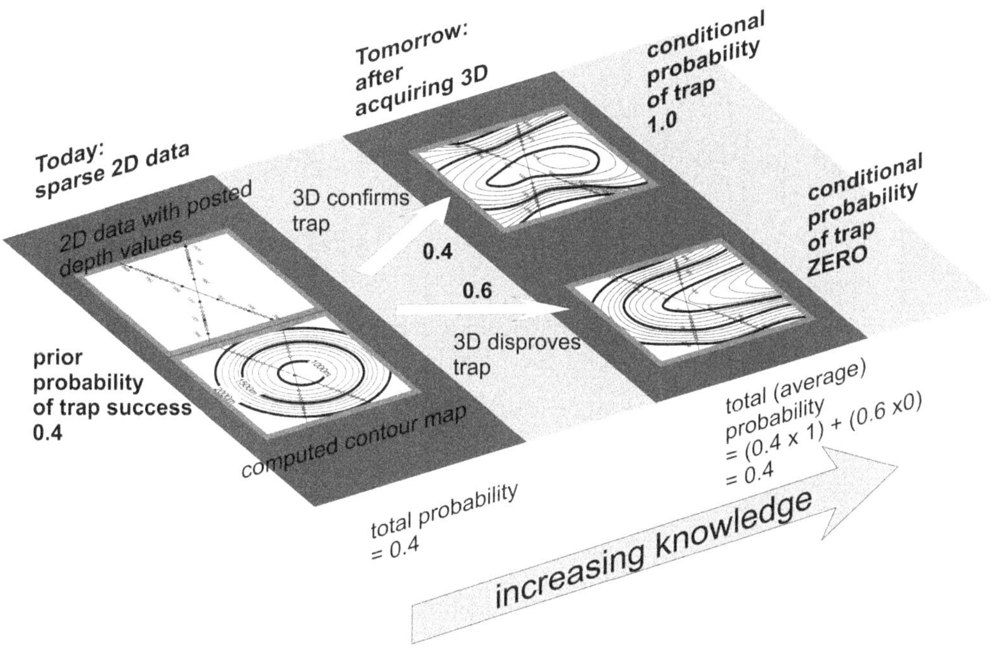

Fig. 3. The Law of Total Probability illustrated by petroleum exploration. The total probability of winning, considering all the possible outcomes, is 0.4 at all stages in the game. Some outcomes give good results, some give bad, but the expected odds (the average of all outcomes) is unchanged.

This article follows Rose (1987, 1992, 2001) in using the symbol *Pg* (Probability of Geological Success). Abbreviations used by other sources for the same quantity include GP (Geological Probability; Ross 1997, 2004), GPoS (Geological Probability of Success), CoS (Chance of Success) and POSg (Quirk & Ruthrauff 2008).

In this article, we quantify the chance of success as numerical probability (0–1, where 0 is no chance of success and 1 is no chance of failure). Other publications quantify it in the form of odds (e.g. 1:10) or as a percentage (0–100%).

How does prospect risk change in the light of new information?

Applying the principles described above to the process of oil and gas exploration, we can derive two important rules for how the prospect chance of success will change, both of which are counter-intuitive, and both of which may initially appear to undermine our rationale for acquiring new information. However, by recognizing that we add value to the exploration cycle by making good decisions (Bratvold & Begg 2008, 2010), not by changing prospect risks, we can demonstrate how new information adds value (Howard 1966).

Rule (i) The expected value of Pg *will be unchanged (i.e. you shouldn't expect to improve* Pg *by acquiring new data)*

If we acquire new information, it helps us make a more reliable estimate of that probability; it may bring good news or bad news, so that our subsequent re-evaluation of the chance of success (i.e. our conditional probability, given the new information) may change for the better or for the worse. The magnitude and direction of the change depend on our prior estimate, on whether the new information appears positive or negative, and how reliable the new information is (Bayes 1764; Laplace 1774; Morgan 1968; for specific application of this method to petroleum exploration, see for example Newendorp 1971, 1975).

Although we expect that our *Pg* estimate for individual prospects will, in most cases, change in the light of new information, we do not know the direction of the change. As described above, the law of Total Probability dictates that the mean value of all the possible outcomes is unchanged.

In other words, we should not expect to reduce risk (i.e. increase the chance of success, *Pg*) by acquiring new information, although we may hope that this is the result. This is illustrated schematically by Figure 4 which shows a prospect for

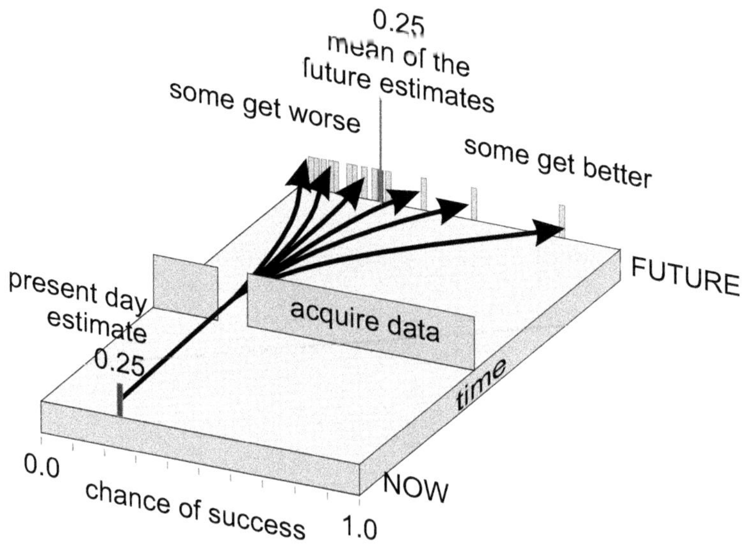

Fig. 4. Schematic diagram showing the effect of new information on a prospect for which our current estimate of *Pg* is 0.25. In some scenarios, *Pg* improves: in others, *Pg* appears worse. The mean of all these 'future world' scenarios is unchanged from the present *Pg* estimate, 0.25.

which the present-day estimated *Pg* is 0.25. After acquiring new information, the new estimate of *Pg* is likely to change, falling within a range of possibilities: but the average of all these possibilities is the same as the prior estimate, 0.25.

Rule (ii): you should expect Pg *to become worse for the majority of prospects*

The majority of real-world exploration prospects have pre-drill chance of success significantly less than 0.5. Quirk & Ruthrauff (2008) showed the average chance of success of exploration wells in 18 oil and gas plays, ranging from 0.2 to 0.48, with a mean of *c.* 0.35.

With additional information, some prospects will see improved *Pg*, and some will see diminished *Pg*, but if the prior *Pg* is below 0.5, the upgraded ones can move further in the positive direction than the downgraded ones can move in the negative direction (see Fig. 5). Therefore, because the expected value of the average prospect risk of all possible outcomes should be unchanged, we should expect that more outcomes will be downgraded than are upgraded.

This is illustrated graphically in Figure 5, which shows the range of change in *Pg* that might result from gaining a particular piece of information whose outcome is a distribution. The new *Pg* estimate is represented by a probability density function (pdf) curve (Ushakov 2001). A typical distribution expected for a prospect with $Pg < 0.5$, is skewed, with the median value smaller than the mean.

Some types of new information are likely to result in a simple skewed distribution, as shown in Figure 5. Other types of new information, which are more diagnostic of success v. failure (e.g. seismic attribute data in a setting where hydrocarbon anomalies are expected to have a strong seismic response), are likely to result in more polarization and a bimodal distribution (Fig. 6).

For our notional prospect, which has a prior (pre-information) *Pg* estimate of 0.25, we would expect that information will improve our *Pg* estimate for about a quarter of the future scenarios, and it will make it worse in about three-quarters of the future scenarios (Fig. 4). The numerical ratio of gainers v. losers in these future scenarios depends on the nature of the new data and the way and the degree to which it can polarize our estimation of *Pg*.

The basis for this is explored in more detail in Peel & Brooks (2015), who use a probabilistic forward model of prospect risk estimation to demonstrate the evolution of *Pg* for a prospect with initial *Pg* of 0.25 through four successive stages of knowledge/data gain. Their model shows that the ratio of gainers (improved *Pg*) to losers (decreased *Pg*) does tend to give more losers than gainers for a prospect with initial $Pg < 0.5$, but that this ratio depends strongly on the nature of the new information. Some types of information may have the opposite effect (more gainers than losers). Their modelling did, however, in all cases support the

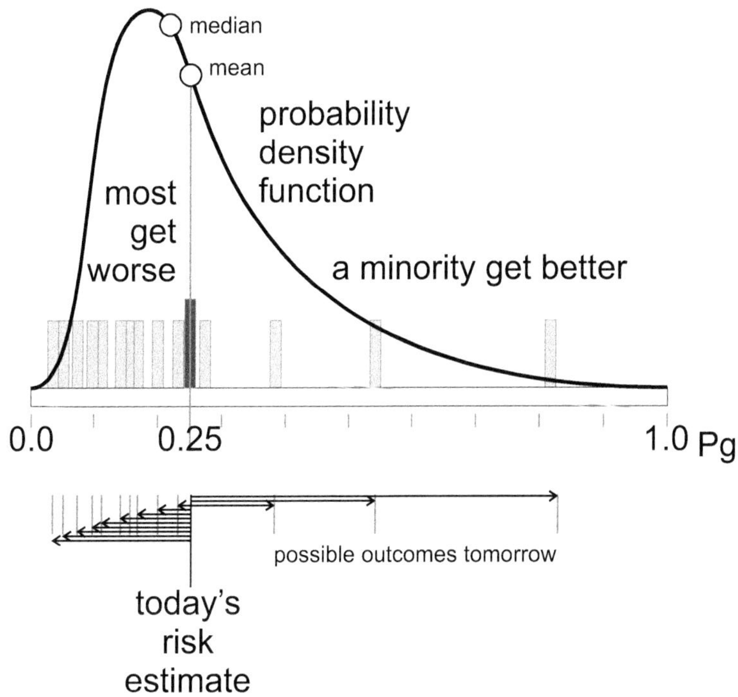

Fig. 5. Probability density function (pdf) showing how our estimate of *Pg* may be distributed tomorrow, after receiving new data unavailable to us today. Black bar shows the present estimate of *Pg*, prior to acquiring the first piece of new information. Grey bars show representative future estimates made after acquiring new information. The average of the grey bars is equivalent to the present estimate of 0.25. Note that the pdf curve has a dimensionless axis.

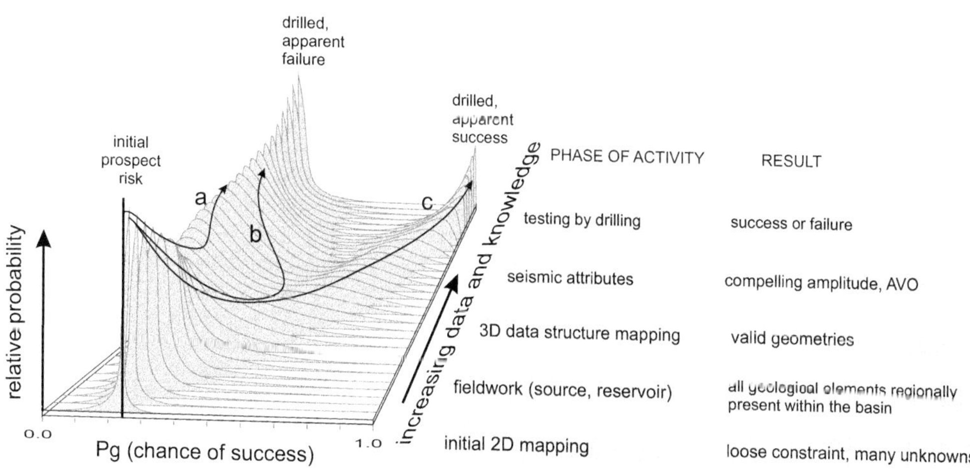

Fig. 6. Future change of the perceived chance of success of the prospect shown in Figure 5, expressed as a probability density function (pdf). Paths a–c represent possible trajectories of the prospect as we gain new information. In trajectory a, initial negative news results in lowered *Pg*, and work on the prospect is abandoned without drilling. In trajectory b, initial encouragement raises the *Pg*, followed by disappointing news that leads to the prospect being abandoned. In trajectory c, a sequence of encouragements progressively improves the *Pg* estimate, and it is drilled.

applicability of the law of Total Probability to prospect risk.

The counter intuitive implication of this is that, if we have an original estimate of the chance of success that was appropriately estimated, and we acquire appropriate data, and we work on it correctly, the expected result will be that the majority of exploration prospects should become riskier (lower *Pg*) with time, and that risk will only improve for a minority of prospects. This outcome does not mean that the original *Pg* estimate was in any way optimistic.

The perceived *Pg* (chance of success) of this prospect may continue to evolve as our state of knowledge and data increase (Fig. 6), but we do not know in advance whether *Pg* will improve or become worse. The distribution of possible outcomes is expressed here as a probability density function (pdf). As more information becomes available, the pdf curve, which represents future outcomes of where *Pg* may move to, as seen and predicted from our current perspective, broadens and changes shape, representing the effect of possible positive or negative evidence that we will gain in the future. Information obtained by fieldwork may provide evidence of some or all of the critical geological elements working regionally (or of them not working). On the schematic example shown in Figure 6, the influence of field studies is shown as moving the pdf curve a short distance up its evolutionary track.

Given enough constraint and suitable geology, the perceived *Pg* may become polarized, giving a bimodal pdf curve, in which the prospect either appears to have very strong evidence in favour of success, or very strong evidence against success. This generally is only achieved with high-quality 3D seismic reflection data containing apparent hydrocarbon indicators (such as AVA anomalies, depth-conforming amplitude anomalies, flat spots, etc.)

We emphasize that the broadening pdf curve shown in Figure 6 represents our uncertainty, from today's perspective, about where we will be in the future, and that our uncertainty about the future estimate increases as we look further down the path. This does not mean that our uncertainty will increase in the future; in each of those future worlds, we will have a single estimate, and our confidence that the estimate is appropriate should be higher.

An exception to Rule (ii) is that, for the minority of prospects for which our prior estimate of *Pg* is higher than 0.5, we should expect that obtaining new information will improve the *Pg* on more prospects than are downgraded, but as before, the expected value of the revised *Pg* is unchanged.

In every case, there is an expectation, in the informal sense, that the estimated chance of success will change from its current (prior) value in the light of new information. We do not know whether this change will be upwards or downwards. The expected (in the formal sense) posterior value of the new, revised estimate made using the new data is equal to the prior value.

Application to a portfolio of prospects

The same logic applies to a portfolio of prospects. Figure 7 shows a map of a hypothetical lease block containing 10 mapped prospects. At present day, each of these has an estimated *Pg* of 0.2, and the expected number of successes if all are drilled is $10 \times 0.2 = 2.0$; if each prospect has a mean success case volume of 100 mmbbl, we can estimate given the information available today that the total risked volume of hydrocarbons on the block is 200 mmbbl. After acquiring new information, our perception of risk may change for the better or for the worse, but the mean of all these possible future scenarios is an expected number of successes which is unchanged at 2.0, and an unchanged total risked volume.

Again, with prospect sets as with individual prospects, if prior $Pg < 0.5$, we expect the information to be disappointing more often than it is encouraging, but the fewer good cases should improve more than the many disappointing cases, so the net effect, on average, is no change. The majority of exploration blocks should become riskier with time, and risk will only improve for a minority of blocks. This outcome does not mean that the original risk was in any way optimistic.

Shared risk, play risk and dependency

In the preceding analysis, we have, for the sake of simplicity, considered the *Pg* of each prospect to be independent from that of the others. It is common for sets of prospects to carry shared risk elements, such that they are not fully independent. If the prospects fall within an unproven play (for a review, see White 1993), we can assign a shared play risk component *P*(play), such that the chance of success of the individual prospect, $Pg = P(\text{play}) * P(\text{prospect}|\text{play})$. This is an effective way of describing and analysing prospect portfolios, but it does not permit the prospect analysis to breach the Law of Total Probability. As before, new information can prove the play, changing *P*(play) to 1, or prove it invalid, changing *P*(play) to zero, or it can change our estimate of *P*(play) to a new intermediate value. However, the expected value of the conditional probability (i.e. the new estimate, given the new information) is equal to the prior estimate; in some cases, it may improve, in some cases, it may worsen, but we should not expect

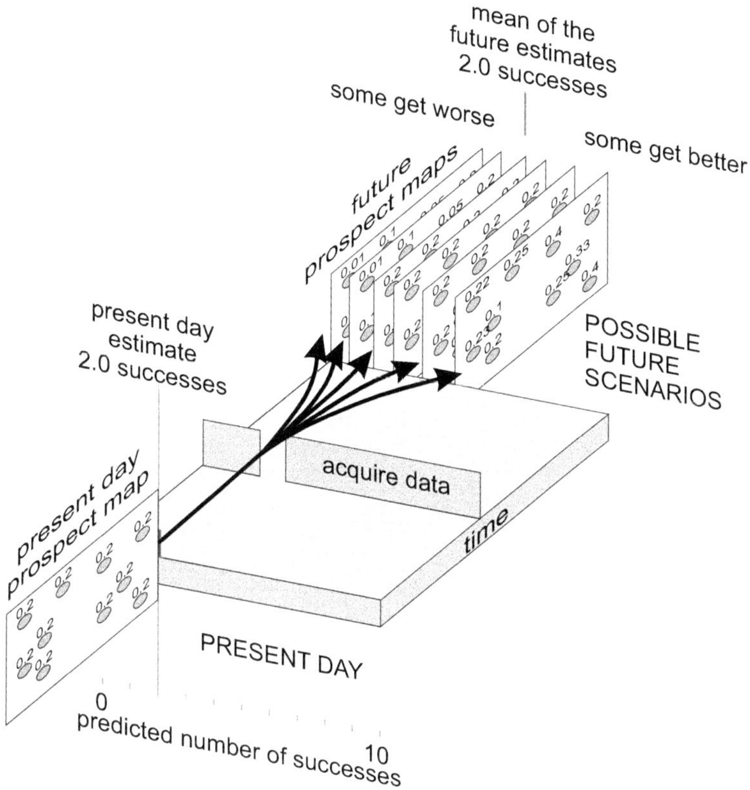

Fig. 7. Schematic diagram showing the effect of new information on a portfolio of 10 prospects within a lease block. The current estimate of *Pg* averages 0.2, and our current expectation of the number of successes is 2.0. In some scenarios, given new information, *Pg* improves; in others, *Pg* appears worse. The mean of all these 'future world' scenarios is equivalent to that of the present (average *Pg* = 0.2, expected number of successes, 2.0).

that by acquiring new information we can improve the chance of success (rule i) and if the play risk is below 0.5, we should expect more outcomes in which it worsens than outcomes in which it improves (rule ii).

If we do not expect the information to improve *Pg*, how does it add value?

Risked volume is a philosophical concept with no concrete existence. Changing the perceived risk on a prospect only has a physical expression in value terms if it causes some action which involves cost or revenue. The principal value effect of refining our pre-drill estimate of *Pg* derives from the money saved by NOT drilling wells. Identifying and avoiding the prospects with least chance of success, and thereby avoiding the sunk cost of a dry hole, is a genuine value addition. Conversely, improving the perceived *Pg* of a prospect which was already on the drilling schedule adds no material value.

The way in which new information adds full cycle value is described by Value of Information Theory. There is extensive literature on this subject, for example Howard (1966), Howard & Matheson (1984), Coopersmith & Cunningham (2002), Coopersmith *et al.* (2006); for a review of the literature, see Bratvold *et al.* (2009) and Bratvold & Begg (2008, 2010).

This is illustrated by the hypothetical lease block shown in Figure 7, which contains 10 prospects, all the same size, with initial *Pg* of 0.2. If we take the cost of an exploration well to be $100 MM and the value of a discovery to be $500 MM, the most likely outcome of drilling out the 10-prospect portfolio is two discoveries. This gives a value of $1000 MM for the outlay of $1000 MM on 10 exploration wells, giving zero net profit.

As shown in Figure 8, we may obtain new information (possibly including field data) at a cost of

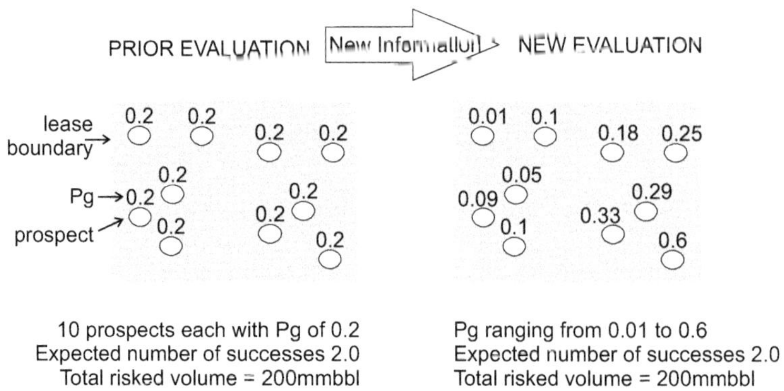

Fig. 8. Changes in the estimated *Pg* of a suite of 10 prospects within an exploration block, all of the same size (mean success case volume 100 MMbbl). On the left, prior to obtaining new information, they are all considered to have the same *Pg* (0.2). On the right is one realization of the change afforded by the new information. The majority of prospects have lower *Pg*; a minority have higher *Pg*. This scenario represents a mean-case realization in which the average *Pg* is unchanged.

$100 MM, which enables us to refine the prospect risk estimates as shown. This refinement has in this case resulted in no net change in total risked volume, and the expected result of drilling out the portfolio still adds up to two discoveries.

The value of obtaining the new information derives from the wells that we choose not to drill. Figure 9 shows the predicted risked value of drilling through the portfolio, starting with the best (highest *Pg*) and working down the list, but stopping at different stages. In this case, value is maximized by choosing to drill only four wells, giving an expected net monetary value of +$235 MM, consisting of an expected value of 1.47 risked discoveries, worth $735 MM, for the cost of four wells ($400 MM) and the cost of obtaining the new information ($100 MM). (The expected value is based on the expected number of discoveries, which is simply the sum of *Pg* of the drilled prospects. The process of exploration drilling will deliver a real number not the expected number – the result of an exploration well is either a discovery or a dry hole. The most likely actual outcomes of this four-well campaign are either one discovery , giving a net value of zero, or two discoveries, giving a net value of +$500 MM, with these two outcomes about equally weighted.)

In this case, if we used the information to optimize the drilling campaign, the information has

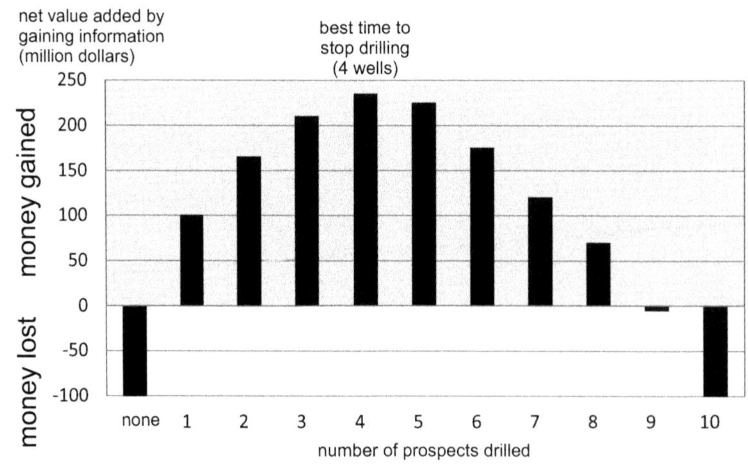

Fig. 9. Predicted net value gained or lost by a drilling campaign, carried out after obtaining new information, in which the wells are drilled in decreasing order of *Pg*. Costs include the cost of drilling ($100 MM/well) plus the cost of new information ($100 MM). Value is calculated as the risked number of successes multiplied by the value of success ($500 MM).

added a value of $235 MM. The key to realizing this value lies in knowing when to stop drilling.

If there is no flexibility in the number of wells to be drilled (e.g. if all 10 are obligatory commitment wells), the information did not, at face value, make any positive contribution. Yet even in this circumstance, some value may still be added from the new information through the time value of money (Capen 1995) by changing the drill order to promote the prospects with highest chance of success, and delaying those with lowest chance of success.

Additional exploration efficiency, and thus value, can also be achieved if we use the principles of play analysis set out by White (1993). We do not acquire new information expecting it to improve our play risk (the expected value is no change). We hope for an improvement in the play risk, ideally changing its status to proven, but if we already planned to drill out the portfolio, the upgrade in the play risk does not lead to added value. Information which allows us to disprove a play has more value impact because it allows us to choose not to drill any of the wells dependent on the play.

Additional sources of value added by new information

The purpose of this article is to show how increased information should not be expected to change our estimate of the average risk on existing prospects, and to show that, for the majority of prospects (those with $Pg < 0.5$), new information will lead to a reduction in our estimated chance of success.

While we believe that these conclusions are important, and that they should inform our decisions, we do not wish to suggest that this is the only reason for acquiring new information, nor that the only source of value lies in the existing prospect portfolio.

In addition to any value change that may occur as a result of changing the *Pg* (risk) estimate for a suite of existing prospects, as discussed above, new information such as that provided by field studies may add substantial value in other ways. These include the following:

- identification of additional prospects in the licensed area, within the same geological play;
- identification of new plays within the same area;
- identification of new opportunities within the region (e.g. follow-on potential in adjacent licence blocks);

Even if the results of the new information are strongly negative, the learnings may result in a positive change relative to the uninformed situation, for example by providing the evidence needed to divest or to farm down the opportunity set, or simply to cut losses and focus attention and resources elsewhere.

Conclusions

There is an expectation (in the informal sense) that the estimated chance of success of a prospect will change from its currently estimated (prior) value in the light of new information. We do not know whether this change will be upwards or downwards. The Law of Total Probability dictates that the expected (in the formal sense) posterior value of the new, revised estimate made using the new data is equal to the prior value. Thus we cannot carry out studies or acquire data for the purpose of increasing probability of success. This does not mean that there is no value in acquiring that new information: on the contrary, it may be vital that we do acquire the new information, because it can be the key to maximizing the value of our overall exploration programme. In some outcomes, we may indeed be rewarded by an improved chance of success.

We do not expect to reduce subsurface risk (i.e. to increase *Pg*) for a given prospect, or for a group of prospects, by carrying out field-based studies, or for that matter, by carrying out any kind of study. In some cases, this desirable improvement may come to pass, but in other cases the risk will worsen; the expected change is, on average, zero.

Given that the typical exploration prospect has $Pg < 0.5$, we expect that more prospects will worsen than will improve, but for the minority of prospects that gain, the positive *Pg* change is likely to be greater than the negative *Pg* change for those that worsen. The net effect is zero average change.

For an existing portfolio of prospects, as with an individual prospect, we expect that the mean result of new information is zero net change in aggregate risk; again we expect numerically that more prospects sets will worsen than improve, but the few winners gain by a larger margin, giving zero net change.

Given improved knowledge to discriminate which prospects are more attractive than others, we are empowered to be selective regarding which subset to drill and in which order to drill them, which can favourably impact the overall economics of a drilling programme.

Obtaining new information adds value to an exploration programme mostly by saving money on the wells we choose not to drill (because their *Pg* has worsened); improving the *Pg* on a well that would be drilled anyway does not add real value. Therefore, ironically, obtaining bad news (for the prospect) before it is drilled is good news (for the company) because it helps us to avoid drilling costly dry holes. Value is also added by accelerating the testing of more favoured prospects for which *Pg* has increased.

New information can be expected to add significant additional value, not by reducing risk on existing prospects, but by identifying new opportunities (new prospects, new plays and new areas into which existing plays can be extended).

The authors gratefully acknowledge support for Frank Peel by NERC, the UK Natural Environment Research Council. We recognize the incisive reviews provided by Peter Carragher and an anonymous reviewer which contributed greatly. We are grateful to David Labonte who provided us with modern references to the Law of Total Probability.

References

Allais, M. 1956. Évaluation des Perspectives Économiques de la Recherche Minière sur de Grands Espaces – Application au Sahara Algérien. *Revue de l'Industrie Minérale*, January, Paris, 329–383.

Aquinas, & T. Saint, 1265–1274. *Summa Theologiae.*

Bayes, T. 1764. An essay toward solving a problem in the Doctrine of Chances. *Philosophical Transactions of the Royal Society of London*, **53**, 370–418.

Bratvold, R. B. & Begg, S. H. 2008. I would rather be vaguely right than precisely wrong: a new approach to decision making in the petroleum exploration and production industry. *AAPG Bulletin*, **92**, 1373–1392.

Bratvold, R. B. & Begg, S. H. 2010. *Making Good Decisions*. Society of Petroleum Engineers, Dallas, TX.

Bratvold, R. B., Bickel, E. J. & Lohne, H. P. 2009. Value of information in the oil and gas industry: past, present, and future. *SPE Reservoir Evaluation & Engineering*, **12**, 630–638, http://doi.org/10.2118/110378-PA

Bean, M. A. 2001. *Probability: The Science of Uncertainty with Applications to Investments, Insurance, and Engineering*. Brooks/Cole, Pacific Grove, CA.

Capen, E. C. 1995. A problem with experts. *In*: *SPE Hydrocarbon Economics and Evaluation Symposium Proceedings*. Society of Petroleum Engineers, Dallas, TX, 22–28 March 1995, 255–266.

Coopersmith, E. M. & Cunningham, P. C. 2002. A practical approach to evaluating the value of information and real option decisions in the upstream petroleum industry. *SPE Annual Technical Conference and Exhibition*, San Antonio, TX, 29 September–2 October 1995, http://doi.org/10.2118/77582-MS

Coopersmith, E. M., Burkholder, M. & Schluze, J. 2006. Value-of-information lookbacks – was the information you gathered really worth getting? Paper SPE 101540 presented at the Society of Petroleum Engineers Annual Technical Conference and Exhibition, San Antonio, Texas, USA, 24–27 September, http://doi.org/10.2118/101540-MS

Cyganowski, S., Kloeden, P. & Ombach, J. 2002. *From Elementary Probability to Stochastic Differential Equations with MAPLE*. Springer Science & Business Media, Berlin.

Hájek, A. 2003. What conditional probability could not be. *Synthese*, **137**, 273–323.

Heinlein, R. A. H. 1959. *Starship Troopers*. Putnam, New York

Howard, R. A. 1966. Information value theory. *IEEE Transactions on Systems Science and Cybernetics (SSC-2)*, **2**, 22–26.

Howard, R. A. & Matheson, J. E. 1984. Influence diagrams. *In*: Howard, R. A. & Matheson, J. E. (eds) *Readings on the Principles and Applications of Decision Analysis*, **II**. Strategic Decisions Group, Menlo Park, CA.

Laplace, P. S. 1774. Mémoire sur la probabilité des causes par les événements. *Mémoires de l'Académie Royale des Sciences de Paris*, **VI**, 621–656.

Marx, K. 1898. Value, price and profit. An address delivered at the General Council of the International Working Men's Association (the First International), on June 20 and 27, 1865. *In*: Marx Aveling, E. (ed.) *Value, Price and Profit*. Swan Sonnenschein, London.

Megill, R. E. 1977. *An Introduction to Exploration Risk Analysis*. PennWell, Tulsa, OK.

Morgan, B. W. 1968. *An Introduction to Bayesian Statistical Processes*. Prentice-Hall, Englewood Cliffs, NJ.

Murkoff, H. E. & Mazel, S. 1984. *What to Expect When You're Expecting*, 1st edn. Workman, New York.

Newendorp, P. D. 1971. Bayesian analysis – a method for updating risk estimates. *Journal of Petroleum Technology*, **24**, 193–198, http://doi.org/10.2118/3463-PA

Newendorp, P. D. 1975. *Decision Analysis for Petroleum Exploration*, 1st edn. PennWell, Tulsa, OK.

Peel, F. J. & Brooks, J. R. V. 2015. What to expect when you're prospecting: how new information changes our estimate of the chance of success of a prospect. *AAPG Bulletin*. First published online 6 August 2015, http://doi.org/10.1306/07061515045

Popper, K. R. 1959. *The Logic of Scientific Discovery*. Basic Books, New York; reprinted, Routledge, London, 1992.

Quirk, D. G. & Ruthrauff, R. G. 2008. Toward consistency in petroleum exploration: a systematic way of constraining uncertainty in prospect volumetrics. *AAPG Bulletin*, **92**, 1263–1291.

Geological Society 2014. *Reducing Subsurface Uncertainty & Risk through Field-based Studies – the Value of Outcrops and Analogues in Hydrocarbon Exploration, Development and Production: Implications for Global Exploration and Production*, 4–8 March. The Geological Society, London.

Rose, P. R. 1987. Dealing with risk and uncertainty in exploration: How can we improve? *AAPG Bulletin*, **71**, 1–16.

Rose, P. R. 1992. Chance of success and its use in petroleum exploration. *In*: Steinmetz, R. (ed.) *The Business of Petroleum Exploration: AAPG Treatise of Petroleum Geology, Handbook of Petroleum Geology*, AAPG, Tulsa, OH, 71–86.

Rose, P. R. 2001. Risk analysis and management of petroleum exploration ventures. *AAPG Methods in Exploration*, **12**, 178.

Ross, J. G. 1997. The philosophy of reserve estimation. *In*: *SPE Hydrocarbon Economics and Evaluation Symposium*, Dallas, TX, 16–18 March 1997, Paper SPE 37960.

Ross, J. G. 2004. Risk and uncertainty in portfolio characterisation. *Journal of Petroleum Science and Engineering*, **44**, 41–53.

Russell, B. 1946. *A History of Western Philosophy*. George Allen & Unwin, London.

Salkind, N. J. (ed.) 2010. *Encyclopedia of Research Design*. SAGE Publications, Thousand Oaks, California, http://doi.org/10.4135/9781412961288

Smith, A. 1776. *An Inquiry into the Nature and Causes of the Wealth of Nations*. W. Strahan and T. Cadell, London.

Ushakov, N. G. 2001. Density of a probability distribution. *In*: Hazewinkel, M. (ed.) *Encyclopedia of Mathematics Supplement III*. Springer, Netherlands.

White, D. A. 1993. Geologic risking guide for prospects and plays. *AAPG Bulletin*, **77**, 2048–2064.

Index

Page numbers in *italic* denote figures. Page numbers in **bold** denote tables.